IEC National
Electrical Apprentice Training Guide

Student Manual
Year Two
2008/2009 Edition

Independent Electrical Contractors, Inc.

Executive Vice President/CEO: Larry Mullins

Writing and development services provided by: Gary Baumgartner, Madeline Borthick, David Hittinger, Cliff Rediger, John Sayne, and George Thess

Desktop Publisher: Sarah Kaducak, Kadu Media

This book was set in Joanna MT by Independent Electrical Contractors, Inc.

The information contained in this manual is general in nature and intended for training purposes only. Actual performance of activities described in this manual requires compliance with all applicable safety procedures under the direction of qualified personnel. References in this manual to patented or proprietary devices do not constitute a recommendation for use.

Copyright © 2008 by the Independent Electrical Contractors, Inc. (IEC), Alexandria, VA 22302.

All rights reserved. Printed in the United States of America. This publication is protected by copyright, and permission should be obtained from IEC prior to any prohibited reproduction, storage in a retrieval system, or transmission in any form or by any means, electronic, mechanical, photocopying, recording, or likewise. For information regarding permission(s), write to: IEC Curriculum, 4401 Ford Avenue, Suite 1100, Alexandria, VA 22302.

In Appreciation

Formal technical training. Experience in the field as electricians, foremen, contractors, inspectors, and electrical engineers. Years of training and experience as instructors and training program administrators. Acknowledged Code experts……..And willing volunteers to boot. If you were to put together a group of individuals to redesign an electrical training curriculum, what more could you ask for?

Last year, when IEC decided to move forward with the development of a whole new version of it's electrical apprenticeship curriculum, the individuals shown below stepped forward to volunteer. Without question, many wonderful, talented, and knowledgeable people work in the electrical contracting industry, but nowhere is there a group of people more talented and capable than the individuals that stepped up to the plate and volunteered to serve on IEC's Electrical Curriculum Development Task Force. IEC National expresses its sincere appreciation for their hard work and dedication to IEC and the electrical industry.

Gary Baumgartner, Sioux Falls, South Dakota

Gary Baumgartner, President of Baumgartner Electric, is a third generation electrical contractor. Gary is heavily involved in the industry—both locally and through IEC National—having served as IEC National Vice President and President, as well as multiple terms as Chairman of IEC National's Apprenticeship and Training Committee. Gary has always strived for continual improvement in IEC's educational offerings and is particularly focused on quality this year as his son Matt, the 4th Baumgartner generation in the business, will start school this fall using this curriculum.

Madeline Borthick, Houston, Texas

Madeline Borthick is the Training Director at the IEC Texas Gulf Coast Chapter, a position she has held since 2001. She brings a wealth of experience to the chapter having also served for ten years as an IEC second year apprenticeship program instructor. In addition, Madeline has served as an IEC representative to Code-Making Panel 10 of the National Electrical Code for over 10 years.

Madeline received a Bachelor of Science Degree in Electrical Engineering from the University of Tulsa in 1976. She worked as an Electrical Draftsman for 8 years and was a Field Engineer for Schlumberger, a leading oil field service provider, for 9 years. From 1985 to 2001, she was an electrician and foreman.

David Hittinger, Cincinnati, Ohio

David Hittinger has worked in the electrical industry for 36 years. In 1987, he started his own electrical contracting company and worked as a part time electrical safety inspector. In 1998, he became the Training Director for the IEC Greater Cincinnati Chapter, and in July of 2004, he became the Executive Director. He holds national certifications as an electrical plans examiner, building and mechanical inspector, and serves as Principal member of NEC® Code Panel One and the Technical Correlating Committee for the National Electrical Code.

Cliff Rediger, Denver, Colorado

Cliff obtained a Bachelor of Science Degree from Colorado State University in Zoology. He has worked as an electrician for 30 years, an electrical contractor for 11 years, and has held a Colorado Master Electrician license for 27 years. Cliff also served as the Electrician Occupations instructor at T.H. Pickens Technical Center for six years. Currently, Cliff is a member of the Colorado State Electrical Board, serves as alternate on NEC® Code Panel Two, and has been the Training Director for Rocky Mountain Chapter IEC for 9 years.

John Sayne, Cincinnati, Ohio

With over 46 years of extensive experience in the electrical field, working as an electrician, foreman, superintendent, project manager and estimator, John Sayne has in-depth commercial and industrial construction background. John holds a State of Ohio Electrical Contractor License, with a license in Columbus, Ohio. He is also a Licensed Master Electrician in Indianapolis, and Master Electrician in Kentucky. He is a Certified Trainer for Aerial Work Platforms, and Material Handling Equipment.

John is a past Chairman of the Independent Electrical Contractors National Apprenticeship and Training Committee. He is also the Chairman of the IEC Apprentice Curriculum Committee. Locally, he is the past Chairman of the Apprenticeship and Training Committee of IEC of Greater Cincinnati, and continues as a valued member. John is also a qualified National Safety Council First Aid, CPR and AED instructor. John has been a valued and trusted employee of Denier Electric Co., Inc. since 1985.

George Thess, St. Louis, Missouri

George Thess has worked in the electrical industry since 1979 and has been a licensed electrical contractor since 1985. He has been a member of the IEC Greater St. Louis Chapter and the International Association of Electrical Inspectors since 1988. George has served as a member of the IEC Greater St. Louis Apprenticeship Committee since 1988 and the IEC National Apprenticeship Committee since 1992. In addition, he is the Chairman of the IEC National Wire-Off Committee and serves on the Technical Committee of the SkillsUSA National Industrial Motor Controls Competition.

Preface

This training manual is intended for use with Electrical Wiring—Residential, 15th Ed. It is the curriculum guide for training the electrical apprentice to become a Journeyman Wireman (electrician). It is primarily intended to be used in an instructor-led course but may also be used as the basis for a course for home study.

The format and outline for each lesson includes: lesson number, title, purpose, objectives, lesson text assignments and worksheets for the NEC® and safety components. The homework assignment at the beginning of each lesson is intended to prepare the student for the lesson prior to coming to class. At the conclusion of the lesson, the instructor should introduce the new subject matter for the following week.

This manual is designed to be interactive with the student. In the ever changing electrical industry, the electrician must strive to be self starting and stay on top of new technology. Functional math, reading and problems solving skills are required to remain on the cutting edge in the electrical trade. It is the intent of this work that the student will develop and hone these skills as well as learn to use the National Electrical Code proficiently.

This curriculum manual was developed and beta tested by the National IEC Apprenticeship and Training Committee's Curriculum Development Task Group. Each Task Group participant is a licensed electrician and experienced instructor. These individuals were inspired to invest their time, energy and money in this project through their own dedication to electrical apprenticeship training.

This document is a work in progress and will provide a pattern and basic style for future curriculum manuals. The Task Group sincerely desires your criticism and comments in order to strive toward perfecting this initial document. If you discover errors (typographical or content) please advise the Task Group immediately, as the curriculum will be revised just as other "Code sensitive" texts are: with each Code cycle.

Please email your comments and editorial corrections to TrainingEditor@ieci.org. Typos and correctly identified errors will not be acknowledged. However, if you identify an error with which the Task Group does not agree, you will get a response and will be encouraged to provide further input, because the question or problem may need to be rewritten if it is not specific enough.

Your assistance is greatly appreciated.

IEC Apprenticeship Curriculum Conditions of Use and Licensing Agreement 2008/2009

1. The publication date for the 2008/2009 edition of IEC's apprenticeship guide is June 2008. The materials included therein are believed to be current and reflective of current regulatory requirements and applicable codes. Users should contact Independent Electrical Contractors, Inc., Attention: Apprenticeship and Training Department, 4401 Ford Avenue, Suite 1100, Alexandria, Virginia 22302 to ascertain that these materials are still current prior to use. The apprenticeship guide and student textbooks ("Materials") are sold with the understanding and agreement of the Purchaser that IEC grants the Purchaser a revocable, nonexclusive, nontransferable license to use the Materials for the purposes of teaching apprenticeship courses conditioned on the Purchaser's continued compliance with the IEC Apprenticeship Curriculum Conditions of Use. IEC does not warrant or represent that the Materials are free from errors or conflicts. Purchaser and any Users of the Materials are encouraged to report any errors, conflicts, or omissions to IEC, Attention: Apprenticeship and Training Department, so that errors can be corrected.

2. The IEC Apprenticeship Program is designed with a number of specific components; each of which must be used if the apprentice is to gain full benefit from the curriculum. Specifically, latest editions of both the student and instructor curriculum guides must be used to ensure that all key components are covered. While lessons may be rearranged within a curriculum year, and requirements of a local nature may be incorporated into the curriculum as required, all material within a curriculum year must be covered within that school year to ensure that the apprentice is properly prepared to progress into the subsequent year's program.

3. Current editions of the curriculum guides must be purchased and implemented annually to help ensure that students are being taught with the most up-to-date materials. Students must be provided or required to obtain the textbooks cited in the curriculum. Teaching from out-of-date editions of either the curriculum or the textbooks may result in students being taught outdated subject matter that might conflict with current practices. The electrical industry is continually changing. New materials and processes are being introduced. Standards are being revised, new standards are being introduced, and local, state and federal requirements are subject to continual revision. Textbooks are updated on an irregular basis.

4. Students should be encouraged to retain all of their texts for reference throughout their apprenticeship and working career, but should be advised to check current codes, regulations, etc. for specific questions or problems.

5. While the apprenticeship training program with its curriculum guides and related texts incorporates a significant degree of safety training, it does not, nor is it intended to, provide all of the types of safety training required for contractor employees under OSHA, EPA, and similar regulations. Contractors and employees are encouraged to refer to applicable local, state, and federal regulations to ensure that they are in compliance with all applicable safety and health guidelines.

6. Upon evidence that Purchaser is not abiding by these conditions, or is no longer in good financial standing with IEC, IEC reserves the right to revoke this License and refuse to provide future curriculum, texts or manpower in support of that individual's/organization's apprenticeship program.

7. Purchaser acknowledges that IEC is the owner of all copyrights in and to the Materials. All rights in and to the Materials are reserved. No part of **the Materials may be reproduced or copied in any form or by any means—graphic, electronic or mechanical, including photocopying, recording, taping, or use on an information retrieval system—without written permission obtained in advance from IEC.**

Approved by IEC National Apprenticeship and Training Committee: January 4, 2007
Approved by IEC National Board of Directors: January 6, 2007

We Need Your Feedback!

You may use this form to report any questions, comments, or corrections you have regarding the IEC Apprenticeship Curriculum. Every question or idea will be considered. (We even want to know about typographical errors, misspelled words, etc.)

Curriculum Year: _____ Edition 2008/2009

Textbook Name: _____

Textbook Page Number(s): _____

Type of error is:

❑ Typo   ❑ Content   ❑ Test Question   ❑ Other

Dear Curriculum Committee:

_____
_____
_____
_____
_____
_____
_____
_____
_____
_____
_____
_____

Signed: _____ Chapter: _____

Mail to: IEC, Inc., 4401 Ford Avenue, Suite 1100
Alexandria, VA 22302
Fax to: (703) 549-7448

Email comments to Trainingeditor@ieci.org

Foreword

This curriculum is application based and has been designed to be taught by electricians. Those instructors who lack field experience may encounter difficulties. Progression through this raining manual should be concurrent with knowledge gained through field instruction.

The material presented in the student manual is presented on the same page, in the same location, asin the instructor manual. Blank spaces appear in the student manual where instructor information was omitted.

Within this curriculum, both proper and industry-accepted terminology is used. It is recognized that regional variations exist and these terms should also be presented by the instructor.

Correct units of measurement include Amperes, kVA, kcmil, AWG, etc. However, please understand that amps, KVA, KCMIL, awg, etc. are frequently used and accepted within the industry. The use of other than scientifically correct units in this manual is deliberate. This will better prepare the students to recognize those commonly used.

Tools and wiring methods are frequently referred to by their trade names. For example lineman's pliers are referred to as "Kleins"; "greenfield" is used instead of flexible metal conduit; "BX" instead of Type AC cable; "Romex®" instead of Type NM cable; etc. These may also be identified by local or regional names. An attempt has been made to expose the students to many variations of nomenclature so that they are better prepared to communicate in the field.

Acknowledgments

This curriculum would not exist were it not for the contributions of the following:

3M
Allied Tube and Conduit
Bosch
Brian Scudder
Capital Safety USA
Chartpak Inc.
Cliff Rediger
David Hittinger
DeWalt
Duracell
Eaton Cutler-Hammer
Energizer
Gary Baumgartner
GE
George Thess
IDEAL Industries, Inc.
IRWIN Industrial Tool Company
John Sayne
Lithonia Lighting
Madeline Borthick
Mercury Marine
Ridge Tool Company
Sonoco
Spectrum Brands
Tim Carney
Vermeer Manufacturing
Yale Materials Handling Corporation

In addition, portions of this course were reprinted with permission from:
American Technical Publishers
Delmar/Cengage Learning
Greenlee/Textron, Inc.
(Greenlee® is a registered trademark of Greenlee Textron Inc., which is a subsidiary of Textron Inc.)
Leviton Manufacturing Company, Inc.
Siemens Energy & Automation, Inc.
Southwire Company

Thanks to all of the many dedicated people who made this project a success.

Table of Contents

Lesson	Content	Class Date	Notes
201	Orientation, Safety, Introduction to Meters, Feeder/Branch Circuits		
202	Service Calculations		
203	Services		
204	Conductor and Overcurrent Protection		
205	Grounding and Bonding Part I		
206	Grounding and Bonding Part II		
207	Wiring Methods		
208	Wiring Materials, Conduit and Box Fill		
209	Mid-Term Review and Exam		
210	Switches, Switchboards and Panelboards		
211	Equipment for General Use		
212	Introduction to AC Theory		
213	AC Theory I		
214	AC Theory II		
215	Single-Phase Transformers Part I		
216	Single-Phase Transformers Part II		
217	Mid-Term Review		
218	Mid-Term Exam		
219	Power Generation — Three-Phase Transformers, Circuits, and Calculations		
220	3Ø Transformers — Delta-Delta		
221	3Ø Transformers — Delta-Wye		
222	Transformers for Non-Linear Loads — 3Ø Fault Currents — Voltage Drop		
223	NEC® Requirements for Transformers		
224	Buck-Boost Transformers: Single- and Three-Phase		

225	Buck-Boost Transformers: Connection and Selection		
226	Generators and Transfer Switches		
227	Mid-term Exam		
228	Electric Motors – DC and 1Ø AC		
229	Electric Motors – Polyphase		
230	Motors: General Knowledge and Sizing Branch Circuit Conductors		
231	Motor Branch Circuit Overcurrent Protective Devices: Short Circuit and Ground Fault Protection		
232	Motor: Branch Circuit Grounding Conductors, Disconnects, Starters, and Overload Protection		
233	Locked Rotor Current and Phase Loss for Motors-A/C and Refrigeration Equipment Generators and Fire Pumps		
234	Motor Feeder Conductors, OCPDs, and Taps – Motor Branch Circuit Conductors and OCPDs		
235	Final Exam Review		
236	Final Exam		

ATTENTION IEC APPRENTICES:

With IEC's 4-year Electrical Apprenticeship program, you can earn up to 37 college credit hours towards an associate's, bachelor's or another degree at colleges and universities that participate in the American Council on Education Recommendation Services (ACE CREDIT) program.

Enter a well-paying career as highly experienced journeymen with up to 37 semester hours towards an undergraduate degree.

Contact your local IEC chapter or IEC National (tcook@ieci.org) to determine if you are eligible to enroll in the ACE Transcript Service.

Lesson 201
Orientation, Safety, Introduction to Meters, Feeder/Branch Circuits

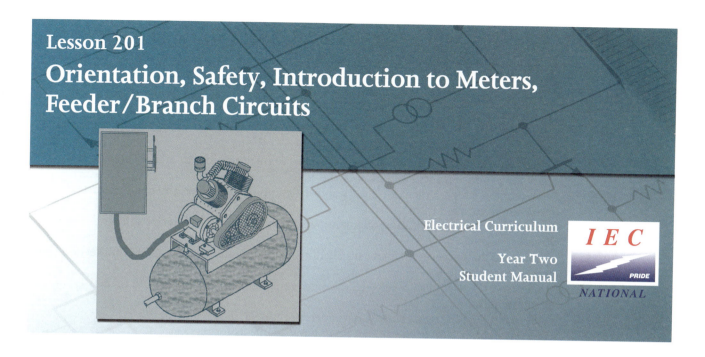

Electrical Curriculum

Year Two
Student Manual

IEC
NATIONAL

Purpose

In this session, students will receive a brief overview of both instructor and student obligations to the apprenticeship training program for Second Year. This lesson will emphasize general electrical and multimeter safety when working around energized circuits. Feeder and Branch circuit descriptions and ratings will be discussed according to NEC requirements.

Homework
(Due at the beginning of this class)

For this lesson, you should:

- Thoroughly read the material contained within Lesson 201..
- Read ES Chapters 1 & 2 in their entirety.
- Complete Objective 201.1 Worksheet.
- Complete Objective 201.2 Worksheet.
- Complete Objective 201.3 Worksheet.
- Complete Objective 201.4 Worksheet.
- Complete Objective 201.5 Worksheet.
- Complete Objective 201.6 Worksheet.
- Complete Objective 201.7 Worksheet.
- Complete Objective 201.8 Worksheet.
- Complete Objective 201.9 Worksheet.
- Complete Lesson 201 NEC Worksheet.
- Read and complete Lesson 201 Safety Worksheet.
- .
- Complete additional worksheets, if available, as directed by your instructor.

Objectives

By the end of this lesson, you should:

201.1 ....Understand the creation, use, format, structure, outline, authority and practical application of the National Electrical Code (NEC).

201.2 ....Understand definitions of electrical terms and equipment in order to prepare you for the study, interpretation and application of the NEC.

201.3 ....Understand the requirement for electrical installations of equipment and conductors according to Article 110 of the NEC.

201.4 ....Be able to identify the circuit wiring, components and voltages in branch circuits of electrical systems.

201.5 ....Understand the purpose, function, installation and NEC requirements for GFCI protection in branch circuits.

201.6 ....Understand the relationship between overcurrent protection, conductors and branch circuit ratings.

201.7 ....Understand the NEC requirements for outlets in residential dwelling units.

201.8 ....Understand basic NEC feeder size and rating requirements for services.

201.9 ....Be able to locate equipment and symbols on the BCES prints in order to solve problems addressed on common construction site scenarios.

> **Attention Students and Instructors:**
> Throughout this curriculum, unless otherwise noted (U.N.O.):
> All conductors are to be THHN/THWN copper and all OCPD terminations are rated at 75°

201.1 National Electrical Code

Common Voltage Systems

This information is to familiarize the 1st year student with the most common voltage systems used by the electrical industry for premises wiring. These voltages are nominal. Nominal means that the voltage can actually vary slightly. Refer to 2008 NEC© 220.5(A). In 2nd year the student will study generators and transformers extensively and learn how these voltage systems are generated and produced.

A symbol that will be used on this worksheet, and extensively on many documents concerning electrical work, is Ø. This symbol is used for "phase". On any voltage system that is represented by xx/yy volt, 120/240 volt for example, it is equally correct for this to be written 240/120 volt. On any of these "dual voltage" systems the lower number is always the voltage measured from one of the ungrounded conductors to the neutral and the higher number is always the voltage measured between any two ungrounded conductors. Voltages between a "hot" and a neutral are called *line-to-neutral* voltages. Voltages between one "hot" and another "hot" are called *line-to-line* voltages.

120/240 volt 1-Ø, 3wire is the voltage system most commonly used in residential and small commercial installations. In this system there are **two** ungrounded conductors ("hots") and **one** grounded conductor ("neutral"). Two plus one equals three. This is therefore a 3-wire system. The voltages, 120 and 240, are *nominal*.

The voltages listed below are the nominal voltages that would be measured between conductors. We will designate one of the "hots" as L1 and the other "hot" will be L2. N will represent the neutral.

L1 to N = 120 volts   L2 to N = 120 volts   L1 to L2 = 240 volts

The electrician needs to be able to distinguish one conductor from another so conventional color codes have been established in your area for the ungrounded conductors and the NEC has established permitted color codes for the grounded conductors. For this system one of the service feeder "hots" is color coded black and the other red or blue. The neutral of the service feeder is color coded white. Generally we use colored electrical tape wrapped around the insulation to identify our conductors. Keep in mind that the NEC does NOT specify any color for ungrounded conductors except in the case of a "high-leg" which we will discuss later in this worksheet. So, unless the local authority having jurisdiction requires particular colors you could actually use any color except white, gray, or green for the ungrounded conductors. The (AHJ) authority having jurisdiction may be the engineer who wrote the specifications on your job.

Cables (MC, AC, NM, etc) that are used for multi-wire branch circuits on this system will routinely contain circuit conductors that have insulation colors of black, red, and white. The grounding conductor is NOT considered a circuit conductor because it will only carry current IF SOMETHING GOES WRONG.

If "pipe and wire" is used as the wiring method the same colors are generally used for the conductors installed in the conduit.

120/240 volt 3Ø, 4-wire is the next voltage system we will consider. This system is commonly used in commercial and large residential installations. Three-phase power may be available to large residential customers because sometimes 3-phase air conditioning/heating equipment is used.

In this system there are **three** ungrounded conductors ("hots") and **one** grounded conductor ("neutral"). Three plus one equals four. This is therefore a **4**-wire system.

The voltages listed below are the nominal voltages that would be measured between conductors. We will designate one of the "hots" as AØ, the second "hot" as BØ, and the third "hot" will be CØ. N will represent the neutral.

    AØ to N = 120 volts
    BØ to N = 208 volts   NO this is not a misprint - This is the **"high-leg"**
    CØ to N = 120 volts
    AØ to BØ = 240 volts   AØ to CØ = 240 volts   BØ to CØ = 240 volts

For this system the color used for the ungrounded service feeder conductor that has the 208 volt line-to-neutral voltage is dictated by the NEC to be orange. While the NEC permits any ungrounded conductor to be orange, or purple, or mauve, or any other color except white, gray, or green the NEC **requires** that the high-leg conductors be colored orange. This is important because, as you can see above, the voltage from BØ to neutral is 208 volts. If a television, or any other load rated to operate at 120 volts, were connected to a circuit breaker installed on BØ it wouldn't last very long at the applied voltage of 208. In many areas the color convention is that AØ is colored black, CØ is colored blue, and white is used for the neutral color. Remember, this may vary from AHJ to AHJ and is NOT required by the NEC. BØ is orange – this IS required by the NEC.

Cables used for multi-wire branch circuits on this system are the same cables used on the 120/240 volt 1-Ø, 3-W system. We can't install a multi-wire branch circuit on this system with 3 "hots" and one neutral because the voltage between the BØ "hot" and neutral would be 208 volts instead of 120 volts. In fact, look again at the definition for multi-wire branch circuit in Article 100 of the NEC. The definition requires that the voltage from **any** ungrounded conductor of the circuit to the neutral be the same. If the high-leg is used the line-to-neutral voltage will be 208 volts. So, the NEC doesn't permit us to use the high-leg as part of a multi-wire branch circuit. We can, however, run a multiwire branch circuit using a circuit connected to AØ along with a circuit connected to CØ.

If "pipe and wire" is used as the wiring method the same colors are generally used for the conductors installed in the conduit.

120/208 volt 3Ø, 4-wire will now be considered. This system is commonly used in industrial, commercial, and large residential installations.

In this system there are **three** ungrounded conductors ("hots") and **one** grounded conductor ("neutral"). Three plus one equals four. This is therefore a **4**-wire system.

The voltages listed below are the nominal voltages that would be measured between conductors. We will designate one of the "hots" as AØ, the second "hot" as BØ, and the third "hot" will be CØ. N will represent the neutral.

AØ to N = 120 volts   BØ to N = 120 volts   CØ to N = 120 volts
AØ to BØ = 208 volts   AØ to CØ = 208 volts   BØ to CØ = 208 volts

For this system the conventional colors used in many areas is: black for AØ, red for BØ, blue for CØ, and white for the neutral. Again, this may vary with the AHJ.

Cables used for multi-wire branch circuits on this system are available with black, red, and white insulation colors where only two circuits are needed. Where 3 circuits are needed cable containing black, red, blue, and white conductors is available. If "pipe and wire" is used as the wiring method the same colors are generally used for the conductors installed in the conduit.

277/480 volt 3Ø, 4-wire will now be considered. This system is commonly used in industrial and commercial installations. Since most facilities require some 120 volt power a transformer is routinely installed to reduce this voltage (480) to 120. Most commonly a transformer that produces a 120/208 volt 3Ø, 4-wire system is used for this purpose.

In this system there are **three** ungrounded conductors ("hots") and **one** grounded conductor ("neutral"). Three plus one equals four. This is therefore a **4**-wire system.

The voltages listed below are the nominal voltages that would be measured between conductors. We will designate one of the "hots" as AØ, the second "hot" as BØ, and the third "hot" will be CØ. N will represent the neutral.

AØ to N = 277 volts   BØ to N = 277 volts   CØ to N = 277 volts
AØ to BØ = 480 volts   AØ to CØ = 480 volts   BØ to CØ = 480 volts

For this system many industrial installations and AHJ's use conventional colors of brown for AØ, orange for BØ, yellow for CØ, and gray for the neutral. Once again, this color coding is NOT an NEC requirement. In some parts of the country the ungrounded conductors are color-coded brown-purple-yellow. Again the NEC does not specify these colors. Local convention or requirements dictate the colors to be used.

The conductors contained in cables available in your area can be re-identified with conventional colors as required. If "pipe and wire" is used as the wiring method conventional colors are generally used for the conductors installed in the conduit.

The gray color used for the grounded (neutral conductor on this voltage system) conductor differentiates this neutral from the white neutral used in a lower voltage system, such as 120/208. The 2008 NEC© requirement at 200.6(D) is that whenever neutrals from different voltage systems are present we must be able to distinguish between them. If one neutral is white and the other gray we meet the NEC requirement.

Summary of 2008 NEC© Insulation Color Requirements

310.12(A): $\Rightarrow$ 200.6

    white or gray is reserved for grounded conductors and can't be used for any other conductor

310.12(B): $\Rightarrow$ 250.119

    green is reserved for the grounding conductors and can't be used for any other conductor

310.12(C):

    any ungrounded conductor can be ANY color (even orange) except white, gray, or green

110.15:

    the hi-leg conductor of a 120/240 volt 3Ø, 4-wire system must be identified by an orange color wherever a grounded conductor is also present

230.56:

    the high-leg service conductor from a 120/240 volt 3Ø, 4-wire system must be identified by an orange color

408.3(E):

    for a 120/240 volt 3Ø, 4-wire system BØ (high-leg) is required to be the middle busbar; also the FPN $\Rightarrow$ 110.15

Voltage System Summary

120/240 volt 1Ø, 3-wire

    L1 to N = 120 volts   L2 to N = 120 volts   L1 to L2 = 240 volts
        1-pole breaker: 120 volt circuit   **2-pole breaker:** 240 volt circuit

120/240 volt 3Ø, 4-wire

    AØ to N = 120 volts   BØ to N = **208** volts   CØ to N = 120 volts
    AØ to BØ = 240 volts   AØ to CØ = 240 volts   BØ to CØ = 240 volts
        1-pole breaker: 120 volt circuit   **2-pole breaker:** 240 volt circuit

120/208 volt 3Ø, 4-wire

    AØ to N = 120 volts   BØ to N = 120 volts   CØ to N = 120 volts
    AØ to BØ = 208 volts   AØ to CØ = 208 volts   BØ to CØ = 208 volts
        1-pole breaker: 120 volt circuit   **2-pole breaker:** 208 volt circuit

Orientation, Safety, Introduction to Meters, Feeder/Branch Circuits

277/480 volt 3Ø, 4-wire

AØ to N = 277 volts   BØ to N = 277 volts   CØ to N = 277 volts
AØ to BØ = 480 volts   AØ to CØ = 480 volts   BØ to CØ = 480 volts
1-pole breaker: 277 volt circuit   **2-pole breaker:** 480 volt circuit

201.2 National Electrical Code Definitions

201.3 Electrical Installation Requirements

201.4 Branch Circuits

201.5 GFCI Protection in Branch Circuits

201.6 Overcurrent Protection, Conductors and Branch Circuit Ratings

201.7 Outlet Requirements for Residential Dwelling Units

201.8 Feeder Sizing and Ratings for Services

201.9 Problem Solving for BCES Blueprints

The abbreviations and symbols that appear on Dwg E1.1 are specific to the electrical plans. Each subset of plans (architectural, plumbing, landscaping, mechanical, etc.) has symbols and abbreviations that are specific to those plans.

Errata on BCES Dwg E1.1: Under **Motors and Controls** the symbol **FSD** represents a Fire Smoke **Damper** (*motorized*). Fire Smoke Detector should be corrected to Fire Smoke Damper on all "E" drawings.

FSDs are installed in mechanical ductwork and firewalls to prevent the spread of fire and smoke when a fire/smoke event occurs. These dampers OPEN when power is applied, and CLOSE when power is removed. The power to these is controlled by the fire control system.

Objective 201.1 Worksheet

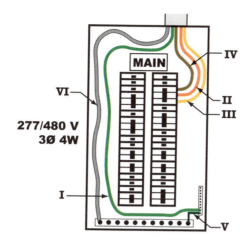

Figure 201.4

_____ 1. The nominal voltage between conductors III and VI as shown above in Figure 201.4 is ___ volts.

a. 277   b. 0   c. 480   d. 240   e. 360

_____ 2. The nominal voltage between conductors II and IV as shown above in Figure 201.4 is ___ volts.

a. 360   b. 0   c. 480   d. 277   e. 240

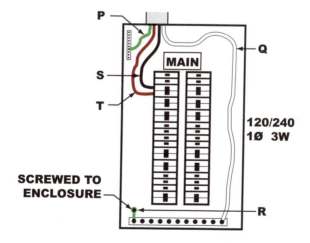

Figure 201.2

_____ 3. Refer to Figure 201.2, shown above. The nominal voltage between conductors P and Q is ___ volts.

a. 208   b. 240   c. 120   d. 0

_____ 4. Refer to Figure 201.2, shown on the previous page. The nominal voltage between conductors P and S is ___ volts.

   a. 120   b. 240   c. 0   d. 208

_____ 5. Annex H in the NEC ___.

   a. is to be used for information only unless specifically adopted by the local AHJ
   b. contains requirements to be applied to all articles

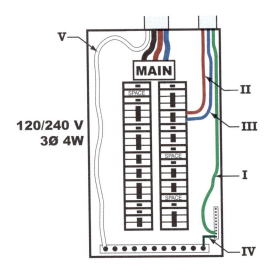

Figure 201.5

_____ 6. Refer to Figure 201.5, shown above. The nominal voltage between conductors II and V is ___ volts.

   a. 208   b. 240   c. 120   d. 0

_____ 7. Refer to Figure 201.5, shown above. The nominal voltage between conductors I and III is ___ volts.

   a. 240   b. 0   c. 120   d. 208

_____ 8. Which of the following installations are NOT covered by the NEC®?

   a. Reliant/HL&P power lines and equipment located in property easements along streets
   b. Sprint towers used for cell phone signal transmission
   c. Bolivar ferry boats
   d. Southwest Airlines jets such as 777's
   e. all of the above

Orientation, Safety, Introduction to Meters, Feeder/Branch Circuits

_____ 9. Where used in the NEC® the term shall be permitted indicates that an action is ___. An example of this useage appears below.

110.53 Conductors. High-voltage conductors in tunnels shall be installed in metal conduit or other metal raceway, Type MC cable, or other approved multiconductor cable. <u>Multiconductor portable cable shall be permitted to supply mobile equipment.</u>

    a. required   b. mandatory   c. permitted   d. not allowed

_____ 10. Protection against ___ is addressed by the requirements in the NEC®?
    I. fault currents   II. thermal effects   III. electric shock

    a. I & II only   b. I only   c. I, II, & III   d. I & III only   e. III only

_____ 11. **Statement I:** All conductors of the same circuit and, where used, the grounded conductor and all equipment grounding conductors and bonding conductors shall be contained within the same raceway, cable, or cord.

Statement II: Equipment grounding conductors shall be permitted to be installed outside a raceway or cable assembly.

    a. Both Statements **I** and **II** are examples of mandatory Code language.
    b. Only Statement **I** is an example of mandatory Code language.
    c. Neither Statements **I** nor **II** are examples of mandatory Code language.
    d. Only Statement **II** is an example of mandatory Code language.

_____ 12. Chapter 8 in the NEC® covers ___ and is independent of the other NEC® chapters and provisions in other chapters do not apply unless directly referenced.

    a. communication systems
    b. formal interpretations
    c. equipment for special use
    d. emergency equipment
    e. Code enforcement

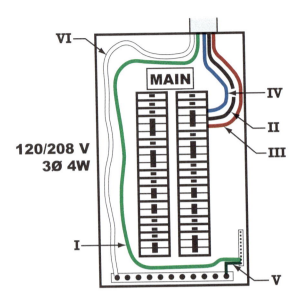

Figure 201.3

_____ 13. Refer to Figure 201.3, shown above. The nominal voltage between conductors IV and V is ___ volts.

   a. 0   b. 120   c. 240   d. 208

_____ 14. Refer to Figure 201.3, shown above. The nominal voltage between conductors I and V is ___ volts.

   a. 0   b. 208   c. 240   d. 120

Objective 201.2 Worksheet

_____ 1. If a motor and the disconnecting means (safety switch) are located in the same room and 60 feet apart, these ___ considered to be within sight (of each other) per the NEC®.

    a. are   b. are not

_____ 2. Which of the installations listed below would be a wet location?

    a. A receptacle installed on the exterior wall of a building
    b. A luminaire installed on a parking lot pole
    c. A run of PVC conduit (RNMC) installed on a fishing pier
    d. All of these
    e. None of these

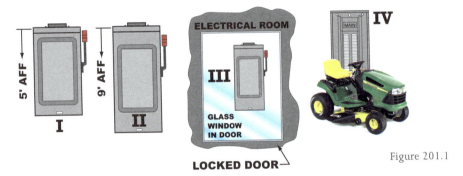

Figure 201.1

_____ 3. The electrical equipment shown above at IV in Figure 201.1 is readily accessible.

    a. True   b. False

Figure 201.8

_____ 4. Refer to Figure 201.8, shown above, where the dimmer is rated for 600 watts at 120 volts and each lamp is rated for 250 watts at 130 volts. If the dimmer is installed as shown ___ will occur.

    a. a short circuit   b. normal operation   c. an overload   d. a ground fault

_____ 5. Which of the drawings below meets the NEC® requirements where equipment 1 and equipment 2 are required to be within sight from each other?

    a. I & III only   b. I only   c. II & III only   d. all of these   e. none of these

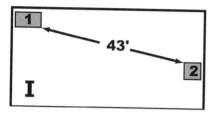 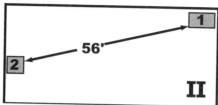 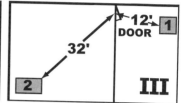

In Sight

_____ 6. A 1900 (4-square or 4″×4″ square) box with a cover that is installed above a sheetrock ceiling is considered ___.

    a. accessible   b. readily accessible   c. either a or b   d. neither a nor b

_____ 7. To assist in the proper application of the NEC® a term that appears in more that one article is defined in ___.

    a. those articles   b. Article 90   c. Article 100   d. Article 80

Equipment Schedule 201.1
    Equipment I: Operates for 3 hours, 20 minutes each time it is started.
    Equipment II: Operates for 1 hour, 15 minutes each time it is started.
    Equipment III: Operates for 45 seconds each time it is started.

_____ 8. Equipment II, shown on the previous page in Equipment Schedule 201.1, is a continuous load.

    a. True   b. False

_____ 9. A fluorescent luminaire is an example of a device.

    a. False   b. True

_____ 10. Which of the following is an example of a bare conductor?

   a. a #12 THHN copper wire
   b. the bare ground wire contained within romex (NM cable) sheathing
   c. the bare aluminum wire twisted with triplex
   d. all of these
   e. none of these

Figure 201.7

_____ 11. The electrical system unit shown above in Figure 201.7 at I is a ___.

   a. device   b. utilization equipment   c. neither of these

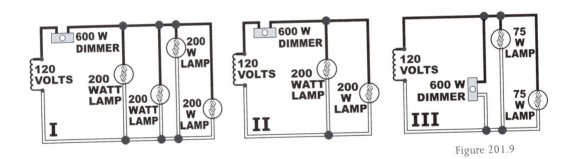

Figure 201.9

_____ 12. If the dimmer is installed as shown above at III in Figure 201.9, ___ will occur.

   a. a short circuit   b. normal operation   c. an overload   d. a ground fault

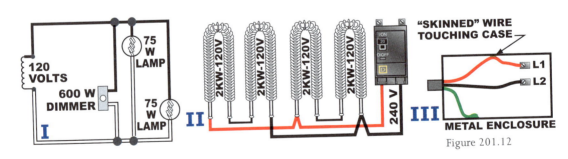

Figure 201.12

_____ 13. Which of the illustrations shown in Figure 201.12 (above) is an example of an overload condition?

   a. II   b. III   c. I   d. All of these   e. None of these

_____ 14. The ampacity of an insulated conductor is ___.

    a. the maximum voltage that can be applied
    b. the amount of current it can carry continuously without melting the contained conductor
    c. the amount of current it can carry without damage to the insulation
    d. all of the above

Objective 201.3 Worksheet

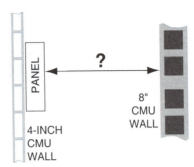

Figure 201.14

_____ 1. The minimum clearance at ? is ___ inches where a 120/208 volt 3Ø, 4-wire panelboard is installed as shown above in Figure 201.14.

   a. 36   b. 42   c. 48

_____ 2. The exact Code reference that requires electrical equipment to be securely fastened to the surface on which it is mounted is ___.

   a. 250.12   b. 110.12(B)   c. 110.27(A)(4)   d. 110.13(A)   e. none of these

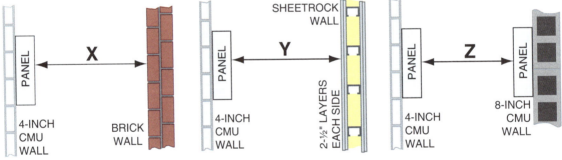

Figure 201.15   Plan Views

_____ 3. Refer to Figure 201.15, shown above. The minimum working clearance at Point Z is ___ feet. The panel is 120/208 3Ø, 4W.

   a. 4   b. 3 1/2   c. 3

_____ 4. Condition ___ clearances would have to be maintained for installation Y, shown above in Figure 201.15.

   a. 3   b. 2   c. 1

_____ 5. Which of the following provides clearance requirements for proper ventilation of transformer enclosures?

a. Job specifications
b. Manufacturer's nameplate
c. The NEC®
d. All of these
e. None of these

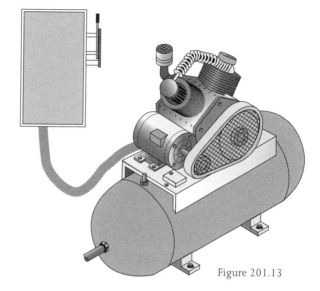

Figure 201.13

_____ 6. Does the NEC® require that the disconnect (safety switch), shown above in Figure 201.13, for the air compressor be marked with a label such as "Air Compressor?"

a. Yes   b. No

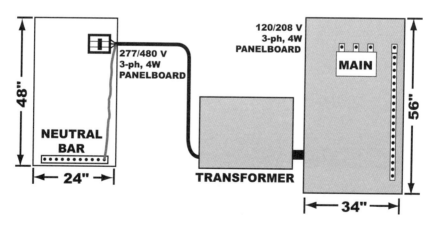

Figure 201.18

_____ 7. For the 277/480 volt panelboard shown above in Figure 201.18, the minimum width of the working space directly in front of the panelboard is ___.

a. 34 inches   b. 64 inches   c. 24 inches   d. 44 inches   e. 30 inches

Objective 201.4 Worksheet

Figure 201.52

_____ 1. Does the NEC® permit the branch circuit conductors of both voltage systems to occupy the same conduit as shown above at I in Figure 201.52?

    a. no   b. yes

1	3 KVA LOAD	4.5 KVA	2-P	2	SCHEDULE
3	SPARE	LOAD		4	
5	3 KVA LOAD	2.5 KVA LOAD		6	PANEL D2
7	2 KVA LOAD	1.5 KVA LOAD		8	
9	SPARE	SPARE		10	120/240 V
11	SPARE	SPARE		12	3Ø, 4W
13	SPARE	SPARE		14	

Schedule Panel

_____ 2. Refer to the panel schedule shown above. The nominal voltage between the breaker in space 5 and the breaker in space 6 is ___ volts.

    a. 0   b. 120   c. 208   d. 240   e. none of these

_____ 3. Refer to the panel schedule shown above. The nominal voltage between the breaker in space 2 and the neutral bar is ___ volts.

    a. 0   b. 208   c. 240   d. 120   e. none of these

_____ 4. A 120/240 volt, 1Ø, 3-wire multiwire branch circuit is connected to a "split-wired" double single-pole switch, as shown below. These are sometimes called "dougle or stack switches". Each of the two switches, that are mounted on the same yoke, will serve a separate 120-volt lighting load. The branch circuit OCPD/disconnect for these circuits is permitted to be ___.

a. a 2-pole fused safety switch
b. a 2-pole circuit breaker
c. two single-pole circuit breakers with approved handle ties
d. any of these
e. none of these

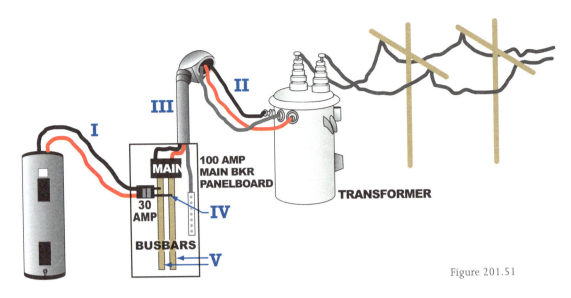

Figure 201.51

_____ 5. The conductors at Point II in Figure 201.51, shown above, meet the definition of ___ conductors.

   a. service drop   b. tap   c. branch circuit   d. service-entrance   e. feeder

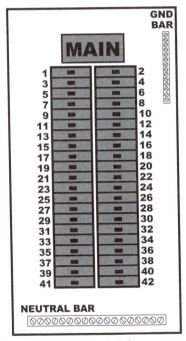

42-Circuit Panelboard

_____ 6. 42-Circuit Panelboard L is shown above. It is fed from a 120/240 volt, 1Ø, 3-wire utility transformer. Circuit number 42 is connected to ___.

   a. CØ   b. AØ   c. BØ

Orientation, Safety, Introduction to Meters, Feeder/Branch Circuits 201-25

_____ 7. 42-Circuit Panelboard L is shown on the previous page. It is fed from a 120/240 volt, 1Ø, 3-wire utility transformer. Circuit number 23 is connected to ___.

   a. AØ   b. BØ   c. CØ

_____ 8. 42-Circuit Panelboard HB is shown on the previous page. It is fed from a 277/480 volt, 3Ø, 4-wire utility transformer bank. Circuit number 25 is connected to ___.

   a. CØ   b. AØ   c. BØ

_____ 9. 42-Circuit Panelboard LC is shown on the previous page. It is fed from a 120/208 volt, 3Ø, 4-wire utility transformer bank. Circuit number 11 is connected to ___.

   a. BØ   b. CØ   c. AØ

_____ 10. 42-Circuit Panelboard HB is shown on the previous page. It is fed from a 277/480 volt, 3Ø, 4-wire utility transformer bank. Circuit number 40 is connected to ___.

   a. CØ   b. BØ   c. AØ

_____ 11. 42-Circuit Panelboard LC is shown on the previous page. It is fed from a 120/208 volt, 3Ø, 4-wire utility transformer bank. Circuit number 39 is connected to ___.

   a. BØ   b. AØ   c. CØ

_____ 12. 42-Circuit Panelboard LC is shown on the previous page. It is fed from a 120/208 volt, 3Ø, 4-wire utility transformer bank. Circuit number 38 is connected to ___.

   a. BØ   b. CØ   c. AØ

_____ 13. 42-Circuit Panelboard HB is shown on the previous page. It is fed from a 277/480 volt, 3Ø, 4-wire utility transformer bank. Circuit number 33 is connected to ___.

   a. AØ   b. CØ   c. BØ

_____ 14. 42-Circuit Panelboard LC is shown on the previous page. It is fed from a 120/208 volt, 3Ø, 4-wire utility transformer bank. Circuit number 18 is connected to ___.

a. AØ   b. BØ   c. CØ

_____ 15. In a perfectly balanced (each ungrounded conductor carries the same current) 3Ø, 4-wire branch circuit where one neutral is "shared" among all three "hots" it is possible for the neutral to carry more current than any one ungrounded conductor. The loads are 277 volt 2 x 4 fluorescent luminaires with electronic ballasts.

a. False   b. True

_____ 16. Per the NEC® (NOT standard practice), which of the following identification colors are acceptable for a 120/240 V, 3-ph, 4-W wiring system?

a. black, orange, blue, gray
b. brown, orange, yellow, gray
c. pink, orange, purple, white
d. any of these
e. none of these

_____ 17. A branch circuit serving a load is protected with 40 A fuses in a 60 A safety switch. This branch circuit has a rating of ___ amps.

a. 40
b. 32
c. 60
d. 48
e. 50

_____ 18. A multiwire branch circuit is to be run from a 120/240 V 3Ø, 4-wire panelboard. The two ungrounded conductors are connected to a 2-pole breaker that is installed in spaces 1 & 3. The voltage between the ungrounded conductors is ___ volts.

a. 208
b. 120
c. 240
d. none of these

Orientation, Safety, Introduction to Meters, Feeder/Branch Circuits

Panelboard HB, shown right, is fed from a 277/480 volt, 3Ø, 4-wire utility transformer bank. Match the voltage (a, b, c, d, e, or f) that would be measured between the points shown in the statements.

    a. 120
    b. 0
    c. 240
    d. 277
    e. 208
    f. 480

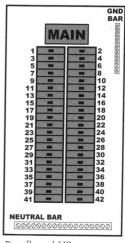

Panelboard HB

_____ 19. Between Breakers 27 & 12 would be ___ volts.

_____ 20. Between the neutral bar and Breaker 37 would be ___ volts.

Panelboard LD, shown right, is fed from a 120/208 volt, 3Ø, 4-wire utility transformer bank. Match the voltage (a, b, c, d, e, or f) that would be measured between the points shown in the questions.

    a. 208
    b. 240
    c. 120
    d. 0
    e. 480
    f. 277

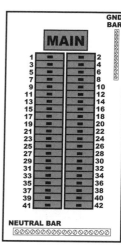

Panelboard LD

_____ 21. Between Breakers 29 & 2 would be ___ volts.

_____ 22. Between the neutral bar and Breaker 10 would be ___ volts.

Panelboard L, shown right, is fed from a 120/240 volt, 3Ø, 4-wire utility transformer bank. Match the voltage (a, b, c, d, e, or f) that would be measured between the points shown in the questions.

    a. 240
    b. 0
    c. 480
    d. 208
    e. 277
    f. 120

Panelboard L

_____ 23. Between Breaker 34 and the neutral bar would be ___ volts.

_____ 24. Between Breakers 38 and 27 would be ___ volts.

Objective 201.5 Worksheet

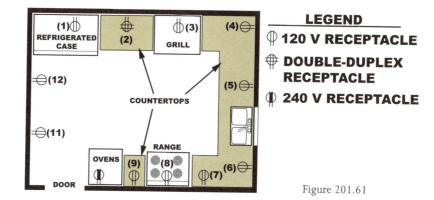

Figure 201.61

_____ 1. In the kitchen plan view, shown above in Figure 201.6$_1$, which receptacles are required by the NEC® to be GFCI protected? This kitchen in located in a restaurant.

a. all twelve
b. 2, 4, 5, 6, 7, 9, 11 & 12
c. 2, 4, 5, 6, 7, & 9
d. all except 10
e. none of these

_____ 2. The table, shown below, has a countertop surface and is against the wall in a school cafeteria kitchen. No electrical appliances will be used on the countertop. The receptacle, below the table, ___ required to be GFCI protected.

a. is   b. is not

Table

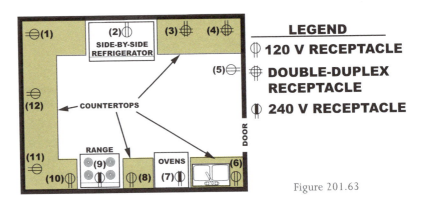

Figure 201.63

_____ 3. The kitchen plan view, shown above in Figure 201.63, is for a kitchen in a house. Receptacles ___ are required by the NEC® to be GFCI protected.

   a. all except 2, 5, 7 & 9
   b. all twelve
   c. 1, 6, 8, 10, 11 & 12 only
   d. all except 7 & 9
   e. all except 2 & 5

_____ 4. Your body has a resistance of about 50,000 ohms between your hands. You complete a 480 volt circuit by holding one "hot" in your left hand and the other "hot" in your right hand. The current flowing in your body is about ___ mA and is sufficient to prevent you from "letting go."

   a. 10   b. 9.6   c. 5   d. .96

Orientation, Safety, Introduction to Meters, Feeder/Branch Circuits

Objective 201.6 Worksheet

_____ 1. What is the minimum size branch circuit conductors permitted to serve a load that is connected with THHN/THWN wire? The branch circuit OCPD is a circuit breaker that is rated for 75° C. The connected load is NOT continuous and will draw 22 amps.

   a. 10 AWG   b. 8 AWG   c. 6 AWG   d. 4 AWG   e. none of these

_____ 2. What is the minimum size THHN/THWN wire permitted to serve a load that will draw 47 amps? The load will operate for about 5 hours at a time. The branch circuit OCPD is a circuit breaker that is rated for 75° C.

   a. 10 AWG   b. 6 AWG   c. 8 AWG   d. 4 AWG   e. none of these

_____ 3. What minimum size conductors are required to feed a 12.5 KW watt 240 volt electric furnace operated at 240 volts? Size the conductors per 424.3(B). The branch circuit OCPD is a circuit breaker that is rated for 75° C.

   a. 6   b. 3   c. 8   d. 10   e. 4

_____ 4. The smallest wire that can be installed to serve a continuous load that has a nameplate rating of 142 amps is ___ AWG. The OCPD is listed to operate at 100% of its rating and has terminations rated at 75° C.

   a. 2/0   b. 3/0   c. 1/0   d. 4/0   e. none of these

_____ 5. Two loads are to be connected to a branch circuit. Load #1 will draw 35 amps and will operate continuously. Load #2 is non-continuous and will pull 42 amps. What is the minimum THWN/THHN copper wire size permitted for the branch circuit? The branch circuit is connected to a circuit breaker with a 75° C rating.

   a. 10 AWG   b. 8 AWG   c. 6 AWG   d. 4 AWG   e. none of these

Year Two (Student Manual)

Orientation, Safety, Introduction to Meters, Feeder/Branch Circuits

Objective 201.7 Worksheet

_____ 1. A hallway in a Holiday Inn is 90 feet in length. The fewest number of convenience receptacles that can be installed per the NEC® is ____.

    a. 2   b. 1   c. 9   d. none are req'd

_____ 2. Where an **air conditioning** (see Index in NEC®) unit is installed on the roof of an office building a convenience receptacle ____.

    a. is not required to be installed but must be GFCI protected if installed
    b. must be installed on the same level within 25′ of the unit and must be GFCI protected
    c. must be GFCI protected, installed within sight of the unit, and not more than 50 ft away

_____ 3. A stairway between the 1st and 2nd floors of a home with ____ to control the stairway lighting will not violate the NEC®.
    I. a 3-way switch at the 1st floor & a 3-way switch at the 2nd floor landing
    II. a single-pole switch at the 1st floor landing
    III. a single-pole switch at the 2nd floor landing

    a. III only   b. II or III only   c. I only   d. I or II or III   e. II only

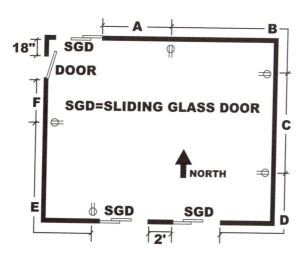

Figure 201.71

_____ 4. Figure 201.71, above, shows the plan view of a room in a house. The maximum dimension for B is ____ feet. Select N/A if the NEC® provides no requirements.

    a. 6   b. 4   c. 2   d. 12   e. N/A

Year Two (Student Manual)

_____ 5. Figure 201.71, on the previous page, shows the plan view of a room in a private club. The maximum dimension for C is ___ feet. Select N/A if the NEC® provides no requirements.

a. 4   b. 2   c. 12   d. 6   e. N/A

Orientation, Safety, Introduction to Meters, Feeder/Branch Circuits

Objective 201.8 Worksheet

_____ 1. A 175 Amp breaker protects a feeder that originates in a 277/480 volt, 3-ph, 4-wire service panel in an office building. Which of the following conductors is the smallest size permitted for the feeder? All terminations are rated for 75° C.

    a. 2/0   b. 1/0   c. 1   d. 3/0   e. 2

_____ 2. A 225 Amp breaker protects a feeder that originates in a 120/240 volt, 1-ph, 3-wire service panel in a residence. Which of the following conductors is the smallest size permitted for the feeder? All terminations are rated for 75° C.

    a. 3/0   b. 1   c. 2/0   d. 1/0   e. 4/0

_____ 3. A 175 Amp breaker protects a feeder that originates in a 120/240 volt, 1-ph, 3-wire service panel in a residence. Which of the following conductors is the smallest size permitted for the feeder? All terminations are rated for 75° C.

    a. 2/0   b. 3   c. 1/0   d. 2   e. 1

_____ 4. A 150 Amp breaker protects a feeder that originates in a 120/208 volt, 3-ph, 4-wire service panel in a school. Which of the following THHN/THWN copper conductors is the smallest size permitted for the feeder? All terminations are rated for 75° C.

    a. 2/0   b. 3/0   c. 2   d. 1/0   e. 1

Objective 201.9 Worksheet

_____ 1. Refer to drawing E1.1 in the BCES set. The minimum wire size for circuits is ___, unless otherwise specified.

    a. #12   b. #10   c. #14   d. #8

_____ 2. The FSD's in the BCES Computer Lab (Room B141) are supplied power from ___.

    a. circuit LE-9   c. FPU 5-4   b. circuit LB2-28   d. the fire alarm panel

_____ 3. On BCES Drawing E1.1, the NEMA configuration for a duplex wall receptacle is ___.

    a. 5-20R   b. 5-30R   c. 6-20R   d. 5-15R

_____ 4. The symbol ∇ shown on the BCES E-drawings, represents a ___ outlet.

    a. telephone   b. lighting   c. power   d. data

_____ 5. "Unless Noted Otherwise" and "Unless Otherwise Noted" are notations used on construction drawings. These are abbreviated UNO and UON, respectively. The abbreviation used on BCES E-drawings is ___.

    a. UNO   b. UON

_____ 6. Circuit ___ serves the entrance lights at the east entrance of BCES Area A.

    a. HL-3   b. HB2-11   c. HA-13   d. HL-5   e. HA-7

_____ 7. Refer to the BCES panel schedule for Panel LB1 Section #1. If we have a homerun conduit with circuits LB1-1, 3, & 5, there will be a total of ___ conductors in this conduit at the panel.

    a. 7   b. 6   c. 3   d. 5   e. 4

Orientation, Safety, Introduction to Meters, Feeder/Branch Circuits

_____ 8. There are ___ GFCI protected receptacles in the BCES Science classroom (D113).

    a. 7   b. 10   c. 13   d. 9   e. 8

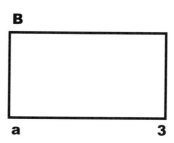

Figure BP1

_____ 9. Refer to Figure BP1, shown above, as shown on BCES drawing E1.1. The number "3" indicates the ___.

    a. circuit number   b. fixture type   c. controlling switch   d. bulb type

_____ 10. FPU 5-4 in BCES Room B141 is connected to circuit ___ in Panel ___.

    a. 19 - LB2   b. 21 - HB   c. 45 - LB2   d. 19 - HE   e. 21 - HE

Lesson 201 NEC Worksheet

_____ 1. Articles 500 through 504 address the electrical requirements for locations where fire or explosion hazards exist due to ___.

   a. flammable gases
   b. flammable liquid–produced vapors
   c. combustible liquid–produced vapors
   d. all of these
   e. none of these

_____ 2. Explosionproof equipment is constructed ___.

   a. to prevent the entry of flammable vapors
   b. to be vaporproof
   c. from soldering, brazing, welding or fusion of glass or metal
   d. to withstand an explosion of gas or vapor and prevent explosion of surrounding flammable atmospheres
   e. none of these

_____ 3. The classification of a location, where ignitable concentrations of flammable liquid–produced vapors or combustible liquid–produced vapors exist under normal operating conditions, is ___.

   a. Class I, Division 1
   b. Class I, Division 2
   c. Class II, Division 1
   d. Class II, Division 2
   e. all of these

_____ 4. The classification of a location, where ignitable concentrations of flammable liquid–produced vapors or combustible liquid–produced vapors above their flash point are normally prevented by positive mechanical ventilation and might become hazardous through failure of abnormal operation of the ventilating equipment, is ___.

   a. Class I, Division 1
   b. Class I, Division 2
   c. Class II, Division 1
   d. Class II, Division 2
   e. all of these

Orientation, Safety, Introduction to Meters, Feeder/Branch Circuits

_____ 5. The classification of a location, where combustible dust due to abnormal operations may be present in the air in quantities sufficient to produce explosive or ignitible mixtures, is ___.

   a. Class I, Division 1
   b. Class I, Division 2
   c. Class II, Division 1
   d. Class II, Division 2
   e. all of these

_____ 6. The classification of a location where combustible dust is in the air under normal operating conditions in quantities sufficient to produce explosive or ignitible mixtures is ___.
   a. Class I, Division 1
   b. Class I, Division 2
   c. Class II, Division 1
   d. Class II, Division 2
   e. Class III, Division 1

_____ 7. The classification of a location where easily ignitible fibers/flyings are stored or handled other than in the process of manufacture is ___.

   a. Class I, Division 1
   b. Class I, Division 2
   c. Class II, Division 1
   d. Class III, Division 2
   e. Class III, Division 1

Lesson 201 Safety Worksheet

Basic Use and Safety for Digital Multimeters

Continuity Tests

Care should be exercised when making continuity checks using a DMM or a tester. Dependent upon the model of the meter (or tester), a continuity test may indicate an open circuit if the resistance of the portion of the circuit under test exceeds the limits of the meter. Some meters and testers have a limit of around 90 ohms, while others have a limit of around 1500 ohms.

For example, the continuity of a small relay coil was tested. The result of the test was that the buzzer failed to sound (or the tester continuity light was not illuminated). This indicated that the coil was open and it should be replaced. However, using the resistance position on a meter, the cold resistance of the coil was 3200 ohms. This is certainly an acceptable reading to indicate that the coil windings are intact.

THE RELAY IS GOOD – DO NOT REPLACE IT!

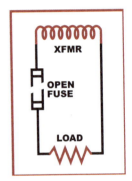

The use of continuity testers on switches and fuses is acceptable. These have very low resistance and a test will yield a correct indication. However, these fuses and switches should be removed from the circuit before testing. In the circuit shown at the right, the transformer has been disconnected from its power source. However, the secondary windings (shown) remain connected to the load and fuse. If the test leads were placed across the open fuse the tester would complete the circuit and **falsely** indicate continuity.

Questions

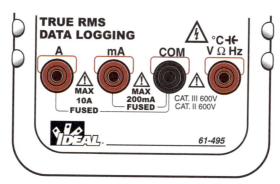

Figure 201.1M

_____ 1. Refer to Figure 201.1M, shown above. You have a 50 watt load operating at 9 volts DC. To most accurately measure the current in this circuit, you should insert the red lead into the jack marked ___. Select "none of these" if the meter capacity will be exceeded.

   a. V Ω Hz   b. COM   c. A   d. mA   e. none of these

_____ 2. Refer to Figure 201.1M, shown above. When inserting the test leads into the meter, the black lead should be inserted in the port panel jack marked ___ for AC voltage measurements.

   a. V Ω Hz   b. mA   c. A   d. COM   e. none of these

_____ 3. Refer to Figure 201.1M, shown above. You have a 20 W load operating at 120 volts AC. To most accurately measure the current in this circuit, you should insert the red lead into the jack marked ___. Select "none of these" if the meter capacity will be exceeded.

   a. mA   b. COM   c. V Ω Hz   d. A   e. none of these

_____ 4. Refer to Figure 201.1M, shown above. You have a 50 watt load operating at 9 volts DC. To most accurately measure the current in this circuit, you should insert the black lead into the jack marked ___. Select "none of these" if the meter capacity will be exceeded.

   a. V Ω Hz   b. A   c. COM   d. mA   e. none of these

_____ 5. Refer to Figure 201.1M, shown above. You have a 2 kW load operating at 120 volts AC. To most accurately measure the current in this circuit, you should insert the red lead into the jack marked ___. Select "none of these" if the meter capacity will be exceeded.

   a. COM   b. mA   c. A   d. V Ω Hz   e. none of these

_____ 6. Refer to Figure 201.1M, shown on the previous page. To check that a switch is "bad" you removed it from the circuit and need to check that the contacts are opening and closing. The red test lead should be inserted in the port panel jack marked ___. Select "none of these" if the meter is not capable of making this measurement.

   a. mA
   b. A
   c. COM
   d. V Ω Hz
   e. none of these

_____ 7. When making voltage measurements, you should ___.

   a. set the meter on a surface
   b. hold the meter in your left hand
   c. hold the meter in your right hand

_____ 8. When setting the range on a DMM for a voltage measurement, if the expected voltage to be measured is 480 volts DC, the range setting should be set on ___.

   a. 2 V ⎓
   b. 600 V ⎓
   c. 600 V AC
   d. 200 V ⎓
   e. 200 V ~
   f. 20 V ⎓

_____ 9. When testing voltage on an AC circuit, the reading on the LCD will ___.

   a. appear as a negative number
   b. appear as a positive number
   c. be a constant number
   d. fluctuate
   e. appear as a sine wave
   f. be blank

_____ 10. Any meter tests on outdoor conductors (overhead or underground) should be performed with a meter that had a CAT ___ rating.

   a. II   b. III   c. I   d. IV

Orientation, Safety, Introduction to Meters, Feeder/Branch Circuits   201-43

_____ 11. Continuity (not resistance readings) tests made on components or equipment, with an expected resistance of 2.2 k Ω, will correctly indicate continuity.

   a. False
   b. True

_____ 12. A DMM is set for a voltage measurement and the range is set to 200 V. If the voltage to be measured is 277 volts, the LCD will display ___.

   a. 277
   b. 2770
   c. 77
   d. OL
   e. 2.77
   f. 27.7

_____ 13. When setting the range on a DMM for a voltage measurement, if the expected voltage to be measured is 18 volts DC, the range setting should be set on ___.

   a. 200 V ⎓
   b. 200m V ~
   c. 200m V ⎓
   d. 200 V ~
   e. 20 V ⎓
   f. 2 V ⎓

_____ 14. Which of these incidents causes more electrical injuries per year?

   a. electric shock
   b. arc blasts

_____ 15. To measure resistance in a circuit using a DMM, the circuit should be ___ and the portion of the circuit under test should ___.

   a. energized – remain in the circuit
   b. energized – be removed from the circuit
   c. deenergized – be removed from the circuit
   d. deenergized – remain in the circuit

Year Two (Student Manual)

_____ 16. When using a Digital Multimeter (DMM) and applying Ohm's Law to measure current, if the voltage is increased, current ___.

   a. goes to zero
   b. is increased
   c. is decreased
   d. remains the same
   e. none of these

_____ 17. When setting the range on a DMM, the precision of measurement is set by moving the decimal point on the LCD screen and is called ___.

   a. action
   b. precision
   c. resolution
   d. definition
   e. delineation

_____ 18. To measure current in a switched circuit using a DMM, the test leads would be placed on the terminal screws of a(an) ___ switch.

   a. open
   b. closed

_____ 19. When a CE marking appears on a piece of equipment, this marking indicates that the equipment conforms to health, safety, environment and consumer protection requirements established by ___.

   a. the European Commission
   b. the China Export consortium
   c. Canadian Electrical testing
   d. the Electrotechnical Committee
   e. none of these

_____ 20. The common voltage for a fire alarm system is ___.

   a. 24 volts DC
   b. 120 volts AC
   c. 9 volts AC
   d. 9 volts DC
   e. 24 volts AC

_____ 21. When testing DC voltage, connect the black test lead to the ___.
  I. positive polarity test point   II. negative polarity test point   III. ground

  a. II or III only
  b. II only
  c. I or II only
  d. I only
  e. I or II or III
  f. I or III only

_____ 22. When continuity tests are made on fuses or switches, these should ___ the circuit.

  a. be removed from
  b. remain connected to

_____ 23. When setting the range on a DMM for a voltage measurement, and you don't know what the expected voltage might be, you should set the voltage range on ___.

  a. best estimated number
  b. the middle number
  c. the lowest number
  d. the highest number
  e. the setting just below the highest number

_____ 24. Which of these meters offers better transient protection?

  a. CAT II - 1000 V
  b. CAT III - 600 V

_____ 25. When testing voltage at switchgear, feeders, short branch circuits, or equipment that are located a short distance from the service equipment, the DMM test equipment should have a CAT rating of ___.

  a. II
  b. III
  c. I
  d. I or II
  e. IV

Lesson 202
Service Calculations

Electrical Curriculum

Year Two
Student Manual

Purpose

This lesson will teach and practice standard and optional loads calculations for branch circuits and feeders for one family and multifamily residential services.

Homework

(Due at the beginning of this class)

For this lesson, you should:

- Thoroughly read the material contained within Lesson 202.
- Read ES Units 3 in its entirety.
- Complete Objective 202.1 Worksheet.
- Complete Objective 202.2 Worksheet.
- Complete Lesson 202 NEC Worksheet
- Read and complete Lesson 202 Safety Worksheet.

- Complete additional assignments or worksheets, if available, and directed by your instructor.

Objectives

By the end of this lesson, you should:

202.1 ....Understand the standard and optional calculation methods for branch circuit and demand load calculations.

202.2 ....Be able to locate equipment and symbols on the BCES prints in order to solve problems addressed on common construction site scenarios.

202.1 Standard and Optional Calculation Methods for Branch Circuit and Demand Load Calculations

Branch circuit and demand calculations are requirements for determining service size and feeder size when wiring One-Family and Multi-Family dwellings. Although it may be argued that the apprentice will not use this knowledge until they become a contractor, this working knowledge is critical to becoming a competent journeyman electrician. The NEC contains numerous examples of branch circuit and service calculations in Annex D. It is necessary for the apprentice to become familiar and proficient in this section of the code in order to be a valued and competent employee.

202.2 Problem Solving for BCES Blueprints

Service Calculations

Objective 202.1 Worksheet

_____ 1. Demand Factor is defined as ___.

    a. the ratio of the maximum demand of a system, or part of a system, to the total connected load of a system or the part of the system under consideration
    b. the difference between the maximum demand of a system, or part of a system, and the total connected load of a system or the part of the system under consideration
    c. 125% of the connected load less the maximum demand of the calculated load
    d. none of these

_____ 2. The feeder neutral load is the maximum unbalance between ___.

    a. any one of the ungrounded conductors and the grounded conductor
    b. any two of the ungrounded conductors
    c. the larger of a or b
    d. the smaller of a or b

_____ 3. The Optional Calculation for dwelling units consists of heating and A/C loads and ___ loads.

    a. lighting   b. largest motor   c. small appliance   d. general

_____ 4. For the calculations required in Article 220, the total area of an occupancy is determined from the ___ dimensions of the building or structure.

    a. inside
    b. outside
    c. either A or B
    d. neither A nor B

_____ 5. Using the standard calculation, four or more appliances which are fastened-in-place in a one-family dwelling are computed at ___ of the total load for all four appliances.

    a. 25   b. 75   c. 80   d. 125

_____ 6. After the minimum ampacity of the service has been calculated, Table ___ and 310.15(B) are used to determine the size of feeder or service-entrance conductors.

    a. 250.122   b. 250.66   c. 310.16   d. 300.5

Year Two (Student Manual)

Service Calculations

Objective 202.2 Worksheet

_____ 1. The mounting locations for the light fixtures in BCES's Room B133 are shown on Drawing A10.2.

   a. True   b. False

_____ 2. As indicated on the BCES food service drawings, the refrigeration rack is rated to pull ___ amps at rated voltage.

   a. 5.8   b. 60.9   c. 16.1   d. 42.5   e. 48.7

_____ 3. On the BCES project, EWH-3 (Electric Water Heater), has a rating of ___ KW.

   a. 1.5   b. 4.0   c. 19.2   d. 12.0   e. 4.5

_____ 4. The maximum number of ungrounded branch circuit conductors (on different phases) that can be installed in a conduit that contains only one neutral is ___ for the BCES project.

   a. 2   b. 1   c. 3   d. unlimited

_____ 5. On the BCES plans the symbol 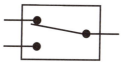 is used for a(an) ___.

   a. mechanically-held lighting contactor
   b. three-way switch
   c. hand-off-auto switch
   d. automatic transfer switch
   e. none of these

_____ 6. On the BCES print MEP1.1, the Telephone Company requires ___ underground conduits.

   a. 3-4″   b. 2-4″   c. 4-3″   d. 3-3″

_____ 7. There are ___ duplex receptacles in the BCES gym that are connected to circuit LD-81.

   a. 6   b. 0   c. 4   d. 3   e. 1

Year Two (Student Manual)

Service Calculations

_____ 8. MSB is located in BCES Room ___.

   a. C126   b. B139   c. D154   d. B125   e. C128

_____ 9. On the BCES project, the demand load for the kitchen equipment, @ 65%, is ___ KVA.

   a. 122   b. 131   c. 1398   d. 164

_____ 10. The light fixtures in BCES's Corridor A110 are to be centered in the corridor.

   a. True   b. False

_____ 11. On the BCES print MEP1.1, the Texas New Mexico Power Company requires ___ underground primary conduits.

   a. 4-3"   b. 4-2"   c. 3-4"   d. 2-4"

_____ 12. On the BCES print MEP1.1, the scale used is ___.

   a. 1/4" = 1'   b. 1/8" = 1'   c. 1/16" = 1'   d. 1/32" = 1'

_____ 13. The transformers on the BCES project have a K rating of K- ___.

   a. 20   b. 3   c. 1   d. 13

_____ 14. On the BCES prints, a panel designation of "LG" represents a ___ panelboard.

   a. 120/208V   b. 277/480V   c. 120/240V

_____ 15. The total calculated load for BCES is ___ kVA.

   a. 493   b. 100,000   c. 1682   d. 2500   e. 1398

_____ 16. There are ___ duplex receptacles to be installed in single-gang boxes in BCES Room B107.

   a. 3   b. 6   c. 1   d. 4   e. 0

Service Calculations

_____ 17. On the BCES project, the connected load for the lighting is ___ KVA.

    a. 122   b. 1398   c. 164   d. 131

_____ 18. On the BCES prints, the electrical service one line diagram is found on drawing ___.

    a. E3.1   b. E4.2   c. E4.1   d. E5.1

_____ 19. On the BCES project, the electrical service size is ___ amperes.

    a. 3000   b. 2500   c. 2000   d. 1000

_____ 20. On the BCES project, there is a total of ___ distribution transformers are located inside the building.

    a. 8   b. 9   c. 7   d. 5   e. 10   f. 4

Lesson 202 NEC Worksheet

_____ 1. The Class and Material Group Classification where acetylene gas is present is ___.

   a. Class I, Group A
   b. Class I, Group B
   c. Class I, Group C
   d. Class I, Group D
   e. Class II, Group B

_____ 2. The Class and Group Classification where combustible metal dusts such as magnesium are present is ___.

   a. Class II, Group A
   b. Class II, Group E
   c. Class II, Group F
   d. Class II, Group G
   e. Class II, Group B

_____ 3. Flour mills and grain elevators contain atmospheres of combustible dusts that require equipment to be classified as ___, in accordance with NEC© Section 500.6.

   a. Class I, Group A
   b. Class I, Group B
   c. Class I, Group D
   d. Class II, Group E
   e. Class II, Group G

_____ 4. Which of the following is an acceptable Protection Technique for electrical and electronic equipment in hazardous locations?

   a. Explosionproof Apparatus
   b. Purged and Pressurized
   c. Intrinsic Safety
   d. Hermetically Sealed
   e. all of these

Lesson 202 Safety Worksheet

Subpart E – Personal Protective and Life Saving Equipment

OSHA CFR1926.95

Criteria for personal protective equipment

(a) Application. Protective equipment, including personal protective equipment for eyes, face, head, and extremities, protective clothing, respiratory devices, and protective shields and barriers, shall be provided, used and maintained in a sanitary and reliable condition wherever it is necessary by reason of hazards of processes or environment, chemical hazards, radiological hazards, or mechanical irritants encountered in a manner capable of causing injury or impairment in the function of any part of the body through absorption, inhalation or physical contact.

(b) Employee-owned equipment. Where employees provide their own protective equipment, the employer shall be responsible to assure its adequacy, including proper maintenance, and sanitation of such equipment.

(c) Design. All personal protective equipment shall be of safe design and construction for the work to be performed.

1926.96

Occupational foot protection

Safety-toe footwear for employees shall meet the requirements and specifications in American National Standard for Men's Safety-Too Footwear, Z41.1-1967.

1926.100

Head protection

(a) Employees working in areas where there is a possible danger of head injury from impact, or from falling or flying objects, or from electrical shock and burns, shall be protected by protective helmets.

(b) Helmets for the protection of employees against impact and penetration of falling and flying objects shall meet the specifications contained in American National Standards Institute, Z89.1-1969, Safety Requirements for Industrial Head Protection.

(c) Helmets for the head protection of employees exposed to high voltage electrical shock and burns shall meet the specifications contained in American National Standards Institute, Z89.2-1971.

The above sections are taken from the U.S. Department of Labor, CFR1926 Safety and Health Regulations for Construction.OSHA CFR1926.101

Hearing protection
(a) Wherever it is not feasible to reduce the noise levels or duration of exposures to those specified in Table D-2. Permissible Noise Exposures, in 1926.52, ear protective devices shall be provided and used.
(b) Ear protective devices inserted in the ear shall be fitted or determined individually by competent persons.
(c) Plain cotton is not an acceptable protective device.

1926.52

(a) Protection against the effects of noise exposure shall be provided when the sound levels exceed those shown in Table D-2 of this section when measured on the A-scale of a standard sound level meter at slow response.
(b) When employees are subjected to sound levels exceeding those listed in Table D-2 of this section, feasible administrative or engineering controls shall be utilized. If such controls fail to reduce sound levels within the levels of the table, personal protective equipment as required in Subpart E, shall be provided and used to reduce sound levels within the levels of the table.
(c) If the variations in noise level involve maxima at intervals of 1 second or less, it is to be considered continuous.
(d) (1) In all cases where the sound levels exceed the values shown herein, a continuing, effective hearing conservation program shall be administered.

TABLE D-2 - PERMISSIBLE NOISE EXPOSURES	
Duration per day, hours	Sound level dBA slow response
8	90
6	92
4	95
3	97
2	100
1 ½	102
1	105
½	110
¼ or less	115

Questions

_____ 1. Where employees provide their own protective equipment, the ___ shall be responsible to assure its adequacy, including proper maintenance and sanitation.

   a. tool man   b. employee   c. employer   d. none of these

_____ 2. Personal Protective Equipment for ___ and extremities shall be used and maintained.

   a. eyes   b. face   c. hands   d. all of these

_____ 3. Safety toe footwear for employees shall meet the requirements and specifications of ___.

   a. ANSI   b. AAA   c. NEMA   d. NHRA

_____ 4. Employees working in areas where there is a possible danger of head injury shall be protected by protective ___.

   a. hoods   b. shields   c. guards   d. helmets

_____ 5. All personal protective equipment shall be of ___ design and construction for the work to be performed.

   a. high quality   b. safe   c. the latest   d. the best

_____ 6. Ear protective devices inserted in the ear shall be fitted or determined by ___ persons.

   a. two   b. competent   c. three   d. any

_____ 7. In all cases where the sound levels ___ the values shown, a continuing, effective hearing conservation program shall be administered.

   a. energizes   b. exceeds   c. exhausts   d. enlarges

_____ 8. Wherever it is not ___ to reduce the noise levels or duration of exposures, ear protective devices shall be used.

   a. fashionable   b. preferred   c. feasible   d. none of these

_____ 9. The maximum permissible noise exposure level, at 97 db, is ___ hours.

    a. 8   b. 3   c. 6   d. 4

_____ 10. When employees are subjected to sound levels exceeding those in table D-2, feasible administrative or ___ controls shall be utilized.

    a. engineering   b. estimating   c. enlarging   d. enriching

_____ 11. Plain cotton can be used as a substitute for other forms of hearing protection.

    a. sometimes   b. yes   c. never

Industrial Clamp Meters

- 660 and 1000 amp clamps for a variety of industrial and commercial environments

- Unique TightSight™ bottom display makes hard-to-reach measurements convenient

- New easy-to-read displays feature large numbers and green glowing backlight

- CAT IV rating for maximum safety

- Selectable audible voltage presence warning on all functions

Exclusive TightSight™ Display

Lesson 203
Services

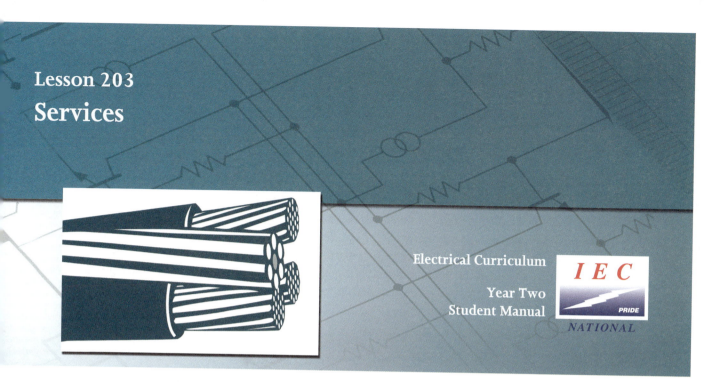

Electrical Curriculum

Year Two
Student Manual

Purpose

This lesson will discuss the equipment, conductors and requirements for services on residential and commercial installations.

Homework

(Due at the beginning of this class)

For this lesson, you should:

- Thoroughly read the material contained within Lesson 203.
- Read ES Chapter 4 in its entirety.
- Complete Objective 203.1 Worksheet.
- Complete Objective 203.2 Worksheet.
- Complete Objective 203.3 Worksheet.
- Complete Objective 203.4 Worksheet.
- Complete Objective 203.5 Worksheet.
- Complete Lesson 203 NEC Worksheet
- Read and complete Lesson 203 Safety Worksheet.

- Complete additional worksheets, if available, as directed by your instructor.

Objectives

By the end of this lesson, you should:

203.1 ....Understand and recognize the key components and the limitations for electrical service entrance conductors.
203.2 ....Understand the requirements and clearances for service drops.
203.3 ....Understand the requirements for service laterals and service entrance conductors.
203.4 ....Understand the requirements for service disconnects.
203.5 ....Be able to locate equipment and symbols on the BCES prints in order to solve problems addressed on common construction site scenarios.

203.1 Key Components and the Limitations for Electrical Service Entrance Conductors

203.2 Requirements and Clearances for Service Drops

Messenger Supported Wiring (multiplex cables) are addressed in Article 396 in the NEC. Service drops of duplex, triplex, or quadruplex messenger cables typically utilize wedge clamps to attach the cable to the point of support. The wedge clamp, shown at the right, is installed on the bare conductor. These are available in ranges that are selected based on the size conductor to be clamped.

Wedge Clamp

Cable cutters should not be used to cut conductors larger than that for which they are rated. Additionally, most cable cutters should not be used to cut ACRS (Aluminum Conductor, Steel Reinforced) cable. The bare conductor in these overhead cables (such as duplex, triplex, or quadruplex) contains a spring steel reinforcement wire in the center (as shown in the following cables) that will damage the cutter.

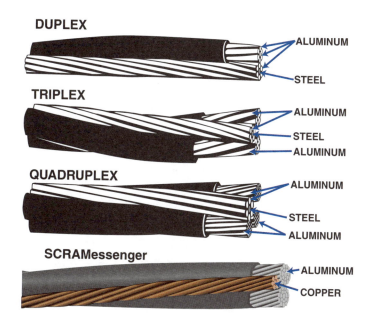

Messenger cable with a bare copper conductor is also available.

Southwire markets their SCRAMessenger, as shown above. It is used to supply power, usually from a pole-mounted transformer, to the user's service head in areas where squirrels are known to damage the bare neutral by chewing. Damage can result in complete loss of the neutral, known to cause voltage stability problems at the service. Used at a maximum voltage of 600 volts phase-to-phase, or 480 volts to ground, and at conductor temperatures not to exceed 75°C for polyethylene conductors (poly) or 90°C for crosslinked polyethylene (XLP) insulated conductors. Conductors are concentrically stranded, compressed 1350-H19 aluminum, insulated with either polyethylene or XLP. Neutral messenger is a bare, hard-drawn concentrically stranded copper SCRAMessenger (Southwire Copper Rodent Adverse Messenger). These cables are available in sizes 4 to 4/0 AWG with a full-size neutral.

203.3 Service Laterals and Service Entrance Conductors

203.4 Service Disconnects.

203.5 Problem Solving for BCES Blueprints

Lesson 203.1 Worksheet

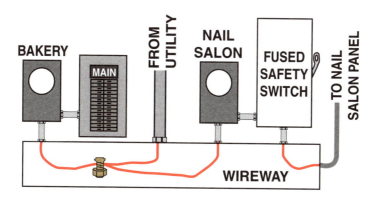

Figure 201.2

Only one conductor is shown in the wireway.

_____ 1. A "gutter service" is built on the back of a strip shopping center, as shown above in Figure 201.2. The service conductors enter a wireway from the utility. These are then tapped in the wireway to feed the meters. The conductors between the safety switch and the nail salon panel are permitted to be be routed through the wireway under what conditions?

   a. when protected by GFPE
   b. when they are are 1/0 or greater
   c. when they are unspliced
   d. any of these
   e. under no condition

_____ 2. The service point is located at the ___. The Power Company owns the pad-mounted transformer and a service lateral has been installed to the meter socket on the customer's building.

   a. weatherhead
   b. pole-mounted transformer
   c. meter socket
   d. pad-mounted transformer

_____ 3. When special permission is granted by the AHJ this is in the form of ___.

   a. verbal permission during an inspection
   b. a written document
   c. a telephone conversation with the chief inspector
   d. City ordinances

_____ 4. The cable connector shown at ___ is a mechanical type connector.

a.

b.

c. both a & b
d. neither a nor b

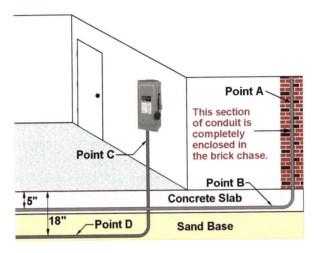

Figure 203.1

_____ 5. Refer to Figure 203.1, above. The conduit at Point B is considered to be ___ the building per the NEC®.

a. outside   b. inside

_____ 6. The cable connector shown at ___ is a compression type connector.

a.

b.

c. both a & b
d. neither a nor b

Services

_____ 7. A building or structure can have a service separate from the main service where it supplies ___.

   a. a fire pump   b. a standby system   c. either a or b   d. neither a nor b

Lesson 203.2 Worksheet

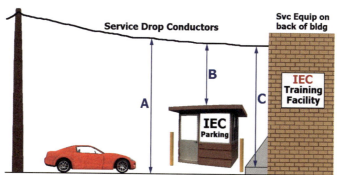

Figure 203.5

_____ 1. Figure 203.5, above, shows a 120/208 volt 3Ø, 4-wire quadruplex drop. The minimum clearance at A is ___ feet. Trucks are prohibited in the parking lot.

   a. 18   b. 8   c. 3   d. 10   e. 12

_____ 2. Refer to Figure 203.5, shown above. The minimum clearance at C, above the sidewalk, is ___ feet for this 120/240 volt 1Ø, 3-wire triplex drop.

   a. 8   b. 18   c. 12   d. 10   e. 3

Year Two (Student Manual)

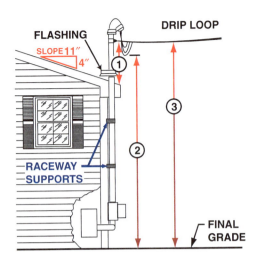

Figure 203.4

_____ 3. Refer to Figure 203.4, shown above. This 480/277 volt 3Ø, 4-wire service supplies an attorney's office. The minimum clearance at ③ is ___. The overhead conductors are quadruplex cable and this area is accessible to automobiles but not trucks.

    a. 10′   b. 12′   c. 18′   d. 15′   e. 8′

_____ 4. Refer to Figure 203.4, shown above. For a 120/240 volt 1Ø, 3-wire service the minimum clearance above the roof at ① is ___.

    a. 3′   b. 8′   c. 18″   d. 12″   e. 4′

_____ 5. The smallest aluminum service drop conductors that can be installed for a service are ___.

    a. 4   b. 2   c. 6   d. 8   e. 3

_____ 6. The smallest THHN/THWN copper grounded conductor that can be installed for a service is ___ when the ungrounded conductors are 1/0 THWN/THHN copper and the load on the neutral is expected to be 120 amps. All terminations are rated at 75° C.

    a. 1   b. 2   c. 3   d. 6   e. 1/0

Lesson 203.3 Worksheet

_____ 1. Drawing ___ meets the requirements of 230.54(C).

a.
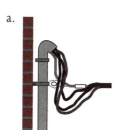

b.
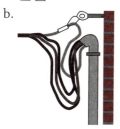

c. either a or b
d. neither a nor b

_____ 2. Which of these wiring methods is NOT permitted to be used for service-entrance conductors?

a. AC cable   b. ENT ("smurf pipe")   c. LFNC   d. Schedule 40 PVC   e. RMC

_____ 3. Underground service laterals that are installed as direct buried cable and covered with 42 inches of dirt shall also have a warning ribbon installed at least ___ inches above the cable.

a. 12   b. 30   c. 24   d. 18   e. 36

_____ 4. The smallest aluminum service lateral conductors that can be installed for a service are ___.

a. 6   b. 8   c. 2   d. 4   e. 3

_____ 5. A service is to be built on the base support pole of a two-sided billboard. Two branch circuits provide power to the lights. Each circuit illuminates one side and will be loaded to 6 amps. The minimum ampacity for the service ungrounded conductors is ___ amps.

a. 15   b. 60   c. 125   d. 30   e. 100

Objective 203.4 Worksheet

_____ 1. A service disconnecting means shall be installed at a(n) ___ location either outside of a building or structure or inside nearest the point of entrance of the service conductors.

   a. accessible   b. readily accessible

_____ 2. The service disconnecting means for a 3-phase service is permitted to consist of ___.

   a. not more than 6 separate circuit breaker enclosures, each with a 3-pole breaker
   b. a fusible switchboard with a maximum of 6 switches
   c. a main breaker panelboard with eight individual 3-pole circuit breakers
   d. any of the above
   e. none of the above

_____ 3. The smallest main breaker that can be installed in a 24-circuit service panel on a commercial building is ___ amps.

   a. 125   b. 100   c. 30   d. 60   e. 15

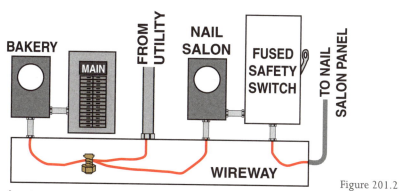

Figure 201.2

Only one conductor is shown in the wireway.

_____ 4. Refer to Figure 201.2, shown above. The main breaker in the Bakery panel provides ___ protection for the ungrounded service conductors between the breaker and the utility transformer.

   a. short circuit   b. ground fault   c. overload   d. any of these

_____ 5. Service equipment shall have an ampere interrupting rating (AIR) ___ the fault current available at the service location.
   I. greater than   II. equal to   III. less than

   a. II or III only   b. I only   c. I or II only   d. II only

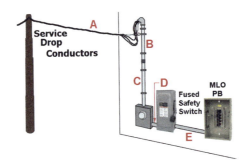

Figure 203.10

_____ 6. The conductors at Point B, in the drawing shown above in Figure 203.10, are protected against ___.

   a. short-circuit   b. overload   c. ground-fault   d. all of these   e. none of these

_____ 7. Copper conductors are installed underground between a utility transformer and a meter can. These conductors continue to a main breaker panel. The conductors between the transformer and the meter are protected against ___ currents.

   a. ground-fault   b. overload   c. short-circuit   d. all of the above   e. none of the above

_____ 8. Which of the following requires that a GFPE device be installed?

   a. 800 A safety switch with 800 A fuses for a 480/277 volt, 3Ø, 4-wire service
   b. 277/480 volt, 3Ø, 4-wire switchgear with a 1200 Amp Main Breaker
   c. both a and b
   d. neither a nor b

Services

Objective 203.5 Worksheet

_____ 1. On BCES drawing E1.1, where there is concealed electrical equipment you are required to ___.

    a. mark the final print "as builts"
    b. provide an access door
    c. notify the owner
    d. all of these

_____ 2. A ___ ampere circuit breaker is to be installed for the BCES dryer outlet.

    a. 40   b. 15   c. 20   d. 25   e. 30

_____ 3. On the BCES project, CHP-1 is fed from Panel ___.

    a. HB2   b. HE   c. HD   d. HL

_____ 4. On the BCES project, the service meter is provided by ___.

    a. the owner
    b. the mechanical contractor
    c. the electrical contractor
    d. Texas New Mexico Power Company

_____ 5. What Trade Size conduits for the BCES service feeder conductors are installed between the pad mount transformer and the main distribution board (MSB)?

    a. 3   b. 2 1/2   c. 6   d. 4   e. 3 1/2

_____ 6. For BCES the minimum size galvanized threaded rod that can be used to support suspended transformers is ___ inch.

    a. 3/4   b. 1/2   c. 5/8   d. 1/4   e. 3/8

_____ 7. On the BCES prints, the parking lot light pole base detail is located on Drawing ___.

    a. E4.4   b. MEP1.1   c. E5.2   d. MEP1.2   e. E5.1

_____ 8. The generator receptacle and manual transfer switch are located ___.

   a. in the boiler room - D156
   b. on the east exterior wall of BCES Area D
   c. on the roof of Area D, as shown on MEP1.2
   d. in the main electrical room - D154

Lesson 203 NEC Worksheet

_____ 1. As addressed in NEC 500.8, generally, equipment for use in hazardous locations shall be marked to show the environment for which it has been evaluated. Equipment that is not marked to indicate a division or is marked as Division 1, is ___.

   a. suitable for Division 1 only
   b. suitable for both Divisions 1 and 2
   c. not suitable for Division 2
   d. allowed to be used for Division 2 in special circumstances
   e. not allowed to be used for Division 1 or

_____ 2. In Article 500.8, Equipment marked for use in Class I locations and has a marked temperature classification of T3B is suitable for use in environments with a maximum surface temperature of ___.

   a. 356°F
   b. 329°F
   c. 180°C
   d. 260°C
   e. 419°F

_____ 3. Except for listed explosionproof equipment with factory threaded entries, all National Standard Pipe Taper threaded conduit entries into explosionproof equipment and threaded in the field shall have at least ___ threads fully engaged.

   a. five
   b. four
   c. six
   d. four and one-half
   e. three

_____ 4. All National Standard Pipe Taper threaded conduit entries into explosionproof equipment shall ___.

   a. have threads that provides a taper of 3/4 inch per foot
   b. be made wrenchtight
   c. ensure the explosionproof integrity of the conduit system
   d. be made up with at least five threads fully engaged
   e. all of these

_____ 5. Where used in hazardous locations, metric threaded conduit entries into equipment ___.

    I. shall not be permitted
    II. shall be permitted for use with the inclusion of listed NPT adapters
    III. shall be permitted where listed cable fittings with metric threads are used in the identified metric entries

    a. II only
    b. III only
    c. either II & III
    d. I only

EYE AND FACE PROTECTION

OSHA CFR1926.102

(a) General.
1. Employees shall be provided with eye and face protection equipment when machines or operations present potential eye or face injury from physical, chemical, or radiation agents.
2. Eye and face protection equipment required by this Part shall meet the requirement specified in American National Standards Institute, Z89.1-1968, Practice for Occupational and Educational Eye and Face Protection.
3. Employees whose vision requires the use of corrective lenses in spectacles, when required by this regulation to wear eye protection, shall be protected by goggles or spectacles of one of the following types:
   i. Spectacles whose protective lenses provide optical correction;
   ii. Goggles that can be worn over corrective spectacles without disturbing the adjustment of the spectacles;
   iii. Goggles that incorporate corrective lenses mounted behind the protective lenses.
4. Face and eye protection equipment shall be kept clean and in good repair. The use of this type equipment with structural or optical defects shall be prohibited.
5. Table E-1 shall be used as a guide in the selection of face and eye protection for hazards and operations noted.

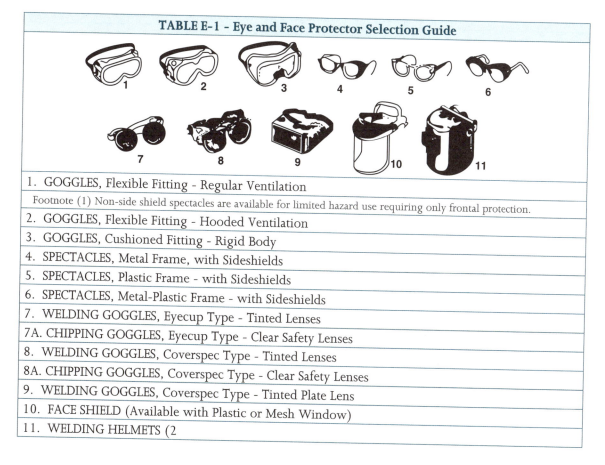

TABLE E-1 - Eye and Face Protector Selection Guide

1.	GOGGLES, Flexible Fitting - Regular Ventilation
	Footnote (1) Non-side shield spectacles are available for limited hazard use requiring only frontal protection.
2.	GOGGLES, Flexible Fitting - Hooded Ventilation
3.	GOGGLES, Cushioned Fitting - Rigid Body
4.	SPECTACLES, Metal Frame, with Sideshields
5.	SPECTACLES, Plastic Frame - with Sideshields
6.	SPECTACLES, Metal-Plastic Frame - with Sideshields
7.	WELDING GOGGLES, Eyecup Type - Tinted Lenses
7A.	CHIPPING GOGGLES, Eyecup Type - Clear Safety Lenses
8.	WELDING GOGGLES, Coverspec Type - Tinted Lenses
8A.	CHIPPING GOGGLES, Coverspec Type - Clear Safety Lenses
9.	WELDING GOGGLES, Coverspec Type - Tinted Plate Lens
10.	FACE SHIELD (Available with Plastic or Mesh Window)
11.	WELDING HELMETS (2

Applications		
Operation	Hazards	Recommended Protectors
Acetylene - Burning, Acetylene – Cutting, Acetylene - Welding	Sparks, harmful rays, molten metal, flying particles	7, 8, 9
Chemical Handling	Splash, acid burns, fumes	2, 10 (For severe exposure add 10 over 2).
Chipping	Flying particles	1, 3, 4, 5, 6, 7A, 8A
Electric (arc) Welding	Sparks, intense rays, molten metal	9, 11 (11 in combination with 4, 5, 6 in tinted lenses advisable)
Furnace Operations	Glare, heat, molten metal	7, 8, 9 (For severe exposure add 10)
Grinding- Light	Flying particles	1, 3, 4, 5, 6, 10
Grinding- Heavy	Flying particles	1, 3, 7A, 8A (For severe exposure add 10)
Laboratory	Chemical splash, glass breakage	2 (10 when in combination with 4, 5, 6)
Machining	Flying particles	1, 3, 4, 5, 6, 10
Molten Metals	Heat, glare, sparks, splash	7, 8, (10 in combination with 4, 5, 6 in tinted lenses)
Spot Welding	Flying particles, sparks	1, 3, 4, 5, 6, 10

6. Protectors shall meet the following minimum requirements:
   (i) They shall provide adequate protection against the particular hazards for which they are designed.
   (ii) They shall be reasonably comfortable when worn under the designated conditions.
   (iii) They shall fit snugly and shall not unduly interfere with the movements of the wearer.
   (iv) They shall be durable.
   (v) They shall be capable of being disinfected.
   (vi) They shall be easily cleanable.
7. Every protector shall be distinctly marked to facilitate identification only of the manufacturer.
8. When limitations or precautions are indicated by manufacturer, they shall be transmitted to the user and care taken to see that such limitations and precautions are strictly observed.
   (b) Protection against radiant energy.
      (1) Selection of shade numbers for welding filter. Table E-2 shall be used as a guide for the selection of the proper shade numbers of filter lenses or plates used in welding. Shades more dense than those listed may be used to suit the individual's needs.
      (2) Laser protection.
         (i) Employees whose occupation or assignment requires exposure to laser beams shall be furnished suitable laser safety goggles which will protect for the specific wavelength of the laser and be of optical density (O.D.) adequate for the energy involved. Table E-3 lists the maximum power or energy density for which adequate protection is afforded by glasses of optical densities from 5 through 8.

Table E-2 – Filter Lens Shade Numbers for Protection Against Radiant Energy	
Welding Operation	**Shade Number**
Shielded metal-arc welding 1/16, 3/32, 1/8, 5/32-inch diameter electrodes	10
Gas-shielded arc welding (nonferrous) 1/16, 3/32, 1/8, 5/32-inch diameter	11
Gas-shielded arc welding (ferrous) 1/16, 3/32, 1/8, 5/32-inch diameter	12
Shield metal-arc welding 3/16, 7/32, ¼-inch diameter electrodes	12
5/16, 3/8-inch diameter electrodes	14
Atomic hydrogen welding	10 – 14
Carbon-arc welding	14
Soldering	2
Torch brazing	3 or 4
Light cutting, up to 1 inch	3 or 4
Medium cutting, 1 inch to 6 inches	4 or 5
Heavy cutting, over 6 inches	5 or 6
Gas welding (light), up to 1/8-inch	4 or 5
Gas welding (medium), 1/8-inch to ½-inch	5 or 6
Gas welding (heavy), over ½-inch	6 or 8

Footnote (2) See Table E-2, in paragraph (b) of this section, Filter Lens Shade Numbers for Protection Against Radiant Energy.

TABLE E-3 - SELECTING LASER SAFETY GLASS		
Intensity, CW maximum power density (watts/cm²)	**Attenuation**	
	Optical density O.D.	Attenuation Factor
10^{-2}	5	10^5
10^{-1}	6	10^6
1.0	7	10^7
10.0	8	10^8

Output levels falling between lines in this table shall require the higher optical density.

(ii) All protective goggles shall bear a label identifying the following data:
    (a) The laser wavelengths for which use is intended;
    (b) The optical density of those wavelengths;
    (c) The visible light transmission.

Questions

_____ 1. Employees whose occupation or assignment requires exposure to laser beams shall be furnished with suitable laser safety ___.

    a. lenses   b. protection   c. goggles   d. all of these

_____ 2. Employees, whose vision requires use of corrective lenses in spectacles, shall be protected by goggles or spectacles which ___.

    a. are colorful
    b. are ventilated
    c. provide optical correction

_____ 3. Face and Eye protection equipment shall be kept ___.

    a. in your truck   b. at home   c. at the shop   d. clean

_____ 4. Flying particles from chipping require type 1, 3, 4, 5, or ___.

    a. 6   b. 7A   c. 8A   d. all of these

_____ 5. Employees shall be provided with eye and face protection when machines or operations present potential eye or face injury from ___.

    a. chemicals   c. physical   b. radiation agents   d. all of these

Lesson 204
Conductor and Overcurrent Protection

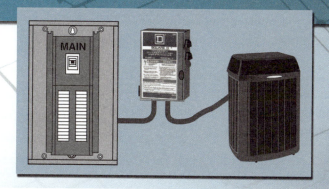

Electrical Curriculum
Year Two
Student Manual

Purpose
This lesson will teach ampacity, insulation ratings and overcurrent protection for electrical systems.

Homework
(Due at the beginning of this class)

For this lesson, you should:

- Thoroughly read the material contained within Lesson 204.
- Read ES Chapter 5 in its entirety.
- Complete Objective 204.1 Worksheet.
- Complete Objective 204.2 Worksheet.
- Complete Objective 204.3 Worksheet.
- Complete Objective 204.4 Worksheet.
- Complete Objective 204.5 Worksheet.
- Complete Objective 204.6 Worksheet.
- Complete Lesson 204 NEC Worksheet
- Read and complete Lesson 204 Safety Worksheet.

- Complete additional worksheets, if available, as directed by your instructor.

Objectives

By the end of this lesson, you should:

    204.1 ....Understand characteristics and installation practices of bare and insulated conductors.

    204.2 ....Understand ampacity, insulation ratings and ambient temperature of circuit conductors.

    204.3 ....Understand the purpose and selection of fuses and circuit breakers for the protection of conductors and equipment.

    204.4 ....Understand the requirements and limitations for the location of OCPD's.

    204.5 ....Understand the operation and application of OCPD's.

    204.6 ....Be able to locate equipment and symbols on the BCES prints in order to solve problems addressed on common construction site scenarios.

Conductor and Overcurrent Protection

204.1 Characteristics and Installation Practices of Bare and Insulated Conductors

204.2 Ampacity, Insulation Ratings and Ambient Temperature of Circuit Conductors

204.3 Purpose and Selection of Fuses and Circuit Breakers for the Protection of Conductors and Equipment

Almost all molded case circuit breakers are to be derated to carry 80% of continuous loads. Although some electronic-trip breakers are rated for 100% of continuous loads, you should assume that a breaker in question should be derated, unless specifically stated otherwise. For example, if a 100 A load is to operate continuously, a 125 A breaker should be selected (125 A x 80% = 100 A). A 125 A breaker can carry a maximum of 100 A of continuous load. Of course, it is rated to carry 125 A of noncontinuous load.

Circuit breakers are listed and labeled for the temperature rating of the wire permitted to be connected to its terminals. If a breaker has a label that shows 75°C then either 75 or 90° wire can be used. Where 90°C wire is used, ampacity corrections beyond the *point of transition** are made in the 90°C column of the ampacity table. However, the ampacity of the 90°C wire at the breaker terminals, and out to the point of transition, is taken from the 75° column of the ampacity table.

Typical installations utilize THHN/THWN conductors. These, of course, are dual-rated at 90/75°C. For application questions presented in worksheets or on exams, you can assume that required ampacity corrections are made beyond the point of transition, unless specifically noted otherwise.

*Point of transition: NEC® 310.60(B)(1) Exception states that this point is the lesser of 10 feet or 10% of the circuit length.

204.4 Requirements and Limitations for the Location of OCPD's

204.5 Operation and Application of OCPD's

Year Two (Student Manual)

204.6 Problem Solving for BCES Blueprints

Conductor and Overcurrent Protection

Objective 204.1 Worksheet

_____ 1. 8 AWG copper branch circuit conductors are installed in PVC to serve a load. The ungrounded conductors are permitted to be ___.
   I. solid   II. stranded   III. Bare

   a. I or II only   b. II or III only   c. II only   d. I only   e. I, II, or III

_____ 2. Two parallel sets of conductors are to be connected to lugs in a transformer and to lugs in a fused safety switch. Does the NEC permit the conductors in the transformer to be connected with single lugs and the conductors in the safety switch to be connected with double lugs?

   a. Yes   b. No

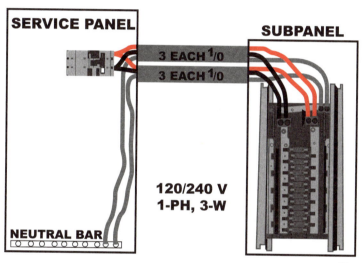

Figure 204.1

The service panel feeds the subpanel with two parallel conduits from a 2P 225 A breaker. There are two "hots" and one neutral in each conduit. AØ loads in the subpanel are pulling 210 amps and BØ loads in the subpanel are pulling 170 amps.

_____ 3. Refer to the illustration and the information shown above in Figure 204.1. The AØ conductor in each conduit should be carrying ___ amps of current.

   a. 20   b. 105   c. 85   d. 0   e. 40

_____ 4. A bare conductor can be used for a grounding electrode conductor where it is installed between a service panel and a ground rod.

   a. True   b. False

Year Two (Student Manual)

_____ 5. A bare conductor can be used for a grounding conductor where it is installed in conduit with branch circuit conductors.

a. True   b. False

Scenario 204.1

Three parallel conduits are installed between a service panel and a subpanel. Each panel is 120/240 volt 1-ph, 3-wire. Three 3/0 THHN/THWN copper conductors are installed in each conduit. There are two ungrounded and one grounded conductor in each conduit. The subpanel is supplying 480 amps of current to all the loads connected to AØ and 450 amps to all the loads connected to BØ.

_____ 6. Refer to Scenario 204.1, shown above. How much current is flowing in each of the three BØ conductors that are run in the conduits?

a. 150   b. 160   c. 260   d. 200   e. 450

_____ 7. Refer to Scenario 204.1, shown above. How much current is flowing in each of the three neutrals that are run in the conduits?

a. 150   b. 200   c. 160   d. 10   e. 30

_____ 8. Which 2 AWG conductor has a lower resistance?

a. copper   b. aluminum

_____ 9. Which of the following are characteristics of THHN/THWN insulation?

a. It is heat-resistant.
b. It is available is sizes as small as 14 AWG and as large as 1000 KCMIL.
c. It can be installed in wet, damp, or dry locations.
d. It is flame-retardant.
e. All (a through d) are true.

_____ 10. When conductors are installed in PVC in a concrete slab the conductors must have an insulation that is rated for installation in ___ locations.

a. hazardous   b. damp   c. dry   d. wet

_____ 11. A bare conductor can be used for an ungrounded branch circuit conductor.

a. False   b. True

1	1.1 KVA load	1.5 KVA load	2	Schedule
3	SPACE	SPACE	4	Panel LB
5	1.4 KVA load	1.4 KVA load	6	
7	SPARE	SPARE	8	
9	SPACE	SPARE	10	120/240 V
11	2.0 KVA load	11.8 KVA load (2-P)	12	1Ø, 3W
13	1.7 KVA load		14	

_____ 12. Refer to the schedule for Panel LB, shown above. Branch circuit LB-11 is run from the panel to the load. The neutral should have to carry ___ amps of current.

   a. 9.61   b. 8.33   c. 4.81   d. 0.96   e. 16.66

1	1.1 KVA LOAD	1.5 KVA LOAD	2	SCHEDULE
3	SPACE	SPACE	4	
5	1.4 KVA LOAD	1.4 KVA LOAD	6	
7	SPARE	SPARE	8	PANEL LD
9	SPACE	11.8 KVA LOAD (3-POLE)	10	
11	2.0 KVA LOAD		12	
13	1.7 KVA LOAD		14	

_____ 13. Refer to the schedule for Panel LB, shown above. This is a 120/240 volt 3Ø, 4-wire panelboard. If circuits 1 and 6 share a neutral there should be ___ amps of current flowing in the neutral.

   a. 0   b. 20.8   c. 10.4   d. 2.5   e. This is a Code violation.

Objective 204.2 Worksheet

_____ 1. For the purpose of determining the number of current carrying conductors in a cable or conduit, the neutral ___ counted when the branch circuit is 120/208 volt 1Ø, 3-wire connected to a 120/208 volt 3Ø wye, 4-wire panelboard. Each of the two circuits carries 16 amps and serves resistive (linear) loads.

   a. is   b. is not

_____ 2. A feeder needs to be installed to supply power to a load that will draw 800 amps. The circuit breaker will be rated at 800 amps. There will be 3 parallel conduits run for the feeder. You will need to install at least ___ AWG or kcmil conductors in each conduit.

   a. 350   b. 300   c. 250   d. 3/0   e. 4/0

_____ 3. Where multiwire branch circuits are connected to breakers in a 277/480 volt 3Ø, 4-wire panel the neutral has to be counted as a current-carrying conductor (when applying ampacity correction factors) where the connected loads consist of ___ lighting.

   a. fluorescent   b. HID   c. both a & b   d. neither a nor b

_____ 4. Where a 2/0 AWG THHN/THWN copper conductor is inside a panelboard and connected to a breaker it can carry ___ amps of current. The breaker has a 75° C rating.

   a. 195   b. 72   c. 175   d. 145   e. 80

Conductor and Overcurrent Protection

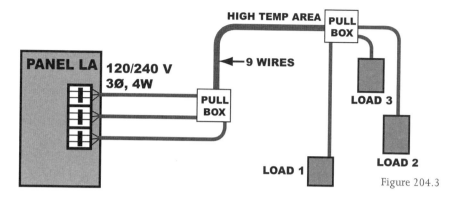

Figure 204.3

_____ 5. Refer to Figure 204.3, shown above. The circuit breakers have a temperature rating of 75° C. The wire is THWN/THHN copper. The wire to Load #3 is 6 AWG, the wire to Load #2 is 1 AWG, and the wire to Load #1 is 2/0 AWG. Each of the loads is 3Ø, 3-wire. How much current can the wire to Load #3 carry in the high temperature area where the temperature is 121° F?

a. 37   b. 43   c. 53   d. 49   e. 34

_____ 6. A 3-pole breaker in a panel is to supply power to a 3Ø load that will pull 53 amps. The wires will be installed in a conduit with other current carrying conductors. The conduit will contain a total of 12 current carrying conductors. The conduit will be run through an area where the temperature is 124 ° F. The smallest size THWN/THHN copper wire that you can install for this load is ___ AWG.

a. 6   b. 4   c. 3   d. 8   e. 1

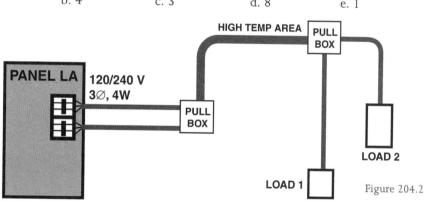

Figure 204.2

_____ 7. Refer to Figure 204.2, shown above. The circuit breakers have a temperature rating of 75° C. The wire is THWN/THHN copper. The wire to Load #2 is 1 AWG and the wire to Load #1 is 2/0 AWG. Each of the loads is 3Ø, 3-wire so there is a total of 6 wires in the conduit between the two pull boxes. How much current can the wire to Load #1 carry in the high temperature area where the temperature is 123° F?

a. 94   b. 117   c. 105   d. 156   e. 119

_____ 8. What minimum size copper THHN/THWN conductors are required to feed a 16 KW 240 volt electric furnace operated at 240 volts? Terminations are rated at 75°C.

    a. 1   b. 3   c. 2   d. 6   e. 4

Conductor and Overcurrent Protection

Objective 204.3 Worksheet

_____ 1. The largest 3-pole circuit breaker that can be used to protect a 12 AWG copper Type SO flexible cord is ___ A.

    a. 35   b. 15   c. 30   d. 25   e. 20

_____ 2. A 15 amp multiwire branch circuit is connected to breakers 1 & 3 in a 120/208 volt 3Ø, 4-wire panelboard. If somewhere in the circuit the "hot" from breaker 3 and the metal conduit touch together___.

    a. an overload will occur and the breaker will trip
    b. the load will operate normally and the breaker will not trip
    c. a ground-fault will occur and the breaker will trip
    d. a short-circuit will occur and the breaker will trip

_____ 3. Which of the following is a type of circuit condition?
      I. selective coordination   II. overload

    a. I only
    b. I & II
    c. II only
    d. neither I nor II

_____ 4. 1 AWG THHN/THWN copper conductors are installed to serve a single load that will not operate more than 3 hours. These branch circuit conductors can be protected by a circuit breaker with a rating no greater than ___ Amps. Breakers have 75°C terminals.

    a. 150
    b. 100
    c. 125
    d. 110
    e. 175

_____ 5. Where a calculation results in an ampacity requirement that does not match the standard rating of an overcurrent protective device, which of the following is permitted?

    a. Select an OCPD that is rated not more than 125% of the ampacity of the conductor.
    b. Select an OCPD with the next higher standard ampere rating.
    c. Select an OCPD that is rated not more than twice the ampacity of the conductor.

6. A 15 amp multiwire branch circuit is connected to breakers 1 and 3 in a 120/208 volt 3Ø, 4-wire panelboard. If somewhere in the circuit the "hot" from breaker 1 and the "hot" from breaker 3 touch together ___.

   a. a ground-fault will occur and the breaker will trip
   b. the load will operate normally and the breaker will not trip
   c. a short-circuit will occur and the breaker will trip
   d. an overload will occur and the breaker will trip

7. A 15 amp multiwire branch circuit is connected to breakers 1 and 3 in a 120/208 volt 3Ø, 4-wire panelboard. ___ if the "hot" from breaker 3 and the neutral are connected to a 2000 watt lighting fixture load.

   a. A ground-fault will occur and the breaker will trip
   b. An overload will occur and the breaker will trip
   c. A short-circuit will occur and the breaker will trip
   d. The load will operate normally and the breaker will not trip

8. 2 AWG conductors are connected to a 3-pole breaker in a panel and serve a 90 amp, 3Ø load. The three wires are installed in a conduit that passes through an area where the temperature is 125° F. The largest breaker that can be installed for this circuit is a ___ Amp.

   a. 90   b. 125   c. 150   d. 110   e. 100

9. For a particular circuit the least current flows when a (an) ___ occurs.

   a. short-circuit   b. overload   c. ground-fault

10. An 8.2 kva load (rated to operate at the applied voltage) is connected to a branch circuit that is fed from a 2-pole, 35 A breaker in a 120/208 volt 3Ø, 4-wire panelboard. Will the breaker trip due to overload?

    a. Yes   b. No

Student Notes

Objective 204.4 Worksheet

_____ 1. 3 AWG THHN/THWN copper feeder conductors are connected to a 2-pole 100 amp circuit breaker in a panelboard. The feeder conductors enter a pull box where several other conductors are connected to them. If 10 AWG conductors are connected to the feeder the 10 AWG conductors are ___.

   a. taps   b. feeders

_____ 2. Where not over 7.5 meters (25 feet) long tap conductors are to have an ampacity of at least ___ the rating of the overcurrent device protecting the feeder conductors per 240.21(B)(2).

   a. one-third   b. one-half   c. one-tenth   d. 10 times

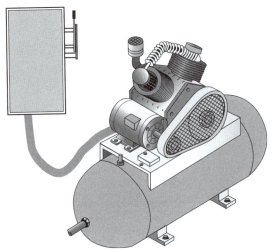

Figure 201.13

_____ 3. The compressor nameplate reads: 240V 3Ø 60 Hz. Per the NEC, the minimum number of fuses that are to be installed in the fused safety switch, shown above in Figure 201.13, is ___.

   a. 1   b. 3   c. 2   d. 0

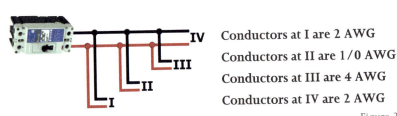

Conductors at I are 2 AWG
Conductors at II are 1/0 AWG
Conductors at III are 4 AWG
Conductors at IV are 2 AWG

Figure 204.5

_____ 4. Refer to Figure 204.5, shown above. Where the circuit breaker is rated at 150 Amps the conductors at Point II are ___.

   a. feeders   b. taps

Conductor and Overcurrent Protection

_____ 5. A circuit breaker enclosure is permitted to be installed in which of the following locations?

   a. napkin and tablecloth storage room in a restaurant
   b. coat closet in a house
   c. mechanical room in a college
   d. any of these
   e. none of these

_____ 6. Tap conductors shall have an ampacity of at least ___ the rating of the overcurrent device protecting the feeder conductors to comply with the requirements of 240.21(B)(1).

   a. one-half   b. 3 times   c. one-third   d. 10 times   e. one-tenth

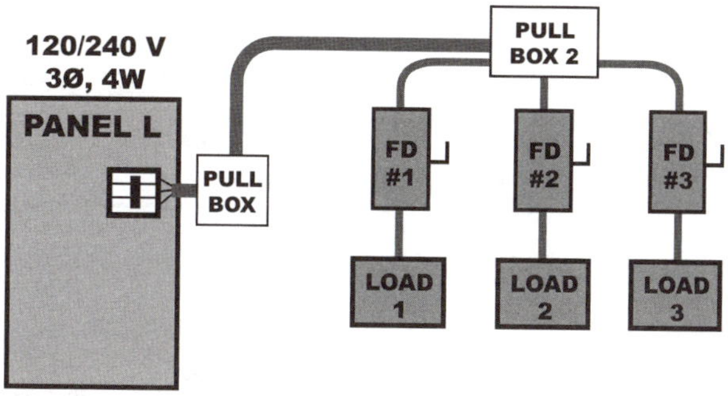

Taps are made in Pull Box 2 to feed the Fused Safety Switches for the three loads.
All terminations are rated at 75° C and all conductors are THHN/THWN copper.

Figure 204.7

_____ 7. Refer to Figure 204.7, shown above. The conductors in the conduit between the panel and the first Pull Box are ___ conductors.

   a. feeder   b. tap   c. branch circuit

_____ 8. Refer to Figure 204.7, shown above, where a 200 A breaker is installed in Panel L. The conductors in the conduit between Pull Box 2 and 30 Amp FD #1 are ___ conductors.

   a. feeder   b. branch circuit   c. tap

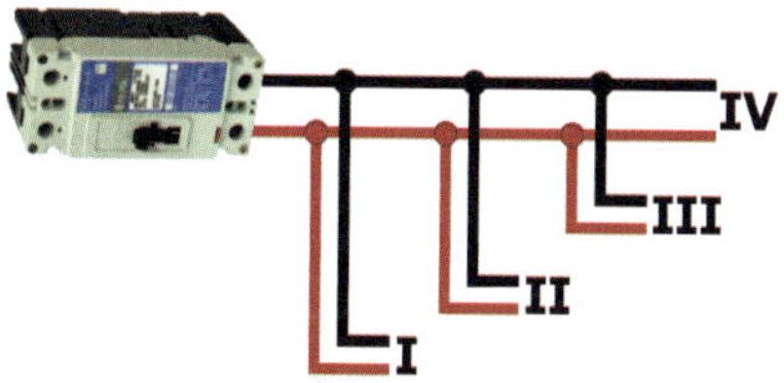

Conductors at I are 3/0 AWG
Conductors at II are 1/0 AWG
Conductors at III are 4 AWG
Conductors at IV are 3/0 AWG

Figure 204.4

_____ 9. Refer to Figure 204.4, shown above. Where the circuit breaker is rated at 200 Amps the conductors at Point III are ___.

   a. taps   b. feeders

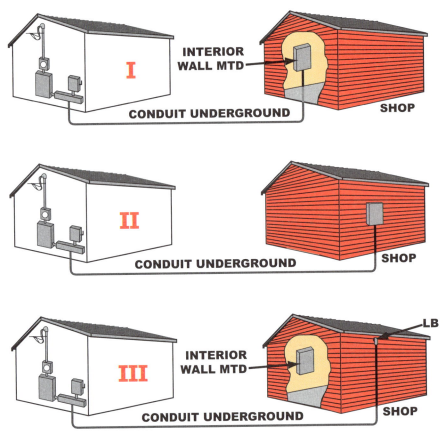

Figure 204.8

Both the service panel and the subpanel on the shop contain MAIN breakers. A feeder breaker in the service panel feeds the gutter. Taps are made in the gutter to feed the shop and the fused switch above the gutter. The shop is located about 120 feet from the gutter

_____ 10. Illustration II, shown above in Figure 204.8, ___ satisfy the requirements of 240.21(B)(5).

    a. does
    b. does not

_____ 11. Where installed in a 3-pole 250-volt fused safety switch, the minimum number of fuses required to protect a 120/240 volt 3-ph, 4-wire branch circuit is ___.

    a. 3
    b. 4
    c. 1
    d. 2

Conductor and Overcurrent Protection

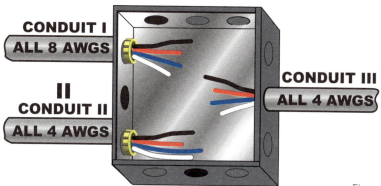

Figure 204.6

All the blue conductors are connected together, all the blacks together, all the reds together, and all the whites together. The conductors in Conduit II originate in a panelboard and are protected with a 3-pole 80 amp circuit breaker. The conductors in Conduit I feed one fused switch and the conductors in Conduit III feed another fused switch.

_____ 12. Refer to Figure 204.6, shown above. The 4 AWG conductors in Conduit II are ___ conductors.

    a. tap
    b. feeder

Objective 204.5 Worksheet

_____ 1. Can a safety switch rated for 250 volts be used as the disconnecting means for a 480 volt circuit?

   a. No   b. Yes

_____ 2. You need to get power to a 208 volt load. Your available power is from a 120/240 volt, 3Ø, 4-wire panelboard. Is it permitted for you to run a branch circuit from the neutral bar and a 1-pole circuit breaker installed in breaker space #4? The CB has a 120 volt rating.

   a. No   b. Yes

_____ 3. Plug fuses and fuseholders can be used to protect circuits on ___ systems.
   I. 120/240 volt 3Ø, 4-wire
   II. 120/240 volt 1Ø, 3-wire
   III. 120/208 volt 3Ø, 4-wire

   a. I only
   b. II only
   c. I or II only
   d. I, II, or III
   e. II or III only

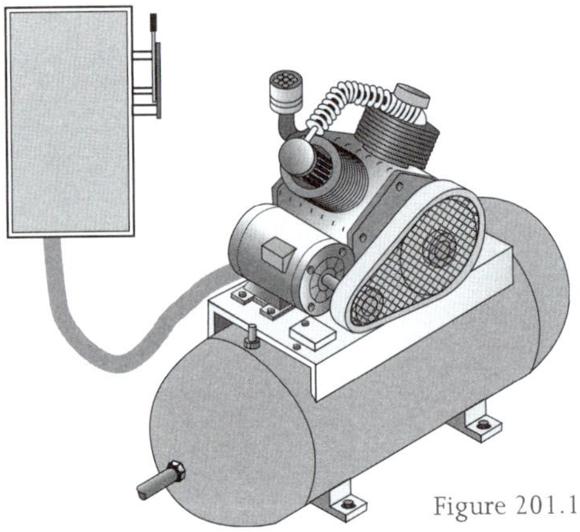

Figure 201.13

_____ 4. The compressor nameplate reads: 240V 1Ø 60 Hz. Which of the following safety switches is (are) permitted to be used as the disconnecting means for the air compressor shown above in Figure 201.13?
   I. 250 volt 2-pole
   II. 600 volt 3-pole
   III. 250 volt 3-pole

   a. II or III only   b. II only   c. I or III only   d. I or II or III   e. III only

_____ 5. The ability of an overcurrent device to safely open under fault conditions is referred to as its ___ rating.

   a. interrupting   b. series   c. coordinated   d. ampere

Conductor and Overcurrent Protection

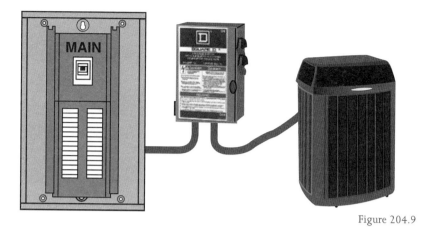

Figure 204.9

_____ 6. The panel shown above in Figure 204.9 is 120/240 volt 3Ø, 4-wire. The disconnect switch is fused with plug fuses and supplied from a 2-pole breaker in space 6-8 in the panel. The use of plug fuses in this installation ___ a Code violation.

   a. is not
   b. is

_____ 7. A circuit breaker was installed so that pushing the handle UP or pulling the handle DOWN operates it. Pulling the handle DOWN turns the breaker ___.

   a. OFF
   b. ON

_____ 8. A fault current calculation requires that the main breaker in a panelboard have a rating of 28k AIC. If all the branch breakers are required to have this same AIC rating then this system is ___ rated.

   a. series
   b. fully

_____ 9. Generally an overcurrent protective device is not permitted to be continuously loaded to more than ___ percent of its rating.

   a. 100
   b. 125
   c. 50
   d. 80

_____ 10. Time-delay fuses are also known as ___ fuses.

    a. single-element   b. dual-element   c. quick-time   d. any of these

_____ 11. Instantaneous-trip circuit breakers are permitted to be installed in a lighting and appliance panelboards.

    a. False   b. True

_____ 12. Which of the following safety switches is (are) permitted to be used as the disconnecting means for a 240 volt 1Ø water heater?
    I. 250 volt 2-pole   II. 600 volt 3-pole   III. 250 volt 3-pole

    a. II or III only   b. I only   c. III only   d. I or II or III   e. I or III only

Lesson 204 NEC Worksheet

_____ 1. In Class I, Division 1 hazardous locations, ___ is a permitted wiring method.

   a. threaded rigid metal conduit
   b. threaded steel intermediate metal conduit
   c. Type MI cable
   d. Type ITC-HL cable, with restricted public access and supervised maintenance by qualified persons,
   e. all of these

_____ 2. In Class I, Division 1 hazardous locations, enclosures containing apparatus require conduit seals. These seals shall be installed within ___ from such enclosures.

   a. 20 inches
   b. 24 inches
   c. 3 feet
   d. 6 feet
   e. 18 inches

_____ 3. In Class I, Division 1 locations, only explosionproof ___ that are not larger than the trade size of the conduit shall be permitted between the sealing fitting and the explosionproof enclosure.

   a. unions and couplings
   b. elbows and capped elbows
   c. conduit bodies similar to L, T and Cross types
   d. reducers
   e. all of these

_____ 4. In each conduit leaving a Class I, Division 1 hazardous location, the sealing fitting shall be permitted on either side of the boundary of such location within ___ of the boundary.

   a. 10 feet
   b. 12 feet
   c. 20 feet
   d. 18 feet
   e. 25 feet

_____ 5. In each conduit, passing from a Class I, Division 2 hazardous location into an unclassified location, ___ are permitted between the conduit seal and the point at which the conduit leaves the Division 2 location.

   a. unions
   b. couplings
   c. boxes
   d. fittings
   e. listed reducers

_____ 6. For application of Article 501, where conduits systems leave a room or enclosure that is unclassified as a result of pressurization into a Class I, Division 2 hazardous location, ___.

   a. seals are required within 18 inches of the boundary
   b. seals are not required at the boundary
   c. seals are required within 10 feet of the boundary
   d. seals are required within 36 inches of the boundary
   e. seals are required to be hermetically sealed

_____ 7. In each conduit seal in Class I, Division 1 and 2 hazardous locations, the compound thickness must be at least ___.

   a. 1 inch
   b. 5/8 inch
   c. 1/4 inch
   d. 6 inches
   e. 3/4 inch

_____ 8. Splices and taps ___ in Class I, Division 1 locations.

   a. shall not be made in fittings which are intended only for sealing compound
   b. may be made in fittings which are intended for sealing compound
   c. may have their junction points filled with sealing compound.
   d. shall have all junction points filled with sealing compound
   e. shall be made in fittings which are intended for sealing compound

_____ 9. For application of Article 501, the conductor fill in a seal shall not exceed ___ percent of the cross-sectional area of a rigid metal conduit of the same trade size unless it is specifically identified for a higher percentage of fill.

   a. 30
   b. 40
   c. 25
   d. 55
   e. 60

_____ 10. Cables in conduit with a gas/vaportight continuous sheath capable of transmitting gases or vapors through the cable core shall be sealed in the Division 1 location ___ so that the sealing compound will surround each individual insulated conductor and the outer jacket.

   a. by splicing within a seal
   b. by applying compound to the cable
   c. by sealing and splicing the cable
   d. after removing the jacket and coverings
   e. none of these

Lesson 204 Safety Worksheet

NFPA70E Standard for Electrical Safety in the Workplace

The NFPA70E Standard for Electrical Safety in the Workplace was first published in 1979. Unlike the NEC (NFPA70), which is primarily to assure the correct installation of equipment, the NFPA70E is all about safety for the electrical worker.

Today the NFPA70E is being used as a tool for OSHA to use to gauge electrical safety in the workplace. The 2004 edition of the publication is divided into four chapters.

- Chapter 1 Safety-Related Work Practices
- Chapter 2 Safety-Related Maintenance Requirements
- Chapter 3 Safety Requirements for Special Equipment
- Chapter4 Installation Safety Requirements

Many electricians have not heard of this publication, primarily because no one has been enforcing the requirements of the document. But all this is changing rapidly. OSHA is using NFPA70E as a benchmark for minimum requirements. If there is an electrical accident, OSHA will be looking at what was not done correctly by the NFPA70E standards, and will then issue citations under the appropriate OSHA standard that applies.

The NFPA70E requires Safety Training, Electrical Safety Programs, Hazard and Risk Evaluation, Electrically Safe Work Condition and other programs to assure worker safety.

There are also requirements to have justification for work on live parts. The program has requirements for all workplaces to perform Arc Flash Analysis and put labels indicating such items as Approach Boundary information, the amount of Incident Energy available and what class of Arc Flash PPE and Insulated tools and Insulated Gloves are required under these conditions. (The labels required in NEC 110.16 do not require this information)

Tables and charts are provided for use in those locations that have not performed the required Arc Flash studies or posted these warning labels. These tables will provide workers with a list of appropriate PPE and tools to help keep them safe under the conditions listed in the tables, providing all the required knowledge and training is followed.

Every person working on any electrical system needs to be familiar with the NFPA70E to be able to protect themselves when working around any energized equipment.

Questions

_____ 1. Tables and ___ are provided for use when Arc Flash studies have not been performed.

      a. chairs   b. tools   c. charts   d. plans

_____ 2. NFPA70E was first published by the National Fire Protection Association in ___.

      a. 1978   b. 1976   c. 1979   d. 1776

_____ 3. Every person working on any ___ system needs to be familiar with the NFPA70E.

      a. computer   b. electrical   c. mechanical   d. solar

_____ 4. There are ___ chapters in the 2004 edition of NFPA70E.

      a. two   b. five   c. four   d. three

_____ 5. NFPA70E rules are enforced by ___.

      a. OSHA   b. police   c. inspectors   d. fire marshall

Lesson 205
Grounding and Bonding Part I

Electrical Curriculum

Year Two Student Manual

Purpose

This lesson will define terms and teach requirements for grounding and bonding of electrical equipment, systems, services, grounded conductors and grounding electrodes.

Homework

(Due at the beginning of this class)

For this lesson, you should:

- Thoroughly read the material contained within Lesson 205.
- Read ES Chapter 6, pages 147 thru 172.
- Complete Objective 205.1 Worksheet.
- Complete Objective 205.2 Worksheet.
- Complete Objective 205.3 Worksheet.
- Complete Objective 205.4 Worksheet.
- Complete Lesson 205 NEC Worksheet
- Read and complete Lesson 205 Safety Worksheet.

- Complete additional worksheets, if available, as directed by your instructor.

Objectives

By the end of this lesson, you should:

205.1 ....Understand grounding terminology and identify grounding components on service equipment.
205.2 ....Understand the requirements for grounded conductors in AC systems.
205.3 ....Understand the bonding requirements for equipment and services.
205.4 ....Be able to locate equipment and symbols on the BCES prints in order to solve problems addressed on common construction site scenarios.

Grounding and Bonding Part I

205.1 Grounding - Terminology and Identification

205.2 Grounded Conductors in AC Systems

205.3 Bonding of Equipment and Services

205.4 Problem Solving for BCES Blueprints

Grounding and Bonding Part I

Objective 205.1 Worksheet

_____ 1. Electrical systems that are grounded shall be connected to earth in a manner that will limit the voltage imposed by ___.
   I. line surges
   II. lightning
   III. unintentional contact with higher-voltage lines

   a. I & II only   b. II & III only   c. I, II, & III   d. I only   e. II only

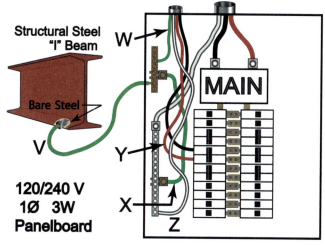

Figure 205.2

_____ 2. Refer to Figure 205.2, shown above. Point ___ is an equipment grounding conductor.

   a. V   b. W   c. X   d. Y   e. Z

_____ 3. Refer to Figure 205.2, shown above. Point ___ is an ungrounded conductor.

   a. Z   b. Y   c. X   d. W   e. V

_____ 4. Refer to Figure 205.2, shown above. Point ___ is the grounding electrode conductor.

   a. Z   b. Y   c. X   d. V   e. W

_____ 5. Refer to Figure 205.2, shown above. Point ___ is the main bonding jumper.

   a. Y   b. W   c. X   d. V   e. Z

_____ 6. Refer to Figure 205.2, shown above. Point ___ is a grounded conductor.

   a. V   b. Z   c. Y   d. W   e. X

_____ 7. On which of the systems listed below is the grounded conductor NOT the neutral?

   a. 277/480 volt, 3-ph, 4-wire wye
   b. 120/240 volt, 3-ph, 4-wire delta
   c. 120/208 volt, 3-ph, 4-wire wye
   d. 480 volt, 3-ph, 3-wire corner-grounded delta

_____ 8. Providing a low impedance ground return path ___ when a ground-fault occurs.

   a. limits the amount of current to a small amount
   b. has no effect on the circuit current
   c. permits a high current to flow

_____ 9. The primary purpose of grounding metal equipment enclosures and metal raceways to the grounding electrode system (earth) is to ___.
   I. cause the quick operation of OCPD's
   II. parts at the same voltage
   III. limit voltage surges

   a. I only
   b. I & III only
   c. II & III only
   d. I & II only   e. I, II, & III

Grounding and Bonding Part I

Objective 205.2 Worksheet

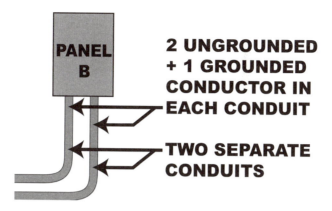

Figure 205.4

____ 1. Panel B, shown above in Figure 205.4, contains a 500 Amp 2-pole main breaker. This is a 120/240 volt 1-ph, 3-wire service. There is one 250 kcmil ungrounded conductor for AØ in each conduit and one in each conduit for BØ. There is also a grounded conductor in each conduit. Most of the loads are 240 volt 1Ø and the neutral load will be 60 amps. The smallest neutral that can be run in EACH of the two conduits is a ___ AWG. Note that the conduits are in PARALLEL and there are special rules to follow!

   a. 10   b. 3   c. 1/0   d. 2/0   e. 2

____ 2. Refer to Figure 205.4, shown above. A 120/240 volt 1-ph, 3-wire service is to be run to serve 400 A main breaker Panel B. As shown two parallel conduits will be installed for the service. Each conduit will contain two 3/0 AWG THHN/THWN copper conductors for the ungrounded conductors. All terminations are 75° C. Most of the loads are 240 volt 1Ø phase without neutrals. The smallest neutral that can be installed in each conduit is a ___ AWG THHN/THWN copper conductor.

   a. 6   b. 3/0   c. 4   d. 1/0   e. 2

____ 3. Refer to Figure 205.4, shown above. A 120/240 volt 1-ph, 3-wire service is to be run to serve 400 A main breaker Panel B. As shown two parallel conduits will be installed for the service. Panel B supplies 200 amps of 240 volt loads. Additionally, 122 amps of 120 volt loads are connected to AØ and 188 amps of 120 volt loads are connected to BØ in Panel B. The BØ conductor in each of the parallel conduits will carry ___ amps.

   a. 144   b. 188   c. 94   d. 194   e. 388

Year Two (Student Manual)

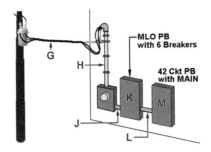

Figure 205.3

_____ 4. The conductors shown above at Point L in Figure 205.3 are ___ conductors.

    a. service entrance   b. feeder   c. branch circuit   d. service drop

_____ 5. The conductors shown above at Point ___ in Figure 205.3 are feeder conductors.

    a. J   b. G   c. H   d. L

_____ 6. The equipment shown above at Point ___ in Figure 205.3 is the subpanel.

    a. K   b. M   c. both K & M   d. neither K nor M

_____ 7. The conductors shown above at Point ___ in Figure 205.3 are service drop conductors.

    a. J   b. G   c. H   d. L

_____ 8. The conductors shown above at Point J in Figure 205.3 are ___ conductors.

    a. branch circuit   b. service drop   c. service entrance   d. feeder

_____ 9. The equipment shown above at Point ___ in Figure 205.3 is the service panel.

    a. K   b. M   c. both K & M   d. neither K nor M

_____ 10. The conductors shown above at Point H in Figure 205.3 are ___ conductors.

    a. service drop   b. feeder   c. branch circuit   d. service entrance

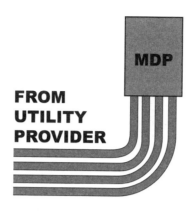

Figure 205.6

_____ 11. As shown above in Figure 205.6, MDP is a 1200 amp, 3Ø, 4-wire service and will have four parallel sets of conductors feeding it. Each set is in a separate conduit. All conductors are THHN/THWN copper. What is the minimum size of the grounded conductor required in each conduit? Most of the loads are 3-phase.

   a. 250   b. 2   c. 4/0   d. 3/0   e. 350   f. none of these

_____ 12. As shown above in Figure 205.6, MDP is a 1200 amp, 3Ø, 4-wire service and will have four parallel sets of conductors feeding it. Each set is in a separate conduit. All conductors are THHN/THWN copper. What is the minimum size of the ungrounded conductors in each conduit?

   a. 3/0   b. 4/0   c. 2   d. 250   e. 350

_____ 13. Where installed as overhead conductors outside of a building, AC systems of less than 50 volts ___ grounded.

   a. shall be   b. shall be permitted to be   c. shall not be

_____ 14. A 350 amp 277/480 volt 3Ø, 4-wire service consists of three 500 kcmil ungrounded conductors and one grounded conductor. The smallest grounded conductor that can be installed for this service is ___ AWG or kcmil where the neutral load is unknown but could be the same as any ungrounded conductor since all the loads are 277 volt lighting.

   a. 1/0
   b. 500
   c. 2
   d. 1
   e. 3

Scenario 205.2

A 400 A 3Ø main lug panel will be fed with one conduit from a 300 Amp 3-pole breaker 120/208 volt 3-ph, 4-wire panelboard. The feeder will consist of three 350 KCMIL conductors (one for AØ, one for BØ, and one for CØ) plus a grounded conductor. All terminations are 75° C. Most of the loads are 120 volt 1-phase and the neutral load is calculated to be 140 amps.

_____ 15. Refer to Scenario 205.2, shown above. The smallest neutral that can be run is a ___ AWG.

    a. 2
    b. 1
    c. 3
    d. 4
    e. 1/0

_____ 16. A 120/240 volt 3-ph, 4-wire service is to be run to serve a 200 A main breaker panel. Each of the ungrounded conductors is a 3/0 AWG THHN/THWN copper conductor. All terminations are 75° C. Most of the loads are 3-phase without neutrals. The smallest neutral that can be run is a ___ AWG THHN/THWN copper conductor.

    a. 1/0
    b. 3/0
    c. 4
    d. 6
    e. 2

_____ 17. The secondary of a transformer delivers 12 volts. One of these secondary conductors must be grounded when the supply voltage to the transformer is from a ___.
    I. 2-pole breaker in a 120/208 volt 3Ø, 4-wire panelboard
    II. 1-pole breaker in a 277/480 volt 3Ø, 4-wire panelboard

    a. I only
    b. either I or II
    c. II only
    d. neither I nor II

_____ 18. Where located on the load side of a service point ___ is considered to be part of a premises wiring system.
    I. factory installed wiring inside a dishwasher
    II. branch circuit conductors to temporary luminaires
    III. feeder conductors to a subpanel

    a. III only
    b. I, II, & III
    c. II & III only
    d. I & II only
    e. I & III only

Grounding and Bonding Part I

Scenario 205.1

A 300 Amp breaker in a 120/240 volt 3-ph, 4-wire panelboard will protect a feeder. The feeder conductors will be installed in one conduit. The feeder supplies a 3Ø subpanel which is expected to be fully loaded. However, the calculated neutral load is 190 amps.

_____ 19. Refer to Scenario 205.1, shown above. The smallest ungrounded conductors that can be installed in the conduit are ___ AWG or kcmil.

   a. 350
   b. 4
   c. 3/0
   d. 2
   e. 250

_____ 20. Refer to Scenario 205.1, shown above. The smallest grounded conductor that can be installed in the conduit is ___ AWG or kcmil.

   a. 2
   b. 350
   c. 3/0
   d. 4
   e. 250

_____ 21. A 120/240 volt 3-ph, 4-wire panelboard will be fed with one conduit. This service panel contains a 350 Amp main breaker. The conductors supplying the service will consist of three 500 kcmil ungrounded conductors (one for AØ, one for BØ, and one for CØ) plus a grounded conductor. The calculated neutral load is 80 amps. The smallest neutral that can be run is a ___ AWG.

   a. 1
   b. 1/0
   c. 4
   d. 2/0
   e. 3

Objective 205.3 Worksheet

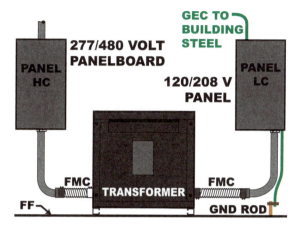

Figure 205.8

_____ 1. Refer to Figure 205.8, shown above. Panel LC has a 400 amp main breaker and is supplied power by the transformer with a set of 600 KCMIL conductors. The smallest permitted system bonding jumper for Panel LC is ___ AWG.

    a. 3   b. 1/0   c. 2/0   d. 2   e. 1

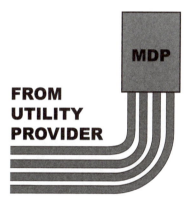

Figure 205.6

_____ 2. Service Panel MDP, shown above in Figure 205.6, has a 1200 amp main breaker. Each conduit contains 350 kcmil conductors. The minimum size main bonding jumper is ___ AWG or kcmil.

    a. 250   b. 2   c. 350   d. 3/0   e. 4/0

_____ 3. Service Panel MDP, shown above in Figure 205.6, has a 1600 amp main breaker. Each conduit contains 600 kcmil conductors. The minimum size main bonding jumper is ___ AWG or kcmil.

    a. 1/0   b. 250   c. 300   d. 4/0   e. 3/0

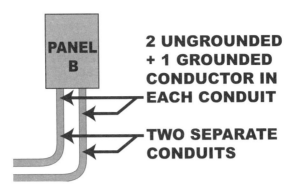

Figure 205.4

_____ 4. Refer to Figure 205.4, shown above. A 120/240 volt 1-ph, 3-wire service is to be run to serve 400 A main breaker Panel B. As shown two parallel conduits will be installed for the service. Each conduit will contain two 3/0 AWG THHN/THWN copper conductors for the ungrounded conductors. All terminations are 75° C. Most of the loads are 240 volt 1Ø phase without neutrals. The smallest MBJ that can be installed in Panel B is a ___ AWG THHN/THWN copper conductor.

a. 1/0
b. 4
c. 3/0
d. 2
e. 6

_____ 5. Panel B, shown above in Figure 205.4, is a 120/240 volt 1-ph, 3-wire service panelboard. Panel B contains a 700 Amp 2-pole breaker Main Breaker. There is one 500 kcmil ungrounded conductor for AØ in each conduit and one in each conduit for BØ. There is also a grounded conductor in each conduit. The smallest main bonding jumper that can be installed is a ___ AWG.

a. 2/0
b. 1/0
c. 1
d. 3
e. 2

_____ 6. PVC water pipe is used underground to connect the Utility's water service to the building. The PVC enters the building, is routed up through the wall, and is extended into the accessible ceiling. In the ceiling the PVC is changed over to copper pipe for the various drops. Since the copper begins more than 5 feet from where the water pipe enters the building does the metal pipe have to be bonded to the electrical service equipment?

a. NO
b. YES

Grounding and Bonding Part I

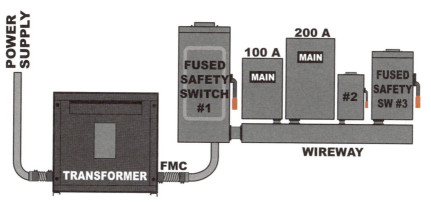

Figure 205.9

_____ 7. Refer to Figure 205.9, above. Safety Switch #1 contains 350 amp fuses. The flex between the transformer and Fused SS #1 contains four 500 KCMIL THWN/THHN copper conductors. The minimum size of the copper bonding jumper for the flex is ___.

   a. 1   b. 2/0   c. 1/0   d. 3/0   e. 2

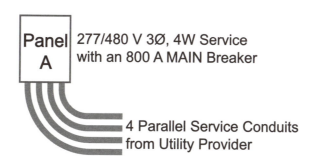

Figure 205.5

_____ 8. Refer to Figure 205.5, shown above. Panel A serves the following loads: 300 amps of 480 volt 3Ø loads. Additionally, 264 amps of 277 volt loads are connected to AØ, 324 amps of 277 volt loads are connected to BØ, and 456 amps of 277 volt loads are connected to CØ. The AØ conductor in each of the parallel conduits will carry ___ amps.

   a. 264   b. 141   c. 437   d. 81   e. 109

_____ 9. The primary purpose of bonding metal equipment enclosures, metal raceways, and the grounded power supply together is to ___.
   I. cause the quick operation of OCPD's
   II. keep all metal parts at the same voltage
   III. limit voltage surges

   a. I only   b. II & III only   c. I & III only   d. I & II only   e. I, II, & III

_____ 10. What is the smallest size copper wire permitted to be used as a main bonding jumper in a 200 A main breaker panelboard? The panel is fed from a transformer with one set of 3/0 THHN/THWN copper wire.

a. 3/0 AWG
b. 2 AWG
c. 1/0 AWG
d. 3 AWG
e. 4 AWG

Grounding and Bonding Part I

Objective 205.4 Worksheet

_____ 1. On the BCES project, each conduit installed between the MSB and Panel HK contains ___ grounding conductor(s).

   a. 1 ea 1 AWG
   b. 2 ea 2 AWG
   c. 2 ea 1 AWG
   d. 1 ea 4 AWG
   e. 1 ea 8 AWG
   f. 1 ea 2 AWG

_____ 2. On the BCES project, the conductors installed between the MSB and Panel HD include ___ ground wire(s).

   a. one - 4 AWG
   b. two - 2 AWG
   c. one - 1 AWG
   d. one - 2 AWG
   e. two - 1 AWG
   f. two - 4 AWG

_____ 3. On the BCES project, each conduit installed between the MSB and the disconnect switch for Chiller CH-2 contains ___ grounding conductor(s).

   a. 1 ea 1 AWG
   b. ea 2 AWG
   c. 2 ea 2 AWG
   d. 2 ea 1 AWG
   e. 1 ea 4 AWG
   f. 1 ea 8 AWG

Lesson 205 NEC Worksheet

_____ 1. There shall be no uninsulated exposed parts that operate at more than ___ volts in Class I, Division 1 and 2 locations.

    a. 125   b. 50   c. 30   d. 277   e. 12

_____ 2. Locknut-bushing and double-locknut types of contacts ___ in Class I, Division 1 and 2 locations.

    a. are the means for electrical bonding purposes
    b. shall not be depended upon for electrical bonding purposes
    c. shall be required for electrical bonding purposes
    d. are the means for electrical bonding purposes if they are listed and labeled
    e. are the means for electrical bonding purposes but are not required to be listed and labeled

_____ 3. Where flexible metal conduit or liquidtight flexible metal conduit is permitted in Class I, Division 1 and 2 locations, external or internal bonding jumpers shall be required except ___.

    a. where liquidtight flexible metal conduit 6 feet or less in length with listed fittings is used
    b. where overcurrent protection in the circuit is limited to 10 amperes or less
    c. where the load is not a power utilization load
    d. all of the these
    e. none of these; a through c are always required

_____ 4. Surge arresters, transient voltage surge suppressors, and capacitors in Class I, Division 1 locations ___.

    a. shall be installed in enclosures identified for such use
    b. shall not be installed in enclosures
    c. shall be installed in vaporproof enclosures
    d. shall be installed in general-purpose type enclosures
    e. are not permitted

_____ 5. Surge arresters (sealed, type MOV), transient voltage surge suppressors (nonarcing type) and surge-protective capacitors designed for specific duty in Class I, Division 2 locations ___.

    a. shall be installed in NEMA 3R enclosures identified for such use
    b. shall not be installed in enclosures
    c. shall be installed in vaporproof enclosures
    d. are permitted to be installed in general-purpose type enclosures
    e. are not permitted

_____ 6. Multiwire branch circuits ___ in Class I, Division 1 and 2 locations.
   I. shall not be permitted
   II. shall be permitted where all ungrounded conductors are opened simultaneously
   III. must be 20 amperes or less

   a. III only
   b. I only
   c. II & III only
   d. I or II, & II
   e. I or II only
   f. I & III only

_____ 7. Where circuit breakers and switches are used in a Class I, Division 1 location, general purpose enclosures ___.

   a. are permitted if the make-and-break contacts are oil-immersed
   b. are not permitted
   c. are only permitted if the enclosure covers are gasketed
   d. are permitted where air exchanges occur at a rate of 50 cfm by forced ventilation
   e. are only permitted when enclosed in a NEMA 3R enclosures

_____ 8. Pendant luminaires used in a Class I, Division 1 locations shall be protected from physical damage by suitable guards or by ___.

   a. tempered glass lenses
   b. location
   c. 1/4 inch thick plexiglass shields
   d. fiber encased lenses
   e. the use of dual-rated lenses

_____ 9. Pendant luminaires used in a Class I, Division 1 location shall be suspended by and supplied through ___.

   a. Type SO cord and jack chain
   b. Type MI cable
   c. FMC
   d. threaded RMC
   e. LFMC

Lesson 205 Safety Worksheet

NFPA70E Standard for Electrical Safety in the Workplace (part 2)

Qualified Persons shall be used to perform any work on live or energized equipment. They must have training and knowledge of the equipment, training and knowledge of the hazards involved, what must be done to work safely.

Written Energized Electrical Work Permits are required if live parts can not be put in an electrically safe condition. These conditions could include increased or additional hazards or infeasibility. These permits must have at least such items as what work is to be performed, why the work must be performed energized, what safe work practices shall be employed, shock hazard analysis, shock protection boundaries, flash protection boundaries, what type of PPE is necessary, how to control the restricted area, job briefing and approval by authority having jurisdiction or responsibility.

Some other precautions are being alert to working near live parts and knowing the dangers related to this work, not being impaired by illness, fatigue, drugs or alcohol. Make sure there is adequate illumination in the work area. Do not wear any type of conductive jewelry or clothing (rings, watches, key chains, necklaces, ear rings, nose rings, peircings, metal frame glasses or any other metal personal items) into the area of work. Be aware of any confined spaces.

Flame Retardant Personal Protective Equipment shall be worn as appropriate for the risks involved. The PPE must be maintained in good condition, laundered according to manufacturer's instructions. PPE is required for head, face, neck, chin, eyes, body, hands, arms, feet and legs. Particular attention should be paid to layering of the clothing to increase the levels of protection provided.

Signs and barricades shall be used to warn employees and other workers in the area about the dangers and to keep out. If the signs and barricades do not provide enough protection for the area to be safe, provide an attendant to warn and protect the employees.

Questions

_____ 1. Personal Protective Equipment shall be worn as ___ for the risks involved.

    a. appropriate   b. necessary   c. specified   d. all of these

_____ 2. Qualified Persons need ___ and knowledge of equipment and hazards to work safely.

    a. training   b. education   c. experience   d. none of these

_____ 3. When working near live parts you must know the ___ related to the work.

    a. situation   b. parts   c. dangers   d. workers

_____ 4. Written Energized Electrical ___ are required if live parts can not be put in an electrically safe condition.

    a. job requests   b. work permits   c. work orders   d. none of these

_____ 5. PPE is required for head, face, neck, ___, hands, arms, feet and legs.

    a. body   b. eyes   c. chin   d. all of these

_____ 6. Do not wear any type of ___ jewelry or clothing.

    a. used   b. costume   c. cheap   d. conductive

_____ 7. ___ persons shall be used to perform any work on live or energized equipment.

    a. Qualified   b. Special   c. Authorized   d. Quality

Lesson 206
Grounding and Bonding Part II

Electrical Curriculum

Year Two
Student Manual

Purpose

This lesson will teach the requirements and installation practices for grounding electrodes and grounding electrode conductors.

Homework
(Due at the beginning of this class)

For this lesson, you should:

- Thoroughly read the material contained within Lesson 206.
- Read ES Chapter 6, pages 172 thru 180.
- Complete Objective 206.1 Worksheet.
- Complete Objective 206.2 Worksheet.
- Complete Objective 206.3 Worksheet.
- Complete Lesson 206 NEC Worksheet
- Read and complete Lesson 206 Safety Worksheet.

- Complete additional worksheets, if available, as directed by your instructor.

Objectives

By the end of this lesson, you should:

206.1 ....Understand the grounding electrode system including installation, types and sizing of grounding electrodes and grounding electrode conductors.

206.2 ....Understand the purpose, installation, and sizing of equipment grounding conductors and load side bonding jumpers to provide safe operation of OCPD's during short circuits and ground faults.

206.3 ....Be able to locate equipment and symbols on the BCES prints in order to solve problems addressed on common construction site scenarios.

The instructor should review next week's material before assigning the worksheets for that lesson to determine if the students will require additional instruction.

206.1 Grounding Electrode System - GES

206.2 Equipment Grounding Conductors and Load-Side Bonding Jumpers (BJs)

206.3 Problem Solving for BCES Blueprints

Grounding and Bonding Part II

Objective 206.1 Worksheet

_____ 1. Ground rods made of aluminum RMC ____.

    I. shall be at least 5/8 inch in diameter   II. shall be at least 10 feet in length

    a. both I & II
    b. neither I nor II
    c. I only
    d. II only
    e. shall not be permitted

_____ 2. Your only available grounding electrode is the metal underground water pipe from the water utility. You are required by the NEC to have one other grounding electrode for your service. Which of the following driven electrodes could you use?

    a. 8´ by 1/2 inch copper-clad ground rod
    b. 10 ft of Trade size 1 rigid metal conduit
    c. either a or b
    d. neither a nor b

_____ 3. The largest size copper wire required to be used as a structure bond from a 400 A main breaker panelboard is ____ AWG. The panel is fed from a transformer with three parallel sets of 1/0 THHN/THWN copper wire.

    a. 1/0   b. 2   c. 2/0   d. 3/0   e. 1

_____ 4. Where run outside on the wall of a building a water bond (grounding electrode conductor to a metal water pipe) shall be ____. A driveway runs along this wall.

    a. at least a 4 AWG and supported
    b. at least a 8 AWG and in properly supported IMC
    c. at least a 6 AWG and supported
    d. none of these

_____ 5. Where a 250 KCMIL copper GEC is NOT subject to physical damage it ____.

    a. can be run exposed if properly supported
    b. can be run inside properly supported Schedule 40 PVC
    c. either a or b
    d. neither a nor b

_____ 6. The minimum size grounding electrode conductor for a service, shown to the right in Figure 205.6, that is fed with four sets of parallel 500 KCMIL service entrance conductors is ___ AWG or kcmil.

   a. 4/0
   b. 2/0
   c. 1/0
   d. 3/0
   e. 250

_____ 7. Service Panel MDP, shown above in Figure 205.6, has a 1200 amp main breaker. Each conduit contains 350 kcmil conductors. The minimum size GEC is ___ AWG or kcmil.

   a. 4/0   b. 350   c. 2   d. 250   e. 3/0

_____ 8. 30 feet of 3/8 inch rebar used to reinforce grade beams within a poured building slab ___ permitted to be used as a grounding electrode for a separately derived system or a service.

   a. is not   b. is

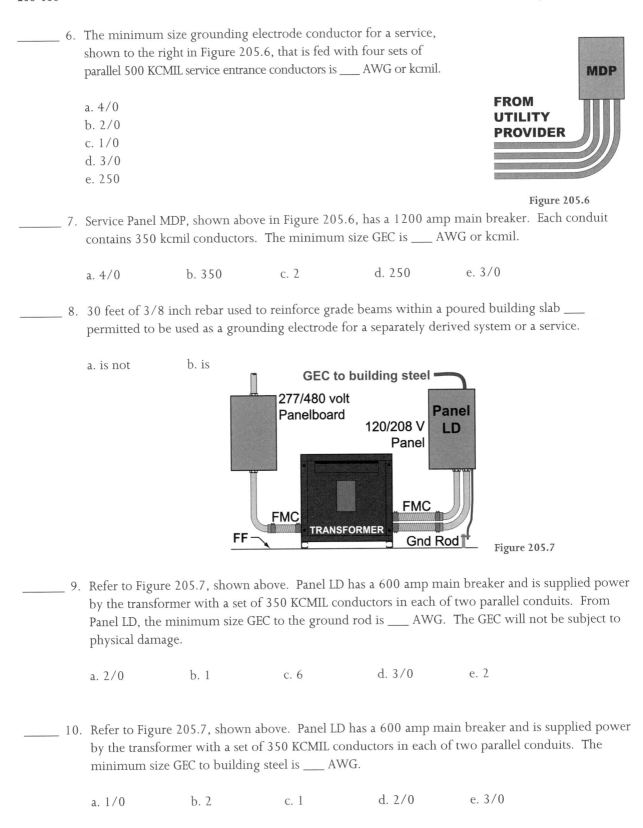

Figure 205.6

Figure 205.7

_____ 9. Refer to Figure 205.7, shown above. Panel LD has a 600 amp main breaker and is supplied power by the transformer with a set of 350 KCMIL conductors in each of two parallel conduits. From Panel LD, the minimum size GEC to the ground rod is ___ AWG. The GEC will not be subject to physical damage.

   a. 2/0   b. 1   c. 6   d. 3/0   e. 2

_____ 10. Refer to Figure 205.7, shown above. Panel LD has a 600 amp main breaker and is supplied power by the transformer with a set of 350 KCMIL conductors in each of two parallel conduits. The minimum size GEC to building steel is ___ AWG.

   a. 1/0   b. 2   c. 1   d. 2/0   e. 3/0

Grounding and Bonding Part II

_____ 11. The smallest size copper wire permitted to be used as a structure bond from 400 A main breaker Panel LD is ___ AWG. As shown on the previous page in Figure 205.7, the panel is fed from a transformer with two parallel sets of 3/0 THHN/THWN copper wire.

   a. 3/0   b. 1/0   c. 2/0   d. 2   e. 1

_____ 12. 25 feet of bare 4 AWG copper that is laid in the grade beam within a poured building slab ___ permitted to be used as a grounding electrode for a separately derived system or a service.

   a. is   b. is NOT

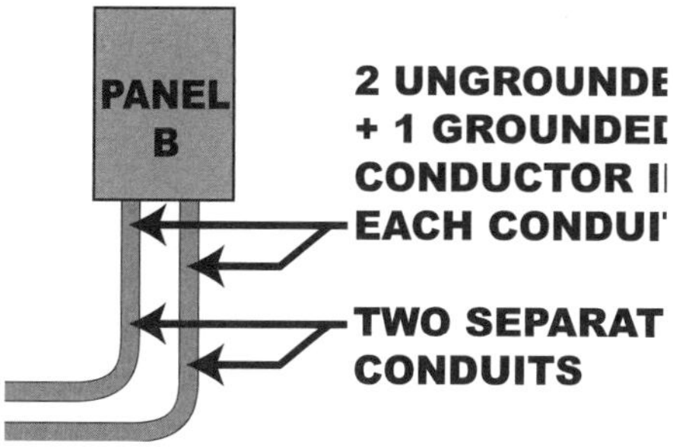

Figure 205.4

_____ 13. A 120/240 volt 1-ph, 3-wire service is to be run to serve 400 A main breaker Panel B. As shown above in Figure 205.4, two parallel conduits will be installed for the service. Each conduit will contain two 3/0 AWG THHN/THWN copper conductors for the ungrounded conductors. All terminations are 75° C. Most of the loads are 240 volt 1Ø phase without neutrals. The smallest GEC that can be installed to the building steel is a ___ AWG THHN/THWN copper conductor.

   a. 4   b. 2   c. 6   d. 1/0   e. 3/0

_____ 14. Ground rods made of galvanized steel ___.
   I. shall be at least 5/8 inch in diameter   II. shall be at least 8 feet in length

   a. both I & II
   c. neither I nor II
   e. shall not be permitted
   b. II only
   d. I only

_____ 15. Ground rods made of galvanized RMC (Rigid Metal Conduit) ___.
   I. shall be at least 3/4 inch in diameter   II. shall be at least 8 feet in length

   a. II only
   b. I only
   c. neither I nor II
   d. both I & II
   e. shall not be permitted

_____ 16. Which of the following grounding electrodes requires that another electrode be utilized to supplement it since it is the only one presently used?

   a. an underground metal water pipe
   b. a ¾" copper-clad ground rod that is 8´ in length
   c. either a or b
   d. neither a nor b

_____ 17. Where a 250 KCMIL copper GEC is subject to physical damage it ___.

   a. can be run inside properly supported Schedule 80 PVC
   b. can be run exposed if properly supported
   c. either a or b
   d. neither a nor b

_____ 18. What is the smallest size copper wire permitted to be used as a structure bond from a 1200 A fused safety switch? The safety switch is fed from a transformer with four parallel sets of 350 KCMIL THHN/THWN copper wire.

   a. 250 KCMIL   b. 3/0 AWG   c. 4/0 AWG   d. 1/0 AWG   e. 2/0 AWG

_____ 19. A copper water pipe run 27 feet underground from the street to the building is permitted to be used as a grounding electrode for a separately derived system or a service.

   a. True
   b. False

Objective 206.2 Worksheet

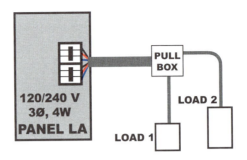

Figure 205.13

The upper breaker feeds Load 1 and has a rating of 250 A. The lower breaker is rated at 400 A and feeds Load 2.

_____ 1. Refer to Figure 205.13, shown above. Where the EGC's are spliced in the pull box are these required to be bonded to the metal pull box?

   a. NO   b. YES

_____ 2. Refer to Figure 205.13, shown above. One conduit is run from Panel LA to a pull box as shown. 250 KCMIL THHN/ THWN copper circuit conductors are connected to the 250 amp breaker, and 500 KCMIL THHN/ THWN copper circuit conductors are connected to the 400 amp breaker. Only one grounding conductor is required in this conduit for all contained circuits. What is the smallest copper EGC permitted to be installed in the conduit between the panel and the pull box?

   a. 4 AWG   b. 3 AWG   c. 1/0 AWG   d. 6 AWG   e. 2 AWG

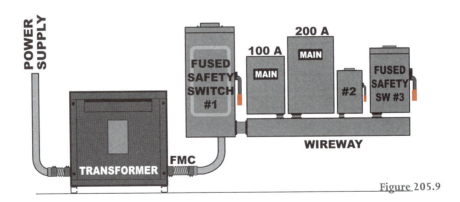

Figure 205.9

_____ 3. Refer to Figure 205.09, above. Safety Switch #1 contains 500 amp fuses. The raceway between the SS #1 and the wireway is an EMT nipple. The nipple contains four 600 KCMIL THWN/THHN copper conductors. The minimum size of the copper bonding jumper for the nipple is ___.

   a. 4   b. 1/0   c. 2/0   d. 3   e. 2

Figure 205.12

_____ 4. Refer to Figure 205.12, shown above. The flex and fittings are listed for grounding. The length of the FMC is 3.5 ft. Does the NEC require a separate equipment grounding conductor to be installed?

    a. no   b. yes

_____ 5. Three 4 AWG THHN/THWN copper conductors are installed in the flex between the switch, containing a 125 A fuses, and the motor. The smallest copper bonding jumper/grounding conductor that can be installed in the flex, shown above in Figure 205.12, is ___.

    a. 6 AWG   b. 10 AWG   c. 8 AWG   d. 4 AWG   e. none of these

_____ 6. Three 1/0 AWG THHN/THWN copper conductors are installed in the flex between the 400 A switch (F225 A), and the motor. The smallest copper bonding jumper/grounding conductor that can be installed in the flex, shown above in Figure 205.12, is ___.

    a. 6 AWG   b. 3 AWG   c. 1/0 AWG   d. 2 AWG   e. none of these

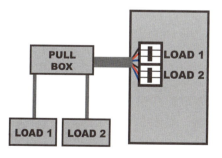

Figure 205.14

_____ 7. Refer to Figure 205.14, shown above. Load 1 is supplied from a 150 A circuit breaker with 1 AWG conductors. Load 2 is suppled from a 300 Amp breaker with 4/0 AWG conductors. Both loads are 3-ph, 3-wire. The smallest grounding conductor that can be installed between the pull box and Load 2 is ___ AWG.

    a. 3   b. 8   c. 6   d. 4   e. 2

_____ 8. Refer to Figure 205.14, shown on the previous page. Load 1 is supplied from a 150 A circuit breaker with 1 AWG conductors. Load 2 is suppled from a 300 Amp breaker with 4/0 AWG conductors. Both loads are 3-ph, 3-wire. The smallest grounding conductor that can be installed between the pull box and Load 1 is ___ AWG.

    a. 6   b. 2   c. 4   d. 3   e. 8

_____ 9. Refer to Figure 205.14, shown on the previous page. Load 1 is supplied from a 150 A circuit breaker with 1 AWG conductors. Load 2 is suppled from a 300 Amp breaker with 4/0 AWG conductors. Both loads are 3-ph, 3-wire. Where only one grounding conductor is installed in the conduit between the panel and the pull box its minimum size is ___ AWG.

    a. 2   b. 6   c. 8   d. 3   e. 4

_____ 10. In a run of conduit, when couplings and connectors are loose this will ___ the amount of time required for a circuit breaker to open when a ground-fault occurs.

    a. increase   b. have no effect on   c. decrease

_____ 11. The Code Reference that requires a separate equipment grounding conductor or permits the FMC to serve as the equipment grounding conductor for the previous question is ___. In other words, which of the following could be a code reference for your answer?

    a. 250.118(6)(e)   b. 348.60   c. either a or b   d. neither a nor b

_____ 12. You are to connect a 3-phase air handler unit (AHU). You decide to use 3 feet of ½ inch FMC (Flexible Metal Conduit or flex) since you realize that it has a motor and as such will tend to vibrate. Neither the flex nor the connectors is listed for grounding. The plans require 3 each #12 THWN/THHN stranded conductors. Per the NEC do you need to install an EGC (Equipment Grounding Conductor) with the FMC between the disconnect switch and the AHU? The disconnect has a 30 amp rating but is fused at 20 amps.

    a. no   b. yes

_____ 13. A wet location is a(n) ___.

    a. installation in concrete slabs contacting earth
    b. underground installation
    c. location saturated with water or other liquids
    d. a, b, and c

Grounding and Bonding Part II

_____ 14. A piece of equipment has a nameplate rating of 23 KW at 240 volts. For some reason the engineer has specified that oversized 4/0 AWG copper THHN/THWN circuit conductors be installed to the equipment. The equipment will operate continuously. The 4/0 circuit conductors are connected to a 2-pole 125-Amp circuit breaker in a 120/240 volt 3Ø, 4-wire panelboard. The smallest EGC that can be installed with the circuit conductors is ___ AWG.

    a. 1/0   b. 4   c. 6   d. 2/0   e. 2

_____ 15. The plans indicate that you should install a 150 amp branch circuit in Schedule 40 PVC with a 6 AWG equipment grounding conductor. However you did not have any #6 wire so you pulled in 10 AWG instead. Compared to the correct circuit (with the 6 AWG EGC) your circuit will take ___ time for the circuit breaker to open when an overload occurs.

    a. more   b. less   c. the same amount of

_____ 16. Electrical service equipment which shall be bonded together includes ___.

    a. metallic raceways protecting the metallic sheath
    b. service raceways with GEC
    c. service conductor enclosures
    d. a, b, and c

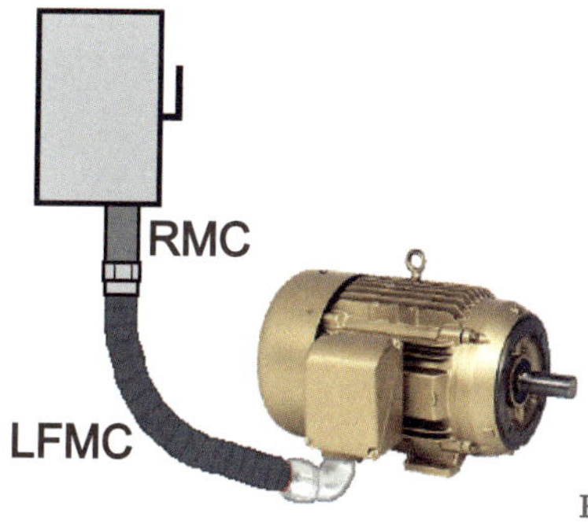

Figure 205.10

_____ 17. Refer to Figure 205.10, shown above. The raceway contains three 2/0 AWG THHN/THWN copper conductors. The minimum size of the copper bonding jumper/grounding conductor between the 400 A switch (fused at 250 A) and the motor is ___.

    a. 3 AWG   b. 8 AWG   c. 6 AWG   d. 4 AWG   e. 2 AWG

_____ 18. Refer to Figure 205.10, shown above. The 400 A safety switch contains 350 amp fuses. The raceways used between the switch and the 3Ø motor contain three 250 KCMIL conductors. The minimum size of the BJ/EGC between the safety switch and the motor is ___ AWG.

    a. 6   b. 4   c. 3   d. 1   e. 2

_____ 19. Refer to Figure 205.10, shown on the previous page. Where the grounding conductor is run on the inside of the RMC/LFMC the length of the LFMC ___.

   a. is unlimited   b. can not exceed 6 feet

_____ 20. Due to voltage drop 6 AWG copper THHN/THWN circuit conductors are installed between some parking lot lights and a 2-pole 20-Amp circuit breaker in a 120/208 volt 3Ø, 4-wire panelboard. The smallest EGC that can be installed with the circuit conductors is ___ AWG.

   a. 8   b. 10   c. 12   d. 6   e. 14

_____ 21. What is the smallest copper equipment grounding conductor that can be installed in EACH conduit for a feeder? The feeder conductors are connected to a 3-pole 500 Amp breaker and are run in two separate conduits. Each conduit contains four 250 KCMIL conductors.

   a. 1/0   b. 2   c. 3   d. 1   e. 4

_____ 22. When couplings and connectors are properly tightened in a run of conduit this will ___ the amount of time required for a circuit breaker to open when a short-circuit occurs.

   a. increase   b. decrease   c. not affect

_____ 23. The circuit installed in a run of PVC conduit requires a 6 AWG equipment grounding conductor. However a 12 AWG was installed instead. Using the 12 instead of the 6 AWG wire will ___ the normal circuit impedance.

   a. have no effect on   b. increase   c. decrease

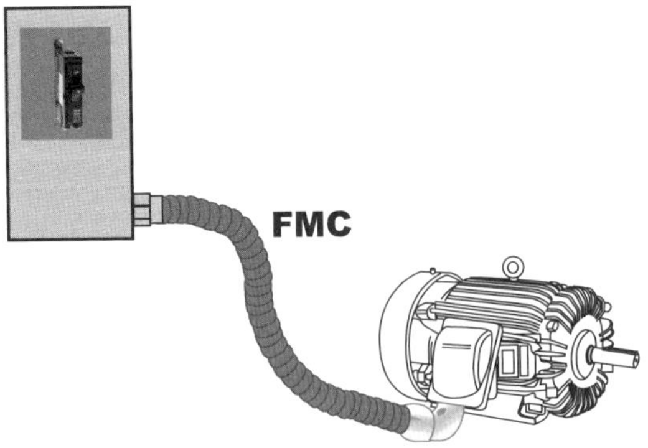

Figure 205.15

_____ 24. Refer to Figure 205.15, shown above. The flex used between the breaker enclosure and the motor contains three 350 kcmil THHN/THWN copper conductors. The minimum size of the copper bonding jumper/grounding conductor between the 700 A breaker and the motor is ___.

   a. 1/0 AWG   b. 3/0 AWG   c. 2 AWG   d. 350 kcmil   e. none of these

Grounding and Bonding Part II

_____ 25. The circuit installed in a run of PVC conduit requires a 6 AWG equipment grounding conductor. However a 12 AWG was installed instead. Using the 12 instead of the 6 AWG wire will ___ the amount of time required for a circuit breaker to open when a ground-fault occurs.

a. have no effect on   b. increase   c. decrease

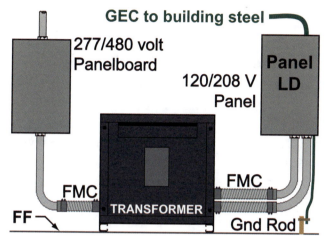

Figure 205.7

_____ 26. Refer to Figure 205.7, shown above. A 250 amp breaker in the 277/480 volt panel supplies power to the transformer. A set of 250 kcmil conductors is installed from this breaker to the transformer. The minimum size bonding jumper for the FMC between the 277/480 volt panel and the transformer is ___ AWG.

a. 3   b. 4   c. 1/0   d. 1   e. 2

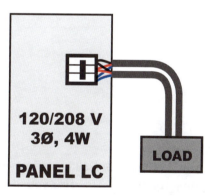

Figure 205.17

_____ 27. A load is fed with 250 kcmil conductors in two parallel conduits from Panel LC as shown above in Figure 205.17. The smallest copper grounding conductor permitted to be installed in each conduit is ___ AWG. The circuit conductors are connected to a 500 amp breaker in the panel.

a. 2   b. 4   c. 1/0   d. 3   e. 1

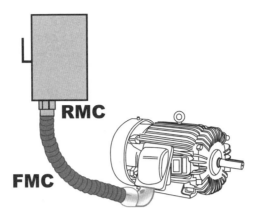

Figure 205.11

_____ 28. Refer to Figure 205.11, shown above. Where the bonding jumper/grounding conductor is installed on the outside of the FMC its length ___.

a. is unlimited
b. can not exceed 6 feet

Scenario 205.1

A 300 Amp breaker in a 120/240 volt 3-ph, 4-wire panelboard will protect a feeder. The feeder conductors will be installed in one conduit. The feeder supplies a 3Ø subpanel which is expected to be fully loaded. However, the calculated neutral load is 190 amps.

_____ 29. Refer to Scenario 205.1, shown above. The smallest equipment grounding conductor that can be installed in the conduit is ___ AWG.

a. 2
b. 4
c. 3/0
d. 6
e. 3

_____ 30. The plans indicate that you should install a 300 amp branch circuit in Schedule 40 PVC with a 4 AWG equipment grounding conductor. However you did not have any #4 wire so you pulled in 10 AWG instead. With a load connected the normal circuit current will be ___ with the #10 EGC (compared to using a 4 AWG EGC).

a. greater
b. less
c. the same

Grounding and Bonding Part II

Match the terms, shown below, to the description.

    a. phase-to-phase voltage   d. effectively grounded
    b. line surge   e. transformer
    c. ground   f. none of these

_____ 31. A device that converts electrical power at one voltage or current to another voltage or current.

_____ 32. The maximum voltage between any two phases of an electrical system.

_____ 33. A temporary increase in the circuit or system voltage or current.

_____ 34. A device that converts electrical power to mechanical power.

_____ 35. A conducting connection between electrical circuits or equipment and the earth.

_____ 36. Grounded with sufficiently low impedance and sufficient current-carrying capacity to prevent hazardous voltage buildups.

Match the terms, shown below, to the description.

    a. grounded conductor   d. bonding
    b. enclosure   e. generator
    c. equipment   f. none of these

_____ 37. Any device, fixture, apparatus, appliance, etc., used in conjunction with electrical installations.

_____ 38. A conductor that has been intentionally grounded.

_____ 39. A device that converts mechanical power to electrical power.

_____ 40. The case or housing of equipment or other apparatus which provides protection from live or energized parts.

_____ 41. Joining metal parts to form a continuous path to conduct safely any current likely to be imposed.

_____ 42. The voltage between any ungrounded conductor and a grounded conductor.

Year Two (Student Manual)

Objective 206.3 Worksheet

_____ 1. A lightning rod must be installed at each BCES pole lighting standard and connected to the pole.

   a. True   b. False

_____ 2. Where lightning rods are installed on the BCES project for pole lighting, the rod can be connected using ___.

   a. cadwelding type connections
   b. acorn type ground clamps
   c. J-type ground clamps

_____ 3. The size grounding electrode conductor that is connected to the steel building structure from the BCES main service is ___ AWG or kcmil.

   a. 4   b. 3/0   c. 500   d. 250   e. 1/0

_____ 4. On the BCES project, each conduit installed between the transformer and Panel LB2 contains ___ grounding conductor(s).

   a. 1 ea 1 AWG
   b. 2 ea 1 AWG
   c. 1 ea 8 AWG
   d. 1 ea 2 AWG
   e. 1 ea 4 AWG
   f. 2 ea 2 AWG

_____ 5. A bare ___ bonding jumper is used for the lightning protection at the BCES parking lot pole bases.

   a. #8 solid copper
   b. #6 copper-clad stranded
   c. #4 stranded copper
   d. none of these

_____ 6. On the BCES project, the electrician is required to provide a bonding jumper on the horizontal cable tray ___.

   a. between horizontal and vertical runs
   b. between each section
   c. where accessible

_____ 7. Refer to the BCES prints. The concrete-encased electrode needs to be installed early in the construction process, before walls have been built or the floor slab poured. The most practical place to install the concrete-encased electrode would be the ___.

   a. West footer in room D154
   b. East footer in room 154
   c. West footer in room D133
   d. South footer in room D156

Lesson 206 NEC® Worksheet

_____ 1. Where lubrication and vehicle servicing is done at places such as Jiffy Lube and Grease Monkey auto service centers, the space below the floor level in a ventilated pit shall be ____. Exhaust entilation is provided at a rate of not less than 0.3 m 3/min/m 2 (1 cfm/ft 2) of floor area at all times that the building is occupied.

   a. classified as Class I, Division 1
   b. unclassified
   c. classified as Class II, Division 2
   d. none of these

_____ 2. Where lubrication and vehicle servicing is done at places such as Jiffy Lube and Grease Monkey auto service centers and the space below the floor and up to 18 inches above the floor level is to be unclassified, the exhaust ventilation air must be taken from a point within ____ of the floor of the pit.

   a. 300 mm
   b. 24 inches
   c. 450 mm
   d. 18 inches
   e. none of these

_____ 3. Where lubrication and vehicle servicing is done at places such as Jiffy Lube and Grease Monkey auto service centers, for lubrication service rooms to be unclassified, there must be mechanical ventilation of ____ per minute of exchanged air for each square foot of floor area.

   a. one cubic foot
   b. three cubic feet
   c. 1.5 cubic feet
   d. five cubic feet
   e. ten cubic feet

_____ 4. Where major repair garages service vehicles that are fueled by lighter-than-air fuels such as natural gas or hydrogen, the area within ____ of the ceiling shall be considered unclassified where ventilation of at least 1 cfm/sq ft of ceiling area is provided.

   a. 24 inches
   b. 12 inches
   c. 300 mm
   d. 60 inches
   e. 18 inches

Grounding and Bonding Part II   206-153

_____ 5. In major repair garages, where vehicles ___ are repaired or stored, the area within 450 mm (18 in.) of the ceiling shall be considered unclassified where ventilation of at least 1 cfm/sq ft of ceiling area taken from a point within 450 mm (18 in.) of the highest point in the ceiling is provided.

   a. that are fueled with lighter-than-air fuels
   b. that have multiple fuel tanks
   c. with propane fueled fleet operations
   d. with oversized fuel tanks

_____ 6. Where lubrication and vehicle servicing is done at places such as Jiffy Lube and Grease Monkey auto service centers, the space below the floor and up to 18 inches above the floor level in an unventilated pit shall be ___.

   a. classified as Class I, Division 1
   b. unclassified
   c. classified as Class I, Division 2
   d. classified Class II, Division 2
   e. none of these

_____ 7. In commercial repair garages where electrical diagnostic equipment, electrical hand tools or portable lighting equipment is used, GFCI protection is required for all ___ receptacles.
   I.   125-volt, single-phase, 15 and 20-ampere
   II. 250-volt, single-phase, 30-ampere
   III. 250-volt, three-phase, 15- and 20-ampere

   a. I, II, & III
   b. II only
   c. I & II only
   d. I & III only
   e. I only

Year Two (Student Manual)

Lesson 206 Safety Worksheet

HAZARD COMMUNICATIONS (Haz Com) INTRODUCTION

The purpose of the Haz Com standard (OSHA CFR1910.1200) is to ensure that the hazards of all chemicals produced or imported are evaluated, and that information concerning their hazards is transmitted to employers and employees. This transmittal of information is to be accomplished by means of comprehensive hazard communication programs, which are to include container labeling and other forms of warning, material safety data sheets (MSDS) and employee training.

The Haz Com standard is intended to address comprehensively the issue of evaluating the potential hazards of chemicals, and communicating information concerning hazards and appropriate protective measures to employees, and to preempt any legal requirements of a state, or political subdivision of a state, pertaining to this subject. Evaluating the potential hazards of chemicals, and communicating information concerning hazards and appropriate protective measures to employees, may include, for example, but is not limited to, provisions for: developing and maintaining a written hazard communication program for the workplace, including lists of hazardous chemicals present; labeling of containers of chemicals in the workplace, as well as of containers of chemicals being shipped to other workplaces; preparation and distribution of material safety data sheets to employees and downstream employers; and development and implementation of employee training programs regarding hazards of chemicals and protective measures.

This standard requires chemical manufacturers or importers to assess the hazards of chemicals which they produce or import, and all employers to provide information to their employees about the hazardous chemicals to which they are exposed, by means of a hazard communication program, labels and other forms of warning, material safety data sheets, and information and training. In addition, this section requires distributors to transmit the required information to employers. (Employers who do not produce or import chemicals need only focus on those parts of this rule that deal with establishing a workplace program and communicating information to their workers.)

This standard applies to any chemical which is known to be present in the workplace in such a manner that employees may be exposed under normal conditions of use or in a foreseeable emergency.

Employers shall ensure that labels on incoming containers of hazardous chemicals are not removed or defaced.

Employers shall maintain any material safety data sheets (MSDS) that are received with incoming shipments of hazardous chemicals, and ensure that they are readily accessible during each work shift to laboratory employees when they are in their work area(s).

Employers shall maintain copies of any material safety data sheets that are received with incoming shipments of the sealed containers of hazardous chemicals, shall obtain a material safety data sheet as soon as possible for sealed containers of hazardous chemicals received without a material safety data sheet if an employee requests the material safety data sheet, and shall ensure that the material safety data sheets are readily accessible during each work shift to employees when they are in their work area(s).

Grounding and Bonding Part II

Questions

_____ 1. The purpose of the Haz Com standard is to ensure that the hazards of all chemicals are ___, and that information concerning their hazards is transmitted to employees.

   a. evaluated
   b. identified
   c. labeled
   d. marked

_____ 2. Employers shall ensure that labels on incoming containers of hazardous chemicals are not ___.
   I. removed   II. outdated   III. Defaced

   a. I only
   b. III only
   c. I & III only
   d. II only
   e. I, II, & III
   f. I & II only

_____ 3. A comprehensive Haz Com program includes ___.

   a. employee training
   b. MSDS training
   c. other forms of warning
   d. all of these

_____ 4. Employers shall maintain any material safety data sheets (MSDS) of ___.

   a. hazardous locations
   b. hazardous chemicals
   c. hazardous energy
   d. hazardous activities

_____ 5. Material safety data sheets must be readily accessible when employees are in their ___.

   a. work vehicle
   b. work area
   c. break area
   d. none of these

Year Two (Student Manual)

Lesson 207
Wiring Methods

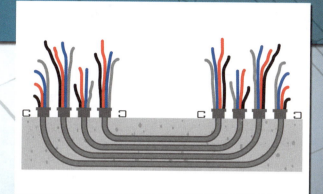

Electrical Curriculum

Year Two
Student Manual

Purpose
This lesson will teach the recognized standards and installation requirements for various wiring methods in electrical systems.

Homework
(Due at the beginning of this class)

For this lesson, you should:

- Thoroughly read the material contained within Lesson 207.
- Read ES Chapter 7 in its entirety.
- Complete Objective 207.1 Worksheet.
- Complete Objective 207.2 Worksheet.
- Complete Objective 207.3 Worksheet.
- Complete Objective 207.4 Worksheet.
- Complete Lesson 207 NEC Worksheet
- Read and complete Lesson 207 Safety Worksheet.

- Complete additional worksheets, if available, as directed by your instructor.

Objectives

By the end of this lesson, you should:

207.1 ....Understand the load considerations applied to conduit and cable wiring methods.
207.2 ....Understand requirements and installation practices for metallic and non-metallic raceways.
207.3 ....Understand requirements and installation practices for underground installations.
207.4 ....Be able to locate equipment and symbols on the BCES prints in order to solve problems addressed on common construction site scenarios.

207.1 Load Considerations Applied to Conduit and Cable Wiring Methods

207.2 Installation Practices for Metallic and Non-metallic Raceways

Repeated reference is made in the NEC concerning ferrous and nonferrous metals and materials. For example, NEC Section 250.64(E) requires that ferrous metals and materials, used to enclose grounding electrode conductors (GECs), be electrically continuous between the cabinet or equipment and the grounding electrode. However, nonferrous GEC enclosures do not have this requirement.

Examples of nonmetallic, nonferrous materials commonly used in electrical materials and equipment construction include:
- Plastic outlet boxes and device covers
- Fiberglass outlet boxes, junction boxes, and equipment enclosures
- PVC outlet boxes and rigid raceways (including Schedule 80 RNMC)
- High Density Polyethylene Conduit (HDPE Conduit)
- The nylon covering the insulation of THHN/THWN conductors
- Plastic flexible raceways: including ENT and LFNC

Although some metals can become magnetized when exposed to various temperatures or when exposed to a magnetic field, for our purposes, metals are classified either ferrous or nonferrous. Ferrous metals consist primarily of iron. Ferrous metals include pure iron and alloys containing iron. Steel is an example of an alloy containing iron. Typically, metals to which a magnet is attracted contain iron. Also typically, if a magnet is not attracted to a piece of metal, the metal is nonferrous. Common ferrous metals include the various irons and common carbon steels. Common steel is an alloy containing various amounts of carbon but primarily iron. Magnetic metals include iron, nickel, and cobalt. These metals are used for permanent magnets.

Examples of ferrous metals commonly used in electrical materials and equipment construction include:

Wiring Methods

- Chrome-plated and common steel device covers
- Steel outlet boxes, junction boxes, and equipment enclosures
- Die-cast steel conduit bodies (LB, LL, T, etc)
- Steel RMC, IMC, and EMT
- Steel flexible metal conduit (FMC)
- Steel-armored Type AC cable

Common types of nonferrous metals include aluminum, tin, copper, silver, lead, zinc, and brass (an alloy of copper and zinc). Some precious metals such as silver, gold, and platinum are also nonferrous.

Austenitic stainless steel is an alloy of chromium, nickel, and iron. Type 304 stainless is the type of austenitic stainless steel used in most electrical equipment and materials. Many device covers, raceways, equipment enclosures, ground rods, etc, are made of this steel because of its high resistance to corrosion. The amount of chromium used in Type 304 stainless makes this steel virtually rust-proof. Type 304 stainless is nonmagnetic because it contains a great deal of chromium but very little iron and nickel. Type 304 stainless has been manufactured through a special process that "cancels" the slight magnetic properties that are present due to the presence of the iron and nickel.

Examples of nonferrous metals commonly used in electrical materials and equipment construction include:
- Stainless steel device covers
- Stainless steel outlet boxes, junction boxes, and equipment enclosures
- Stainless steel ground rods
- Brass device covers
- Brass RMC – typically used in swimming pool installations
- Aluminum RMC and fittings
- Die-cast aluminum EMT set-screw fittings
- Aluminum junction boxes
- Aluminum conduit bodies (LB, LL, T, etc)
- Type MC cable with aluminum interlocked armor
- Aluminum flexible metal conduit (FMC)
- Copper and aluminum wire and busbars

207.3 Requirements and Practices for Underground Installations

207.4 Problem Solving for BCES Blueprints

Wiring Methods

Objective 207.1 Worksheet

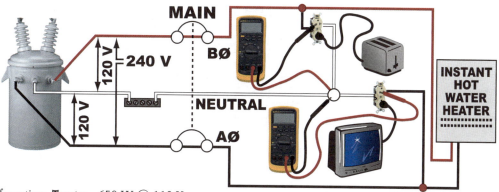

Nameplate Information: **Toaster** - 650 W @ 110 V
Television - 3.2 A @ 130 V   **Water Heater** - 2500 W @ 250 V

Figure 207.5

_____ 1. Refer to the wiring diagram shown above in Figure 207.5. When all three equipments are operating and the neutral is removed from the neutral bar the voltage across the toaster will be ___ volts. Select the closest answer.

   a. 240   b. 120   c. 165   d. 60   e. 75

_____ 2. Refer to the wiring diagram shown above in Figure 207.5. When all three equipments are operating the current through AØ of the MAIN breaker will be ___ amps. Select the closest answer.

   a. 13.2   b. 9.6   c. 16.0   d. 19.1   e. 12.6

_____ 3. Refer to the wiring diagram shown above in Figure 207.5. When all three equipments are operating and the neutral is removed from the neutral bar the voltage across the water heater will be ___ volts. Select the closest answer.

   a. 165   b. 75   c. 240   d. 120   e. 60

_____ 4. Refer to the wiring diagram shown above in Figure 207.5. When all three equipments are operating the current through BØ of the MAIN breaker will be ___ amps. Select the closest answer.

   a. 16.0   b. 19.1   c. 9.6   d. 13.2   e. 12.6

_____ 5. Refer to the wiring diagram shown above in Figure 207.5. When all three equipments are operating and the neutral is removed from the neutral bar the voltage across the television will be ___ volts. Select the closest answer.

   a. 120   b. 165   c. 240   d. 75   e. 60

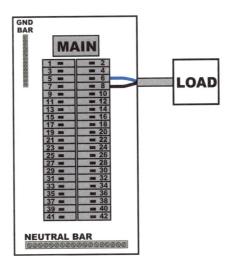

Figure 207.6

_____ 6. The panel, shown above in Figure 207.6, is 120/240 volt 1Ø, 3-wire. The load has a rating of 32.4 KW at 240 volts. Each ungrounded conductor will carry ___ amps.

a. 78
b. 135
c. 156
d. 270
e. 292

_____ 7. The panel, shown above in Figure 207.6, is 120/208 volt 3Ø, 4-wire. The load has a rating of 22.4 KW at 240 volts. Each ungrounded conductor will carry ___ amps.

a. 107.7
b. 116.7
c. 186.7
d. 80.9
e. 93.3

_____ 8. Among the conduits shown below, Conduit(s) ___ will have a magnetic field induced around it. All circuits originate in a 120/240 volt 1-ph, 3-wire panelboard. Each ungrounded conductor carries 12 amps of current.

a. I only   b. II and II only   c. I and II only   d. III only   e. II only

Wiring Methods

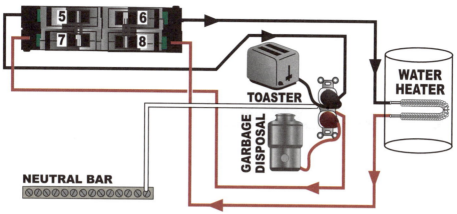

Figure 207.2

The circuit breakers (5-7 & 6-8) shown above are installed in MLO 120/240 V, 1ph, 3-W Panel K. Panel K is fed with Circuit LD-1&3. LD-1 is connected to the left lug and LD-3 is connected to the right lug in Panel K. Feeder Circuit LD-1&3 is 120/240 volt 1Ø, 3-wire. The nameplate info is as follows:

Toaster: 700 W at 110 V   **Disposer:** 12.9 A at 130 V   **Water Heater:** 4.7 KW at 250 V

_____ 9. Refer to the wiring diagram shown above in Figure 207.2. How much current will flow through FEEDER conductor LD-1 with all the loads operating at the panel voltage? Select the closest answer.

    a. 25   b. 30   c. 31   d. 32   e. 16

_____ 10. Refer to the wiring diagram shown above in Figure 207.2. How much current will flow through FEEDER conductor LD-3 with all the loads operating at the panel voltage? Select the closest answer.

    a. 16   b. 32   c. 31   d. 30   e. 25

_____ 11. The symbol, or abbreviation, used for the unit of measurement for the answer to the previous question is ___.

    a. E   b. W   c. A   d. V   e. I

_____ 12. Refer to the drawing shown below. These circuits are from a 120/240 volt 1Ø, 3-wire panel. Each ungrounded conductor carries 16 amps. The neutral in Conduit B will carry ___ amps.

    a. 0   b. 32   c. 8   d. 16

_____ 13. Refer to the drawing shown below. These circuits are from a 120/240 volt 1Ø, 3-wire panel. Each ungrounded conductor carries 16 amps. The neutral in Conduit A will carry ___ amps.

a. 0   b. 16   c. 8   d. 32

Objective 207.2 Worksheet

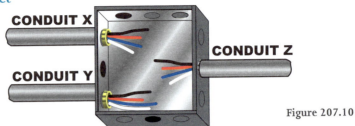

Figure 207.10

_____ 1. Refer to Figure 207.10, shown above. Conduit X contains 8 AWG, Conduit Y contains 6 AWG, and Conduit Z contains 4 AWG conductors. Although bushings have been installed on all the conduits which of the conduits actually are required to have bushings per the NEC®?
I. Conduit X   II. Conduit Y   III. Conduit Z

a. I, II, & III   b. I & II only   c. II & III only   d. III only   e. none of these

_____ 2. Which of the following conductor arrangements is permitted by the NEC®?

a.
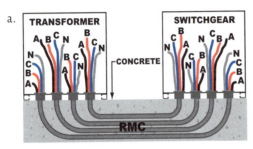

b.
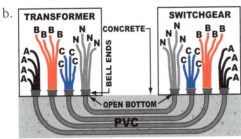

c.
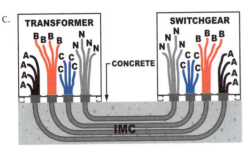

d. a or b only
e. a or b or c

Wiring Methods

_____ 3. Which of the following metals are considered to be ferrous as referenced at 300.6(A) in the NEC®?

   a. carbon steel   b. copper   c. aluminum   d. stainless steel   e. all of these

_____ 4. As referenced at 250.52(A)(7) in the NEC® ___ is a nonferrous metal.
   I. stainless steel   II. carbon steel   III. Aluminum

   a. I only   b. II only   c. III only   d. I & III only   e. none of these

_____ 5. Which of the drawings, shown below in Figure 207.12, satisfies the requirements of the NEC® for conductor length in boxes?

   a. II, III, & IV only
   b. I & IV only
   c. I, II, III, & IV
   d. IV only
   e. II & IV only

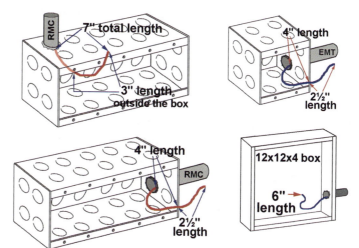

Figure 207.12

_____ 6. The 12" × 12" junction box, shown below in Figure 207.13, has a depth (d) of 6". From where the conductors emerge from the conduit, the minimum length of the conductors is ___".

   a. 6   b. 12   c. 9   d. 3   e. 15

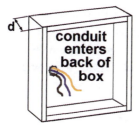

Figure 207.13

_____ 7. The depth (d) of the 1-gang masonry box shown below is 3½ inches. From where the conductors emerge from the conduit the minimum length of the conductors is ___".

   a. 4   b. 2½   c. 6½   d. 3   e. 7

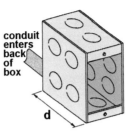

_____ 8. 547.5(C)(3) in the NEC® includes an informational FPN that indicates that in an agricultural environment enclosures made of ___ material may corrode.

   a. aluminum   b. magnetic ferrous   c. either a or b   d. neither a nor b

_____ 9. Six foot lighting "tails" made from 3/8 inch LFNC (Carlon® flex) are used to connect 2′ × 4′ luminaires above a lay-in grid ceiling. The space between the ceiling and the roof deck is used as an air return for the air conditioning/heating system. In this plenum ceiling space, is the use of lighting tails made of LFNC a violation of the NEC®?

   a. Yes   b. No

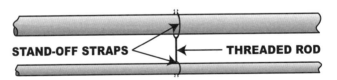

Figure 207.11

_____ 10. Under which conditions, listed below, does the NEC permit one conduit to be supported from another as shown above in Figure 207.11? Both conduits contain current carrying conductors.
   I. Where the upper wiring method is EMT and contains circuit conductors to a heating unit and the lower wiring method is Class 2 thermostat cable for the heating unit.
   II. Where the upper conduit contains 4 AWG circuit conductors and the lower conduit contains circuit conductors not larger than 6 AWG.
   III. Where the upper wiring method is RMC, the lower wiring method is Type MC cable, and the threaded rod is at least 3/8 inch in size.

   a. I only   b. neither I nor II nor III   c. II or III only   d. III only   e. I or II only

Wiring Methods

_____ 11. Where the upper conduit trade size is at least twice (200%) the trade size of the lower conduit, the lower conduit is permitted to be supported as shown on the previous page in Figure 207.11.

   a. False   b. True

_____ 12. Where installed through wooden studs as shown below in Figure 207.9, which of the installations is permitted by the NEC®?

   a. II & III only   b. III only   c. I & III only   d. II only   e. I, II, & III

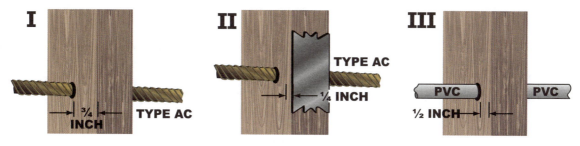

Figure 207.9

_____ 13. Article 300 covers wiring that is factory-installed in ___.

   a. air-handlers
   c. refrigeration condenser units
   b. water heaters
   d. all of the above
   e. none of the above

_____ 14. Rigid Metal Conduit (RMC) is generally made of galvanized steel or aluminum but special applications, such as for swimming pools or in chemical plants, may require the use of ___ conduit.

   a. stainless steel   b. bronze   c. either a or b   d. neither a nor b

_____ 15. As referenced in the NEC at 300.6(A) and 300.20(B) the Code frequently distinguishes between ferrous and nonferrous metals because ferrous metals ___.

   a. have magnetic properties
   b. are more likely to corrode than nonferrous metals
   c. both a and b
   d. neither a nor b

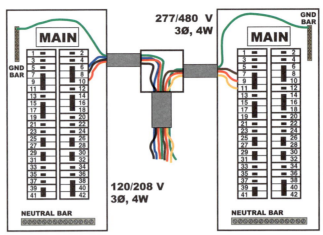

Figure 207.8

_____ 16. Refer to Figure 207.8, shown above. The conductors from Conduit X are permitted to occupy the junction box with the conductors from Conduit Z.

a. False   b. True

_____ 17. Refer to Figure 207.8, shown above. The conductors from Conduit X are permitted to occupy Conduit Y with the conductors from Conduit Z.

a. False   b. True

_____ 18. In vertical conduit runs, such as up through a high-rise building, conductors made of aluminum have support requirements that are ___ than for copper conductors.

a. farther apart   b. closer together

_____ 19. Two 12 AWG circuit conductors are permitted to be spliced using a wire nut inside ___.
I. a 3″ run of EMT   II. a 4 square box   III. an LR conduit body

a. I or II or III   b. I or III only   c. II only   d. II or III only   e. I or II only

_____ 20. An electrician has been using steel AC cable with single screw connectors. The instructions on the box the connectors came in states "for use with steel cable only". His next delivery of cable has an aluminum sheath. Would using the same connectors be a violation of the Code?

a. Yes   b. No

Wiring Methods

_____ 21. As referenced at 250.64(E) in the NEC, ___ is a ferrous metal enclosure.
   I. aluminum RMC   II. steel EMT   III. aluminum IMC

   a. I & III only   b. I & II only   c. II only   d. III only   e. all of these

_____ 22. 10 AWG conductors can be spliced inside a run of conduit using ___.
   I. irreversible connections   II. exothermic welding   III. wire nuts

   a. I or II or III   b. I or II only   c. III only   d. I only   e. neither I nor II nor III

Student Notes

Objective 207.3 Worksheet

_____ 1. RMC is installed in the slab floor of a psychiatric hospital. There must be a minimum of ___ inches of concrete covering the conduit.

　　a. 18　　　b. 4　　　c. 0　　　d. 12　　　e. 6　　　f. 24

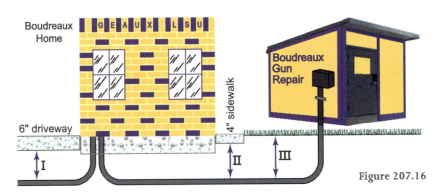

Figure 207.16

_____ 2. A 120-volt circuit is fed from a 15-Amp GFCI breaker in the panel in Boudreaux' home and is routed underground to the gun shop, as shown above in Figure 207.16. The minimum dimension permitted at III is ___ inches where the raceway is Schedule 40 PVC.

　　a. 6　　　b. 18　　　c. 24　　　d. 12　　　e. 0　　　f. 4

_____ 3. Schedule 80 PVC is the wiring method installed beneath the driveway shown above in Figure 207.16. The 120-volt circuit is from the house panel and feeds the garage. The minimum dimension permitted for I is ___ inches.

　　a. 6　　　b. 12　　　c. 0　　　d. 14　　　e. 18　　　f. 24

_____ 4. Type UF cable is the wiring method installed beneath the grassy area shown above in Figure 207.16. The circuit is from a low voltage transformer in the home and feeds 24-volt equipment in the gun shop. The minimum dimension permitted for III is ___ inches.

　　a. 12　　　b. 24　　　c. 4　　　d. 14　　　e. 6　　　f. 18

_____ 5. A 120-volt circuit is fed from a 15-Amp GFCI breaker in the panel in Boudreaux' home and is routed under the driveway, as shown above in Figure 207.16. The minimum dimension permitted at I is ___ inches where the wiring method is Type UF cable.

　　a. 18　　　b. 6　　　c. 12　　　d. 24　　　e. 0　　　f. 4

Wiring Methods

_____ 6. A 120-volt circuit is fed from a 15-Amp GFCI breaker in the panel in Boudreaux' home and is routed underground to the gun shop, as shown on the previous page in Figure 207.16. The minimum dimension permitted at III is ___ inches where the wiring method is Type UF cable.

    a. 6   b. 4   c. 24   d. 18   e. 12   f. 0

_____ 7. Schedule 80 PVC is the wiring method installed beneath the sidewalk shown on the previous page in Figure 207.16. The 120-volt circuit is from the house panel and feeds lighting in the gun shop. The minimum dimension permitted for II is ___ inches.

    a. 12   b. 0   c. 14   d. 24   e. 6   f. 18

_____ 8. Type UF cable is the wiring method installed beneath the sidewalk shown on the previous page in Figure 207.16. The 120-volt circuit is from the house panel and feeds lighting in the gun shop. The minimum dimension permitted for II is ___ inches.

    a. 0   b. 6   c. 12   d. 18   e. 24   f. 14

_____ 9. A half-way house for former guests of the State of California requires a circuit from a 24-volt transformer to feed the hot tub lights. This circuit is installed in Schedule 80 PVC and is routed beneath the lawn. The minimum burial depth permitted for the conduit is ___ inches.

    a. 18   b. 12   c. 0   d. 6   e. 24   f. 4

_____ 10. A 1 1/4-inch PVC conduit is to be run underground below an area that will be used as a parking lot. The conduit will contain circuit conductors feeding a sign for this hospital. The minimum depth the conduit must be below the present dirt surface level is ___ inches. After the trench is backfilled concrete will be poured on top of the dirt and will be 6″ thick.

    a. 18   b. 0   c. 24   d. 12   e. 6

_____ 11. As shown on the next page in Figure 207.18 luminaires have been installed along the sides of a parking lot at a university and Schedule 40 PVC was installed under the 6″ thick concrete parking lot between the luminaires. The minimum dimension permitted for Y is ___ inches.

    a. 24   b. 0   c. 6   d. 18   e. 12

Year Two (Student Manual)

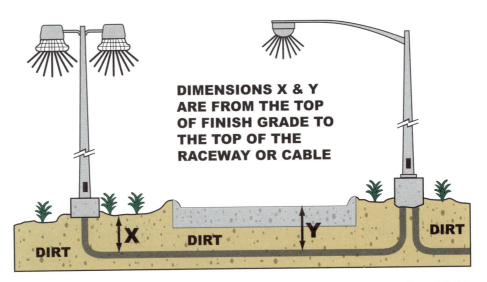

Figure 207.18

_____ 12. Figure 207.18, above, shows luminaires along a driveway to a single-family ranch home. RMC is the wiring method installed under the 7″ thick concrete drive between the luminaires. The minimum dimension permitted for Y is ___ inches.

    a. 12   b. 24   c. 6   d. 0   e. 18

_____ 13. Refer to Figure 207.18, shown above, where luminaires have been installed along a public street. RMC is the wiring method installed along the side of the street and between the luminaires. The minimum dimension permitted for X along the side of the street is ___ inches.

    a. 24   b. 12   c. 18   d. 0   e. 6

_____ 14. Refer to Figure 207.18, shown above, where luminaires have been installed along a driveway to a single-family ranch home. Where Schedule 40 PVC is the wiring method installed along the side of the drive and between the luminaires the minimum dimension permitted for X is ___ inches.

    a. 6   b. 24   c. 18   d. 12   e. 0

_____ 15. Refer to Figure 207.18, shown above, where luminaires have been installed along a driveway to a single-family ranch home. Type UF cable is the wiring method installed between the luminaires and along the side of the drive. The minimum dimension permitted for X along the side of the driveway is ___ inches.

    a. 0   b. 6   c. 12   d. 18   e. 24

Wiring Methods

_____ 16. Luminaires have been installed along the sides of a parking lot at a university as shown on the previous page in Figure 207.18. Type UF cable was installed under the 6″ thick concrete parking lot between the luminaires. The minimum dimension permitted for Y is ___ inches.

a. 12
b. 18
c. 0
d. 6
e. 24

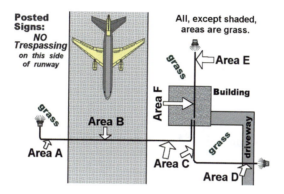

Figure 207.17

_____ 17. Refer to Figure 207.17, shown above. The cover requirements for the underground conduit in Area C are taken from the row in Table 300.5 entitled ___.

a. All locations not specified below
b. In or under airport runways, including adjacent areas where trespassing prohibited
c. Under a building
d. One- and two-family dwelling driveways and outdoor parking areas, and used only for dwelling-related purposes
e. Under streets, highways, roads, alleys, driveways, and parking lots

_____ 18. Refer to Figure 207.17, shown above. The cover requirements for the underground conduit in Area E are taken from the row in Table 300.5 entitled ___.

a. All locations not specified below
b. Under streets, highways, roads, alleys, driveways, and parking lots
c. Under a building
d. One- and two-family dwelling driveways and outdoor parking areas, and used only for dwelling-related purposes
e. In or under airport runways, including adjacent areas where trespassing prohibited

_____ 19. Refer to Figure 207.17, shown on the previous page. The cover requirements for the underground conduit in Area A are taken from the row in Table 300.5 entitled ___.

   a. In or under airport runways, including adjacent areas where trespassing prohibited
   b. One- and two-family dwelling driveways and outdoor parking areas, and used only for dwelling-related purposes
   c. Under a building
   d. Under streets, highways, roads, alleys, driveways, and parking lots
   e. All locations not specified above

_____ 20. Sharp, or heavy, rocks ___ permitted to be used to backfill a trench containing RMC.

   a. are   b. are not

_____ 21. Where covered with stabilized sand to prevent physical damage, Type ___ cable is permitted to be directly buried in a trench underground.
   I. UF   II. MI   III. AC

   a. II only   b. I & III only   c. I, II, & III   d. I & II only   e. I only

_____ 22. "S" loops ___ allowed for ground movement where service lateral conductors emerge from an underground raceway adjacent to a pad mounted transformer.

   a. are   b. are not

_____ 23. Which of the following are permitted to be buried in a trench underground? The wiring will be covered with stabilized sand to prevent physical damage.

   a. THHN/THWN
   b. Type MI cable
   c. Type AC cable
   d. a, b, and c
   e. neither a, nor b, nor c

_____ 24. Type USE cable is routed underground and then attached to the side of a building. Protection against physical damage ___ required on the side of the building above 8 feet.

   a. is not   b. is

Objective 207.4 Worksheet

_____ 1. What type raceway for the BCES service feeder conductors is installed between the pad mount transformer and the Texas New Mexico power distribution pole?

    a. RMC   b. EMT   c. IMC   d. cable tray   e. PVC

_____ 2. Refer to the BCES food service drawings. A slicer (Item 7) requires power. The power for this equipment is provided by ___.

    a. direct connection to a stubbed-up conduit
    b. cord-and-plug connection to a wall-mounted duplex receptacle
    c. flex connection to a stubbed-up conduit
    d. cord-and-plug connection to a floor-mounted duplex receptacle

_____ 3. Refer to drawing E1.1 in the BCES set. The maximum number of fixture whips that can be connected to a single junction box is ___.

    a. two   b. unlimited   c. four   d. three

_____ 4. Heavy wall metal conduit is frequently referred to as Heavy Wall, Rigid, Rigid Galvanized Steel (RGS), Galvanized Rigid Conduit (GRC) or Rigid Metal Conduit (RMC). The NEC® utilizes the term Rigid Metal Conduit (RMC). In the BCES prints, this conduit is referred to as ___.

    a. RMC   b. Rigid   c. RGS   d. HW

_____ 5. On the BCES project, the minimum permitted trade size rigid metal conduit is ___ inch.

    a. $1^1/_4$ b. $^1/_2$ c. 1   d. $^3/_4$ e. $1^1/_2$

_____ 6. As indicated on the BCES food service drawings, the fryer is connected to electrical power through a ___.

    a. wall-mounted J-Box
    b. 3Ø, 4-wire receptacle
    c. stubbed-up conduit
    d. 20A duplex receptacle

_____ 7. As indicated on the BCES food service drawings, the booster heater is gas fired.

   a. True
   b. False

_____ 8. As indicated on the BCES food service drawings, the fryer is gas-fired.

   a. True
   b. False

_____ 9. How many conduits for the BCES service feeder conductors are installed between the pad mount transformer and the Texas New Mexico power distribution pole?

   a. 3
   b. 4
   c. 7
   d. 9
   e. 1

Lesson 207 NEC Worksheet

_____ 1. Electrical wiring that is installed in and under aircraft hangar floors shall comply with the requirements for a ___ location.

   a. Class I, Division 1
   b. Class I, Division 2
   c. Class I, Zone 0
   d. Class I, Zone 1
   e. Class I, Zone 2

_____ 2. In aircraft hangars where electrical diagnostic equipment, electrical hand tools or portable lighting equipment are used, GFCI protection is required for all ___ receptacles.
   I. 125-volt, single-phase, 15 and 20-ampere
   II. 250-volt, single-phase, 30-ampere
   III. 250-volt, three-phase, 15- and 20-ampere

   a. I & II only
   b. I only
   c. I, II, & III
   d. I & III only
   e. II only

_____ 3. Where Schedule 80 PVC conduit is used for underground wiring in a bulk storage plant, an equipment grounding conductor ___ electrical continuity of the raceway system and the grounding of all non-current carrying metal parts.

   a. shall be included to provide
   b. shall not be included so as to provide isolation and to prohibit

_____ 4. Where non-metallic sheathed cable is used for underground wiring in a bulk storage plant, an equipment grounding conductor ___ electrical continuity of the raceway system and the grounding of all non-current carrying metal parts.

   a. shall be included to provide
   b. shall not be included so as to provide isolation and to prohibit

_____ 5. Buried raceways and cables under Class I locations in bulk storage plants shall be considered to be within a ___ location.

   a. Class I, Zone 1
   b. Class I, Division 2
   c. Class I, Zone 0
   d. Class I, Zone 2
   e. Class II, Zone 2

Student Notes

Lesson 207 Safety Worksheet

HAZARD COMMUNICATIONS (Haz Com) part 2

Employers shall ensure that employees are provided with information and training to the extent necessary to protect them in the event of a spill or leak of a hazardous chemical.

Written hazard communication program.

Employers shall develop, implement, and maintain at each workplace, a written hazard communication program which at least describes how the labels and other forms of warning, material safety data sheets, and employee information and training will be met. It also requires a list of the hazardous chemicals known to be present in the workplace, and the methods the employer will use to inform the employees of the hazards associated with these chemicals.

Multi-employer worksites.

Employers who produce, use, or store hazardous chemicals at a workplace in such a way that the employees of another employer may be exposed shall additionally ensure that the hazard communication programs include, the methods the employer will use to provide the other employers on-site access to material safety data sheets. The method the employer will use to inform the other employer of any precautionary measures that need to be taken during the workplace's normal operation conditions and in foreseeable emergencies and what method of labeling will be used.

Labels and other forms of warning.

The chemical manufacturer, importer, or distributor shall ensure that each container of hazardous chemicals leaving the workplace is labeled, tagged or marked with the identity of the hazardous chemical, appropriate hazard warnings, name and address of the manufacturer or importer, the identity of the hazardous chemical contained, and the appropriate warnings regarding the hazards of the chemical, including any health information necessary.

MSDS sheets (Material Safety Data Sheets)

Chemical manufacturers and importers shall obtain or develop MSDS sheets for each hazardous chemical they produce or import. These MSDS sheets shall be in English (although they may be available in other languages). The MSDS sheets shall contain the following information, the identity of the chemical, the chemical and common name of all ingredients which have been determined to be a health hazard, physical and chemical characteristics of the hazardous chemical, the health hazards, the primary routes of entry, the OSHA permissible exposure limits, emergency 1st aid procedures, the name, address and telephone number of the chemical manufacturer, importer or other responsible party and more.

Questions

_____ 1. Employers shall ensure that employees are provided with information and ___ to the extent necessary to protect them in the event of a spill or leak of a hazardous chemical.

   a. training
   b. knowledge
   c. skills
   d. none of these

_____ 2. A written Haz Com program requires a list of ___ known to be present in the workplace, and the methods the employer will use to inform the employees of the hazards.

   a. chemicals
   b. ingredients
   c. compounds
   d. all of these

_____ 3. Chemical manufacturers and importers shall ___ MSDS sheets for each hazardous chemical they produce or import.
       I. design   II. purchase   III. propose

   a. I or II only
   b. I only
   c. neither I nor II nor III
   d. II only
   e. I or III only
   f. either I or II or III

_____ 4. The chemical ___ shall ensure that each container of hazardous chemical leaving the workplace is labeled, tagged or marked with the identity of the hazardous chemical.

   a. manufacturer
   b. distributor
   c. importer
   d. all of these

_____ 5. Written Haz Com programs shall describe how the ___ and other forms of warning, MSDS sheets, and employee information and training will be met.

   a. barricades
   b. signs
   c. labels
   d. signals

_____ 6. ___ shall develop, implement, and maintain at each workplace, a written hazard communication program.

    a. Employees
    b. Military
    c. Government
    d. Employers

Lesson 208
Wiring Materials, Conduit and Box Fill

Electrical Curriculum

Year Two
Student Manual

IEC
PRIDE
NATIONAL

Purpose
This lesson will teach installation and practices for raceways, cables and boxes in electrical systems and box/raceway fill calculations.

Homework
(Due at the beginning of this class)

For this lesson, you should:

- Thoroughly read the material contained within Lesson 208.
- Read ES Chapter 8 in its entirety.
- Complete Objective 208.1 Worksheet.
- Complete Objective 208.2 Worksheet.
- Complete Objective 208.3 Worksheet.
- Complete Objective 208.4 Worksheet.
- Complete Objective 208.5 Worksheet.
- Complete Lesson 208 NEC Worksheet.
- Read and complete Lesson 208 Safety Worksheet.

- Complete additional worksheets, if available, as directed by your instructor.

Objectives

By the end of this lesson, you should:

208.1 ....Understand support methods and limitations for various raceway and cable wiring.
208.2 ....Understand pull box and conduit volume calculations for equipment enclosures, raceways, wireways and cable systems.
208.3 ....Understand box fill calculations.
208.4 ....Understand and apply the NEC requirements for conduit installation and pull boxes.
208.5 ....Be able to locate equipment and symbols on the BCES prints in order to solve problems addressed on common construction site scenarios.

Read the Summary for this Lesson!

208.1 Support methods and limitations for various raceway and cable wiring

Article 352 in the NEC addresses Rigid Nonmetallic Conduit (RNC). Section 352.10(F) permits RNC to be installed in locations subject to physical damage if identified for such use. Refer to the Carlon inserts located in this manual. Schedule 80 PVC conduit manufactured by Carlon is listed for use in locations subject to physical damage. However, Carlon's Schedule 40 is not identified for such use. Schedule 80 from most manufacturers is listed for use in locations subject to physical damage. A comparison of the wall thicknesses of Schedule 40 and Schedule 80 reveals a substantial difference between the two. Consequently, Sch 80 can withstand more abuse than Sch 40.

Carlon Rigid Nonmetallic Conduit (RNC), Fittings & Accessories

Carlon manufactures the most complete line of nonmetallic conduits and fittings in the electrical industry. Carlon Schedule 40 and Schedule 80 conduits are designed for use aboveground and underground as described in the National Electrical Code Specify only Carlon conduits and fittings to insure raceway system integrity.

Features

Ease of Installation
Nonmetallic conduits are 1/4 to 1/5 the weight of metallic systems, can be installed in less than half the time, and are easily fabricated on the job.

Safety
Nonmetallic conduits are nonconductive, assuring a safe system.

Impact Resistant
Carlon Schedule and Schedule 80 nonmetallic conduits are resistant to sunlight and are listed for exposed or outdoor usage. The use of expansion fittings allows the system to expand and contract with temperature variations.

Corrosion Resistant
Carlon conduits and fittings are nonmetallic and will not rust or corrode.

Carlon nonmetallic Schedule 40 and Schedule 80 conduits and elbows are manufactured to NEMA TC-2, Federal specification WC1094A and UL 651 specification WC1094A and UL514B. Both conduit and fittings carry respective UL or ETL Listings and UL or ETL labels.

Support of Carlon Rigid Nonmetallic Conduit in Aboveground Installations

Table 352.30(B) NEC shows the support requirements for Schedule 40 and Schedule 80 rigid PVC nonmetallic conduit. Plastic conduit should alwuas be installed away from steam lines, etc. Support straps should allow for lineal movemenr caused by expansion and contraction. Maximum ambient temperature is 122°F (50°C)

Table 352.30(B), NEC	
Trade Size	Maximum Spacing Between Supports (feet)
1/2 - 1	3
1 1/4 - 2	5
2 1/2 - 3	6
3 1/2 - 5	7
6	8

Acceptable Dimensions in Inches of Integral Bell per UL 651

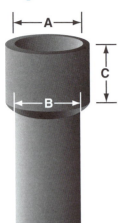

TRADE SIZE	A AT ENTRANCE (in.)		B AT BOTTOM (in.)		C NOMINAL BELL
	Maximum	Minimum	Maximum	Minimum	Depth (in.)
1/2	0.860	0.844	0.844	0.828	1.375
3/4	1.074	1.054	1.056	1.036	1.500
1	1.340	1.320	1.320	1.300	1.750
1 1/4	1.689	1.665	1.667	1.643	1.875
1 1/2	1.930	1.906	1.906	1.882	2.750
2	2.405	2.381	2.381	2.357	3.250
2 1/2	2.905	2.875	2.883	2.853	3.250
3	3.530	3.500	3.507	3.477	3.875
3 1/2	4.065	3.965	4.007	3.977	3.875
4	4.565	4.465	4.506	4.476	4.625
5	5.643	5.543	5.583	5.523	5.625
6	6.708	6.608	6.644	6.584	6.375

Schedule 40 PVC Rigid Nonmetallic Conduit (RNC)
(Heavy Wall EPC)

ETL Listed to UL 651 in compliance to the NEC

LISTED E35297

RUS Listed

Listed for underground applications encased in concrete or direct burial. Also for use in exposed or concealed applications aboveground.

- Sunlight resistant
- Rated for use with 90°C conductors
- Superior weathering characteristics

Schedule 40 Heavy Wall

With Intregal Bell*

Part No.		Nom Size	Std. Crate Qty.		Wt. Per 100'	Dimensions		Wall
10'	20'		10'	20'		O.D.	I.D.	
49005-010		1/2"	6000'		17	.840	.622	.109
49007-010	49007-020	3/4"	4400'	8800'	23	1.050	.824	.113
49008-010	49008-020	1"	3600'	7200'	34	1.315	1.049	.133
49009-010	49009-020	1 1/4"	3300'	6600'	46	1.660	1.380	.140
49010-010	49010-020	1 1/2"	2250'	4500'	55	1.900	1.610	.145
49011-010	49011-020	2"	1400'	2800'	73	2.375	2.067	.154
49012-010	49012-020	2 1/2"	930'	1860'	124	2.875	2.469	.203
49013-010	49013-020	3"	880'	1760'	163	3.500	3.068	.216
49014-010	49014-020	3 1/2"	630'	1260'	196	4.000	3.548	.226
49015-010	49015-020	4"	570'	1140'	232	4.500	4.026	.237
49016-010	49016-020	5"	380'	760'	315	5.563	5.047	.258
49017-010	49017-020	6"	260'	520'	409	6.625	6.065	.280

Rigid nonmetallic conduit is normally supplied in standard 10´ lengths, with one belled end per length. For specific requirements, it may be produced in lengths shorter or longer than 10´ with or without belled ends.

Use RNC Fittings with Schedule 40 and Schedule 80 Conduit

Notes:
1. Special fittings and conduit sizes will be quoted on request.
2. DON'T FORGET TO ORDER CEMENT.
3. Carlon reserves the right: to ship to the nearest unitized quantity.

Wiring Materials, Conduit and Box Fill

Schedule 80 PVC Rigid Nonmetallic Conduit (RNC)
(Extra Heavy Wall EPC-80)

Listed for use in aboveground and belowground applications that are subject to physical damage.
- Sunlight resistant
- Rated for use with 90C
- Superior weathering characteristics
- For use in areas subject to physical damage

RUS Listed

Schedule 80 Extra Heavy Wall
With Intregal Bell*

Part No.		Nom Size	Std. Crate Qty.		Wt. Per 100'	Dimensions		
10'	20'		10'	20'		O.D.	I.D.	Wall
49405-010	49405-020	1/2"	6000'	12000	21	.840	.546	.147
49407-010	49407-020	3/4"	4400'	8000'	30	1.050	.742	.154
49408-010	49408-020	1"	3600'	7200'	44	1.315	.957	.179
49409-010	49409-02	1 1/4"	3300'	6600'	60	1.660	1.278	.191
49410-010	49410-02	1 1/2"	2250'	3600'	72	1.900	1.500	.200
49411-010	49411-020	2"	1400'	2800'	101	2.375	1.939	.218
49412-010	49412-020	2 1/2"	930'	1880'	154	2.875	2.323	.276
49413-010	49413-020	3"	880'	1760'	210	3.500	2.900	.300
49415-010	49415-020	4"	570'	1140'	308	4.500	3.826	.337
49416-010		5"	380'		428	5.563	4.813	.375
49417-010	49417-020	6"	260'	520'	588	6.625	5.761	4.32

Rigid nonmetallic conduit is normally supplied in standard 10´ lengths, with one belled end per length. For specific requirements, it may be produced in lengths shorter or longer than 10´ with or without belled ends.

Use RNC Fittings with Schedule 40 and Schedule 80 Conduit

Notes:
1. Special fittings and conduit sizes will be quoted on request.
2. DON'T FORGET TO ORDER CEMENT.
3. Carlon reserves the right: to ship to the nearest unitized quantity.

Year Two (Student Manual)

208.2 Pull Box and Conduit Volume Calculations for Equipment Enclosures, Raceways, Wireways and Cable Systems

Wireways and auxiliary gutters are either metallic or nonmetallic. Metallic types are available in stainless steel, powder coated steel, or galvanized steel. Nonmetallic types are typically UV resistant PVC or fiberglass. Auxiliary gutters (wiring troughs) have end caps and are available in specified lengths. Wireways consist of straight sections, angle sections, end pieces, support brackets, and couplings.

Power distribution blocks can be installed in wireways and gutters to make splices and taps in conductors. One of these is shown at the right. Although individual single-pole blocks are available, these are typically ordered as 3- or 4-wire. The size of the block to be ordered is based on the size and number of conductors to be placed under the lugs.

3-Pole Power Distribution Block

208.3 Box Fill Calculations

208.4 NEC Requirements for Conduit Installation and Pull Boxes

208.5 Problem Solving for BCES Blueprints

Wiring Materials, Conduit and Box Fill

Objective 208.1 Worksheet

I II III IV

Figure 208.15

_____ 1. The screw holes for mounting a cover on Box I, shown above in Figure 208.15, are tapped for ___ screws.

    a. 10-32   b. 6-32   c. 10-24   d. 8-32   e. none of these

_____ 2. The screw holes for mounting a cover on Box IV, shown above in Figure 208.15, are tapped for ___ screws.

    a. 10-32   b. 8-32   c. 6-32   d. 10-24   e. none of these

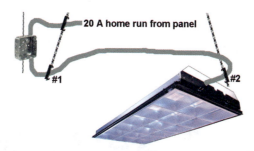

Figure 208.2

_____ 3. MC cable is used to connect a 2 x 4 lay-in fluorescent luminaire to a "home run" box as shown above in Figure 208.2. The length of the cable between the box and the luminaire is 26 feet. The maximum distance between the luminaire and support #2 is ___.

    a. 12 inches   b. 10 feet   c. 6 feet   d. 36 inches   e. 4½ feet

_____ 4. MC cable is used to connect a 2 x 4 lay-in fluorescent luminaire to a "home run" box as shown above in Figure 208.2. The length of the cable between the box and the luminaire is 26 feet. Where supports #1 & #2 are as far from the connectors as permitted ___ additional supports are required between #1 and #2.

    a. 3   b. 0   c. more than 3   d. 1   e. 2

5. MC cable is used to connect a 2 x 4 lay-in fluorescent luminaire to a "home run" box as shown on the previous page, in Figure 208.2. The length of the cable between the box and the luminaire is 26 feet. The maximum distance between the box and support #1 is ___.

   a. 10 feet   b. 3 feet   c. 12 inches   d. 6 feet   e. 4½ feet

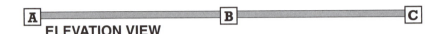

Figure 208.12

6. Refer to Figure 208.12, shown above. You are to install Trade Size 3 IMC using threadless (compression type) couplings and connectors between enclosures A, B, and C. The distance from enclosure A to B is 93 feet. 84 feet of IMC will be required between enclosures B and C. You will need at least ___ straps to secure and support the conduit.

   a. 11   b. 19   c. 12   d. 17   e. 18

7. Refer to Figure 208.12, shown above. You will need at least ___ straps to support the conduit between enclosures A & B. Trade Size 2½ IMC with threaded couplings and connectors are used. You will cut and thread the conduit as required. The distance from enclosure A to B is 123 feet.

   a. 11   b. 12   c. 10   d. 9   e. 14

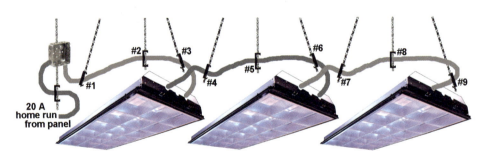

Figure 208.3

A string of 2´ × 4´ lay-in fluorescent luminaires is to be connected by "looping" the wiring method. Between luminaires, the raceway entries are 6´ apart. The length of the cable (or raceway) from the box to the first luminaire is 7 feet.

8. AC cable is the wiring method to be used above in Figure 208.3. The maximum distance permitted between support #7 and support #8 is ___ feet.

   a. 1   b. 10   c. 6   d. 4½   e. 3   f. support not req'd

Wiring Materials, Conduit and Box Fill

_____ 9. ENT is the wiring method to be used on the previous page, in Figure 208.3. The maximum distance permitted between the "home run" box and support #1 is ___ feet.

    a. 4½   b. 3   c. 10   d. 6   e. 1   f. support not req'd

_____ 10. MC cable is the wiring method to be used on the previous page, in Figure 208.3. The maximum distance permitted between support #5 and support #6 is ___ feet.

    a. 10   b. 4½   c. 6   d. 3   e. 1   f. support not req'd

_____ 11. LFNC is the wiring method to be used on the previous page, in Figure 208.3. The maximum distance permitted between support #1 and support #2 is ___ feet.

    a. 4½   b. 1   c. 10   d. 3   e. 6   f. support not req'd

_____ 12. According to the NEC® an 8 AWG bare solid copper conductor has a cross-sectional area of ___ square inches.

    a. .017   b. .0366   c. .051   d. .013   e. none of these

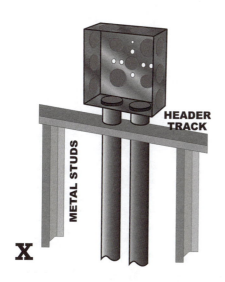

Figure 208.1

_____ 13. Does the NEC® permit the installation shown above at Y in Figure 208.1 where the box is supported only by the conduits? The distance from ground level to the bottom of the "bell" box is 17 inches. Both conduits are RMC and are threaded into the box.

    a. Yes   b. No

_____ 14. Does the NEC® permit the installation shown on the previous page, at X in Figure 208.1 where the box is supported only by the conduits? The distance from the top of the head track to the bottom of the "2100" box is 2 inches. Both conduits are EMT and are attached to the box with steel compression connectors (not visible in drawing).

   a. Yes
   b. No

_____ 15. Does the NEC® permit the installation shown on the previous page, at X in Figure 208.1 where the box is supported only by the conduits? The distance from the top of the head track to the bottom of the "2100" box is 2 inches. Both conduits are threaded RMC and are attached to the box with steel locknuts.

   a. No
   b. Yes

Figure 208.19

_____ 16. The conduit body shown above at ___ in Figure 208.19 is a Type LB.

   a. IV
   b. I
   c. III
   d. II
   e. none of these

_____ 17. The conduit body shown above at ___ in Figure 208.19 is a Type LL.

   a. IV
   b. III
   c. II
   d. I
   e. none of these

Wiring Materials, Conduit and Box Fill

Figure 208.14

_____ 18. Refer to Figure 208.14, shown above. The flexible wiring method between the time clock and the panel is permitted to be ___.

   I. 1/2" LFNC   II. 1/2" LFMC   III. 3/8" FMC

   a. III only
   b. II or III only
   c. I only
   d. I, II, or III
   e. II only
   f. I or II only

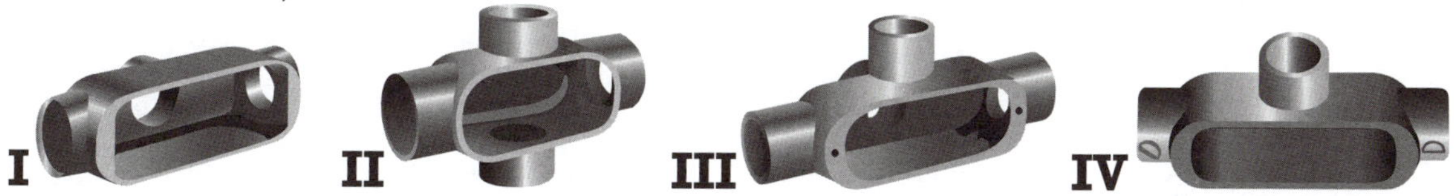

Figure 208.18

_____ 19. The conduit body shown above at ___ in Figure 208.18 is a Type TA.

   a. III   b. IV   c. I   d. II   e. none of these

_____ 20. The conduit body shown above at ___ in Figure 208.18 is a Type TB.

   a. II   b. I   c. IV   d. III   e. none of these

_____ 21. Schedule 40 PVC is to be installed in a straight run between two junction boxes on the back exterior wall of a strip shopping center. The summertime temperature is 100° F and the wintertime temperature is −5° F. An expansion fitting would be required in this 80′ run of PVC because it would expand ___ inches between winter and summer.

   a. 3.85   b. 3.408   c. 4.26   d. 5.325   e. 3.08

_____ 22. Trade size ¾ IMC and RMC use the same fittings. Which of them has the greatest wall thickness?

   a. ¾ RMC   b. ¾ IMC   c. Both are equal

Year Two (Student Manual)

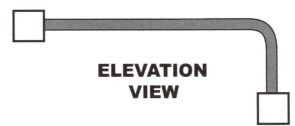

Figure 208.11

_____ 23. You are to surface mount 128′ of EMT on a wall between two boxes as shown above in Figure 208.11. You will use 1-hole straps to securely fasten the EMT to the wall. The minimum number of straps required is ___.

    a. 15   b. 14   c. 13   d. 11   e. 12

_____ 24. The characteristics of Schedule 80 PVC, versus Schedule 40, is that Sch 80 ___.
    I. has a thicker wall
    II. can be used where subject to physical damage
    III. has a smaller cross-sectional area for fill purposes

    a. I only
    b. I & III only
    c. II & III only
    d. III only
    e. I & II only
    f. I, II, & III

_____ 25. 3/4-inch flex is used to feed power into the 1900 box shown below. The box will be mounted behind a sheetrock wall with a 2-gang plaster ring. Two duplex receptacles will be mounted on the box assembly. The NEC® permits ___ to be used.
    I. the 90° flex connector   II. the straight flex connector

    a. neither I nor II
    b. II only
    c. I only
    d. either I or II

_____ 26. Conductors that are enclosed in Schedule 80 PVC underground are ___.
    I. protected from physical damage
    II. capable of being removed
    III. protected from exposure to water

    a. II only   b. I only   c. I, II, & III   d. II & III only   e. I & II only

Wiring Materials, Conduit and Box Fill

Match the metric designator to the Trade Size for each item listed below.

a. 12	d. 27	g. 53	j. 91	m. 155
b. 15	e. 35	h. 63	k. 103	
c. 21	f. 41	i. 78	l. 129	

_____ 27. A Trade Size 2 EMT connector = _____

_____ 28. A Trade Size 3 weatherhead = _____

_____ 29. Trade Size 6 RMC = _____

_____ 30. A Trade Size ¾ threaded rigid coupling = _____

_____ 31. Trade Size 4 2-hole EMT strap = _____

_____ 32. Trade Size 1½ electrical nonmetallic tubing = _____

Wiring Materials, Conduit and Box Fill

Objective 208.2 Worksheet

_____ 1. The maximum number of $3/0$ AWG conductors that can be installed in an 18 foot long Trade Size 3 section of FMC that is installed between two equipment enclosures is ___.

   a. 11 b. 8 c. 13 d. 10 e. 15

_____ 2. The minimum trade size Schedule 80 PVC that can be installed for an underground run between two buildings is ___. The conduit will contain five each $1/0$, six each $3/0$, and one 2 AWG conductors.

   a. 4 b. 5 c. 3 d. 2½ e. 3½

_____ 3. What is the minimum trade size Schedule 40 PVC permitted to be installed between two junction boxes that are 20 feet apart? The PVC will contain thirty-four 8 AWG, twenty 12 AWG, nine 6 AWG, and one 10 AWG THHN/THWN copper conductors.

   a. 2 b. 3½ c. 3 d. 2½ e. 4

_____ 4. What is the minimum trade size RMC permitted to be installed between two panelboards that are 5 inches apart? The RMC will contain four 4/0 AWG conductors and one 4 AWG ground. All conductors are THHN/THWN copper.

   a. 2 b. 3 c. 3½ d. 4 e. 2½

_____ 5. The smallest trade size RMC available that can be installed between two big pull boxes is ___.

   a. 1 b. $3/8$ c. ¾ d. ½ e. none of these

_____ 6. What is the maximum number of #4 THHN/THWN conductors that can be installed in a run of 2 inch IMC?

   a. 28 b. 26 c. 7 d. 11 e. 17

_____ 7. What is the minimum trade size RMC permitted to be installed between two panelboards that are 5 feet apart? The RMC will contain four 4/0 AWG conductors and one 4 AWG ground. All conductors are THHN/THWN copper.

   a. 2½ b. 2 c. 3½ d. 3 e. 4

Year Two (Student Manual)

8. Which of these Trade Size 1¼ raceways has the largest cross-sectional area?

   a. LFMC b. FMC c. Schedule 80 PVC d. EMT e. RMC

9. You are to install IMC between two equipment enclosures. You don't know how many or what size wires will have to be pulled into the flex. If you install the largest available this is trade size ___.

   a. 5 b. 3½ c. 3 d. 6 e. 4

10. The smallest trade size Schedule 80 PVC available that can be installed between two big pull boxes is ___.

    a. ¾ b. 1 c. ³⁄₈ d. ½ e. none of these

11. The smallest trade size IMC available that can be installed between two big pull boxes is ___.

    a. ½ b. 1 c. ³⁄₈ d. ¾ e. none of these

12. A 42′ run of 2″ ENT is installed in a poured concrete slab for several circuits. The maximum area that can be occupied by the circuit conductors is ___ sq. in.

    a. 1.282 b. 1.342 c. 0.993 d. 3.205 e. 1.923

13. What is the minimum trade size EMT permitted to be installed between two junction boxes that are 5 inches apart? The EMT will contain forty 8 AWG, twenty 12 AWG, nine 6 AWG, and one 10 AWG THHN/THWN copper conductors.

    a. 2 b. 2½ c. 4 d. 3 e. 3½

14. In a 30 foot run of conduit containing two conductors the maximum cross-sectional area that can be occupied by conductors is ___ %.

    a. 100 b. 60 c. 31 d. 40 e. 53

15. You are to install FMC between two equipment enclosures. You don't know how many or what size wires will have to be pulled into the flex. If you install the largest flex available this is trade size ___.

    a. 3½ b. 3 c. 6 d. 5 e. 4

_____ 16. When installing a 19 inch long piece of conduit between two panelboards____.

    a. the conduit can be filled with wire to 60% of its area
    b. the ampacity of the contained conductors must be derated if there are 4 or more
    c. both a and b
    d. neither a nor b

_____ 17. A raceway has a total cross-sectional area of 6.258 square inches according to the table in Chapter 9 of the NEC®. Where this conduit run will contain four conductors the maximum area that can be occupied by the conductors is 40% or ____ in2.

    a. .2503   b. 8.761   c. 2.343   d. 2.503   e. 15.645

_____ 18. Which of the following 2-inch raceways has the smallest cross-sectional area?

    a. Schedule 80 PVC (nonmetallic conduit)
    b. LFMC (liquidtight flexible conduit)
    c. RMC (rigid metal conduit)
    d. EMT (electrical metallic tubing)

_____ 19. What minimum size schedule 80 PVC conduit is required for a 65´ run? The conduit will contain three 350 kcmil, one 3/0, and one #2 THHN/THWN conductors.

    a. 3½   b. 2½   c. 5   d. 4   e. 3

_____ 20. The maximum number of 250 kcmil conductors that can be installed in an 8 inch long Trade Size 2½ section of FMC that is installed between two equipment enclosures is ____.

    a. 9   b. 6   c. 7   d. 8   e. 5

_____ 21. The maximum cross-sectional area that can be occupied by conductor in a 40-foot run of conduit is 53% where the conduit contains ____ conductor(s).

    a. 1   b. 2   c. 5

_____ 22. The largest trade size EMT available that can be installed between two big pull boxes is ____.

    a. 3   b. 5   c. 2   d. 6   e. 4

Objective 208.3 Worksheet

_____ 1. A 4″ square box (1900) with a 2-gang plaster ring contains:
   (one)   12/2 with ground AC cable (BX) with exterior connector
   (contains hot & neutral + isolated grounding conductor)
   (one)   14/2 MC cable (contains a ground) with exterior connector (contains hot & neutral)
   (one)   14/2 MC cable (contains a ground) with exterior connector (contains sw. leg & neutral)
   (one)   toggle switch (connected to the 14/2)
   (one)   isolated ground receptacle (connected to the 12/2 BX and its isolated ground)

   The plaster ring has 4.3 cu. in. stamped into the metal. The smallest "1900" box you can use and not violate the NEC is one that is ___ inches deep.

   a. 1½
   b. 1¼
   c. 2⅛
   d. none of these

_____ 2. A switch box will contain the following:
   (one)   isolated ground receptacle
   (one)   12/2 with ground AC cable (BX) with exterior connector

   The shallowest switch box you can use for this installation is 3″ × 2″ × ___″.

   a. 3½   b. 2   c. 2½   d. 2¼   e. 2¾

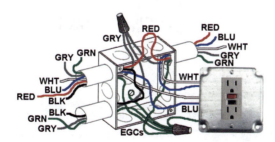

Figure 208.36

_____ 3. Refer to Figure 208.36, shown above. Which industrial cover/box combination will NOT violate the NEC? All the conductors are #12 AWG.
   I.   4 x 1½ square with 5.0 in3 industrial cover
   II. 4¹¹/₁₆ x 1¼ square box with a 4.2 in3 industrial cover
   III. 4 x 2⅛ square box with a 3.6 in3 industrial cover

   a. I only   b. III only   c. II & III only   d. II only   e. I & II only

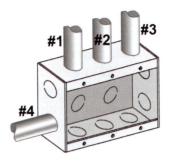

Figure 208.38

_____ 4. The shallow masonry box, shown above in Figure 208.38, will contain two 4-way switches and one three-way switch. Conduit #1 contains five wires, #2 contains four wires, #3 contains four wires, and #4 contains two wires. All conductors are 12 AWG THHN/THWN copper and are spliced in the box or terminated on the switches. Will this installation violate the Code?

a. yes
b. no

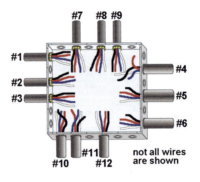

Figure 208.34

_____ 5. Refer to Figure 208.34, shown above. The smallest (volume) junction box that can be used in the installation shown below is a____. Conduits #1 and #4 each contain twenty-four 10 AWG wires. Conduits #2 and #5 each contain nine 6 AWG wires. Conduits #3 and #6 each contain sixteen 8 AWG wires. Conduits #7, #10, #9, and #12 each contain twenty-eight 12 AWG wires. Conduits #8 and #11 each contain six 8 AWG wires. All conductors are "wire nutted" inside the box.

a. 10″ × 10″ × 8″ deep
b. 12″ × 12″ × 4″ deep
c. 12″ × 12″ × 6″ deep
d. 10″ × 10″ × 6″ deep

Wiring Materials, Conduit and Box Fill

_____ 6. A 4-square box with a 3.9 cu. in. plaster ring will contain one isolated ground receptacle and one four-way switch. A 3/4 inch Schedule 40 PVC conduit is installed from the panel to the box and then extended to a three-way switch. The receptacle is on a 20 Amp dedicated circuit and the switch is for lighting connected to a 15 Amp circuit. The shallowest 4-square box you can use for this installation is 4″ × 4″ × ____″.

   a. 1 ½
   b. 1 ¼
   c. 2 ⅛

Figure 208.33

_____ 7. The smallest (volume) junction box that can be used in the installation, shown above in Figure 208.33, is a ____. Conduits #1 and #4 each contain twenty-four 10 AWG wires. Conduits #2 and #5 each contain twelve 6 AWG wires. Conduits #3 and #6 each contain sixteen 8 AWG wires. All conductors are "wire nutted" inside the box. Additionally, each conduit contains a ground wire.

   a. 8″ × 8″ × 4″ deep
   b. 12″ × 12″ × 4″ deep
   c. 10″ × 10″ × 4″ deep
   d. 6″ × 6″ × 4″ deep
   e. these are all too small

_____ 8. Among the possible box assemblies the smallest permitted to contain the following is ____. Exterior MC cable connectors are used on the cables.

   One – 30 amp 240-volt single receptacle   and   One – 15 amp 120-volt single receptacle
   Two – 10/2 with ground MC cables   plus   One – 14/2 with ground MC cable

   a. two – 3 x 2 x 2 switch boxes (ganged to form a 2-gang box)
   b. 4 x 2⅛ square box with a 3.6 cu. in. two-gang plaster ring
   c. 4 x 1½ square box with a 5.0 cu. in. two-gang plaster ring
   d. two – 3 x 2 x 2½ switch boxes (ganged to form a 2-gang box)

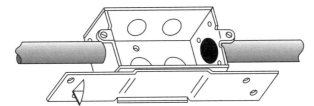

Figure 208.32

_____ 9. Each conduit, connected to the switch box shown above in Figure 208.32, will contain one "hot", one neutral, and one EGC. A duplex receptacle will be installed in the box. The shallowest box permitted for this installation is ___″ deep. All conductors are 14 AWG THHN/THWN.

   a. 2¾
   b. 2½
   c. 2¼
   d. 2
   e. 3½

_____ 10. A switch box, shown above in Figure 208.32, will contain a duplex receptacle and each of the connected conduits will contain 12 AWG copper THHN/THWN conductors (one "hot", one neutral, and one EGC). The shallowest switch box permitted for this installation is ___ mm deep.

   a. 50
   b. 57
   c. 65
   d. 70
   e. 90

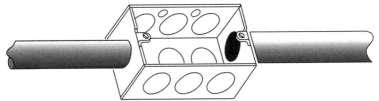

Figure 208.31

_____ 11. A handy box (shown above in Figure 208.31) will contain a duplex receptacle and each connected conduit will contain 12 AWG copper THHN/THWN conductors (one "hot", one neutral, and one EGC). The shallowest handy box permitted for this installation is ___.

   a. 1⁷⁄₈ inches deep
   b. 1¹⁄₂ inches deep
   c. 2¹⁄₈ inches deep
   d. none of these - this is a Code violation

Objective 208.4 Worksheet

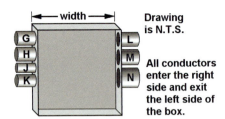

Figure 208.55

_____ 1. The minimum CALCULATED dimension for **width** shown above in Figure 208.55 is ___ inches.
   Conduit **G** = 4″
   Conduit **H** = 4″
   Conduit **J** = 2″
   Conduit **K** = 3″
   Conduit **L** = 4″
   Conduit **M** = 4″
   Conduit **N** = 5″

   a. 40   b. 61   c. 41   d. 48   e. none of these

_____ 2. The minimum CALCULATED dimension for **width** shown above in Figure 208.55 is ___ inches.
   Conduit **G** = 3½″
   Conduit **H** = 2″
   Conduit **J** = 2½″
   Conduit **K** = 3″
   Conduit **L** = 3″
   Conduit **M** = 3″
   Conduit **N** = 2″

   a. 31   b. 35   c. 28½   d. 21   e. none of these

Figure 208.42

_____ 3. As shown above in Figure 208.42, you are to surface mount EMT between two equipment enclosures as shown. You will make 10° offsets to get from each box to the wall surface. To avoid violating the NEC you can make ___ ° saddles.
   I. 22½   II. 30   III. 45

   a. I or II or III   b. I or II only   c. I only   d. none of these

Wiring Materials, Conduit and Box Fill

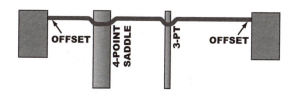

Figure 208.45

_____ 4. The 4-point saddle shown above in Figure 208.45 will be a 15° saddle. This saddle has a total of ___ degrees of bend.

　a. 45　　　　b. 90　　　　c. 60　　　　d. 30　　　　e. none of these

_____ 5. The 3-point saddle shown above in Figure 208.45 will be a 30° saddle. This saddle has a total of ___ degrees of bend.

　a. 45　　　　b. 30　　　　c. 90　　　　d. 60　　　　e. none of these

_____ 6. Although indicated in Figure 208.45, NO OFFSETS will be required to enter the boxes because you will use stand-off straps for the EMT between the two boxes. You **will** have to saddle over the obstacles shown. The NEC® will permit you to make ___ degree saddles.
　　I. 30　　　II. 45　　　III. 60

　a. I or II or III　　b. I or II only　　c. I only　　d. none of these

Figure 208.61

_____ 7. A section of 8″ × 8″ wireway is shown above in Figure 208.61. The maximum number of 2/0 AWG THHN/THWN copper conductors that can be installed in the wireway is ___.

　a. 12　　　b. 36　　　c. 30　　　d. 13　　　e. 57　　　f. 115

_____ 8. What minimum standard size wireway, such as shown above in Figure 208.61, is required to contain one hundred #8 THHN/THWN conductors?

　a. 2″ × 2″
　b. 10″ × 10″
　c. 8″ × 8″
　d. 4″ × 4″
　e. 6″ × 6″
　f. none of these

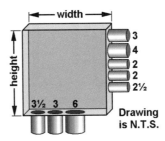

_____ 9. For the pull box shown above in Figure 208.52 the minimum calculated width is ___ inches.

All conductors enter the bottom and exit the right side of the box.

a. 33½   b. 54½   c. 48   d. 41½   e. none of these

_____ 10. For the pull box shown above in Figure 208.52 the minimum calculated height is ___ inches.

All conductors enter the bottom and exit the right side of the box.

a. 33½   b. 41½   c. 48   d. 54½   e. none of these

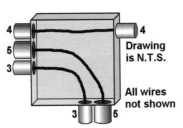

Figure 208.53

_____ 11. The calculated minimum **height** (top-to-bottom) of the pull box, shown above in Figure 208.53, is ___ inches. The conduits considered for this calculation contain conductors that enter this side and exit the adjacent side.

a. 37   b. 33   c. 30   d. 40   e. 32

_____ 12. The calculated minimum **width** (left-to-right) of the pull box, shown above in Figure 208.53, is ___ inches. Select the *larger* dimension required: straight pull dimension versus angle pull dimension.

a. 30   b. 33   c. 40   d. 37   e. 32

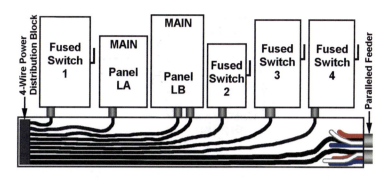

Figure 208.64

Only the AØ tap and feeder conductors are shown connected to the distribution block. BØ and CØ taps are run from the block into each switch and panelboard. Additionally, neutral taps will be run to the panels, and to the switches, where required.

_____ 13. What is the smallest metal wireway that can be used for the service "gutter" shown above in Figure 208.64? All conductors are THHN/THWN copper. The 3Ø, 4-W feeder consists of eight 600 kcmil conductors. The following taps will be made in the distribution block:

Panel LA: four each 1/0 AWG   Switch 1: four each 2/0 AWG
Panel LB: eight each 2/0 AWG   Switch 2: three each 1 AWG
(4 in ea conduit)   Switch 3: four each 3/0 AWG
                                      Switch 4: three each 1/0 AWG

a. 12 × 12
b. 8 × 8
c. 10 × 10
d. 6 × 6
e. none of these - this is a Code violation

_____ 14. Per the NEC® which of the following bends are permitted in a conduit run between two junction boxes?

a. **one**-back-to-back 90 *plus* **two**-60° 3-point saddles
b. **two**-30° offsets *plus* **three** 90's
c. **two**-90's *plus* **one**-45° three-point saddle *plus* one-12° offset
d. any one of these
e. none of these

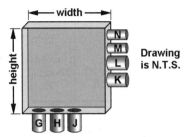

Figure 208.51

_____ 15. Refer to Figure 208.51, shown above. The minimum CALCULATED dimension for **width** is ___ inches. All conductors enter the right size of the pull box and exit the bottom.
  Conduit **G** = 3″
  Conduit **H** = 3½″
  Conduit **J** = 6″
  Conduit **K** = 3″
  Conduit **L** = 4″
  Conduit **M** = 2″
  Conduit **N** = 2½″

  a. 32
  b. 39½
  c. 48
  d. 42½
  e. none of these

_____ 16. Refer to Figure 208.51, shown above. The minimum CALCULATED dimension for **height** is ___ inches. All conductors enter the right size of the pull box and exit the bottom.
  Conduit **G** = 3″
  Conduit **H** = 3½″
  Conduit **J** = 6″
  Conduit **K** = 3″
  Conduit **L** = 4″
  Conduit **M** = 2″
  Conduit **N** = 2½″

  a. 39½
  b. 48
  c. 32
  d. 42½
  e. none of these

_____ 17. Refer to Figure 208.51, shown on the previous page. Conduit **N** is on the side that is ___ the left side of the box.

   a. opposite
   b. adjacent to

_____ 18. Refer to Figure 208.51, shownon the previous page. Conduit **G** is on the side that is ___ the top of the box.

   a. adjacent to
   b. opposite

_____ 19. Refer to Figure 208.51, shown on the previous page. Conduit **L** enters the side that is ___ the side of the box where Conduit **J** enters.

   a. opposite
   b. adjacent to

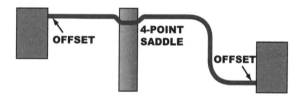

Figure 208.41

_____ 20. As shown above in Figure 208.41, you will make 15° offsets to get from each box to the wall surface where you will surface mount EMT. To avoid violating the NEC you can make a ___ ° saddle.
   I. 30
   II. 45
   III. 60

   a. I or II only
   b. I or II or III
   c. I only
   d. none of these

Objective 208.5 Worksheet

_____ 1. On the BCES project, ___ conduit(s) is/are installed between the MSB and Panel HK.

   a. seven
   b. one
   c. two
   d. three

_____ 2. Which of the following raceways are permitted to be installed in pole bases on BCES?
   I. Schedule 80 PVC   II. RMC   III. IMC

   a. II only
   b. I & II only
   c. II & III only
   d. I only
   e. I, II, & III

_____ 3. From the telephone outlets, the EC (Electrical Contractor) is to provide a 3/4-inch conduit from the wall box to an accessible ceiling. The EC shall provide ___

   a. a 4" × 4" box on conduit in the accessible ceiling
   b. a pull string in each conduit
   c. a pull wire in each conduit
   d. both a and b

_____ 4. On the BCES project, there will be a total of ___ 3" conduits connected to the main switchboard.

   a. 4
   b. 12
   c. 6
   d. 2

_____ 5. On the BCES project, the conduits for the parking lot lighting pole bases must be at a minimum depth of ___.

   a. 18"
   b. 12"
   c. 36"
   d. 24"

Wiring Materials, Conduit and Box Fill

_____ 6. Which of the following types of cable tray are permitted on the BCES job?

   a. plastic
   b. steel
   c. aluminum
   d. fiberglass

_____ 7. On the BCES project, ____ conduit(s) is/are installed between the transformer and Panel LB2.

   a. seven   b. one   c. three   d. two

_____ 8. On the BCES project, ____ conduit(s) is/are installed between the MSB and the disconnect switch for Chiller CH-2.

   a. one   b. three   c. seven   d. two

_____ 9. Trade Size ____ conduit(s) is installed between BCES Transformer XE and Panel LE.

   a. 1½   b. 2½   c. 3   d. 3½   e. 4

_____ 10. On the BCES prints, the engineer ____ permit lay-in light fixtures to be connected fixture-to-fixture.

   a. does   b. does not

_____ 11. On the BCES prints, the engineer specifies the use flexible metal conduit for the connection of light fixtures.

   a. False   b. True

_____ 12. Trade Size ____ conduit(s) is installed between BCES's MSB and Panel HE.

   a. 3½   b. 2½   c. 1½   d. 3   e. 4

_____ 13. On the BCES project, Trade Size ____ conduit is required to the parking lot poles.

   a. ½"   b. ¾"   c. 1-¼"   d. 1"

Wiring Materials, Conduit and Box Fill

Lesson 208 NEC® Worksheet

_____ 1. An upward-discharge vent for an underground gasoline tank and the area within 3 feet, in all directions, of the open end of the vent is a ___ location.

   a. Class II, Division 1
   b. Class I, Division 1
   c. Class I, Division 2
   d. Class I, Zone 2
   e. Class I, Zone 0

_____ 2. Motor fuel dispensing areas for dispensing gasoline are classified as ___ locations.

   a. Class III, Division 1
   b. Class II, Division 2
   c. Class I, Division 2
   d. Class I, Division 1
   e. Class II, Division 1

_____ 3. Where compressed natural gas dispensers are located under a canopy or enclosure that prevents the accumulation or entrapment of ignitible vapors, electrical equipment ___.

   a. shall be suitable for Class I, Division 2 hazardous locations
   b. shall be suitable for Class I, Division 1 hazardous locations
   c. shall not be required to meet Class I, Division 2 hazardous location requirements
   d. shall be suitable for Class II, Division 1 hazardous locations
   e. shall be suitable for Class I, Group G requirements

_____ 4. Locations within ___, in all directions, from a compressed natural gas dispenser enclosure are classified as Class I, Division 2 locations.

   a. 450 mm (18 in.)
   b. 900 mm (3 ft)
   c. 1.5 m (5 ft)
   d. 3 m (10 ft)
   e. 6 m (20 ft

_____ 5. Locations within ___, in all directions, from a liquefied natural gas dispenser enclosure are classified as Class I, Division 1 locations.

   a. 450 mm (18 in.)
   b. 900 mm (3 ft)
   c. 1.5 m (5 ft)
   d. 3 m (10 ft)
   e. 6 m (20 ft)

Year Two (Student Manual)

_____ 6. A location that is 9 feet from a liquefied natural gas dispenser enclosure is ___ location.

   a. a Class I, Div 1
   b. a Class I, Div 2
   c. an unclassified

_____ 7. Where Class I, Division 1 underground wiring emerges from the ground, the electrical wiring shall be sealed within ___ of the point of emergence above grade at motor fuel dispensing facilities.

   a. 3 feet
   b. 5 feet
   c. 6 meters
   d. 1 meter
   e. 10 feet

_____ 8. Where a conduit emerges from the earth or concrete under a gasoline pump, a seal must be installed ___ after the conduit emerges for the earth or concrete.

   a. within 2 feet
   b. within 12 inches
   c. as the first fitting
   d. within 18 inches

Lesson 208 Safety Worksheet

HAZARD COMMUNICATIONS (Haz Com) part 3

Employers shall maintain in the workplace copies of the required MSDS sheets for each hazardous chemical, and shall ensure that they are readily accessible during each work shift to employees when they are in their work areas. If employees must travel between workplaces during a work shift, or their work is carried out at more than one geographical location, the MSDS sheets may be kept at the primary workplace facility. In this situation the employer shall ensure that employees can **immediately** obtain the required information in an emergency.

Employee information and training
Employers shall provide employees with effective information and training on hazardous chemicals in the work area at the time of their initial assignment, and when a new physical or health hazard the employees have not been previously trained about is introduced into their work area. Chemical-specific information must always be available through labels and MSDS. Employees must be informed of any operations in their work area where hazardous chemicals are present, and the location and availability of the written hazard communication program, including the required list(s) of hazardous chemicals and MDSD sheets.

Training
Employee training shall include the methods and observations that may be used to detect the presence or release of a hazardous chemical in the work area, the physical and health hazards of the chemicals, the measures employees can take to protect themselves from these hazards and details of the employers Haz Com program including an explanation of the labeling system and MSDS sheets and how employees can obtain any of the information.

Additional information is available from many sources. Some very good information is online at some of these websites:

www.hazard.com/msds/index.php

www.ilpi.com/msds/index#Internet

www.ehs.cornell.edu/

http://hazard.com/msds/index.php

Questions

_____ 1. Employee training shall include the ___ and observations that may be used to detect the presence or release of a hazardous chemical in the work area.

    a. information   b. methods   c. type   d. all of these

_____ 2. Employers shall provide employees with ___ information and training on hazardous chemicals in the work area at the time of their initial assignment.

    a. timely   b. occasional   c. effective   d. worthwhile

_____ 3. Employers shall maintain in the ___, copies of the required MSDS sheets for each hazardous chemical.

    a. trailer   b. workplace   c. truck   d. office

_____ 4. Chemical-specific information must always be ___ through labels and MSDS sheets.

    a. forwarded   b. ready   c. on-hand   d. available

_____ 5. Training must also include the physical and ___ hazards of the chemicals and the measures that employees can take to protect themselves from these hazards.

    a. stress   b. emotional   c. health   d. none of these

_____ 6. Employees must be informed of any operations in their work area where hazardous chemicals are present, and the ___ and availability of the written Haz Com program, including the required list(s) of hazardous chemicals and MSDS sheets.

    a. cost   b. purpose   c. location   d. none of these

Lesson 209
Mid-Term Review and Exam

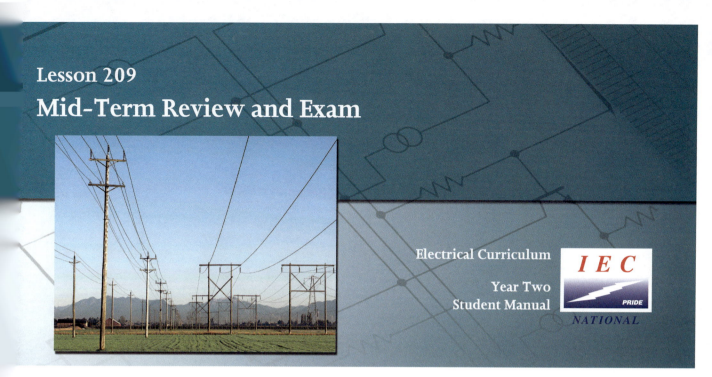

Electrical Curriculum

Year Two
Student Manual

Purpose
Mid-Term Exam is scheduled for this class.

Homework
For this lesson, you should:

- Review all your worksheets, homework, and quizzes.
- Complete your "Practice Exam".
- Be prepared to clarify any questions or problem areas you encountered during your review.

Lesson 210
Switches, Switchboards and Panelboards

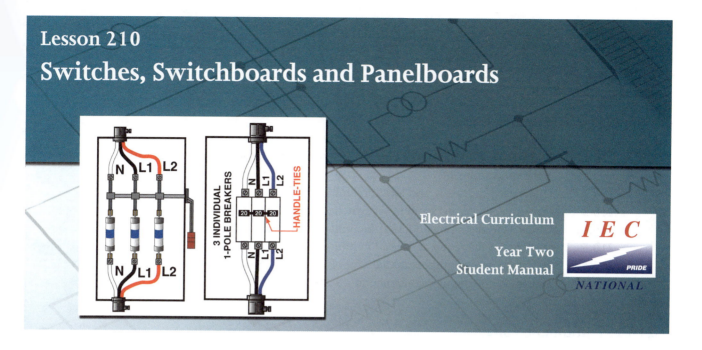

Electrical Curriculum

Year Two
Student Manual

Purpose
This lesson will teach requirements and limitations for switches, switchboards and panelboards.

Homework
(Due at the beginning of this class)

For this lesson, you should:

- Thoroughly read the material contained within Lesson 210.
- Read ES Chapter 9 in its entirety.
- Complete Objective 210.1 Worksheet.
- Complete Objective 210.2 Worksheet.
- Complete Objective 210.3 Worksheet.
- Complete Objective 210.4 Worksheet.
- Complete Objective 210.5 Worksheet.
- Complete Objective 210.6 Worksheet.
- Complete Lesson 210 NEC Worksheet
- Read and complete Lesson 210 Safety Worksheet.

- Complete additional worksheets, if available, as directed by your instructor.

Objectives

By the end of this lesson, you should:

210.1 ....Know the commonly available voltage systems and the voltages between various points in electrical equipment.
210.2 ....Understand the rating and location limitations for switches.
210.3 ....Have a general understanding of switches and panelboards and the NEC requirements.
210.4 ....Understand the rating and location limitations for switchboards.
210.5 ....Understand the rating and location limitations for panelboards.
210.6 ....Be able to locate equipment and symbols on the BCES prints in order to solve problems addressed on common construction site scenarios.

210.1   Equipment Voltages

210.2   Rating and Location Limitations for Switches

210.3   Switchboards and Panelboards

210.4   Rating and Location Limitations for Switchboards

210.5   Rating and Location Limitations for Panelboards

210.6   Problem Solving for BCES Blueprints

Switches, Switchboards and Panelboards

Objective 210.1 Worksheet

_____ 1. You are sent on a service call to install some equipment. The main service equipment is a 42 circuit panelboard that does not have the system voltage stamped on it. There is a sticker on the cover that reads: **Proudly Installed by BOOTLEG BROTHERS ELECTRIC**

After removing the panelboard cover and deadfront you see that the main breaker is a 3 pole 200 Amp and that the feeder consists of three conductors colored brown, red and blue. The feeder also has a conductor colored gray that is connected to the neutral bar.

To verify the system voltage, you check with your voltmeter and get the following readings:

AØ to neutral = 120 volts   BØ to neutral = 208 volts   CØ to neutral = 120 volts

You have now determined that the system is ___.

a. 277/480 volt 3-ph, 4-W
b. 120/240 volt 3-ph, 4-W
c. 120/240 volt 1-ph, 3-W
d. 480 volt 3-ph, 3-W
e. 120/208 volt 3-ph, 4-W

_____ 2. You are sent on a service call to install some equipment. The main service equipment is a 42 circuit panelboard that does not have the system voltage stamped on it. There is a sticker on the cover that reads: **Proudly Installed by SIDE JOB ELECTRIC**

After removing the panelboard cover and deadfront you see that the main breaker is a 3 pole 175 Amp and that the feeder consists of three "hot" conductors and another conductor, with gray phase tape, that is connected to the neutral bar.

You see that the service conductors have been marked with phase tape as follows:

AØ has black tape; BØ has orange tape; and CØ has blue tape. You then check with your voltmeter and get the following readings:

AØ to BØ = 208 volts   AØ to neutral = 120 volts   CØ to neutral = 120 volts

You have now determined that the system is ___.

a. 277/480 volt 3-ph, 4-W
b. 120/240 volt 3-ph, 4-W
c. 120/240 volt 1-ph, 3-W
d. 480 volt 3-ph, 3-W
e. 120/208 volt 3-ph, 4-W

Year Two (Student Manual)

3. You are sent on a service call to install some equipment. The main service equipment is a 42 circuit panelboard that does not have the system voltage stamped on it. There is a sticker on the cover that reads:
Proudly Installed by SIDE JOB ELECTRIC

After removing the panelboard cover and deadfront you see that the main breaker is a 3 pole 200 Amp and that the feeder consists of three "hot" conductors and another conductor that is connected to the neutral bar. Each of the "hot" conductors has orange colored "phase tape" on it and the neutral has gray tape on it. To verify the system voltage, you check with your voltmeter and get the following readings:

AØ to neutral = 277 volts   BØ to neutral = 277 volts   CØ to neutral = 277 volts

You have now determined that the system is ___.

a. 277/480 volt 3-ph, 4-W
b. 120/240 volt 3-ph, 4-W
c. 120/240 volt 1-ph, 3-W
d. 480 volt 3-ph, 3-W
e. 120/208 volt 3-ph, 4-W

4. Shown here is a 42 Circuit Panelboard. In a 277/480 volt, 3Ø, 4-W panel a multiwire circuit connected to breakers 6 & 29 would have ___ volts between the neutral and the wire connected to Breaker 29.

a. 0
b. 240
c. 480
d. 277
e. none of these

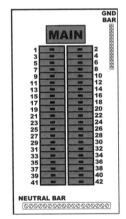

The properly bonded and grounded panelboard, shown at the right, is supplied from a 120/240 volt, 3Ø, 4-wire system. All breakers are in the ON position.

Select the correct voltage (a, b, c, d, e, or f) to complete the statements.

a. 0
b. 120
c. 208
d. 240
e. 277
f. 480

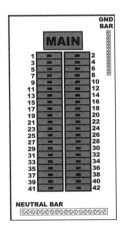

Switches, Switchboards and Panelboards 210-233

_____ 5. The voltage between breakers 29 and 34 is ___ volts.

_____ 6. The voltage between breaker 22 and the neutral bar is ___ volts.

_____ 7. The voltage between breaker 17 and the ground bar is ___ volts.

The properly bonded and grounded panelboard, shown at the right, is supplied from a 120/208 volt, 3Ø, 4-wire system.
All breakers are in the ON position. **Select the correct voltage (a, b, c, d, e, or f) to complete the statements.**

a. 0   b. 120   c. 208
d. 240   e. 277   f. 480

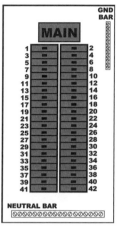

_____ 8. The voltage between breakers 14 and 27 is ___ volts.

_____ 9. The voltage between breaker 22 and the neutral bar is ___ volts.

_____ 10. The voltage between breaker 17 and the ground bar is ___ volts.

The panel shown at the right is fed from a 120/240 volt, 1Ø, 3-wire system and is installed on a house. The service is properly bonded and grounded. All breakers are in the ON position. **Select the correct voltage (a, b, c, d, e, or f) to complete the statements.**

a. 0   b. 120   c. 208
d. 240   e. 277   f. 480

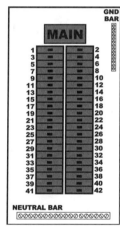

_____ 11. The voltage between breakers 9 and 28 is ___ volts.

_____ 12. The voltage between breakers 4 and 23 is ___ volts.

_____ 13. The voltage between breaker 22 and the neutral bar is ___ volts.

_____ 14. The voltage between breaker 17 and the ground bar is ___ volts.

Year Two (Student Manual)

Objective 210.2 Worksheet

_____ 1. Which of the following would be considered to be a continuous load?
  I.   service station canopy lighting
  II. a clothes dryer in a laundromat
  III. a boiler in a hotel

   a. I only   b. I & II only   c. I, II, & III   d. I & III only   e. III only

_____ 2. A ___ is a device with a current and voltage rating used to open or close an electrical circuit.

   a. switch   b. receptacle   c. either a or b   d. neither a nor b

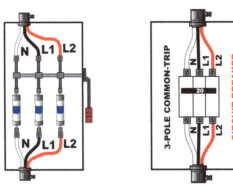
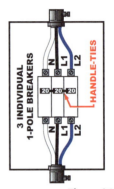

Figure 210.1

The switch or breaker supplies a 120/240 V, 3-wire load.

_____ 3. Which of the illustrations shown above in Figure 210.1 would satisfy the Code requirements of 404.2(B) Exception?

   a. I only   b. II only   c. III only   d. II or III only   e. I or II or III

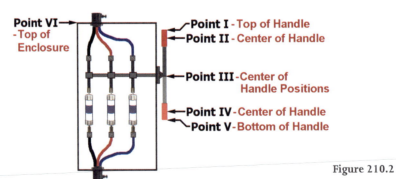

Figure 210.2

_____ 4. When the NEC® states the maximum height for a safety switch, as shown above in in Figure 210.2, the maximum height is at Point ___.

   a. III   b. IV   c. II   d. VI   e. V   f. I

Switches, Switchboards and Panelboards   210-235

_____ 5. Double-throw knife switches mounted vertically shall be equipped with a locking device to prevent ___ from closing the blades.

   a. vandals   b. children   c. gravity   d. unauthorized persons

_____ 6. Does the NEC® permit a breaker to be mounted in a breaker enclosure such that to turn the breaker "ON" you pull the breaker handle down?

   a. Yes   b. No

_____ 7. Refer to Figure 210.3, to the right. The NEC® has limitations for the maximum mounting heights above a floor or working platform. This maximum height refers to ___.

   a. the bottom of the panel
   b. the center of breaker 23
   c. the top of the panel
   d. the center of breaker 2
   e. none of these

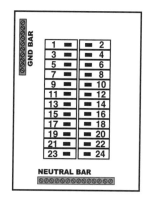

Figure 210.3

_____ 8. A 5-gang box and a partition are shown below in Figure 210.5. The box will contain single-pole switches A, B, C, D, & E that are fed from the following circuits that originate in a 277/480 volt, 3Ø, 4 wire panelboard:
Switches A and B are fed from Circuit 1
Switches C and D are fed from Circuit 2
Switch E is fed from Circuit 3

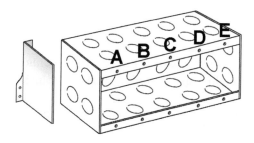

Figure 210.5

The NEC® requires that partitions be installed between switches ___.
   I. A & B   II. B & C   III. C & D   IV. D & E

   a. II & IV only   b. II only   c. IV only   d. I, II, III, & IV   e. No partitions are required

Year Two (Student Manual)

_____ 9. You are installing a double-pole (single-throw) toggle switch to use as the disconnect for a motor. The information stamped on the switch is "AC only   20 A   250 volts". You can use this switch for a 230 volt motor that draws ___ amps at 240 volts.

   a. 11.4   b. 12.9   c. 15.7   d. any of these   e. none of these-this is a Code violation

_____ 10. 30 A toggle switches that are stamped CU can be connected to ___ conductors.

   a. copper   b. aluminum   c. copper-clad aluminum   d. any of these   e. none of these

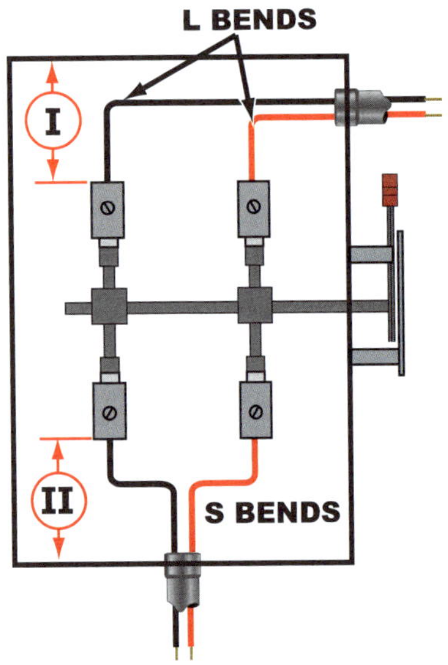

Figure 210.6

_____ 11. Refer to Figure 210.6, shown above. The conductors connected to the terminal lugs are 350 kcmil. The minimum wire bending space required at **I** is ___ inches. These lugs are of the lay-in type.

   a. 12   b. 5   c. 10   d. 8   e. none of these

_____ 12. Refer to Figure 210.6, shown above. The conductors connected to the terminal lugs are 350 kcmil. The minimum wire bending space required at **II** is ___ inches. These lugs are of the lay-in type.

   a. 12   b. 5   c. 10   d. 8   e. none of these

_____ 13. Refer to Figure 210.6, shown above. The conductors connected to the terminal lugs are 4/0 AWG. The minimum wire bending space required at **I** is ___ inches. These lugs are NOT removable.

   a. 8   b. 7   c. 4   d. 6   e. none of these

Switches, Switchboards and Panelboards   210-237

_____ 14. Refer to Figure 210.6, shown above. The conductors connected to the terminal lugs are 4/0 AWG. The minimum wire bending space required at **II** is ___ inches. These lugs are NOT removable.

   a. 8
   b. 7
   c. 4
   d. 6
   e. none of these

_____ 15. Your journeyman says that, according to the NEC®, the top of a panelboard should be mounted no higher than 6´-0˝ above the floor. He is ___.

   a. correct again
   b. wrong, this time

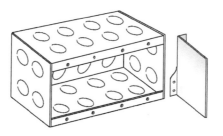

Figure 210.4

_____ 16. A 4-gang box and a partition are shown above in Figure 210.4. The box will contain 4 devices in Positions 1 through 4 (*left-to-right*). A three-way switch is in Position 1 on the far left. This switch controls 120-volt recessed lighting and is fed from Circuit LA-1. Adjacent to (next to) this switch, a receptacle will be installed in Position 2 and is fed from Circuit LA-3. Panel LA is 120/208 volt, 3Ø, 4-wire panel. In Position 3 will be a single-pole switch fed from Circuit HA-14. Position 4 will contain a four-way switch that is connected to Circuit HA-13. Panel HA is a 277/480 volt, 3Ø, 4 wire panelboard. The NEC® requires partitions between Positions ___.

   a. 1 and 2
   b. 2 and 3
   c. 3 & 4
   d. all of these
   e. none of these

Year Two (Student Manual)

Objective 210.3 Worksheet

_____ 1. Refer to the drawing shown below. Which phase conductor is required to be identified as the "high-leg"?

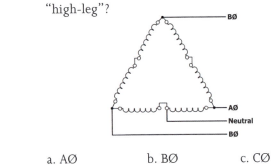

   a. AØ   b. BØ   c. CØ

_____ 2. A ___ is a single panel or group of assembled panels with buses and overcurrent devices, which may have switches to control light, heat, or power circuits.

   a. switchboard   b. panelboard   c. neither a nor b

_____ 3. A ___ is accessible from the front or the rear.

   a. switchboard   b. panelboard   c. both a & b

Figure 210.7

Switches, Switchboards and Panelboards

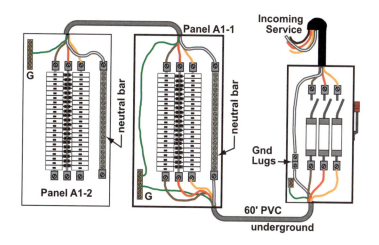

_____ 4. Refer to Figure 210.7, shown above. The conductors between Panel A1-1 and the switch are ___ conductors.

   a. service lateral   b. feeder   c. branch circuit   d. tap

_____ 5. Refer to Figure 210.7, shown above. Panel ___ is a subpanel.

   a. A1-1 only   b. neither A1-1 nor A1-2   c. both A1-1 & A1-2   d. A1-2 only

_____ 6. A circuit breaker is an overcurrent protection device that is designed to automatically open the circuit when a(an) ___ condition occurs.

   a. overload   b. short-circuit   c. ground-fault   d. any of these

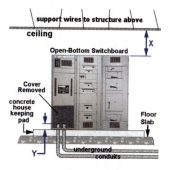

Figure 210.9

_____ 7. Refer to Figure 210.9, shown above, where the conduits enter the bottom of the equipment. The maximum height permitted for dimension Y in the drawing is ___ mm per the NEC®.

   a. 150   b. 75   c. 200   d. 600   e. 250

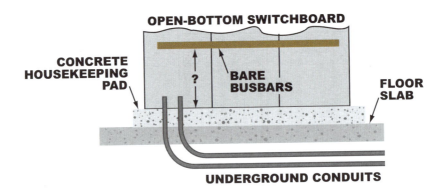

Figure 210.10

_____ 8. Refer to Figure 210.10, shown above. The maximum height permitted for dimension "?" in the drawing is ___ mm per the NEC®.

    a. 150   b. 75   c. 200   d. 600   e. 250

_____ 9. The service entrance conductors are connected to a 200 amp fused safety switch. The switch supplies power to MLO Panel HD1-1. Panel HD-1 has feed-through lugs and supplies power to MLO Panel HD-2. Panel ___ is a subpanel. MLO is the designation for Main Lugs Only.

    a. HD-2 only   b. both HD-1 & HD-2   c. HD-1 only   d. neither HD-1 nor HD-2

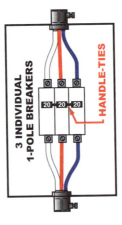

Figure 210.11

_____ 10. Figure 210.11, above, shows the breakers handle-tied to meet the requirements of the NEC. All three breakers under the handle-tie will trip when a ground-fault occurs between the load and the breaker.

    a. False   b. True

Switches, Switchboards and Panelboards

_____ 11. You need to move some breakers around in a panel to make room for a 3-pole breaker. You can move breaker #1 to an empty space near the bottom but the "hot" will then be too short. Does the NEC® permit you to just splice a short piece of wire onto the "hot" with wire nuts so you can then reach the relocated breaker?

   a. yes   b. no

_____ 12. If 2 individual single-pole breakers are handle-tied and used for a 1Ø 240 volt circuit then ___.
   I. both "hots" will open when a short-circuit or ground-fault occurs
   II. both "hots" can be manually opened

   a. I only   b. II only   c. neither I nor II   d. both I and II

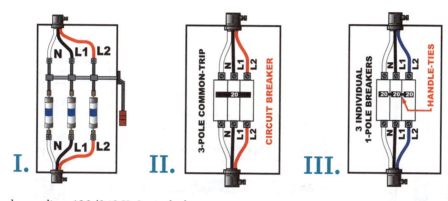

The switch or breaker supplies a 120/240 V, 3-wire load.

Figure 210.1

_____ 13. Which of the illustrations, shown above in Figure 210.1, fails to satisfy the Code requirements of 240.22(1)?

   a. I only
   b. II only
   c. III only
   d. I & III only
   e. I & II & III
   f. none of these

_____ 14. Refer to NEC® 240.22(1) and 404.2(B) exception. 240.22 addresses overcurrent device operation while 404.2 addresses switch disconnection. Which of the illustrations shown on the previous page in Figure 210.1 would satisfy both Code requirements?

   a. I only   b. II only   c. III only   d. II or III only   e. I or II or III

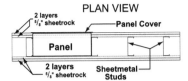

Figure 210.12

_____ 15. The Code permits a ___" maximum space between the panel enclosure ("can") and the panel cover as shown above in Figure 210.12.

a. 1/8
b. ½
c. 3/8
d. ¼
e. none of these

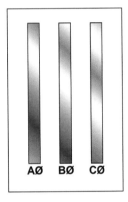 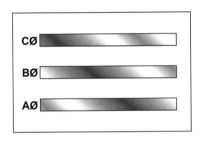 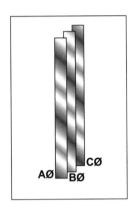

Figure 210.13

_____ 16. Among those shown above in Figure 210.13, Drawing ___ shows the correct busbar arrangement.

a. II only
b. II & III only
c. I, II, & III
d. I only
e. I & II only
f. I & III only

Year Two (Student Manual)

Objective 210.4 Worksheet

_____ 1. A space of not less than ___ feet shall be provided between the top of a switchboard and a combustible ceiling.

    a. 3   b. 8   c. 6   d. none of these

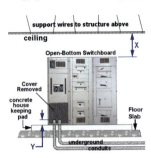

Figure 210.9

_____ 2. Refer to Figure 210.9, shown above. The 3-section switchboard assembly has removable covers on the front, the back, and both ends. The minimum dimension X permitted by the NEC® is ___ feet. The ceiling is a grid system with 15-minute rated ceiling tiles.

    a. 0   b. 3   c. 3.5   d. 4

_____ 3. Refer to Figure 210.9, shown above. Where the ceiling is plywood the minimum clear space required between the top of the open-bottom switchboard and the ceiling is ___.

    a. 3 ½′   b. 4′   c. 3′   d. 6′   e. none of these

_____ 4. A service switchboard is installed in an electrical equipment room to supply a commercial building. The room has a fire-rated sheetrock ceiling. The switchboard is setting on a 4″ high housekeeping pad that has been poured on top of the finished floor. The switchboard is 6′-3″ in height. The ceiling height in this room must be at least ___ above the finished floor.

    a. 6′-6″   b. 6′   c. 6′-7″   d. 25′   e. none of these

Switches, Switchboards and Panelboards

Objective 210.5 Worksheet

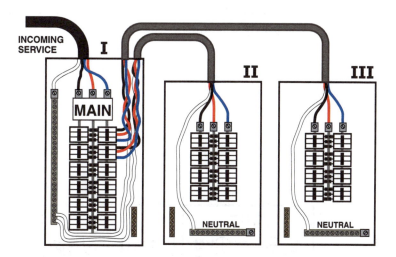

Figure 210.8

_____ 1. In Figure 210.8 the busbars in Panel II are ___ conductors.

   a. branch circuit
   b. service
   c. tap
   d. feeder

_____ 2. In Figure 210.8 the conductors that supply power to Panel I are ___ conductors.

   a. tap
   b. feeder
   c. branch circuit
   d. service

_____ 3. In Figure 210.8 the conductors that supply power to Panel III are ___ conductors.

   a. feeder
   b. tap
   c. service
   d. branch circuit

_____ 4. All the equipment shown on the previous page in Figure 210.8 is rated at 200 Amps and is 277/480 volt 3Ø, 4-wire. Your helper has installed bonding jumpers in all three panels. You should now tell him to remove MBJ's in Panel ___.

a. I only
b. III only
c. I, II, & III
d. II & III only
e. I & II only

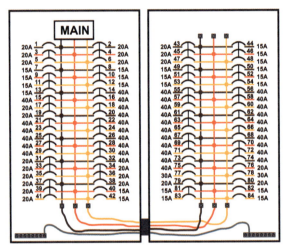

Figure 210.14

_____ 5. Refer to Figure 210.14, shown above. The panel on the left is fed from a 277/480 volt, 3Ø, 4-wire system. The load connected to the breaker installed in spaces 3-5 is a ___ load.

a. 1Ø
b. 3Ø

_____ 6. Refer to Figure 210.14, shown above. You have been told to feed the 225 amp MAIN in the MBO panel on the left with 4/0 AWG THHN/THWN. 4/0 conductors are connected between the bottom lugs on the MBO panel and the bottom lugs on the 225 amp MLO panel. Does the NEC permit 4/0 conductors to be connected between the MLO panel top lugs and a 400 amp switch, fused at 225 amps?

a. Yes
b. No

Figure 210.15

Switches, Switchboards and Panelboards   210-247

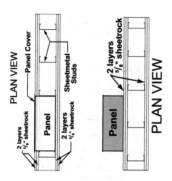

Figure 210.15

_____ 7. Which of the panels, shown above in Figure 210.15 (looking down on the wall), is "surface" mounted?

a. I only
b. both I & II
c. II only
d. neither I nor II

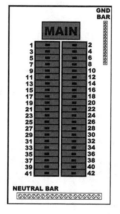

42-Circuit Panelboard

_____ 8. A circuit conductor that is connected to breaker 12 is permitted by the NEC® to be ___ in color. This 42 Circuit 480/277 volt, 3Ø, 4-W panelboard is shown above.
I. brown   II. orange   III. yellow

a. III only   b. I or II only   c. II only   d. I or II or III   e. none of these

_____ 9. The 42 Circuit panelboard shown above is supplied from a 120/208 volt, 3Ø, 4-wire system. Where connected to breaker 21 a conductor is permitted by the NEC to be ___ in color.
I. black   II. purple   III. orange

a. III only   b. I or II or III   c. I or II only   d. I only   e. none of these

_____ 10. In a 120/240 volt 3-ph, 4-wire panelboard the ungrounded conductor for ___-phase can be marked orange. A 42 Circuit Panelboard is shown above.

a. A only   b. B only   c. C only   d. A, B, and C   e. neither A nor B nor C

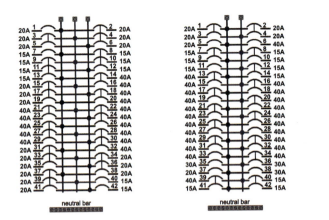
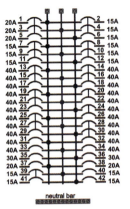

Figure 210.17

_____ 11. Refer to Figure 210.17, shown above. Which of the panels is a lighting and appliance branch circuit panelboard? Assume multi-pole breakers serve loads that do NOT require a neutral.

a. II only   b. III only   c. II & III only   d. I, II, & III   e. none of these

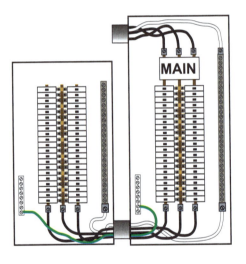

Figure 210.18

_____ 12. Refer to Figure 210.18, shown above. The panel on the left is a 42-circuit 200 Amp MLO (main lug only) panelboard that is fed from the 42-circuit Main Breaker panelboard on the right. The main is a 200 Amp breaker. The service conductors and the feeder conductors to the MLO panel are 3/0 THHN/THWN copper. A total of 84 circuits are tapped from the busbars. The busbars are protected by the main breaker. Is this installation a violation of 408.35 in the 2008 NEC®?

a. yes   b. no

_____ 13. Refer to Figure 210.18, shown above. The EGC between the two panels carries current during ___.

a. normal operation   b. a short-circuit   c. a ground fault   d. either a or b

Switches, Switchboards and Panelboards   210-249

_____ 14. Refer to Figure 210.18, shown on the previous page. The grounded conductor between the MBO panel and the source of the power can carry current during ___.

    I. normal operation   II. a ground-fault   III. a short-circuit

    a. I only   b. II only   c. I, II, & III   d. I & II only   e. II & III only

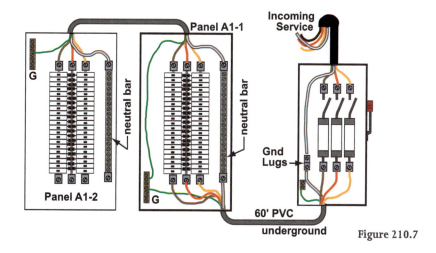

Figure 210.7

_____ 15. Refer to Figure 210.7, shown above. Panels A1-1 and A1-2 are **200** Amp panelboards. There are **225** amp fuses in the switch. Is this installation a violation of 408.36?

    a. yes   b. no

_____ 16. Refer to Figure 210.7, shown above. Panels A1-1 and A1-2 are 120/**208** volt 3Ø, 4-wire panelboards. The switch is rated for 250 volts. If the incoming service is 120/**240** volt 3Ø, 4-wire is using these panels with their voltage ratings a violation of the NEC®?

    a. yes   b. no

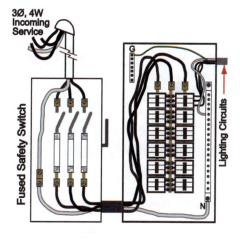

Figure 210.19

Year Two (Student Manual)

_____ 17. Refer to Figure 210.19, shown on the previous page. If the incoming service is 277/480 the NEC® permits the panelboard to be rated for ___ volts.
    I. 250   II. 600

    a. II only   b. neither I nor II   c. either I or II   d. I only

_____ 18. Refer to Figure 210.19, shown on the previous page. If the incoming service is 120/208 the NEC® permits the switch to be rated for ___ volts.
    I. 250   II. 600

    a. II only   b. neither I nor II   c. either I or II   d. I only

_____ 19. Refer to Figure 210.19, shown on the previous page. If the incoming service is 120/240 the NEC® permits the panelboard to be rated for ___ volts.
    I. 120/208   II. 120/240

    a. II only   b. neither I nor II   c. either I or II   d. I only

Switches, Switchboards and Panelboards

Objective 210.6 Worksheet

_____ 1. The luminaires in BCES classroom A104 are controlled by ___.

   a. a two-pole switch
   b. two single-pole switches
   c. a motion sensor
   d. a photocell
   e. none of these

_____ 2. A total of ungrounded conductors are installed from BCES Transformer XD to Panel LD.

   a. 8   b. 4   c. 6   d. 3

_____ 3. On the BCES project, each conduit installed between the transformer and Panel LB2 contains ___ grounded circuit conductor(s).

   a. 2 ea 3/0 AWG
   b. 2 ea 4/0 AWG
   c. 1 ea 3/0 AWG
   d. 1 ea 4/0 AWG
   e. 1 ea 350 kcmil
   f. 2 ea 350 kcmil

_____ 4. At the entrance to each BCES classroom there should be a ___ mounted on the strike side of the door.
   I. switch   II. strobe

   a. both I & II   b. I only   c. II only   d. neither I nor II

_____ 5. On the BCES project, the main switchboard is designed to be ___ accessible.

   a. side   b. front   c. front and rear   d. rear

_____ 6. The symbol **FSD**, shown on the BCES E-drawings, represents a ___.

   a. Field Supplied Disconnect
   b. Fire Smoke Damper
   c. Floor Seismic Detector
   d. Fused Switching Device

_____ 7. In room B114 on the BCES prints, the type of light switch shown is a ___.

   a. single-pole, key-type
   b. dimmer
   c. single-pole
   d. three-way

_____ 8. In BCES's main switchboard there is/are ___ branch circuit breaker(s).

   a. 6   b. 3   c. 12   d. 15   e. 1   f. 16

_____ 9. In BCES's main switchboard there are ___ spare breakers for future use.

   a. 12   b. 1   c. 3   d. 6   e. 0

_____ 10. On the BCES prints, ___ 208/120-volt panels are marked to have feed-thru lugs.

   a. 7   b. 8   c. 0   d. 9   e. 6

_____ 11. On the BCES project, Panel HK is a ___ panelboard.

   a. two-section   b. single-section

_____ 12. On the BCES prints, the light fixtures in corridor D114 are controlled through the BMCS (Building Management Control System) and an override switch(es). The override switch(es) is/are located in ___.

   a. Main Electric Room D154
   b. Mechanical Room B133
   c. Principal's Office A128
   d. Foyer B101, and Vestibule D142

_____ 13. Refer to BCES plans. Exhaust Fan 7 (EF-7) provides ventilation for Rooms C131/132 (toilets). The motor-rated switch for the fan is shown on Drawing E3.3 and is the disconnecting means for the fan. The switch is installed ___.

   a. on the roof with the fan
   b. on the wall adjacent to the light switch in Room C132
   c. above the ceiling of Room C132
   d. on the wall adjacent to the light switch in Room C133

Switches, Switchboards and Panelboards

_____ 14. On the BCES project, Panel LB2 contains ___.

   a. a 100 A main breaker
   b. no main breaker
   c. a 400 A main breaker
   d. a 225 A main breaker

_____ 15. On the BCES prints, there are ___ electrical panels that are marked to be flush mounted.

   a. 7   b. 9   c. 0   d. 8   e. 6

_____ 16. A total of grounded conductors are installed from BCES Transformer XD to Panel LD.

   a. 4
   b. 8
   c. 6
   d. 3

_____ 17. On the BCES project, the disconnect switch for Chiller CH-1 has a NEMA ___ enclosure.

   a. 1
   b. 4X
   c. 12
   d. 3R

_____ 18. On the BCES project, Panel LB2 contains busbars that are rated at ___ amps.

   a. 225
   b. 400
   c. 250
   d. 100

_____ 19. Excluding the MSB, there are ___ 480/277-volt Main Circuit Breaker panels shown on the BCES prints.

   a. 0
   b. 15
   c. 8
   d. 9

Year Two (Student Manual)

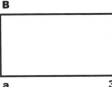

Figure BP1

_____ 20. Refer to Figure BP1, shown above, as shown on BCES drawing E1.1. The letter "a" indicates the ___.

  a. fixture type
  b. bulb type
  c. circuit number
  d. controlling switch

_____ 21. In BCES's main switchboard there is/are ___ feeder breaker(s).

  a. 16
  b. 12
  c. 15
  d. 11
  e. 1
  f. 6

_____ 22. Panel LB1 is located in BCES Room ___.

  a. B145
  b. B125
  c. B138
  d. B139
  e. B141

_____ 23. There are ___ 112.5 KVA transformers on the BCES project.

  a. 7
  b. 0
  c. 1
  d. 4

_____ 24. As indicated on the BCES food service drawings, the circuit breakers supplying power to the convection ovens are to trip when the fire system is activated.

  a. False
  b. True

Lesson 210 NEC Worksheet

_____ 1. All conductors of circuits, including grounded conductors, are required to be ____ on circuits leading to or through dispensing or remote pumping equipment.

    a. simultaneously disconnected
    b. bonded
    c. be rated 20 ampere, 250 volt
    d. be independently disconnected
    e. locked out

_____ 2. Each circuit, leading to or through motor fuel dispensing equipment for remote pumping systems, shall be provided with a disconnecting means that will simultaneously disconnect all conductors of the circuits, including the grounded conductor. ____ are not permitted for such required disconnecting means.

    a. Two-pole breakers
    b. Single-pole breakers
    c. Single-pole breakers utilizing handle ties
    d. Emergency stop buttons
    e. Double-throw switches

_____ 3. Emergency controls to simultaneously disconnect motor fuel dispensing equipment at attended self-service facilities shall be located not more than ____ from dispensers.

    a. 30 m (100 ft)
    b. 20 m (66 ft)
    c. 10 m (33 ft)
    d. 5 m (16 1/2 ft)
    e. 40 m (133 ft)

_____ 4. Emergency controls to simultaneously disconnect motor fuel dispensing equipment at unattended motor fuel dispensing facilities shall be located ____ from dispensers.

    a. more than 6 m (20 ft) but less than 30 m (100 ft)
    b. more than 6 m (20 ft) but less than 20 m (66 ft)
    c. more than 10 m (33 ft) but less than 20 m (66 ft)
    d. more than 5 m (16 1/2 ft) but less than 30 m (100 ft)
    e. more than 40 m (133 ft)

Switches, Switchboards and Panelboards

_____ 5. At an unattended motor fuel dispensing facility, emergency controls are required to be located not less than ___ and not more than ___ from the dispensers.

   a. 900 mm (3 feet), 2.0 m (6 1/2 ft)
   b. 1.8 m (6 ft), 3.0 m (10 ft)
   c. 6 m (20 ft), 30 m (100 ft)
   d. 7.5 m (25 ft), 15 m (50 ft)
   e. 2 m (6 1/2 ft), 5 m (16 ft)

_____ 6. Motor fuel pumps and dispensing equipment shall be provided with a means to remove all external voltage sources during maintenance and servicing operations. The disconnecting means shall ___.
   I. be required to be within the dispensing equipment
   II. be permitted to be located in other than inside or adjacent to the dispensing device
   III. be capable of being locked in the open position

   a. I & II only
   b. I only
   c. III only
   d. II only
   e. II & III only
   f. I & III only

_____ 7. During maintenance and services events, fuel dispensing devices are required to provide means to remove ___ voltage sources.

   a. power
   b. control
   c. communication
   d. external
   e. high

Lesson 210 Safety Worksheet

Switches, Switchboards and Panelboards

Article 110.16 of the National Electric Code (NEC®), and article 400.11 of the NFPA70E, requires that switchboards, panelboards, and industrial motor control panels, that are in other than dwelling occupancies and are likely to require examination, adjustment, servicing, or maintenance while energized shall be field marked to warn qualified persons of arc flash hazards. The marking shall be located so as to be clearly visible to qualified persons before examination, adjustment, servicing, or maintenance of the equipment. This would also include such tasks as removing covers, testing or troubleshooting.

The reasons for these requirements are obvious. This equipment is designed to contain energized parts. The voltages are hazardous and must be treated with respect. The term **qualified persons** here means one who has knowledge, training and understanding of the hazards involved and what precautions must be taken to prevent any personal harm or equipment damage. They must have knowledge in the selection and proper care and use of Personal Protective Equipment (PPE) and other safety equipment and tools.

The power supplying this equipment must be turned off and all Lock-out/Tag-out procedures must be followed to assure an electrically safe work condition. Refer to safety lessons 204 and 205 (NFPA70E) for more information.

This equipment must be installed following all of the manufacturer's installation instructions and all items of Listing and Labeling included with the equipment. These items will include information regarding the correct procedures for inspecting and mounting of the equipment.

There can be no open knock-out holes or open breaker spaces that would expose the busbars or breaker mounting hardware. This would create an unsafe condition when the equipment is energized. These openings must be closed using manufactured closures or fillers. **You cannot use electrical tape or duct tape or similar items to close these openings.**

Switches, Switchboards and Panelboards

Questions

_____ 1. Switchboards and panelboards are designed to contain ___ parts.

   a. foreign
   b. many
   c. energized
   d. cheap

_____ 2. A qualified person is one who has ___ of the hazards involved and what precautions must be taken to prevent any personal harm or equipment damage.

   a. understanding
   b. knowledge
   c. training
   d. all of these

_____ 3. The qualified person must also have knowledge in the selection and proper care and use of ___ and other safety equipment and tools.

   a. NFPA
   b. PPE
   c. NEC®
   d. all of these

_____ 4. The power supplying switchboards or panelboards must be turned off and all ___ procedures must be followed to assure an electrically-safe work condition.

   a. look-out
   b. lock-out
   c. lock-tite
   d. none of these

_____ 5. The NEC® and ___ require that all switchboards, panelboards and industrial motor control panels shall be field marked to warn qualified persons of arc flash hazards.

   a. NFPA70E
   b. ANSI
   c. NFPA72
   d. OSHA

Year Two (Student Manual)

_____ 6. Unused openings in panelboards can be closed using ___.

    a. duct tape
    b. electrical tape
    c. Cardboard
    d. blanks

_____ 7. When installing panelboards or switchboards, you must follow all ___ requirements.

    a. owner
    b. Company
    c. Manufacturer
    d. none of these

_____ 8. Listing and labeling of equipment contains information regarding the correct ___.
    I. inspection
    II. mounting

    a. neither I nor II
    b. I only
    c. both I & II
    d. II only

Lesson 211
Equipment for General Use

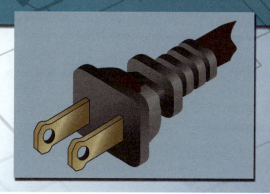

Electrical Curriculum

Year Two
Student Manual

IEC
PRIDE
NATIONAL

Purpose

This lesson will teach the rating, limitations and requirements for flexible cords, flexible cables and receptacles that supply luminaries, appliances and other portable general use equipment.

Homework

(Due at the beginning of this class)

For this lesson, you should:

- Thoroughly read the material contained within Lesson 211.
- Read ES Chapter 10 in its entirety.
- Complete Objective 211.1 Worksheet.
- Complete Objective 211.2 Worksheet.
- Complete Objective 211.3 Worksheet.
- Complete Objective 211.4 Worksheet.
- Complete Objective 211.5 Worksheet.
- Complete Lesson 211 NEC Worksheet.
- Read and complete Lesson 211 Safety Worksheet.

- Complete additional worksheets, if available, as directed by your instructor.

Objectives

By the end of this lesson, you should:

211.1 ....Understand the ratings, installation and over current protection for flexible cords and cables.
211.2 ....Understand the location, support, grounding and installation requirements for luminaries.
211.3 ....Understand the type, grounding and installation requirements for receptacles, cord connectors and attachment plugs.
211.4 ....Understand branch circuit ratings, over current protection, disconnect means, markings and installation requirements for appliances.
211.5 ....Be able to locate equipment and symbols on the BCES prints in order to solve problems addressed on common construction site scenarios.

Equipment for General Use   211-263

211.1 Ratings, Installation and Over Current Protection for Flexible Cords and Cables

211.2 Location, Support, Grounding and Installation Requirements for Luminaries

211.3 Type, Grounding and Installation Requirements for Receptacles, Cord Connectors and Attachment Plugs

211.4 Branch Circuit Ratings, Over Current Protection, Disconnect Means, Markings and Installation Requirements for Appliances

211.5 Problem Solving for BCES Blueprints

One of the lamps required on the BCES Fixture Schedule is a Phillips MS1000/BU. This lamp, shown at the right, is a Metal Halide type. The full product name is Switch Start MH Std 1000W WH Mog BT56.

Year Two (Student Manual)

Equipment for General Use

Objective 211.1 Worksheet

_____ 1. The maximum length of a flexible cord used to connect data processing equipment to a branch circuit outlet box is ____.

    a. 6 feet   b. 4.5 meters   c. 3.15 m   d. 10 feet   e. 3.0 m

_____ 2. The damaged power cord on a 120 volt appliance can be replaced with a 6-ft length of flexible SJO cord not smaller than ____ AWG where the appliance is to be connected to a 15 A branch circuit.

    a. 16   b. 12   c. 14   d. 18   e. none of these

_____ 3. Does the NEC® permit 12 AWG SO cord to be run inside a wall between 15 amp receptacle outlet boxes? CGB fittings that are listed for use with the cord will be used.

    a. yes   b. no

_____ 4. A listed appliance is manufactured with a 14 AWG flexible power cord that is used for connection to a 30 amp receptacle. The nameplate on the appliance contains the information: "Maximum 30 A". This appliance is considered to be protected by a 30 amp branch circuit breaker.

    a. False   b. True

_____ 5. Where LS appears on a flexible cable or cord this is an abbreviation that designates that the cord or cable ____.

    a. is low smoke producing
    b. is light sensitive
    c. has low stretch characteristics
    d. has larger shielding

_____ 6. The "O" designation in any of the "S" type flexible cords designates ____ insulation and/or outer covering.

    a. circular shaped
    b. an oil-resistant
    c. oxygen impregnated   d. none of these

_____ 7. A 3Ø 480 volt piece of equipment needs to be connected with flexible cord. Which of the cord types listed below could you use?
I. SJO   II. SO   III. STO

a. I or III only
b. I, II, or III
c. II only   d. II or III only
e. I or II only
f. none of these

_____ 8. Which of the flexible cables or cords listed below can be used in wet locations?

a. PD
b. SO
c. EV
d. any of these   e. none of these

_____ 9. Which of the flexible cords or cables listed below could be used to connect a 120 volt piece of equipment that draws 14 amps?
I. SVO   II. SPE-2   III. SO

a. I or II only
b. II or III only
c. I or III only
d. III only
e. I only
f. none of these

_____ 10. Where 14 AWG SJO flexible cord is field connected between a fused safety switch and a vibrating piece of equipment the largest fuses that can be installed in the switch are ___ amp. The equipment is 1-phase 240 volt.

a. 10
b. 15
c. 20
d. 25
e. 30

_____ 11. When connected to a 120 volt heater a 12 AWG HPD flexible cord can carry ___ amps.

a. 20   b. 25   c. 30   d. 35   e. 16

Equipment for General Use	211-267

_____ 12. A 208 volt 1-ph piece of equipment is to be connected with SOO flexible cord. What is the smallest AWG size cord that can be used to connect the equipment? The maximum load on any one conductor is 18 amps but the equipment will operate continuously.

a. 12   b. 6   c. 14   d. 8   e. 10

_____ 13. A certain piece of equipment is to be connected with SOO flexible cord. The equipment requires 2 ungrounded and 1 grounded conductors. What is the smallest AWG size cord that can be used to connect the equipment? All conductors should be counted as a current-carrying. The maximum load on any one conductor is 31 amps. The equipment will operate continuously.

a. 12   b. 6   c. 14   d. 8   e. 10

_____ 14. A certain piece of equipment is to be connected with STO flexible cord. The equipment requires 6 ungrounded and 4 grounded conductors. What is the smallest AWG size cord that can be used to connect the equipment? All conductors should be counted as a current-carrying. The maximum load on any one conductor is 30 amps.

a. 10
b. 6
c. 4
d. 8
e. 2

_____ 15. When connected to a 3Ø piece of equipment that operates at 240 volts a 12 AWG SJO flexible cord can carry ___ amps.

a. 20
b. 25
c. 30
d. 35
e. 16

_____ 16. Where 6 AWG/7-conductor Type G flexible cable is used to supply fastened-in-place utilization equipment it has an ampacity of ___ amps. See T. 400.4 for available sizes. This equipment has a metal enclosure and the connection terminals have 75° C ratings.

a. 36
b. 62
c. 45
d. 77
e. 54

Year Two (Student Manual)

Equipment for General Use

Objective 211.2 Worksheet

_____ 1. A(n) ___ is a complete unit consisting of the lamp or lamps, reflector, lamp guards, and lamp power supply.

   a. lighting outlet
   b. lighting fixture
   c. assembly
   d. luminaire
   e. none of these

_____ 2. An electric-discharge luminaire utilizes a ___ for the operation of the lamp.

   a. thermal protector
   b. starter
   c. ballast
   d. none of these

_____ 3. ___ is a type of lamp used in an HID luminaire.
   I. Mercury-vapor   II. Metal halide   III. Nickel cadmium

   a. I only
   b. I & III only
   c. II only
   d. I & II only
   e. I, II, & III
   f. II & III only

_____ 4. Per 220.43(B) in the NEC®, ___ VA must be allocated for the lighting track when determining the calculated load on the service where a 12′ length of lighting track is installed in the conference room in an office building.

   a. 1800   b. 150   c. 0   d. 900

_____ 5. The 8-32 screws in an octagonal 4/0 box is permitted to support a luminaire provided the luminaire weight does not exceed ___ kg.

   a. 6   b. 23   c. 50   d. 9   e. 3

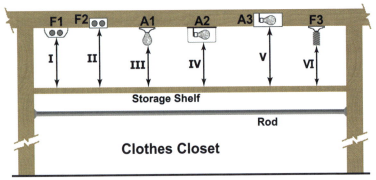

Fixture Schedule

F1 - surface-mounted wrap-around fluorescent
F2 - recessed fluorescent with lens
F3 - keyless lampholder with spiral fluorescent lamp
A1 - keyless lampholder with incandescent lamp
A2 - totally enclosed surface-mounted incandescent
A3 - recessed incandescent with lens

Figure 211.1

_____ 6. Refer to Figure 211.1, shown above. The minimum clearance at Point I is ___.

   a. 6"   b. 3"   c. 1'   d. 8'   e. 2'   f. not permitted

_____ 7. Refer to Figure 211.1, shown above. The minimum clearance at Point III is ___.

   a. 6"   b. 3"   c. 1'   d. 8'   e. 2'   f. not permitted

_____ 8. Refer to Figure 211.1, shown above. The minimum clearance at Point V is ___.

   a. 6"   b. 3"   c. 1'   d. 8'   e. 2'   f. not permitted

_____ 9. Where a luminaire is listed for installation in a damp location it can be installed in a ___ location.
   I. wet   II. dry   III. damp

   a. I, II, or III   b. I or III only   c. III only   d. I or II only   e. II or III only

_____ 10. The horizontal bathtub zone in which luminaires are not permitted is ___´ from the bathtub rim.

   a. 2.5   b. 8   c. 3   d. 5   e. none of these

_____ 11. An EGC is installed with the circuit conductors in underground PVC and the circuit supplies power to luminaires mounted on the top of a metal parking lot pole. The EGC is required to be connected to the ___.

   a. luminaries   b. pole   c. both of these   d. neither of these

Equipment for General Use

_____ 12. A wall-mounted luminaire, such as a sconce, that weighs 9 pounds is permitted to be mounted to and supported by a ___.

  a. switch box
  b. 1900 box with a single-gang plaster ring
  c. 4/0 box
  d. any of these
  e. none of these

_____ 13. A wall-mounted luminaire, such as a sconce, that weighs 4 pounds is permitted to be mounted to and supported by a ___.

  a. switch box
  b. 1900 box with a single-gang plaster ring
  c. 4/0 box
  d. any of these
  e. none of these

_____ 14. The NEC® requires that lay-in type fixtures be supported or attached to ___.

  a. the ceiling grid system   b. the building structure   c. both a and b   d. neither a nor b

_____ 15. Outdoor luminaires ___ be attached and supported by trees.

  a. are permitted to   b. shall not

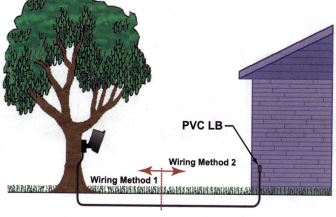

Figure 211.2

_____ 16. The installation shown above in Figure 211.2 violates the provisions of ___ in the NEC®. Wiring Method 1 is UF cable and Wiring Method 2 is Schedule 40 PVC.

  a. 225.26   b. 300.5   c. both a & b   d. neither a nor b

Year Two (Student Manual)

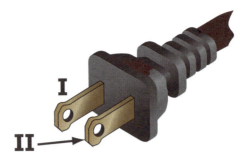

Figure 211.3

_____ 17. The attachment cord cap, shown in Figure 211.3 (above), is properly connected to an incandescent luminaire. When the cord cap is properly plugged into a wall receptacle the ___ conductor of the branch circuit is connected to the screw shell of the lampholder.

　　a. grounded　　b. grounding　　c. ungrounded

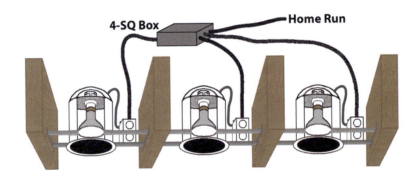

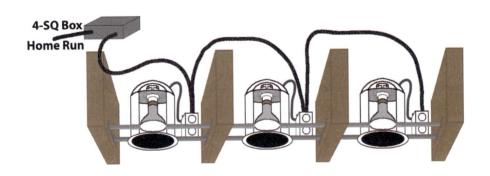

Figure 211.4

_____ 18. Which of the cable connections to the luminaires, shown above in Figure 211.4, can be used where the luminaires are not listed for thru-wiring?

　　a. I only　　b. either I or II　　c. II only　　d. neither I nor II

Equipment for General Use

_____ 19. Fluorescent luminaires are connected end-to-end as shown below. The "Home Run" conduit contains circuits HA-16, 23, 32 (3 "hots" + 1 neutral) and nightlight (NL) circuit HA-2. All of these circuits originate in a 277/480 volt 3Ø, 4-W panelboard. These conductors pass through the luminaires and the circuits serve the luminaires as indicated on the illustration.

"Home Run"
conduit ▬▬▬| HA-16 | HA-23 | NL | HA-16 | HA-32 | NL |

Which of the following statements is true?

a. This installation is permitted by NEC® 410.32.
b. This installation is not permitted by the NEC® because the voltage between conductors exceeds 150 volts.
c. This installation is not permitted by the NEC® because the normal lighting circuits do not meet the definition of a multiwire branch circuit.

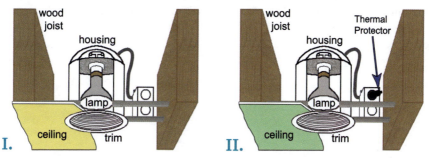

Figure 211.5

_____ 20. Refer to Figure 211.5, shown above. Which of the luminaires can be installed in the sheetrock ceiling in an office building?

a. II only   b. either I or II   c. neither I nor II   d. I only

_____ 21. Refer to Figure 211.5, shown above. The housing of luminaire II can be ____ the joists where the luminaire is NOT a listed Type IC.

a. in direct contact with
b. no closer than 6" from
c. no closer than 3" from
d. no closer than 1/2" from

_____ 22. Luminaire II, shown above in Figure 211.5, is listed and identified as Type IC. Thermal insulation ____ be installed above this luminaire.

a. can   b. can not

_____ 23. A recessed incandescent luminaire is to be installed in a ceiling grid system. The shortest 14 AWG "fixture whip" between a 4-square 20 amp branch circuit box and the wiring compartment of the luminaire is ___. Note that the 14 AWG wires are smaller than the 20 amp circuit breaker can protect and are therefor "tap conductors" by definition.

　　a. 1 foot　　　　b. 6 feet　　　　c. 1.5 feet　　　　d. no minimum limit

_____ 24. An electrical installation requires that recessed HID luminaires are to be installed in a ceiling of poured concrete decks. The HID luminaires are not provided with thermal protection. Does this installation violate the NEC® in Article 410?

　　a. Yes　　　　b. No

_____ 25. A customer wants to install 120-volt track lighting on the wall 2 feet above a countertop. The top of the countertop is 32 inches AFF. Does the NEC permit this installation?

　　a. Yes　　　　b. No

Equipment for General Use

Objective 211.3 Worksheet

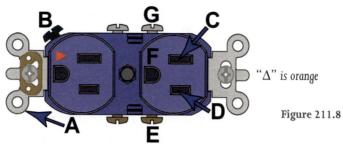

Figure 211.8

"Δ" is orange

_____ 1. Refer to Figure 211.8, shown above. The receptacle is properly connected and is energized. If the leads of a voltmeter are placed between Points E and G the meter should indicate ___ volts.

   a. 120   b. 0

_____ 2. Refer to Figure 211.8, shown above. The receptacle is properly connected and is energized. If the leads of a voltmeter are placed between Points A and D the meter should indicate ___ volts.

   a. 120   b. 0

_____ 3. Refer to Figure 211.8, shown above. The receptacle is properly connected and is energized. If the leads of a voltmeter are placed between Points B and F the meter should indicate ___ volts.

   a. 120   b. 0

_____ 4. Refer to Figure 211.8, shown above. The receptacle is properly connected and is energized. If the leads of a voltmeter are placed between Points D and G the meter should indicate ___ volts.

   a. 120   b. 0

_____ 5. The receptacle shown above in Figure 211.8 should be installed ___.
   I. for computer equipment requiring an isolated ground
   II. in areas that are subject to excessive abuse
   III. in hospitals

   a. II or III only
   b. II only
   c. III only
   d. I or III only
   e. I only

211-276 Equipment for General Use

_____ 6. Refer to Figure 211.8, shown above. After removing the receptacle from its package you are going to check it with an ohmmeter. If the leads of the ohmmeter are placed between Points A and B the meter should indicate ___ ohms.

   a. ∞   b. 0

_____ 7. Refer to Figure 211.8, shown above. The receptacle is properly connected and is energized. If the leads of a voltmeter are placed between Points C and E the meter should indicate ___ volts.
   a. 0   b. 120

_____ 8. Refer to Figure 211.8, shown above. The receptacle is properly connected and is energized. If the leads of a voltmeter are placed between Points D and E the meter should indicate ___ volts.

   a. 120   b. 0

_____ 9. Refer to Figure 211.8, shown on the previous page. The receptacle is properly connected and is energized. If the leads of a voltmeter are placed between Points B and C the meter should indicate ___ volts.

   a. 120   b. 0

_____ 10. A convenience receptacle installed ___ is required by the NEC to be GFCI protected.

   a. in the executive restroom in an office building

   b. in a bathroom in a house

   c. for temporary power on a construction site
   d. a and b only
   e. a, b, and c

Year Two (Student Manual)

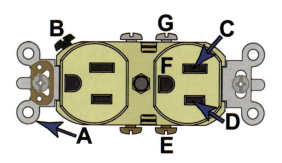

Figure 211.6

_____ 11. Refer to Figure 211.6, shown above. After removing the receptacle from its package you are going to check it with an ohmmeter. If the leads of the ohmmeter are placed between Points D and G the meter should indicate ___ ohms.

    a. ∞   b. 0

_____ 12. Refer to Figure 211.6, shown above. After removing the receptacle from its package you are going to check it with an ohmmeter. If the leads of the ohmmeter are placed between Points A and B the meter should indicate ___ ohms.

    a. ∞   b. 0

_____ 13. You are going to pull in five dedicated (individual) branch circuits into one metal conduit. Each circuit supplies power to a 120 volt load. The ungrounded conductors will be connected to breakers 2, 4, 6, 8, & 10 in a 120/240 volt 1Ø, 3-wire panelboard. Per the NEC the fewest number of neutral conductors you can install for the branch circuits is ___.

    a. 4   b. 2   c. 5   d. 3   e. 1

Figure 211.3

_____ 14. The attachment cord cap, shown in Figure 211.3 (above), is properly connected to an appliance. When the cord cap is properly plugged into a wall receptacle, blade ___ is inserted into the ungrounded slot of the receptacle.

    a. either I or II   b. I only   c. neither I nor II   d. II only

_____ 15. Five isolated ground 120 volt receptacles are to be supplied power from a 120/240 volt 1Ø, 3-wire panelboard. Each receptacle is on a dedicated (individual) branch circuit. You installed one metal conduit and will connect the ungrounded conductors to breakers 2, 4, 6, 8, & 10. Per the NEC® the fewest number of grounding conductors you can install in the conduit is ___.

   a. 1
   b. 3
   c. 5
   d. 2
   e. 0

_____ 16. You are going to pull in four dedicated (individual) branch circuits into one conduit. Each circuit supplies power to 120 volt loads. The ungrounded conductors will be connected to breakers 3, 5, 7, & 32 in a 120/208 volt 3Ø, 4-wire panelboard. Per the NEC® the fewest number of neutral conductors you can install for the branch circuits is ___.

   a. 3
   b. 4
   c. 0
   d. 1
   e. 2

_____ 17. Class A GFCI devices (those currently manufactured) are designed to operate when the ground fault current flow is between 4 mA and 6 mA.

   a. True   b. False

Figure 211.9

_____ 18. The receptacle, shown above at II in Figure 211.9, will be installed in a metal box. The box will be connected to the panelboard with PVC. The minimum number of conductors that are required to be pulled into the PVC is ___.

   a. four   b. five   c. three   d. two

Equipment for General Use   211-279

_____ 19. Receptacle ___, shown below, is rated at 15 amps?
   a.

   b.

   c. both a and b
   d. neither a nor b

_____ 20. ___-grade receptacles are the highest grade receptacles manufactured by the electrical industry.

   a. industrial   b. Hospital   c. commercial   d. none of these

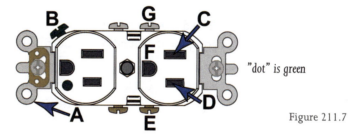

Figure 211.7

_____ 21. Refer to Figure 211.7, shown above. After removing the receptacle from its package you are going to check it with an ohmmeter. If the leads of the ohmmeter are placed between Points A and B the meter should indicate ___ ohms.

   a. ∞   b. 0

_____ 22. The receptacle shown above in Figure 211.7 should be installed ___.
   I. for computer equipment requiring an isolated ground
   II. in areas that are subject to excessive abuse
   III. in hospitals

   a. I only
   b. III only
   c. I or III only
   d. II or III only
   e. II only

_____ 23. Receptacle ___, shown below, is rated at 20 amps?

a.

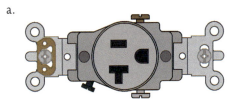

b.

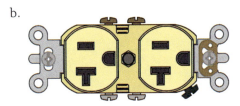

c. both a and b
d. neither a nor b

_____ 24. You are going to pull in four dedicated (individual) branch circuits into a PVC conduit. Per the NEC the fewest number of equipment grounding conductors you can install with the branch circuits is ___.

a. 4
b. 1
c. 2
d. 3
e. 0

_____ 25. A 15 A GFCI convenience receptacle is installed outdoors on the exterior wall of an office building. The receptacle is exposed to the direct effects of the weather. The receptacle is for supplying power to vending machines that are "unplugged" only for servicing. The enclosure for the receptacle is weatherproof only when the receptacle cover is closed. Does this installation violate the requirements of Article 406 in the NEC®?

a. Yes
b. No

Equipment for General Use

Complete the statements with the possible answers shown below.

a.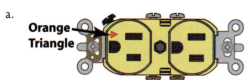

d.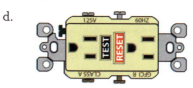

b.

e.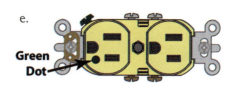

c.

f. None of these

_____ 26. ___ is a GFCI receptacle.

_____ 27. ___ is an isolated-ground receptacle.

_____ 28. ___ is a grounding receptacle.

_____ 29. ___ is a hospital grade receptacle.

_____ 30. ___ is a non-grounding receptacle.

Equipment for General Use

Objective 211.4 Worksheet

_____ 1. A(n) ___ is any piece of utilization equipment that performs one or more functions.

    a. luminaire
    b. appliance
    c. controller
    d. none of these

_____ 2. The minimum branch-circuit rating for an individual branch circuit which is used to supply a fastened-in-place appliance is ___ amps. The marked rating (equipment nameplate) on the appliance is 35 A. The appliance is not continuously loaded.

    a. 20
    b. 40
    c. 50
    d. 30
    e. none of these

_____ 3. A 60 gallon water heater is to be installed to supply hot water to the restrooms in an office building. The water heater is rated for 8.2 KW at 240 volts 1-phase. The minimum size THHN/THWN copper wire that can be installed between the required disconnect switch and the water heater is ___ AWG.

    a. 12
    b. 10
    c. 8
    d. 6
    e. 4

_____ 4. The smallest circuit breaker that can be installed in a panel is ___ Amp. The circuit supplies power to a 240-volt, single-phase, 5.8 KW, 50-gallon water heater.

    a. 30
    b. 35
    c. 40
    d. 45
    e. 50

Figure 211.10

_____ 5. Refer to Figure 211.10, shown above. You are to connect the dishwasher. A dedicated circuit has been run to a receptacle that is located under the sink. The 2-pole, 3-wire receptacle is flush-mounted in the sheetrock wall under the sink. Your two installation options are:

I. Install a flexible cord with a grounding-type attachment plug. There will be 33 inches of cord between the back of the dishwasher and the receptacle.
II. Replace the receptacle with a handy box extension ring and a single-pole switch with a cover, and run Type MC cable from the e-ring to the dishwasher.

Which installation method, shown above, violates the provisions of NEC Article 422?

a. neither I nor II   b. I only   c. II only   d. both I and II

_____ 6. A ceiling fan weighing ___ pounds or less can be supported by an outlet box that is identified for such use. The outlet box does not have a marking that indicates the maximum weight that can be supported.

a. 32   b. 70   c. 6   d. 16   e. 35

_____ 7. A 100 gallon boiler is to be installed to supply hot water to the restrooms in an office building. The boiler is rated for 9.8 KW at 240 volts 1-phase. The minimum size THHN/THWN copper wire that can be installed between the required disconnect switch and the boiler is ___ AWG.

a. 12   b. 10   c. 8   d. 6   e. 4

_____ 8. A 100 gallon boiler is to be installed to supply hot water to the restrooms in an office building. The boiler is rated for 9.8 KW at 240 volts 1-phase. The circuit for the boiler originates in a 120/208 volt 3Ø, 4-wire panelboard. The minimum size of the boiler branch circuit breaker that can be installed in the panel is ___.

a. 50   b. 30   c. 45   d. 35   e. 40

Equipment for General Use

_____ 9. If connecting a garbage disposal with flexible cord the cord must be at least 18″ long.

   a. False   b. True

_____ 10. A unit switch is the ON/OFF switch on an appliance that disconnects all ungrounded conductors.

   a. True   b. False

Objective 211.5 Worksheet

_____ 1. What type lamps are used in BCES's site lighting fixtures?

   a. metal halide   b. high pressure sodium   c. mercury vapor   d. low pressure sodium

_____ 2. The BCES fixture schedule can be used to determine ___ used on a project.

   a. the location of each type of luminaire
   b. the type of luminaries
   c. the shipping weight of each type of luminaire
   d. the quantity of each type of luminaire

Figure BP2

_____ 3. The symbol, shown above in Figure BP2, appears in BCES Corridors A110 and A153 as well as Commons A107 and represents a luminaire. The luminaire is a 2x4, 4-lamp lay-in fluorescent fixture with a one switch for the outboard lamps and a separate switch for the inboard lamps.

   a. True   b. False

_____ 4. The symbol, shown above in Figure BP2, appears in BCES Corridors A110 and A153 as well as Commons A107 and represents a luminaire. The luminaire is a 2x4, 3-lamp, lay-in fluorescent fixture with a battery backup for the center lamp.

   a. True   b. False

_____ 5. The BCES parking lot lights are controlled using a ___.
   I. photocell
   II. time clock
   III. lighting contactor

   a. I & II only
   b. II only
   c. I only
   d. I & III only
   e. II & III only

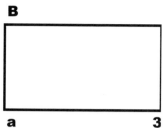

Figure BP1

_____ 6. Refer to Figure BP1, shown above, as shown on BCES drawing E1.1. The letter "B" indicates the ___.

    a. controlling switch   b. bulb type   c. circuit number   d. fixture type

_____ 7. The FSD's in the BCES Computer Lab (Room B141) contain ___ motors.

    a. no   b. 12 VAC   c. 12 VDC   d. 120 VAC

_____ 8. As indicated on the BCES food service drawings, the fryer is to be disconnected from the power when the fire system is activated.

    a. False   b. True

_____ 9. The motorized projection screen is located in BCES Room ___.

    a. A139   b. D147   c. B136   d. D138   e. D108

_____ 10. Refer to BCES plans. Exhaust Fan 3 (EF-3) provides ventilation for Room A131 (toilet). The fan runs when ___.

    a. AHU-9 is running   b. the lights are ON in Room A131   c. neither a nor b

_____ 11. The luminaires used in BCES Room A126 are Type ___.

    a. A7   b. C   c. A   d. B   e. B

_____ 12. On the BCES plans, RCP refers to Drawings ___.

    a. A10.4   b. A10.1   c. A10.2   d. A10.3   e. all of these   f. none of these

_____ 13. As indicated on the BCES food service drawings, the fryer is rated to pull ___ amps at rated voltage.

   a. 15.0
   b. 10
   c. 1.0
   d. 20.0

_____ 14. The FSD's in the BCES Computer Lab (Room B141) are to ___ when any smoke detector in the building is activated.

   a. open
   b. close

_____ 15. Wall boxes for BCES emergency lighting fixtures should be roughed in so that the fixture will be ___" AFF to center.

   a. 96   b. 72   c. 78   d. 84   e. 80

_____ 16. On the BCES project, AHU-10 is fed from circuits ___.

   a. HD-2,4,6   b. HK-1,3,5   c. HD-31,33,35   d. HE-2,4,6

_____ 17. Circuit ___ supplies the motorized projection screen on the BCES project.

   a. LD-73   b. LD-75   c. LD-71   d. LD-69   e. LD-67

_____ 18. The BCES parking lot light poles (Types PA and PB) are ___ feet in height.

   a. 30   b. 28   c. 18   d. 20   e. 24

_____ 19. Refer to BCES plans. Exhaust Fan 7 (EF-7) provides ventilation for Rooms C131/132 (toilets). The fan runs when ___.

   a. the lights are ON in Room C132
   b. the lights are ON in Room C131
   c. AHU-9 is running
   d. either a or b

_____ 20. Refer to the BCES food service drawings. In addition to the grounding conductor, ___ conductors are required for each circuit to the heated cabinets.

    a. three ungrounded
    b. two ungrounded
    c. two ungrounded and one grounded

Lesson 211 NEC Worksheet

_____ 1. For application of Article 516, the interior of any open or closed container of a flammable liquid is considered to be ___ location.

   a. Class I, Zone 1
   b. Class I, Division 2
   c. Class I, Zone 0
   d. Class I, Zone 2

_____ 2. The location classification for spray application for the interior of any open or closed container of a flammable liquid is ___.

   a. Class I, Division 1
   b. Class I, Division 2
   c. Class II, Zone 0
   d. Class I, Zone 2

_____ 3. The interior of any dip or coating tank of a flammable liquid is considered to be ___ location.

   a. a Class I, Division 1
   b. a Class I, Division 2
   c. an unclassified
   d. a Class II, Division 1
   e. a Class III, Division 2
   f. none of these

_____ 4. The interior of spray booths and rooms except as specifically provided in 516.3(D) is considered to be a ___ location.
   I.   Class I, Zone 1
   II. Class II, Division 1
   III. Class I, Division 1

   a. III only
   b. I or III only
   c. II only
   d. I or II only
   e. I or II or III

_____ 5. The interior of exhaust ducts of a spray booth or spray application area is considered to be a ___ location.
   I. Class I, Zone 1
   II. Class II, Division 1
   III. Class I, Zone 0

   a. I or II only
   b. II only
   c. III only
   d. I only
   e. II or III only

_____ 6. Any spray area in direct path of spray operations is considered to be a ___ location.
   I. Class I, Zone 1
   II. Class II, Division 1
   III. Class I, Zone 0

   a. I only
   b. II or III only
   c. III only
   d. I or II only
   e. II only

_____ 7. In sumps, pits and belowgrade channels, within areas used for spray applications, within 7.5 m (25 ft) horizontally of a flammable vapor source without a vapor stop shall be classified as a ___ location.

   a. Class I, Division 1
   b. Class II, Division 1
   c. Class I, Zone 1
   d. any of these
   e. none of these

Lesson 211 Safety Worksheet

Introduction to OSHA

OSHA (Occupational Safety and Health Administration) is a federal agency which is a department of the U.S. Department of Labor (DOL). OSHA was created by an act of congress in 1970 and signed into law by then president Richard M. Nixon. The OSHA regulations went into effect in early 1971, and have been in use since then.

The purpose of OSHA is "To assure safe and healthful working conditions for working men and women; by authorizing enforcement of the standards developed under the Act; by assisting and encouraging the States in their efforts to assure safe and healthful working conditions; by providing for research, information, education, and training in the field of occupational safety and health; and for other purposes."

"Congress finds that personal injuries and illnesses arising out of work situations impose a substantial burden upon, and are a hindrance to, interstate commerce in terms of lost production, wage loss, medical expenses, and disability compensation payments"

Congress went on to state they would so far as possible, provide every working man and women in the Nation safe and healthful working conditions and preserve our human resources.

They created what is often referred to as the "General Duty clause". Section 5(a) reads "Each employer (1) shall furnish to each of his employees, employment and a place of employment which are free from recognized hazards that are causing or are likely to cause death or serious harm to his employees. And (2) shall comply with occupational safety and health standards promulgated under the Act." Section 5 (b) further states "Each employee shall comply with occupational safety and health standards and all rules, regulations, and orders issued pursuant to this Act which applicable to his own actions and conduct."

To implement these directives OSHA has the right to enter workplaces, after first presenting official credentials at an opening conference, and inspect and determine the safety and health conditions. If they find violations during their inspection, they will note these and will discuss them at a closing conference.

The OSHA area director, not the compliance officer, establishes the amount of the proposed fines to be paid. The employer has the option to accept the proposed fines, or contest the citation if they feel it is not a valid violation. A copy of the citation must be posted at the location of the violation for three days or until the violation is corrected, whichever is longer. The proposed penalty amount does not need to be shown. The fines can range from $5000.00 to as much as $70,000.00 for a willful or repeat violation. It is also possible to get sentenced to jail with some of these conditions as well.

Remember that OSHA compliance officers are federal employees, so you do not want to subject them to threats, intimidation or physical harm.

Equipment for General Use

Questions

_____ 1. OSHA was created in ___ by an act of the ___.

    a. 1970 – congress   b. 1971 - supreme court   c. 1971 - senate

_____ 2. The "General Duty clause" states each employer shall furnish to each of his employees employment and a place of employment that are free from ___ hazards that are causing or are likely to cause death or serious harm to his employees.

    a. occasional   b. recognized   c. usual   d. many

_____ 3. Congress finds that personal injuries and illnesses arising out of work situations impose a substantial burden upon, and are a hindrance to, interstate commerce in terms of ___, and disability compensation payments.

    a. wage loss
    b. medical expenses
    c. lost production
    d. all of these
    e. none of these

_____ 4. The purpose of OSHA is to assure ___ working conditions for working men and women.

    a. safe and healthful   b. warm and friendly   c. neat and clean

_____ 5. OSHA will provide ___, and training in the field of occupational safety and health and for other purposes.

    a. education   b. research   c. information   d. all of these

_____ 6. Congress went on to state they would so far as possible, provide every working man and woman in the Nation ___ working conditions and preserve our human resources.

    a. safe and healthful
    b. clean and warm
    c. safe and warm
    d. clean and neat
    e. none of these

Lesson 212
Introduction to AC Theory

Electrical Curriculum

Year Two
Student Manual

IEC
PRIDE
NATIONAL

Purpose

This lesson will introduce basic AC theory which includes electrical charges, magnetism, AC generation, frequency, voltage and current, and a basic understanding of inductance and capacitance.

Homework

(Due at the beginning of this class)

For this lesson, you should:

- Thoroughly read the material contained within Lesson 212.
- Complete Objective 212.1 Worksheet.
- Complete Objective 212.2 Worksheet.
- Complete Objective 212.3 Worksheet.
- Complete Lesson 212 NEC Worksheet
- Read and complete Lesson 212 Safety Worksheet.

- Complete additional worksheets, if available, as directed by your instructor.

Objectives

By the end of this lesson, you should:

212.1 ....Understand the terms used for electrical charges, magnetism, and electromagnetism.
212.2 ....Understand the basics of AC power, AC generation, frequency, voltage and current.
212.3 ....Understand inductance and capacitance.

212.1 Electrical Charges, Magnetism, and Electromagnetism

Electric Charges

Neutral State of an Atom

Elements are often identified by the number of electrons in orbit around the nucleus of the atoms making up the element and by the number of protons in the nucleus. A hydrogen atom, for example, has only one electron and one proton. An aluminum atom has 13 electrons and 13 protons. An atom with an equal number of electrons and protons is said to be electrically neutral.

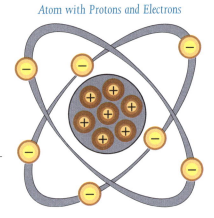

Atom with Protons and Electrons

Positive and Negative Charges

Electrons in the outer band of an atom are easily displaced by the application of some external force. Electrons which are forced out of their orbits can result in a lack of electrons where they leave and an excess of electrons where they come to rest. The lack of electrons is called a positive charge because there are more protons than electrons. The excess of electrons has a negative charge. A positive or negative charge is caused by an absence or excess of electrons. The number of protons remains constant.

Neutral Charge

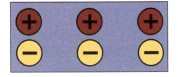

Negative Charge

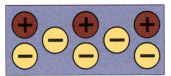

Positive Charge

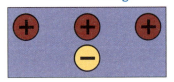

Attraction and Repulsion of Electric Charges

The old saying, "opposites attract"; is true when dealing with electric charges. Charged bodies have an invisible electric field around them. When two like-charged bodies are brought together, their electric field will work to repel them. When two unlike-charged bodies are brought together, their electric field will work to attract them. The electric field around a charged body is represented by invisible lines of force. The invisible lines

Unlike Charges Attract

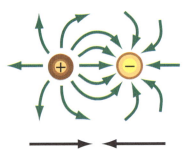

UNLIKE CHARGES ATTRACT

Like Charges Repel

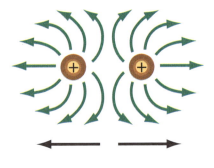

LIKE CHARGES REPEL

of force represent an invisible electrical field that causes the attraction and repulsion. Lines of force are shown leaving a body with a positive charge and entering a body with a negative charge.

Coulomb's Law

During the 18th century a French scientist, Charles A. Coulomb, studied fields of force that surround charged bodies. Coulomb discovered that charged bodies attract or repel each other with a force that is directly proportional to the product of the charges, and inversely proportional to the square of the distance between them. Today we call this Coulomb's Law of Charges. Simply put, the force of attraction or repulsion depends on the strength of the charged bodies, and the distance between them.

Magnetism

The principles of magnetism are an integral part of electricity. In fact, magnetism can be used to produce electric current and vice versa.

Types of Magnets

When we think of a permanent magnet, we often envision a horseshoe or bar magnet or a compass needle, but permanent magnets come in many shapes. However, all magnets have two characteristics. They attract iron and, if free to move (like the compass needle), a magnet will assume a north-south orientation. This is due to the molecular alignment within the magnetized

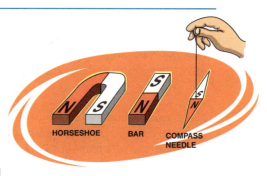

material. Soft iron is a commonly used material for permanent magnets because it is easily magnetized. Permanent magnets retain their magnetism for long periods of time. Permanent magnets are used in the construction of DC motors such as used in cordless drills.

Most commonly used metals contain iron. These are called ferrous or ferromagnetic materials. Specialty metals, such as stainless steel, copper, and aluminum are nonferrous because these do not contain iron. Many nonferrous materials can be temporarily magnetized by passing current through them or placing them within a magnetic field. When the magnetizing force is removed from these temporary magnets, they lose their magnetism. Temporary magnets are used in the construction of transformers, door chimes, and AC motors.

Permanent magnets use ferromagnetic materials while temporary magnets use nonferrous materials.

Magnetic Lines of Flux

Every magnet has two poles, one north pole and one south pole. Invisible magnetic lines of flux leave the north pole and enter the south pole. While the lines of flux are invisible, the effects of magnetic fields can be made visible. When a sheet of paper is placed on a magnet and iron filings loosely scattered over it, the filings will arrange themselves along the invisible lines of flux. By drawing lines the way the iron filings have arranged themselves, the lines of flux can be visualized. The field lines exist outside and inside the magnet. The magnetic lines of flux always form closed loops. Magnetic lines of flux leave the north pole and enter the south pole, returning to the north pole through the magnet.

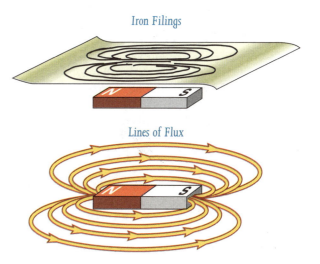

Interaction between Two Magnets

When two magnets are brought together, the magnetic flux field around the magnets causes some form of interaction. Two unlike poles brought together cause the magnets to attract each other. Two like poles brought together cause the magnets to repel each other.

Unlike Charges Attract *Like Charges Repel*

Magnetism is used to produce most of the world's electricity. Magnetism is the force between two magnets or a magnet and ferromagnetic materials. Magnetism causes motor shafts to rotate and solenoids to operate. Solenoids use linear motion (or force) to operate. Electrically operated valves, magnetic door locks, and door chimes utilize solenoids to perform their function. When the magnetic field (usually, current) is removed the solenoid returns to its "out-of-the-box" state.

Solenoids were studied in the first year of the curriculum. The door chimes studied in first year made use of solenoids. These door chimes operated when a circuit voltage was applied to the coil, producing a magnetic field. The magnetic field caused a plunger to move and strike a flat metal bar. Striking the bar produced a tone.

Solenoid valves operate by using the same principles as a door chime. With a solenoid valve, the action of the plunger opens or closes the valve. Solenoid valves are purchased to perform the function required, to open – or to close. Most solenoid valves open when power is applied, and close when power is removed. Your instructor should be familiar with solenoids should you have further questions concerning their operation or connection.

Magnetic locks on doors also make use of solenoids. Magnetic door locks are either fail-safe or fail-secure. Fail-safe locks hold the door locked when power is present. However, when power is lost or interrupted, the lock releases. This is so that, in the event of a power failure, people can exit through the door that is normally locked. Fail-secure locks hold the door locked unless power is applied to the solenoid. These are typically used on doors that limit access to authorized personnel. For example, the door to a room containing valuable jewelry would utilize a fail-secure lock. If a thunderstorm caused a power loss, the door would remain locked.

Electromagnetism

Current-Carrying Coil

A coil of wire, carrying a current, acts like a magnet. Individual loops of wire act as small magnets. The individual fields add together to form one magnet. The strength of the field can be increased by adding more turns to the coil or by wrapping the coils around an iron core. The strength can also be increased by increasing the current.

Electromagnets

An electromagnet is composed of a coil of wire wound around a core. The core is usually a soft iron which conducts magnetic lines of force with relative ease. When current is passed through the coil, the core becomes magnetized. The ability to control the strength and direction of the magnetic force makes electromagnets useful. As

Introduction to AC Theory

with permanent magnets, opposite poles attract. An electromagnet can be made to control the strength of its field which controls the strength of the magnetic poles.

When DC is passed through a coil the strength of the magnetic field is constant. However, when AC is passed through the coil the strength varies with the changing AC current levels.

A large variety of electrical devices such as motors, circuit breakers, contactors, relays and motor starters use electromagnetic principles.

212.2 Basics of AC power, AC generation, Frequency, Voltage and Current

Introduction to AC

The supply of current for electrical devices may come from a direct current source (DC), or an alternating current source (AC). In direct current electricity, electrons flow continuously in one direction from the source of power through a conductor to a load and back to the source of power. Voltage in direct current remains constant. DC power sources include batteries and DC generators. In alternating current an AC generator is used to make electrons flow first in one direction then in another. Another name for an AC generator is an alternator.

AC Current Flow

Current Flow Left-to-Right   Current Flow Right-to-Left

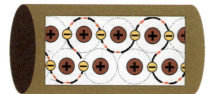 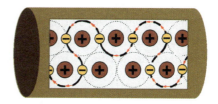

The AC generator reverses terminal polarity many times a second. Electrons will flow through a conductor from the negative terminal to the positive terminal, first in one direction then another.

AC Sine Wave

Alternating voltage and current vary continuously. The graphic representation for AC is a sine wave. A sine wave can represent current or voltage. There are two axes. The vertical axis represents the direction and magnitude of current or voltage. The horizontal axis represents time. When the waveform is above the time axis, current is flowing in one direction.

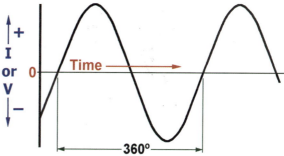

This is referred to as the positive direction. When the waveform is below the time axis, current is flowing in the opposite direction. This is referred to as the negative direction. The magnetic field around a conductor collapses (zero strength) each time the current level goes to zero. Note on the sine wave that the current level goes to zero twice in each cycle. A sine wave moves through a complete rotation of 360 degrees, which is referred to as one cycle. Alternating current goes through many of these cycles each second. The unit of measurement of cycles per second is hertz. In the United States alternating current is usually generated at 60 hertz.

Single-Phase and Three-Phase AC Power

Alternating current is divided into single-phase and three-phase types. Single-phase power is used for small electrical demands such as found in the home. Three-phase power is used where large blocks of power are required, such as found in commercial applications and industrial plants. Single-phase power is shown in the above illustration. Three-phase power, as shown at the right, is a continuous series of three overlapping AC cycles. Each wave represents a phase, and is offset by 120 electrical degrees.

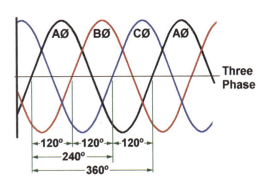

AC Generator
Basic Generator

A basic generator consists of a magnetic field, an armature, slip rings, brushes and a resistive load. The magnetic field is usually an electromagnet. An armature is any number of conductive wires, wound in loops, which rotate through the magnetic field. For simplicity, one loop is shown. When a conductor is moved through a magnetic field, a voltage is induced in the conductor. As the armature rotates through the magnetic field, a voltage is generated in the armature. This causes current to flow through the armature. Slip rings are attached to the armature and rotate with it. Carbon brushes ride against the slip rings to conduct current from the armature to the resistive load.

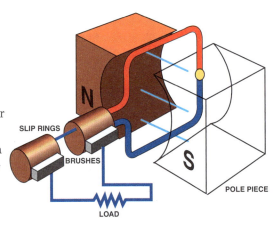

Basic Generator Operation

An armature rotates through the magnetic field. At an initial position of zero degrees, the armature conductors are moving parallel to the magnetic field and not cutting through any magnetic lines of flux. No voltage is induced. At 0°, on the sine wave drawing, voltage = 0 V.

Generator Operation from Zero to 90 Degrees

The armature rotates clockwise from zero to 90 degrees. The conductors cut through more and more lines of flux, building up to a maximum induced voltage in the positive direction. At 90°, on the sine wave drawing, voltage = positive maximum V.

Sine Wave Drawing

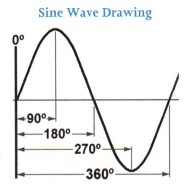

Introduction to AC Theory

Generator Operation from 90 to 180 Degrees

The armature continues to rotate from 90 to 180 degrees, cutting less lines of flux. The induced voltage decreases from a maximum positive value to zero. At 180°, on the sine wave drawing, voltage = 0 V.

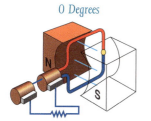
0 Degrees

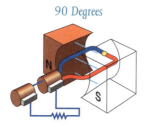
90 Degrees

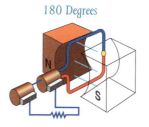
180 Degrees

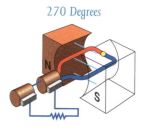
270 Degrees

Generator Operation from 180 to 270 Degrees

The armature continues to rotate from 180 degrees to 270 degrees. The conductors cut more and more lines of flux, but in the opposite direction, voltage is induced in the negative direction building up to a maximum at 270 degrees. At 270°, on the sine wave drawing, voltage = negative maximum V.

Generator Operation from 270 to 360 Degrees

The armature continues to rotate from 270 to 360 degrees. Induced voltage decreases from a maximum negative value to zero. At 360°, on the sine wave drawing, voltage = 0 V. This completes one cycle. The armature will continue to rotate at a constant speed. The cycle will continuously repeat as long as the armature rotates.

AC Generators

The number of cycles per second made by voltage induced in the armature is the frequency of the generator. If the armature rotates at a speed of 60 revolutions per second, the generated voltage will be 60 cycles per second. The accepted term for cycles per second is hertz. The standard frequency in the United States is 60 hertz.

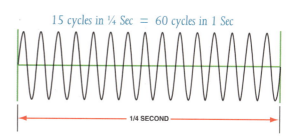
15 cycles in ¼ Sec = 60 cycles in 1 Sec
1/4 SECOND

Four-Pole AC Generator

The frequency equals the number of rotations per second if the magnetic field is produced by only two poles. An increase in the number of poles would cause an increase in the number of cycles completed in a revolution. A two-pole generator would complete one cycle per revolution and a four-pole generator would complete two cycles per revolution. An AC generator produces one cycle per revolution for each pair of poles.

4-Pole Generator

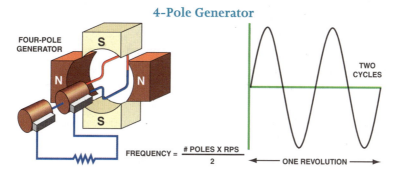

FREQUENCY = # POLES X RPS / 2

Year Two (Student Manual)

Voltage and Current

Peak Value

The sine wave illustrates how voltage and current in an AC circuit rise and fall with time. The peak value of a sine wave occurs twice each cycle, once at the positive maximum value and once at the negative maximum value.

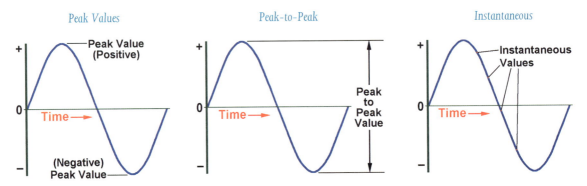

Peak-to-Peak Value

The value of the voltage or current between the peak positive and peak negative values is called the peak-to-peak value.

Instantaneous Value

The instantaneous value is the value at any one particular time. It can be in the range of anywhere from zero to the peak value. The exact value can be determined using trigonometry if the time (in degrees) and the peak value are known.

Effective Value of an AC Sine Wave

Since the instantaneous values of an alternating voltage and current are constantly changing, a method of translating the varying values into an equivalent constant value is needed. The effective value of voltage and current is the common method of expressing the value of AC. This is also known as the RMS (root-mean-square) value. If the voltage in the average home is said to be 120 volts, this is the RMS value. The effective value figures out to be 0.707 times the peak value. Typical voltmeters and ammeters read RMS voltage and current.

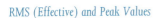
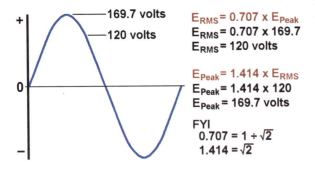

For purpose of circuit design, the peak value may also be needed. For example, insulation on meter leads must be designed to withstand the peak value, not just the effective value. It may be that only the effective value is known. To calculate the peak value, multiply the effective value by 1.414. Using the RMS effective value used previously, the peak value is 169.7 volts.

Introduction to AC Theory

The effective value of AC is defined in terms of an equivalent heating effect when compared to DC. One RMS ampere of current flowing through a resistance will produce heat at the same rate as a DC ampere.

212.3 Inductance and Capacitance

Inductance

The circuits studied to this point have been resistive. Resistance and voltage are not the only circuit properties that affect current flow, however. Inductance is the property of an electric circuit that opposes any change in electric current. Resistance opposes current flow while inductance opposes change in current flow. Inductance is designated by the letter "L". The unit of measurement for inductance is the henry (h).

Current Flow and Field Strength

Current flow produces a magnetic field in a conductor. The amount of current determines the strength of the magnetic field. As current flow increases, field strength increases, and as current flow decreases, field strength decreases.

Any change in current causes a corresponding change in the magnetic field surrounding the conductor. Current is constant in DC, except when the circuit is turned on and off, or when there is a load change. Current is constantly changing in AC, so inductance is a continual factor. A change in the magnetic field surrounding the conductor induces a voltage in the conductor. This self-induced voltage opposes the change in current. This is known as counter EMF. This opposition causes a delay in the time it takes current to attain its new steady value. If current increases, inductance tries to hold it down. If current decreases, inductance tries to hold it up. Inductance is somewhat like mechanical inertia which must be overcome to get a mechanical object moving or to stop a mechanical object from moving. A vehicle, for example, takes a few moments to accelerate to a desired speed, or decelerate to a stop.

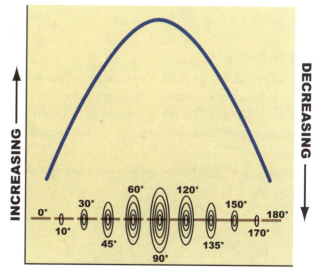

MAGNETIC FIELD STRENGTH

Inductors

Inductance is usually indicated symbolically on an electrical drawing by using a curled, or coiled, line as shown at the right.

Inductors are coils of wire. They may be wrapped around a core. The inductance of a coil is determined by the number of turns in the coil, the spacing between the turns, the coil diameter, the core material, the number of layers of windings, the type of winding, and the shape of the coil. Examples of inductors are transformers, chokes, and motors.

Simple Inductive Circuit

In a resistive circuit, current change is considered instantaneous. If an inductor is used in the circuit, the current does not change as quickly. In inductive circuits, the current LAGS behind the applied voltage.

Series Inductors

The same rules for calculating total series resistance can be applied to series inductive circuits. In the following circuit, an AC generator is used to supply electrical power to four inductors. A resistor is shown in the circuit because there will always be some amount of resistance and inductance in any circuit. The electrical wire used in the circuit and the inductive loads both have some resistance and inductance. Total inductance in a series circuit is calculated using the following formula:

$$L_T = L_1 + L_2 + L_3 + L_4$$

For the circuit shown here:

$$L_T = (2 + 2 + 1 + 1) \text{ mh} = 6 \text{ mh}$$

Parallel Inductors

In the following circuit, an AC generator is used to supply electrical power to three inductors. Note that this is a series-parallel circuit but, as with any circuit of this type, the equivalent of the paralleled portion can be determined first. Total inductance can be found using a formula that mimics that for resistance. The total inductance, paralleled portion of the circuit, is calculated using the following formula:

$$\frac{1}{L_T} = \frac{1}{L_1} + \frac{1}{L_2} + \frac{1}{L_3}$$

or

$$L_T = 1 \div \left(\frac{1}{L_1} + \frac{1}{L_2} + \frac{1}{L_3} \right)$$

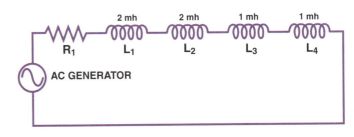

For this example: $L_T = 1 \div \left(\frac{1}{5} + \frac{1}{10} + \frac{1}{20} \right) \text{ mh} = 1 \div 0.35 = 2.857 \text{ mh}$

Introduction to AC Theory

Capacitance

Capacitance and Capacitors

Capacitance is a measure of a circuit's ability to store an electrical charge. A device manufactured to have a specific amount of capacitance is called a capacitor. A capacitor is made up of a pair of conductive plates separated by a thin layer of insulating material. Another name for the insulating material is dielectric material. When a voltage is applied to the plates, electrons are forced onto one plate. That plate has an excess of electrons while the other plate has a deficiency of electrons. The plate with an excess of electrons is negatively charged. The plate with a deficiency of electrons is positively charged.

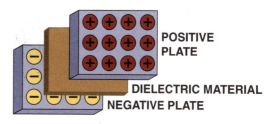

Direct current cannot flow through the dielectric material because it is an insulator; however it can be used to charge a capacitor. Capacitors have a capacity to hold a specific quantity of electrons. The capacitance of a capacitor depends on the area of the plates, the distance between the plates, and the material of the dielectric. The unit of measurement for capacitance is farads (F). Capacitors usually are rated in µF (microfarads), or pF (picofarads).

Capacitor Circuit Symbols

Capacitance is usually indicated symbolically on an electrical drawing by a combination of a straight line with a curved line, or two straight lines.

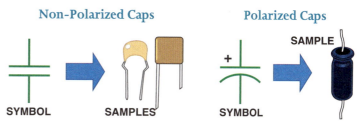

Simple Capacitive Circuit

In a resistive circuit, voltage change is considered instantaneous. If a capacitor is used in the circuit, the voltage across the capacitor does not change as quickly. In capacitive circuits, the current LEADS the applied voltage.

Series Capacitors

Connecting capacitors in series decreases total capacitance. The effect is like increasing the space between the plates. The formula for series capacitors is similar to the formula for parallel resistors. In the following circuit, an AC generator supplies electrical power to three capacitors. Total capacitance is calculated using the following formula:

$$\frac{1}{C_T} = \frac{1}{C_1} + \frac{1}{C_2} + \frac{1}{C_3}$$

or $\quad C_T = 1 \div \dfrac{1}{C_1} + \dfrac{1}{C_2} + \dfrac{1}{C_3}$

For this example: $C_T = 1 \div \left(\dfrac{1}{C_1} + \dfrac{1}{C_2} + \dfrac{1}{C_3} \right)$

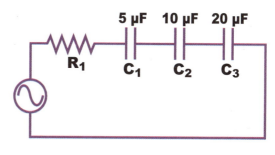

Year Two (Student Manual)

Parallel Capacitors

In the following circuit, an AC generator is used to supply electrical power to three capacitors. Total capacitance is calculated using the following formula:

$C_T = C_1 + C_2 + C_3$

For the circuit shown here:

$C_T = (5 + 10 + 20)\ \mu F$

$C_T = 35\ \mu F$

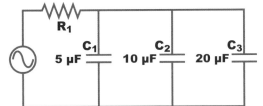

Introduction to AC Theory

Objective 212.1 Worksheet

_____ 1. Charged bodies attract or repel each other with a force that is ___.
   I. directly proportional to the product of the charges
   II. inversely proportional to the square of the distance between them

   a. neither I nor II   b. II only   c. I only   d. both I & II

_____ 2. An atom with more electrons than protons is said to be electrically ___.

   a. positive   b. charged   c. neutral   d. negative

_____ 3. An atom with an equal number of protons and electrons is said to be electrically ___.

   a. charged   b. positive   c. neutral   d. negative

_____ 4. The discovery that charged bodies attract or repel each other is called ___.

   a. Thomas Paine's Common Sense
   b. Coulomb's Law of Charges
   c. Darwin's Survival of the Fittest
   d. Newton's Law of Gravity

_____ 5. When two like-charged bodies are brought together, their electrical field will work to ___ them.

   a. report   b. attack   c. repel   d. attract   e. distort

_____ 6. Magnetism can be used to produce electric current.

   a. False   b. True

_____ 7. The three most common types of permanent magnets include all except ___.

   a. bar magnets
   b. horseshoe magnets
   c. electromagnets
   d. needle compass magnets

Year Two (Student Manual)

_____ 8. When power is ___ a solenoid valve controlling the flow of water into a clothes washing machine the valve opens and allows water to flow.

   a. applied to   b. removed from

_____ 9. Magnetic flux is the invisible lines of force that make up the magnetic field.

   a. False   b. True

_____ 10. As a safety consideration, when power is ___ a solenoid valve controlling the flow of natural gas to the gas-fired cooking appliances in a restaurant the valve <u>closes</u> and stops the gas flow to the appliances.

   a. applied to   b. removed from

_____ 11. Temporary magnets lose their magnetism as soon as the magnetizing force is removed.

   a. False   b. True

_____ 12. As the distance between two magnetic poles <u>increases</u>, the force of attraction (strength of the magnetic field) ___.

   a. increases   b. decreases

_____ 13. Temporary magnets are used in the construction of ___.
   I. cordless drills   II. AC motors

   a. both I & II   b. I only   c. II only   d. neither I nor II

_____ 14. Every magnet has ___ pole.

   a. one north and one west
   b. one south and one north
   c. one east and one west
   d. two north and two south
   e. only one

Introduction to AC Theory

_____ 15. _____ is a force that interacts with other magnets and ferromagnetic materials.

    a. Induction   b. Magnetism   c. Transformation   d. Reaction

_____ 16. The invisible magnetic lines that leave the north pole and enter the south pole of a magnet are _____.

    a. iron filings
    b. lines of flux
    c. north-south orientation
    d. electron flow

_____ 17. Permanent magnets are used in the construction of _____.
    I. cordless drills   II. AC motors

    a. both I & II   b. neither I nor II   c. II only   d. I only

_____ 18. The basic law of magnetism states that <u>unlike</u> poles (south and north, for example) _____ each other.

    a. attract   b. repel

_____ 19. Nonferrous materials, such as stainless steel, _____ attracted by the magnetic field around a magnet.

    a. are not   b. are

_____ 20. A coil of wire, when carrying current, acts like a _____.

    a. capacitor   b. magnet   c. motor   d. resistor

_____ 21. Production of most of the electricity consumed as well as the development of linear motion in solenoids and the rotary motion in motors is based on the use of magnetism.

    a. True   b. False

_____ 22. The strength of the magnetic field around a coil (or winding) _____ when DC voltage is applied.

    a. is constant   b. varies

_____ 23. The strength of the magnetic field around a conductor can be increased by ___.
   I. increasing the amount of current flow through the conductor
   II. forming the conductor into a coil
   III. wrapping the conductor in a spiral around an iron rod (or core)

   a. I only
   b. II or III only
   c. I or II only
   d. II only
   e. I, II, & III

_____ 24. The strength of the magnetic field around a coil is constant when AC current is flowing through the conductor.

   a. True   b. False

_____ 25. ___ is the magnetic field produced when electricity passes through a conductor.

   a. Electromagnetism
   b. Conveyance
   c. Galvanic action
   d. Displacement

Introduction to AC Theory

Objective 212.2 Worksheet

_____ 1. _____ current flows continuously in one direction from a power source to a load and then back to the power source, maintaining constant magnitude.

    a. MC   b. DC   c. AC   d. BC

_____ 2. _____ current constantly changes direction of flow and magnitude as it completes the circuit from a power source to a load and then back to the power source.

    a. MC   b. AC   c. BC   d. DC

_____ 3. The two axes of alternating voltage or current on a sine wave are a graphic representation of _____.

    a. a time continuum
    b. direction and space
    c. magnitude and time
    d. space and time

_____ 4. The graphic representation of AC voltage or current values over a period of time is a _____.

    a. sine wave   b. frequency modulation   c. bar graph   d. none of these

_____ 5. Each phase of three phase AC power is offset by _____ degrees.

    a. 90   b. 100   c. 180   d. none of these

_____ 6. A(n) _____ is any number of conductive wires, wound in loops, which rotate through a magnetic field.

    a. slip ring   b. armature   c. brush   d. pole piece

_____ 7. _____ is/are attached to the armature and rotate with the armature.

    a. Slip rings
    b. Pole pieces
    c. Brushes
    d. A resistive load

_____ 8. ___ do not rotate in a generator but, instead, ride against the slip rings to conduct current from the armature to a load.

   a. Brushes
   b. Magnetic fields
   c. Pole pieces

_____ 9. ___ is the output of an AC generator measured in cycles per second.

   a. Frequency   b. Voltage   c. Amperage   d. Resistance

_____ 10. ___ would be an equivalent term for cycles per second.

   a. Hertz   b. Volts   c. Sparks   d. Amps

_____ 11. Normally, AC power in the United States operates at ___ cps (cycles per second).

   a. 100   b. 120   c. 50   d. 60

_____ 12. An AC generator produces ___ cycle(s) per revolution for each pair of poles.

   a. two   b. one   c. three   d. four

_____ 13. A sine wave will hit its peak value ___ time(s) during each cycle.

   a. 1   b. 3   c. 60   d. 120   e. 2

_____ 14. The peak-to-peak value of a sine wave is ___ the maximum positive peak value.

   a. twice   b. half   c. three times   d. none of these

_____ 15. ___ is an abbreviation for RMS.

   a. **R**equired **M**echanical **S**tandards
   b. **R**esonance to **M**inimal **S**pikes
   c. **R**elative **M**agnetic **S**tability
   d. **R**oot **M**ean **S**quare
   e. **R**egressive **M**anaged **S**electivity

Introduction to AC Theory

_____ 16. _____ value is the value at any point in time on the sine wave.

    a. Instantaneous   b. Peak-to-Peak   c. Peak   d. RMS

_____ 17. Effective value of an AC sine wave is also known as _____ value.

    a. peak   b. instantaneous   c. peak-to-peak   d. RMS

_____ 18. One RMS ampere of current flowing through a resistor will produce heat at the same rate as _____.

    a. one DC ampere
    b. two peak amperes
    c. two DC amperes
    d. one instantaneous ampere

_____ 19. Which of the conversion formulas listed below is correct?

    a. RMS = Peak ÷ 0.707
    b. RMS = Peak × 1.414
    c. RMS = Peak × 0.707
    d. none of these

_____ 20. Which of the conversion formulas listed below is correct?

    a. Peak = RMS × 0.707
    b. Peak = RMS ÷ 1.414
    c. Peak = RMS × 1.414
    d. none of these

_____ 21. Where the RMS voltage is 100 the peak value is _____ volts.

    a. 70.7   b. 100   c. 141.4   d. none of these

_____ 22. The peak voltage of a circuit is _____ the RMS voltage.

    a. less than
    b. equal to
    c. greater than

_____ 23. Where the peak value is 169.7 the RMS value is ___ volts.

    a. 120   b. 238.6   c. 169.2   d. none of these

_____ 24. What is the peak value if the RMS value is 240 volts?

    a. 338.4   b. 170   c. 240   d. none of these

_____ 25. Meter leads must have an insulation that is rated for the ___ voltage of a circuit.

    a. average   b. effective   c. positive-to-negative sum   d. peak

Introduction to AC Theory

Objective 212.3 Worksheet

_____ 1. The unit of measure for capacitance is ___.

    a. Amps   b. Farads   c. Henrys   d. Volts

_____ 2. In an inductive circuit, the current ___ the applied voltage.

    a. leads b.   lags

_____ 3. Which of the symbols, shown below, is used to represent an AC power source in a schematic circuit?

a.
b.

c.
d.

e. none of these

_____ 4. Capacitors in parallel add like resistors in ___.

    a. vacuums   b. parallel   c. glass tubes   d. series

_____ 5. Which of the formulas, shown below, is correct to find the equivalent of 3 caps in a parallel circuit?

    a. $L_T = 1 \div \left(\dfrac{1}{L_1} + \dfrac{1}{L_2} + \dfrac{1}{L_3} \right)$ b. $C_T = 1 \div \left(\dfrac{1}{C_1} + \dfrac{1}{C_2} + \dfrac{1}{C_3} \right)$

    c. $C_T = C_1 + C_2 + C_3$ d. $L_T = L_1 + L_2 + L_3$

_____ 6. A capacitor is made up of a pair of conductive plates separated by a thin layer of ___ material.

    a. conductive
    b. insulating
    c. metallic
    d. liquid

Year Two (Student Manual)

_____ 7. Which of the symbols, shown below, is used to represent a DC power source in a schematic circuit?

a.  b.

c.   d.

e. none of these

_____ 8. _____ is the measure of a circuit's ability to store an electrical charge.

a. Inductance   b. Resistance   c. Impedance   d. Capacitance

_____ 9. Capacitors in series add like resistors in _____.

a. series   b. parallel   c. vacuums   d. glass tubes

_____ 10. A change in the current in the magnetic field surrounding the conductor induces a voltage in the conductor. This induced voltage is known as _____.

a. EMF (ElectroMotive Force)
b. voltage
c. amperage
d. Counter EMF

_____ 11. Which of the formulas, shown below, is correct to find the equivalent of 3 inductors in a series circuit?

a. $C_T = C_1 + C_2 + C_3$

b. $L_T = L_1 + L_2 + L_3$

c. $C_T = 1 \div \left(\dfrac{1}{C_1} + \dfrac{1}{C_2} + \dfrac{1}{C_3} \right)$

d. $L_T = 1 \div \left(\dfrac{1}{L_1} + \dfrac{1}{L_2} + \dfrac{1}{L_3} \right)$

_____ 12. Inductance is represented by the letter _____.

a. L   b. E   c. R   d. C

Introduction to AC Theory

_____ 13. Which of the formulas, shown below, is correct to find the equivalent of 3 caps in a series circuit?

a. $L_T = 1 \div \left(\dfrac{1}{L_1} + \dfrac{1}{L_2} + \dfrac{1}{L_3} \right)$

b. $L_T = L_1 + L_2 + L_3$

c. $C_T = 1 \div \left(\dfrac{1}{C_1} + \dfrac{1}{C_2} + \dfrac{1}{C_3} \right)$

d. $C_T = C_1 + C_2 + C_3$

_____ 14. In a capacitive circuit, the current ____ the applied voltage.

a. leads   b. lags

_____ 15. A coil of wire with or without a core is a(n) ____.

a. capacitor   b. resistor   c. inductor   d. none of these

_____ 16. Inductance is measured in ____.

a. Henrys   b. Volts   c. Ohms   d. Farads

_____ 17. Which of the symbols, shown below, is used to represent a resistor in a schematic circuit?

a.

b.

c.

d.

e. none of these

_____ 18. ____ is the property of an electrical circuit that opposes any change in current.

a. Capacitance
b. Impedance
c. Resistance
d. Inductance

_____ 19. Which of the symbols, shown below, is used to represent an inductor in a schematic circuit?

a.   b.

c.  d. ⌒⌒⌒⌒

e. none of these

_____ 20. Which of the formulas, shown below, is correct to find the equivalent of 3 inductors in a parallel circuit?

a. $L_T = 1 \div \left(\dfrac{1}{L_1} + \dfrac{1}{L_2} + \dfrac{1}{L_3} \right)$

b. $L_T = L_1 + L_2 + L_3$

c. $C_T = 1 \div \left(\dfrac{1}{C_1} + \dfrac{1}{C_2} + \dfrac{1}{C_3} \right)$

d. $C_T = C_1 + C_2 + C_3$

_____ 21. Inductors in series add like resistors in _____.

a. parallel   b. glass tubes   c. a vacuum   d. series

_____ 22. The magnetic field strength around a conductor is at its maximum when the AC current is at ___ degrees.

a. 0   b. 180   c. 120   d. 90

_____ 23. As current flow through a conductor increases, the magnetic field strength _____.

a. remains the same
b. goes to zero
c. decreases
d. increases

_____ 24. Inductors in parallel add like resistors in _____.

a. glass tubes   b. parallel   c. a vacuum   d. series

_____ 25. Which of the symbols, shown below, is used to represent a capacitor in a schematic circuit?

a.    b.

c.    d.

e. none of these

Lesson 212 NEC Worksheet

_____ 1. All space in all directions outside of but within 900 mm (3 ft) of open containers, supply containers, spray gun cleaners and solvent distillation units containing flammable liquids are considered to be a ___ location.

   a. Class I, Division 1
   b. Class II, Division 1
   c. Class I, Zone 1
   d. any of these
   e. none of these

_____ 2. When spray application equipment cannot be operated unless the exhaust ventilation system is operating and functioning properly and spray application is automatically stopped if the exhaust ventilation system fails the system is said to be ___.

   a. interlocked
   b. interconnected
   c. autoconnected
   d. interactively supported
   e. electrically integrated

_____ 3. For spraying operations conducted within an open top spray booth, the space ___ vertically above the booth and within ___ of other booth openings shall be considered a Class I, Division 2 location.

   a. 900 mm (3 ft)   900 mm (3 ft)
   b. 100 mm (1 ft)   900 mm (3 ft)
   c. 2 m (6 1/2 ft)   5 m (15 ft)
   d. 5 m (15 ft)   2 m (6 1/2 ft)
   e. 10 m (30 ft)   2 m (6 1/2 ft)

_____ 4. For spraying operations confined to an enclosed spray booth or room, the space within ___ in all directions from any openings shall be considered Class I, Division 2 or Class II, Division 2 locations.

   a. 900 mm (3 ft)
   b. 100 mm (1 ft)
   c. 2 m (6 1/2 ft)
   d. 5 m (15 ft)
   e. 10 m (30 ft)

Lesson 212 Safety Worksheet

OSHA Facts

Since its inception, OSHA has helped to reduce the number of workplace fatalities by more than 60 percent and the rate of occupational injury and illness by over 40 percent. During the same years the total number of worksites grew from roughly 3.5 million to over 7.2 million. The number of workers also grew from 58 million to over 115 million.

For the fiscal year 2005, OSHA had more than 2200 employees, including 1100 inspectors, also know as compliance officers. The agency is financed by appropriations from the congress (not monies collected by fines. For FY2005 their budget was $468.1 million.

Under the current administration, their focus is on: 1) strong, fair and effective enforcement; 2) outreach, education and compliance assistance; and 3) partnerships and cooperative programs.

OSHA has an extensive website at WWW.OSHA.GOV There are many sections with quite a large number of articles, information, standards, interpretations and much more.

They are offering education to small businesses to help them provide a safe environment for their workers. There are partnerships available to large companies that are already providing safety in the workplace, but want more.

In FY2004, they performed 39,167 inspections nation wide, 57% of which were construction. There were 86,708 violations found during those inspections. The total of proposed penalties was $85,192,940.00.

Not all states use the federal OSHA system. A number of the states use their own OSHA systems for safety and health enforcement.

Questions

_____ 1. Since OSHA's inception, the number of fatalities has been reduced by more than ___ percent and the rate of occupational injury and illnesses by over ___ percent.

    a. 40 and 60   b. 66 and 40   c. 60 and 40   d. 60 and 44

_____ 2. Today there are over ___ million worksites with over ___ million workers.

    a. 3.5 and 115   b. 7.2 and 58   c. 7.2 and 58   d. 7.2 and 115

_____ 3. For FY2005, OSHA had more than ___ employees and ___ inspectors.

    a. 2200 and 1500
    b. 2200 and 1100
    c. 1100 and 2200
    d. 1800 and 1100

_____ 4. The website for OSHA is ___.

    a. WWW.OSHA.COM
    b. WWW.OSHA.EDU
    c. WWW.OSHA.GOV

_____ 5. OSHA provides ___ to small businesses to help them provide a safe environment for their workers.

    a. citations   b. education   c. harassment   d. none of these

_____ 6. Of the 39,167 inspections OSHA performed in FY2004, ___% were on construction sites.

    a. 57   b. 75   c. 67   d. 59

_____ 7. All of the states use the federal OSHA program for safety enforcement in their states.

    a. yes   b. no

Lesson 213
AC Theory I

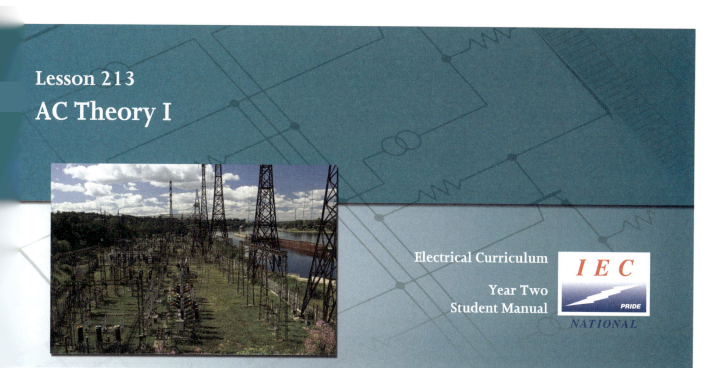

Electrical Curriculum

Year Two
Student Manual

Purpose

This lesson will introduce the effects Inductive Reactance and Capacitive Reactance have on an AC circuit.

Homework

(Due at the beginning of this class)

For this lesson, you should:

- Thoroughly read the material contained within Lesson 213.
- Complete Objective 213.1 Worksheet.
- Complete Lesson 213 NEC Worksheet
- Read and complete Lesson 213 Safety Worksheet.
- Complete additional worksheets, if available, as directed by your instructor.

Objectives

By the end of this lesson, you should:

    213.1 ....Understand the leading and lagging effects, between the voltage and current relationship, when inductive reactance and capacitive reactance are introduced to an AC circuit.

213.1 Inductive and Capacitive Reactance

In a purely resistive AC circuit, opposition to current flow is called resistance. In an AC circuit containing only inductance, capacitance, or both, opposition to current flow is called reactance. Total opposition to current flow in an AC circuit that contains both reactance and resistance is called **impedance**, designated by the symbol Z. Both inductive and capacitive reactance are measured in ohms. Impedance is a combination of resistance and reactance and is also measured in ohms. In AC circuits, all the Ohm's Law formulas remain the same with one substitution. R is replaced with Z. Remember that both resistance and impedance are measured in ohms.

Inductive Reactance

Inductance only affects current flow when the current is changing. Inductance produces a self-induced voltage (counter EMF) that opposes changes in current. In an AC circuit, current is changing constantly. Inductance in an AC circuit, therefore, causes a continual opposition. This opposition to current flow is called **inductive reactance** and is designated by the symbol **XL**.

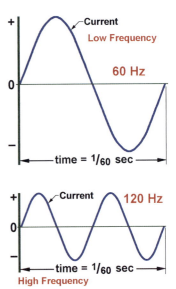

Inductive reactance is dependent on the amount of inductance and frequency. If frequency is low, current has more time to reach a higher value before the polarity of the sine wave reverses. If frequency is high, current has less time to reach a higher value. In the following illustration, voltage remains constant. Current rises to a higher value at a lower frequency than a higher frequency.

Where frequency is in Hz and inductance is in henrys, the formula for inductive reactance is:

$$X_L = 2\pi fL = 2 \times 3.142 \times \text{Frequency} \times \text{Inductance}$$

As shown in the upper illustration for a 60 hertz, 10 volt circuit containing a 10 mh inductor, the inductive reactance would be:

$$X_L = 2\pi fL = 2 \times 3.142 \times 60 \times 0.010 = 3.77 \, \Omega = X_L$$

Since this is a purely inductive circuit, R = 0 ohms. So, Z = X_L. Now that inductive reactance is known, Ohm's Law can be used to calculate reactive current.

$$I_{at\ 60\ Hz} = E \div Z = 10 \text{ volts} \div 3.77 \text{ ohms} = 2.65 \text{ Amps}$$

For a higher frequency 120 Hz, 10 V circuit with a 10 mh inductor, as shown in the upper illustration, the inductive reactance would be:

$$X_L = 2\pi fL = 2 \times 3.142 \times 120 \times 0.010 = 7.54 \, \Omega = X_L$$

Again, for this purely inductive circuit, R = 0 ohms. So, Z = X_L.

$$I_{at\ 120\ Hz} = E \div Z = 10 \text{ volts} \div 7.54 \text{ ohms} = 1.33 \text{ Amps}$$

Comparison of the current at 60 Hz to that at 120 Hz verifies that frequency does affect current levels in an inductive circuit.

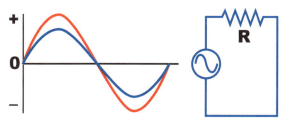
Purely Resistive Circuit

Phase Relationship between Current and Voltage in an Inductive Circuit

Current does not rise at the same time as the source voltage in an inductive circuit. Current is delayed depending on the amount of inductance. In a purely resistive circuit, current and voltage rise and fall at the same time. They are said to be "in phase." In this circuit there is no inductance. Resistance and impedance are equal.

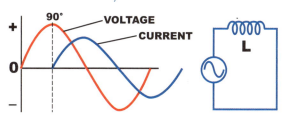
Purely Inductive Circuit

In a purely inductive circuit, current lags behind voltage by 90 degrees. Current and voltage are said to be "out of phase". In this circuit, impedance and inductive reactance are the same.

All inductive circuits have some amount of resistance. AC current will lag somewhere between a purely resistive circuit, and a purely inductive circuit. The exact amount of lag depends on the ratio of resistance and inductive reactance. The more resistive a circuit is, the closer it is to being in phase. The more inductive a circuit is, the more out of phase it is. In the illustration shown above, resistance and inductive reactance are equal. Current lags voltage by 45°.

Resistive-Inductive Circuit

Calculating Impedance in an Inductive Circuit

When working with a circuit containing elements of inductance, capacitance, and resistance, impedance must be calculated. Because electrical concepts deal with trigonometric functions, this is not a simple matter of subtraction and addition. The following formula is used to calculate impedance in a resistive-inductive circuit that does not contain capacitance:

$$Z = \sqrt{R^2 + X_L^2}$$

For the circuit illustrated above, where R = 10 Ω and X_L = 10 Ω:

$$Z = \sqrt{10^2 + 10^2} = \sqrt{200} = 14.142 \ \Omega$$

Now that total circuit impedance is known, a simple application of Ohm's Law can be used to find total circuit current.

Vectors

Another way to represent impedance is with a vector. A vector is a graphic representation of a quantity that has direction and magnitude. A vector on a map might indicate that one city is 50 miles southwest from another. The magnitude is 50 miles and the direction is southwest. Vectors are also used to show electrical relationships. As mentioned earlier, impedance (Z) is the total opposition to current

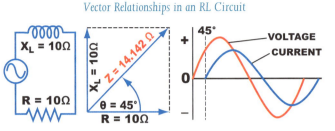

Vector Relationships in an RL Circuit

flow in an AC circuit containing reactance, inductance, and capacitance. The illustration above shows the vector relationship between resistance and inductive reactance in a circuit containing equal values of each. The angle between the vectors is the phase angle represented by the symbol θ. When inductive reactance is equal to resistance the resultant angle is 45 degrees. It is this angle that determines how much current will lag voltage.

Capacitive Reactance

Capacitance also opposes AC current flow. Capacitive reactance is designated by the symbol X_C and, like X_L, is measured in ohms. Capacitive reactance decreases as the size of the capacitor increases. The phase relationship between current and voltage in a capacitive circuit are opposite to the phase relationships in an inductive circuit. In a purely capacitive circuit, current leads voltage by 90 degrees.

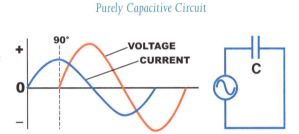

Purely Capacitive Circuit

Calculating Impedance in a Capacitive Circuit

Current flow in a capacitive AC circuit is also dependent on frequency. Where frequency is in Hz and capacitance is in farads, the following formula is used to calculate capacitive reactance.

$$X_C = \frac{1}{2\pi f C}$$

Capacitive reactance is equal to 1 divided by 2 times pi, times the frequency, times the capacitance. The capacitive reactance for a 60 hertz circuit with a 10 microfarad capacitor is:

$$X_C = \frac{1}{2\pi f C} = \frac{1}{(2) \times (3.142) \times (60) \times (0.000010)} = 265.2 \, \Omega$$

Once capacitive reactance is known, Ohm's Law can be used to calculate reactive current. The following calculation is for a circuit with a 10 volt source.

For this purely inductive circuit, R = 0 ohms. So, Z = X_C.

I = E ÷ Z = 10 volts ÷ 265.2 ohms = 0.0377 Amps = 37.7 mA

AC Theory I

Phase Relationship between Current and Voltage in a Capacitive Circuit

All capacitive circuits have some amount of resistance. AC current will lead somewhere between a purely resistive circuit and a purely capacitive circuit. The exact amount of lead depends on the ratio of resistance and capacitive reactance. The more resistive a circuit is, the closer it is to being in phase. The more capacitive a circuit is, the more out of phase it is. In the following illustration, resistance

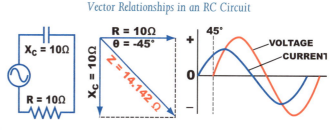

Vector Relationships in an RC Circuit

and capacitive reactance are equal. Current leads voltage by 45 degrees. The vector illustration above shows the relationship between resistance and capacitive reactance of a circuit containing equal values of each. The angle between the vectors is the phase angle represented by the symbol θ. When capacitive reactance is equal to resistance the resultant angle is -45 degrees. It is this angle that determines how much current will lead voltage.

Calculating Impedance in a Capacitive Circuit

The following formula is used to calculate impedance in a capacitive circuit:

$$Z = \sqrt{R^2 + X_C^2}$$

In the circuit illustrated above, resistance and capacitive reactance are each 10 ohms. Impedance is 1 4.1421 ohms.

$$Z = \sqrt{10^2 + 10^2} = \sqrt{200} = 14.142 \, \Omega$$

AC Theory I   213-329

Objective 213.1 Worksheet

_____ 1. In a circuit with 10 ohms resistance and 15 ohms inductive reactance, the impedance is ___ Ω.

      a. 25   b. 18.0277   c. 150   d. 5

_____ 2. The presence of capacitive reactance in a circuit affects current flow. What is the designated symbol for capacitive reactance?

      a. X_C b. X_L c. Z   d. C   e. L

_____ 3. In a purely capacitive circuit, current is said to ___ the applied source voltage.

      a. be in phase with   b. Lead   c. lag

_____ 4. When a resistive AC circuit also contains inductance and/or resistance, the total opposition to current flow is called ___.

      a. vectors
      b. reactance
      c. frequency
      d. impedance
      e. resistance

_____ 5. In a purely resistive circuit, current is said to ___ the applied source voltage.

      a. be in phase with   b. lag   c. lead

_____ 6. In an AC circuit containing resistance *plus* inductance, or resistance *plus* capacitance, or resistance *plus* inductance and capacitance, the opposition to current flow is called ___.

      a. reactance
      b. Impedance
      c. resistance
      d. vectors
      e. frequency

_____ 7. In a circuit with 25 ohms resistance and 50 ohms capacitive reactance, the impedance is ___ Ω.

    a. 2   b. 55.9017   c. 1250   d. 75

_____ 8. In either a purely inductive or purely capacitive circuit, the most the phase angle between the current and the voltage can be out of phase with each other is ___.

    a. 270°   b. 90°   c. 180°   d. 0°

_____ 9. In a 60 Hz circuit, containing a 12 µf capacitor, the capacitive reactance is ___ ohms. (Remember to change microfarads to farads.)

    a. 221.04   b. .2   c. .00022   d. 720

_____ 10. In a 50 Hz circuit, containing a 25 µf capacitor, the capacitive reactance is ___ ohms. (Remember to change microfarads to farads.)

    a. 1250   b. .01273   c. 2   d. 127.32

_____ 11. Inductive reactance is measured in ___.

    a. hertz   b. ohms   c. amperes   d. phasors   e. vectors

_____ 12. In a purely inductive circuit, current is said to ___ the applied source voltage.

    a. lag   b. lead   c. be in phase with

_____ 13. In a 100 Hz circuit, containing a 15 mh inductor, the inductive reactance is ___ ohms. (Remember to change millihenrys to henrys.)

    a. 1500   b. 9424.77   c. 9.42477   d. .15

_____ 14. In an AC circuit containing only inductance and/or capacitance, the opposition to current flow is called ___.

    a. impedance   b. vectors   c. frequency   d. reactance   e. resistance

15. In a 60 Hz circuit, containing a 1.3 mh inductor, the inductive reactance is ___ ohms. (Remember to change millihenrys to henrys.)

    a. 78 b. 490.088 c. .0217 d. .49

16. The presence of inductive reactance in a circuit affects current flow. What is the designated symbol for inductive reactance?

    a. X_C b. C c. Z d. L e. X_L

17. Impedance is represented by the letter ___.

    a. L b. C c. R d. Z

Lesson 213 NEC Worksheet

_____ 1. All branch circuits serving patient care areas shall be provided with an effective path for ground-fault current by installation in ___.

      I. nonmetallic cable   II. metal raceway   III. metallic armored cable

    a. II only
    b. I or II only
    c. I only
    d. I or II or III
    e. III only
    f. II or III only

_____ 2. In patient care areas, all receptacles and all non-current carrying conductive surfaces of fixed electric equipment likely to become energized and subject to personal contact, operating over 100 volts, shall be connected to ___.

    a. an insulated copper equipment grounding conductor
    b. a bare or insulated copper equipment grounding conductor
    c. a continuous ground path established by the circuit raceway
    d. a continuous bonding strap included within the armored cable
    e. a gray insulated conductor included within the armored cable

_____ 3. Luminaires more than 2.3 m (7 1/2 ft) above the floor and switches located outside of the patient vicinity ___.

    a. shall be grounded by a an insulated copper conductor
    b. are not required to be grounded by an insulated equipment grounding conductor
    c. are required to be non-conductive
    d. are require to be GFCI protected
    e. shall not be required to be grounded

_____ 4. In patient care areas, equipment grounding terminal buses of normal and essential branch- circuit panelboards serving the same individual patient vicinity shall be connected together with an insulated continuous copper conductor not smaller than ___ AWG.

    a. 12   b. 10   c. 8   d. 6   e. 4

_____ 5. In patient care areas, ___ receptacles are permitted if these are identifiable after installation. Such receptacles are most often identified by an orange triangle on the face of the receptacle.

    a. insulated grounding
    b. uninsulated grounding
    c. dedicated
    d. single, 125 volt, 20 ampere
    e. duplex, 125 volt, 20 ampere

Lesson 213 Safety Worksheet

Employee emergency action plans

- Emergency action plans shall be in writing and shall cover the designated actions employers and employees must take to ensure employee safety from fire and other emergencies.

- Emergency procedures and emergency escape routes must be included. Any procedures employees, who are to remain to operate any critical equipment, must follow and procedures to account for all employees after an emergency evacuation has been completed must be included in the plan.

- Any rescue or medical duties (that are part of your job duties) must be outlined in the plan.

- A preferred means of reporting the fire or other emergency must be covered as well.

- You must also list the names and titles and contact information of persons who can be contacted should additional information or explanation be necessary.

- An alarm system must comply with OSHA CFR1926.159 (see OSHA website for information).

- Before initiating the plan, the employer shall designate and train employees to assist in the safe and orderly emergency evacuation of employees.

- The employer shall review the plan with each employee at the particular site, when the plan is developed, whenever the employee's responsibilities change and whenever the plan is changed or modified.

- The employer shall review the plan with employees upon initial assignment. Employees must know how to protect themselves in the event there is an emergency. The plan shall be kept at the worksite
- and made available for review by any employees.

- Quite often the GC (General Contractor) will have an emergency action plan in effect for the jobsite. You may be able to get a copy of his plan and modify it for use by your company. Then be sure that you have covered all of the required elements outlined above.

Questions

_____ 1. The employer shall ___ an emergency action plan with employees upon their initial assignment.

    a. discuss   b. show   c. review   d. draw

_____ 2. Employees must know how to ___ themselves in the event there is an emergency.

    a. warn   b. evacuate   c. protect   d. all of these

_____ 3. Before initiating an emergency action plan, the employer shall designate and train ___ to assist in the safe and orderly evacuation of employees.

    a. employees   b. workers   c. everyone   d. people

_____ 4. An emergency action plan should include a list of the ___ and contact information of persons who can be contacted should additional information or explanation be necessary.

    a. names and addresses   b. names and titles   c. names and numbers

_____ 5. Emergency action plans must be ___ and shall cover the designated actions employers and employees must take to ensure employee safety from fire and other emergencies.

    a. in the trailer   b. in writing   c. in the office   d. in the file

_____ 6. Quite often the ___ contractor will have an emergency action plan in effect for the jobsite.

    a. masonry   b. HVAC   c. sheet metal   d. general

_____ 7. Emergency procedures and emergency ___ routes must be included in an emergency action plan.

    a. safety   b. exit   c. escape   d. entrance

Lesson 214
AC Theory II

Electrical Curriculum

Year Two
Student Manual

Purpose

This lesson will continue to discuss the effects Inductive Reactance and Capacitive Reactance have on series and parallel AC circuits. True Power and Apparent Power will be discussed and Power Factor will be calculated.

Homework

(Due at the beginning of this class)

For this lesson, you should:

- Thoroughly read the material contained within Lesson 214.
- Complete Objective 214.1 Worksheet.
- Complete Objective 214.2 Worksheet.
- Complete Lesson 214 NEC Worksheet.
- Read and complete Lesson 214 Safety Worksheet.

- Complete additional worksheets, if available, as directed by your instructor.

Objectives

By the end of this lesson, you should:

214.1 ....Understand the leading, lagging, and corrective effects, between the voltage and current relationship, when inductive reactance and capacitive reactance are introduced to series and parallel AC circuits.

214.2 ....Understand true versus apparent power, and how power factor affects an AC circuit.

214.1 Series and Parallel RLC Circuits

Series RLC Circuits

Circuits often contain elements of resistance, inductance, and capacitance. In an inductive AC circuit, current *lags* voltage by 90 degrees. In a capacitive AC circuit, current *leads* voltage by 90 degrees. It can be seen that inductance and capacitance are 180 degrees apart. Since they are 180 degrees apart, one element will cancel out all or part of the other element. If the inductive and capacitive components of the circuit are equal, they cancel each other totally. A vector diagram of a series AC circuit that has elements of inductive reactance, capacitive reactance, and resistance is shown at the right.

Vectors for a Series RLC Circuit

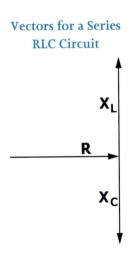

An AC circuit is:

- Resistive if X_L and X_C are equal
- Inductive if X_L is greater than X_C
- Capacitive if X_C is greater than X_L

Calculating Total Impedance in a Series RLC Circuit

The following formula is used to calculate total impedance of a circuit containing resistance, capacitance, and inductance:

$$Z = \sqrt{R^2 + (X_L - X_C)^2}$$

In the case where inductive reactance is greater than capacitive reactance, subtracting XC from XL results in a positive number. The positive phase angle θ (theta) is an indicator that the net circuit reactance is inductive, and current lags voltage.

In the case where capacitive reactance is greater than inductive reactance, subtracting XC from XL results in a negative number. The negative phase angle θ is an indicator that the net circuit reactance is capacitive and current leads voltage. In either case, the value squared will result in positive number.

Calculating Reactance and Impedance in a Series RLC Circuit

The circuit shown at the right has an applied (source) voltage of 120 volts at 60 hertz. The resistance is 1000 Ω, inductance is 5 mh, and capacitance is 2 µF. To calculate total impedance, first calculate the values of XL and XC, then impedance can be calculated.

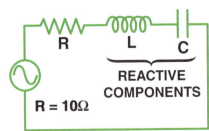

$X_L = 2\pi fL = 2 \times 3.142 \times 60 \times 0.005 = 1.885\ \Omega = X_L$

AC Theory II

$$X_C = \frac{1}{2\pi f C} = \frac{1}{(2) \times (3.142) \times (60) \times (0.000002)} = 1326 \, \Omega$$

Then: $Z = \sqrt{(1000)^2 + (1.885 - 1326)^2} = \sqrt{(1000)^2 + (-1324.115)^2}$

$$Z = \sqrt{1,000,000 + 1,753,281} = \sqrt{(2,753,281} = 1659.3 \, \Omega$$

Calculating Circuit Current in a Series RLC Circuit

Ohm's Law can now be applied to calculate total circuit current.

$$I = E_a \div Z = 120 \div 1659.3 = .0723 \text{ Amps}$$

Parallel RLC Circuits

Calculating Impedance in a Parallel RLC Circuit

Total impedance (Z_T) can be calculated in a parallel RLC circuit if values of resistance and reactance are known. One method of calculating impedance involves first calculating total current, then using the following formula:

$$Z_T = E_a \div I_T$$

Total current is the vector sum of current flowing through the resistance plus, the difference between inductive current and capacitive current. To determine total current in a parallel RLC circuit the following formula is used:

$$I_T = \sqrt{I_R^2 + (I_L - I_C)^2}$$

The circuit shown at the right has an applied (source) voltage of 120 volts at 60 hertz. The capacitive reactance has been calculated to be 25 Ω and inductive reactance 50 Ω. Resistance is 1000 Ω. A simple application of Ohm's Law will find the branch currents. Remember, voltage is constant throughout a parallel circuit.

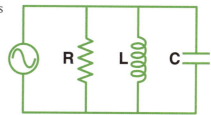

$I_R = E_a \div R = 120 \div 1000 = 0.12$ Amps

$I_L = E_a \div X_L = 120 \div 50 = 2.4$ Amps

$I_C = E_a \div X_C = 120 \div 25 = 4.8$ Amps

Once the branch currents are known, total current can be calculated.

$$I_T = \sqrt{I_R^2 + (I_L - I_C)^2} = \sqrt{(.12)^2 + (2.4 - 4.8)^2} = \sqrt{5.7744} = 2.403 \text{ Amps}$$

Impedance is now found with an application of Ohm's Law.

$$Z_T = E_a \div I_T = 120 \div 2.403 = 49.9 \text{ Ohms}$$

214.2 Power and Power Factor in AC Circuits

Power consumed by a resistor is dissipated in heat and not returned to the source. This is true power. True power is the rate at which energy is used.

Current in an AC circuit rises to peak values and diminishes to zero many times a second. The energy stored in the magnetic field of an inductor, or plates of a capacitor, is returned to the source when current changes direction.

Although reactive components do not consume energy, they do increase the amount of energy that must be generated to do the same amount of work. The rate at which this non-working energy must be generated is called reactive power. The symbol used for total reactive power is P_X. The reactive power due to the presence of capacitance is represented by P_{XC}. The reactive power due to the presence of inductance is represented by P_{XL}. Reactive power is measured in volt-amps reactive (VAR).

Power in an AC circuit is the vector sum of true power and reactive power. This is called apparent power. True power is equal to apparent power in a purely resistive circuit because voltage and current are in phase. Voltage and current are also in phase in a circuit containing equal values of inductive reactance and capacitive reactance. If voltage and current are 90 degrees out of phase, as would be in a purely capacitive or purely inductive circuit, the average value of true power is equal to zero. There are high positive and negative peak values of power, but when added together the result is zero.

True Power and Apparent Power Formulas

The formula for apparent power is:

$P_A = I_T \times E_a$ Apparent power is measured in volt-amps (VA).

True power is calculated from another trigonometric function, the cosine of the phase angle (Cos θ). The formula for true power is:

$P_T = I_T \times E_a \times \text{Cos } \theta$ True power is measured in watts (W).

AC Theory II

In a purely resistive circuit, current and voltage are in phase. There is a zero degree angle displacement between current and voltage. The cosine of zero is one. Multiplying a value by one does not change the value. In a purely resistive circuit the cosine of the angle is ignored.

In a purely reactive circuit (either inductive or capacitive) current and voltage are 90 degrees out of phase. The cosine of 90 degrees is zero. Multiplying a value times zero results in a zero product. No power is consumed in a purely reactive circuit.

Power Factor

Power factor is the ratio of true power to apparent power in an AC circuit. Power factor can be determined using either of the following formulas:

$$PF = \frac{\text{True Power}}{\text{Apparent Power}} = \frac{P_T}{P_A} \quad \text{or} \quad PF = \frac{I_T \times E_a \times \cos\theta}{I_T \times E_a} = \cos\theta$$

If PF is known and θ is to be determined, use this formula: $\theta = \cos^{-1} P$

For example, if the power factor is .82, enter .82 into your calculator and then press the **Cos-1** button. The result should be 34.9°.

In a purely resistive circuit, where current and voltage are in phase, there is no angle of displacement between current and voltage. The cosine of a zero degree angle is one. The power factor is 1, or 100%. This means that all energy delivered by the source is consumed by the circuit and dissipated in the form of heat.

In a purely reactive circuit, voltage and current are 90 degrees apart. The cosine of a 90 degree angle is zero. The power factor is 0, or 0%. <u>This means the circuit returns all energy it receives from the source to the source.</u>

In a circuit where reactance and resistance are equal, voltage and current are displaced by 45 degrees. The cosine of a 45 degree angle is .7071. The power factor is .7071, or 70.71%. This means the circuit uses approximately 70% of the energy supplied by the source and returns approximately 30%.

Summary and Comparison of True Power to Apparent Power

Watts (W) is the unit of measure for **True Power** (PT) and volt-amps (VA) is the unit of measure for **Apparent Power** (PA). So, what is the difference between the two and why do we care?

Let's start with the difference between true and apparent power. Some loads are rated in W or KW while others are rated in VA or KVA. Any load will have some resistance, some inductance (L), and some capacitance (C). In a DC circuit inductance and capacitance have no effect on the current. However in AC circuits all three of these oppose the flow of current. Some loads (those you've studied so far) are almost purely resistive and we don't really concern ourselves with the inductance or capacitance. When inductance and/or capacitance are present in a circuit the opposition to current flow is referred to as inductive *reactance* (XL) and capacitive *reactance* (XC). Just as with resistance (R) these are measured in ohms. The effect on current caused by the presence of *reactance* in a circuit is due to the fact that XL and XC don't like changes in voltage and oppose the flow of current in the circuit. Since in a DC circuit voltage doesn't change DC current is not affected by reactance because DC is constant.

AC is another matter entirely. AC voltage is a sine wave as shown at the right. This is called a sine wave because of its sinusoidal shape. AC voltage is constantly changing. It is either increasing toward its + peak value or decreasing toward its − peak value. An AC electrical cycle is measured in degrees where 1 cycle equals 360°. The frequency of the power source determines the number of these cycles completed in a second of time. 60 cycles-per-second AC is used in the United States. Cycles per second is commonly abbreviated as hertz or hz.

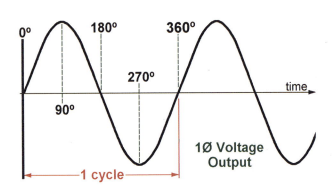

In a purely resistive circuit Ea and I are in phase with each other – that is, at the same time voltage starts to increase toward its + peak the current immediately starts to increase toward its + peak. Both Ea and I reach their peaks at the same time. When inductance is present in an AC circuit it causes a phase shift by slowing down the current. On the other hand, when capacitance is present it speeds up the current. When either or both inductance and capacitance are present in an AC circuit these cause the current to be *out-of-phase* (or phase shifted) with the **applied voltage** (Ea). Inductance and capacitance directly oppose each other. This is why capacitors are used to "correct" the power factor in circuits that supply power to induction motors. Power factor will be addressed shortly.

Another term you should be aware of is **Reactive Power** or P_X. **Reactive power** (P_X) is simply the product of current times the *reactance*. $P_{XC} = I \times X_C$ and $P_{XL} = I \times X_L$. **Reactive power** is measured in volt-amps-reactive or VAR's.

- An incandescent lamp with a 200 W rating is an example of a "resistive load".
- A capacitor bank is an example of a load that is almost all capacitance with very, very little resistance or inductance. Resistance and inductance are such small quantities that we don't have to consider them. Capacitor banks are rated in kilovolt-amperes-reactive or KVAR's.
- An induction motor is a type of load that, although does have a tiny resistive and capacitive component, is considered to be (for all practical purposes) an inductive load.
- Any circuit will have some resistance, some capacitance, and some inductance.

Summary: in an AC circuit the presence of these components (resistance, capacitance, and inductance) oppose the flow of current in the circuit. That is:

resistance + inductive reactance + capacitive reactance = total opposition to current flow

AC Theory II

Now we could call this TOTCF (total opposition to current flow) but an English engineer, Oliver Heaviside, came up with another name. He decided that this should be called **impedance** and he abbreviated it with a **Z**. Why didn't he use I? Because I was already used to indicate the Intensity of electron flow (current). We are stuck with using **Z** for **impedance** and since it is the opposition to current flow it is measured in ohms. Now R, X_L, and X_C don't simply add together because each affects current at a different point in time. Because they occur at different times on the sine wave time line these, as well as the associated voltages (E_R, E_{XL}, & E_{XC}), and powers (P_T, P_{XL}, & P_{XC}) actually add vectorially which is point-by-point sine wave addition. Electricians are generally concerned only with determining line current in order to size conductors and OCPD's and in fact are not routinely given enough nameplate information to determine the various voltages and components of impedance. Complete analysis of circuits using sophisticated equipment is required to determine these factors. Using this detailed information, a power quality analysis is made, and the installation of corrective equipment is assigned to the electricians.

To best illustrate the difference between P_T and P_A let's consider a motor load. The nameplate, on this 1Ø 10 hp motor, shows that it will pull 40 amps at 240 volts. Horsepower is the amount of actual work (output power) delivered by a motor. This can be converted to "watts" if we remember that 1 hp = 746 watts.

$$P_T = 10 \text{ hp} \times 746 \text{ W/hp} = 7460 \text{ watts} = 7.46 \text{ KW} = \text{true power}$$

If it's measured in watts — then it is True Power or Output Power

Remember though that for this motor to deliver 10 hp it requires 40 amps of current at 240 volts. E_a is the *applied voltage* of the circuit. This input power required is called the apparent power or P_A. For this motor:

$$P_A = E_a \times I = 240 \text{ volts} \times 40 \text{ amps} = 9600 \text{ VA} = 9.6 \text{ KVA} = \text{apparent power}$$

If it's measured in volt-amps — then it is Apparent Power or Input Power

Now, as you can see, we have to "put in" 9.6 KVA of power to "get out" 7.46 KW of power (work) from this motor. This is not a good thing because we will have to pay for the power "put in". We will address this when we get to the "why do we care" part. The amount of power "put in" to a piece of equipment (or apparatus) is called INPUT POWER. The amount of power delivered (horsepower for example) is called OUTPUT POWER. The difference is due to two things. The first is efficiency and the second is called **power factor**.

For now we'll look only at **power factor** (or PF). When pf = 1.0 or 100% we can be very happy because this is perfection. Simply stated:

pf = $P_T \div P_A$ = watts ÷ volt-amps

Power factor can never be greater than 1.0 because true power can never be larger than apparent power. Just remember, you can't get more out than what you put in. When watts equals volt-amps then pf will be 1.0 or 100%. For example, if a 2400 watt heating element (remember this is a purely resistive load) is putting out

2400 watts it will pull 10 amps of current at 240 volts. So, here in this example:
input power = 240 volts × 10 amps = 2400 VA = P_A
& the output power is the nameplate rating (2400 W) = P_T

then: **pf = P_T ÷ P_A** = 2400 W ÷ 2400 VA = 1.0 = 100%

Let's go back to our 10 hp motor (ignoring the efficiency for now – assume efficiency is 100%) and put a number to power factor for the motor. Remember our P_A or input power was 9600 VA and our P_T or output power was 7460 W.

then: **pf = P_T ÷ P_A** = 7460 W ÷ 9600 VA = .7771 = 77.71%

77.71% is a horrible power factor! As the difference between true and apparent power increases the power factor gets smaller. A small power factor is bad because we're going to have to pay for the apparent (input) power used and not the actual work (true or output power) we get out of it. Why is there such a difference? Almost all motors are induction motors and are considered to be inductive loads (negligible resistance and capacitance). Remember impedance (Z)? This Z is what opposes current flow in our circuit. If we go back to our Ohm's Law formula: I = E/R we remember that R was our opposition to current flow in the purely resistive circuits we studied before. Now that we realize that many loads have not only resistance but also capacitance and inductance we need to "plug" Z into the Ohm's Law formula:

$I = E_a/Z$

You won't see Z or Ω listed on the nameplate of a piece of equipment. Have you yet to see R on the nameplate of any machinery you have connected thus far? You simply need to be aware that each and every load we connect has impedance that opposes current flow. Even our lowly little "light bulb" has impedance. It's just that with a purely resistive load Z = R because there is no inductance or capacitance. We don't need to calculate how much of Z is due to resistance or how much is due to inductance or how much is due to capacitance. FROM NOW ON we can "plug" Z into any Ohm's Law or Power formula. The only time we will need to find Z is if we're going to operate a load at other than its nameplate rated voltage. Sound familiar?

One of our objectives as electricians is to size wire to carry the current that a load will pull. If we have a piece of equipment that has a nameplate rating of 18 KVA @ 240 volts and we are going to operate it on a 240 volt circuit then we can find I using:

I = P_A/E_a = 18000 ÷ 240 = 75 amps

We will install circuit conductors that can carry 75 amps.

If you have to run a circuit to a piece of equipment with a nameplate that gives you:

pf = .8   14.4 KW   240 volts   (not uncommon for foreign made equipment)

Then you will use the formula:

I = (P_T) ÷ (pf × E_a) = (14400) ÷ (.8 × 240) = 75 amps

Again you will install wire that can carry 75 amps.

AC Theory II

Why Do We Care?

Let's proceed with the "why do we care" part of the question because we all have to pay "the electric bill" whether we pay it ourselves directly or it's included in the price of our rent or the price of the goods and services we receive or enjoy.

A residential customer's electric bill is based on the number of kilowatt-hours (KW-hr) used during a billing cycle multiplied times the rate charged per KW-hr. A residential electric meter measures KW-hrs and not KVA-hrs. A residential customer could actually be using 3000 KVA-hrs of electricity but if the pf is such that only 1000 KW-hrs are shown on the meter then the residential customer pays only for the true power used and not the apparent power. Are the utilities stupid? No, they simply realize that the big differences are due to fluorescent and HID lighting as well as motor loads in commercial and industrial installations and that residential customers just don't have enough of these inductive loads to mess with.

Let's use our 10 hp motor again to illustrate the cost for a residential versus a commercial customer to operate the motor. In both cases we will operate this motor @ 240 volts for 100 hours in a monthly billing cycle.

For the residential customer the billing rate is $0.15 per KW-hr so we will need to find true power before we begin:

$$P_T = P_A \times pf = 9.6 \text{ KVA} \times .7771 = 7.46 \text{ KW}$$

Then: $\dfrac{\$0.15}{\text{KW hr}} \times (7.46 \text{ KW}) \times (100 \text{ hrs}) = \$111.90$ for the residential customer.

For the commercial customer the billing rate is $0.15 per KVA-hr.

$$\dfrac{\$0.15}{\text{KVA hr}} \times (9.6 \text{ KVA}) \times (100 \text{ hrs}) = \$144.00$$

Think large scale now. What if this business had 100 of these motors? For 100 motors his cost for the month would be **$14,400** versus **$11,190** if these motors were billed on a residential meter that measured only KW (true power).

Power Factor Correction

There **IS** a way to make true power = apparent power (or watts = volt amps)! Capacitor banks are routinely connected to motor loads because the current opposition due to inductive motors can be cancelled by adding capacitance to the circuit. This is called **power factor** correction. Journeymen electricians aren't required to determine the size of the cap bank required to accomplish this so we won't go into it here. However, you need to realize the reason these banks are added and be able to calculate the current that a capacitor bank will draw. The nameplate on a cap bank will be in KVAR's or KVA-Reactive. We're going to use this nameplate value and plug it into our power formula (replacing P) to find the current that the bank will draw:

$$I_{cap} = \text{KVAR} \div E_a$$

For example, if you had to install a 25 KVAR, 240-volt capacitor bank you could calculate the current that the bank would draw:

$$I_{cap} = 25 \text{ KVAR} \div 240 \text{ V} = 104.17 \text{ amps}$$

You would then install wire that could carry 104 amps. There are some special Code rules for sizing the wire based on this 104 amps but this will be addressed later during your study of motors.

The Bottom Line or Follow the Money

Using power factor correction equipment the power factor can be kept very close to 100% or unity. This enables the electrician to install smaller wire, pipe, and OCPDs because the line current is smaller than it would be without the correction equipment. Additionally, the customer's electrical bill is smaller than it would be without the equipment. Even the utility provider is happy because he can install smaller transformers and wire for the customer's KVA needs. **Everybody wins!**

AC Theory II

Objective 214.1 Worksheet

_____ 1. In a series circuit where inductive reactance is less than capacitive reactance, resulting in a negative phase angle, current will ___ voltage.

   a. lag   b. lead   c. be in phase with

_____ 2. In a series circuit where the inductive and capacitive loads are identical, what interaction do they have on each other?

   a. They will mix like oil and water.
   b. They will cancel each other out.
   c. They will add together
   d. none of these

_____ 3. The impedance of a series circuit is ___ ohms. The circuit has a resistance of 25 ohms, an inductive reactance of 14 ohms and a capacitive reactance of 10 ohms.

   a. 7   b. 29   c. 27   d. 25.3   e. 49

_____ 4. In a series circuit where inductive reactance is greater than capacitive reactance, resulting in a positive phase angle, current will ___ voltage.

   a. be in phase with   b. lag   c. lead

_____ 5. The total current in a series circuit is ___ Amps. The circuit is supplied from a 2-pole breaker in a 120/208 V 3-ph panel. The circuit has a resistance of 25 ohms, an inductive reactance of 14 ohms and a capacitive reactance of 10 ohms.

    a. 10.22   b. 25.32   c. 8.2   d. 9.48   e. 14.26

_____ 6. The total current in a parallel circuit is ___ Amps. The circuit is supplied from a 2-pole breaker in a 120/208 V 3-ph panel. The circuit has a resistance of 25 ohms, an inductive reactance of 14 ohms and a capacitive reactance of 10 ohms.

    a. 10.22   b. 8.2   c. 9.48   d. 14.26   e. 25.32

AC Theory II

Objective 214.2 Worksheet

_____ 1. Total opposition to current flow in an AC circuit is referred to as ___.

    a. Impedance   b. Reactance   c. Resistance   d. none of these

_____ 2. The phase angle of a circuit is 28 degrees. The power factor is ___%.

    a. 28   b. 46.9   c. 72   d. 88.3

_____ 3. Capacitor banks are rated in ___.

    a. KW   b. KVA   c. KVAR

_____ 4. Normally AC power in the United States operates at ___ cps (cycles per second).

    a. 50   b. 60   c. 120   d. 100

_____ 5. Apparent power is measured in ___.

    a. Watts
    b. Volt-Amps
    c. Volt-Amps Reactive
    d. Ohms
    e. Hertz

_____ 6. Apparent power can be ___ true power.
    I. more than   II. equal to   III. less than

    a. II only   b. I or II only   c. I only   d. I or II or III   e. II or III only

_____ 7. Power is measured in ___.
    I. watts   II. volt-amperes

    a. I only   b. neither I nor II   c. either I or II   d. II only

8. Which of the following are measured in ohms?
    I. inductive reactance (X_L)
    II. impedance (Z)
    III. capacitive reactance (X_C)
    IV. resistance (R)

    a. IV only   b. I & III only   c. II & IV only   d. I, II, III, & IV   e. II & III only

9. Ideally the power factor (pf) should be kept as close to ___ as possible.

    a. .5   b. 0   c. 1

10. To an electrician the drawing at the right represents ___.

    a. a sine wave
    b. ½ of a figure 8 race track
    c. half of an infinite (∞) value
    d. a roller coaster

11. ___ power is power that passes through reactive components, which do not consume energy, and returns to the source.

    a. Reactive   b. True   c. Apparent

12. Poor power factor could occur as the result of the presence of ___.
    I. HID lighting
    II. 120 volt incandescent lighting
    III. refrigeration equipment using induction motors

    a. I & II only
    b. I only
    c. I & III only
    d. III only
    e. I, II, & III

13. True power is measured in ___.

    a. Volt-Amps Reactive   b. Hertz   c. Ohms   d. Volt-Amps   e. Watts

AC Theory II

_____ 14. If a circuit breaker protecting a 60 Hz circuit can operate on a short-circuit in one-half of an electrical cycle this is ___ mS.

    a. 33.33   b. 50   c. 8.33   d. 16.67   e. none of these

_____ 15. The power factor for a load is ___ where true power = 7.5 kW & apparent power = 9800 VA.

    a. .841   b. .644   c. .765   d. 1.31   e. .631

_____ 16. Where operating in the United States, AC voltage reaches a positive peak ___ times during 1 second.

    a. 100   b. 120   c. 30   d. 50   e. none of these

_____ 17. If the power factor of a load is 0.73, the phase angle, θ, is ___ degrees.

    a. 43.1   b. 95.6   c. 73   d. 46.9   e. 29.2

_____ 18. ___ power is power consumed by a resistor, dissipated as heat, and not returned to the generation source.

    a. Reactive   b. Apparent   c. True

_____ 19. Reactive power is measured in ___.

    a. Volt-Amps
    b. Watts
    c. Hertz
    d. Ohms
    e. Volt-Amps Reactive

_____ 20. ___ power is the power required to supply both resistive and reactive components of a load or loads.

    a. Reactive   b. Apparent   c. True

_____ 21. AC circuits are in phase only when they are connected to purely resistive circuits.

    a. False   b. True

_____ 22. Which of these power factors is the best?

    a. 100%   b. 75%   c. 50%   d. 25%

_____ 23. One AC voltage cycle is completed in ___ degrees.

    a. 120   b. 30   c. 180   d. 60   e. none of these

AC Theory II

Lesson 214 NEC Worksheet

_____ 1. In patient care areas, patient bed location branch circuits from the normal system shall originate ___.

   a. from the same panelboard
   b. from adjacent panelboards
   c. from different panelboards
   d. from load centers having different locations
   e. through the same conduit or raceway

_____ 2. In patient care areas, each patient bed location shall be supplied by at least ___.

   a. one branch circuit
   b. two branch circuits
   c. three branch circuits
   d. one normal branch circuit and one emergency branch circuit
   e. two essential branch circuits

_____ 3. In general care areas, each patient bed location shall be provided with a minimum of ___ receptacles.

   a. one
   b. two
   c. three
   d. four
   e. six

_____ 4. In general care areas, each patient bed location shall be provided with a minimum number of receptacles and these shall be listed ___ and so identified.

   a. spec grade
   b. standard grade
   c. hospital grade
   d. isolated ground
   e. industrial grade

_____ 5. In general care areas, receptacles in pediatric patient care areas shall be listed ___.

   a. spec grade
   b. weatherproof and covered
   c. tamper resistant or employ a tamper resistant cover
   d. ungrounded
   e. GFCI protected

Year Two (Student Manual)

_____ 6. In ___ care areas, patient bed location emergency system branch circuit receptacles shall be identified and shall also indicate the panelboard and circuit number supplying them.

   a. critical
   b. pediatric
   c. emergency
   d. surgical
   e. outpatient

Lesson 214 Safety Worksheet

Walking-Working Surfaces (slips, trips and falls)

Many slips, trips and falls are the result of poor housekeeping practices. Any materials or trash left carelessly on the floor or stairs can cause slips, trips and falls.

Covers and/or guardrails shall be provided to protect personnel from the hazards of open pits, tanks, vats, ditches and the like.

Floor openings, floor holes and wall openings pose hazards for employees working in areas where they exist. These must be guarded or covered to prevent employees from falling into or through these openings.

Open sides of upper levels shall be guarded to protect employees from falling to the lower level. Stairs shall have stair rails and handrails along any open sides. Remember to follow all the steps of ladder safety, for both straight ladders and portable ladders. Do not use a ladder that is defective or broken or not set-up properly.

Generally be aware of your surroundings. Look down and ahead of where you will be walking. Watch for the unexpected person or piece of equipment to move suddenly without any warning. Don't let this cause you to be injured. **BE ALERT !!!!!   DON'T GET HURT !!!!!!**

Questions

_____ 1. Open sides of upper levels shall be ___ to protect ___ from falling to the lower level of the working or walking surface.

   a. grounded - workers
   b. guarded - employees
   c. walled - people

_____ 2. Do not use a ladder that is ___ or ___ or set-up improperly.

   a. defective - broken
   b. cracked - dirty
   c. painted - labeled

_____ 3. Look down and ___ of where you will be walking.

   a. ahead
   b. behind
   c. over
   d. none of these

_____ 4. Many ___ are the result of poor housekeeping practices.

   a. falls
   b. trips
   c. slips
   d. all of these

_____ 5. Floor openings, floor holes and wall openings shall be ___ to protect employees from falling into or through these openings.

   a. grounded
   b. blocked
   c. guarded
   d. fenced

_____ 6. Materials or trash left carelessly on the floor or stairs can cause ___.

    a. slips
    b. trips
    c. falls
    d. any of these

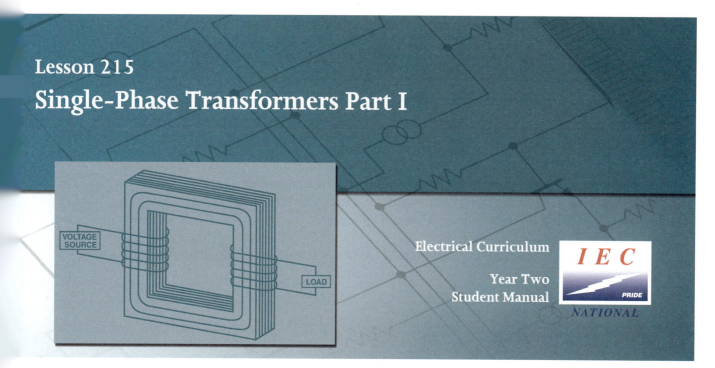

Lesson 215
Single-Phase Transformers Part I

Electrical Curriculum

Year Two
Student Manual

Purpose

This lesson will introduce the basic construction, theory, operation and types of single phase transformers. Primary and Secondary turns ratio will be discussed as applied to step-up and step-down transformer applications.

Homework

(Due at the beginning of this class)

For this lesson, you should:

- Thoroughly read the material contained within Lesson 215.
- Thoroughly read Yr-2 Annex A (Located at the back of this manual).
- Locate and use as a reference: Yr-2 Annex B (Located at the back of this manual).
- Complete Objective 215.1 Worksheet.
- Complete Objective 215.2 Worksheet.
- Complete Objective 215.3 Worksheet.
- Complete Objective 215.4 Worksheet.
- Complete Objective 215.5 Worksheet.
- Complete Lesson 215 NEC Worksheet
- Read and complete Lesson 215 Safety Worksheet.

- Complete additional worksheets, if available, as directed by your instructor.

Objectives

By the end of this lesson, you should:

215.1 ....Understand the terms used for transformers and the basic theory of operation.
215.2 ....Understand the principles of operation for transformers.
215.3 ....Know the various types of transformers and where they are used.
215.4 ....Know typical features, general information, and special considerations for dry-type transformers.
215.5 ....Understand the connection of transformers with single-voltage primaries that have either single- or dual-voltage secondaries.

215.1 Basic Operation

While studying transformers, you should frequently refer to the *Glossary of Transformer Terms* to make certain that you understand the material. This glossary is located in Yr-2 Annex A at the back of this manual.

What is a Transformer?

A transformer is a device that transfers electrical energy from one electrical circuit to another, without changing the frequency. This is accomplished using the principles of electromagnetic induction. Electromagnetic induction is the voltage provided by the lines of magnetic flux cutting a conductor. Usually, the energy transfer takes place with a change of voltage from the input to the output. It either increases (steps up) or decreases (steps down) AC voltage.

A transformer does not generate electrical power. It transfers electrical power from one AC circuit to another through magnetic coupling. This method is when one circuit is linked to another circuit by a common magnetic field. Magnetic coupling is used to transfer electrical energy from one coil to another. The transformer core is used to provide a controlled path for the magnetic flux generated in the transformer by the current flowing through the windings (also called coils).

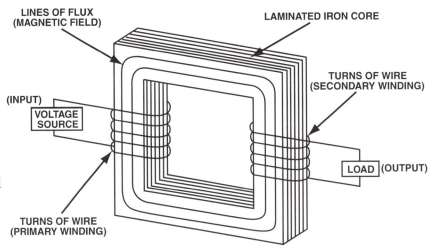

In order to understand the advantage and use of a transformer, let's first look at the basic transformer. There are four basic parts:

- Input connections
- Output connections
- Windings or coils
- Core

Input Connections

The input side is called the primary side of the transformer because this is where the main electrical power to be changed is connected.

Output Connections

The output side is called the secondary side of the transformer. This is where the electrical power is sent to the load. Depending upon the requirement of the load, the incoming electric voltage is either increased or decreased.

Windings

The simplest transformer has two windings. These two windings are called the primary winding and the secondary winding. Both of these windings are wound around the same iron core. The primary winding is connected to the supply circuit, and the secondary winding is connected to the load.

The primary and secondary windings of practically all transformers are subdivided into several coils. This is to reduce the creation of flux that does not link both primary and secondary. The transforming action can only exist when flux (mutual flux) couples both the primary and the secondary. Flux that does not do so is, in effect, leakage flux.

The windings are also subdivided to reduce the voltage per coil. This is important in high voltage transformers, in which insulation thicknesses make up a considerable part of the construction. In practice it is customary to subdivide a winding so that the voltage across each coil does not exceed about 5,000 volts.

Core

The transformer core is used to provide a controlled path for the magnetic flux generated in the transformer. The core is not a solid bar of steel, but is constructed of many layers (laminations) of thin sheet steel. Since heating creates power losses, cores are laminated to reduce these losses. Because the primary and secondary circuits are not electrically connected, the core serves to transfer (induce) the primary coil's voltage into the secondary coil. This occurs through the principle of electromagnetic (or magnetic) induction. There are two general types of cores: core type, and shell type. These are distinguished from each other by the manner in which the pri-

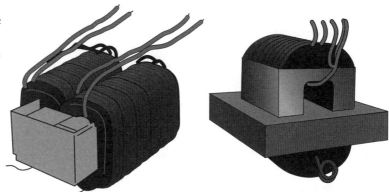

mary and secondary coils are placed around the laminated steel core. The core type is shown on the left and is generally used in the construction of distribution transformers. The shell type is shown on the right and is generally used in the construction of high voltage transformers. Note that each illustration shows a transformer with one primary winding (with two primary leads), and one secondary winding (with two secondary leads).

How a Transformer Works

Induced Voltage

When an input voltage is applied to the primary winding, alternating current starts to flow in the primary winding. As the current flows, a changing magnetic field is set up in the transformer core. As this magnetic field cuts across the secondary winding, alternating voltage is produced in the secondary winding. In short, a voltage is being *induced* on the secondary winding.

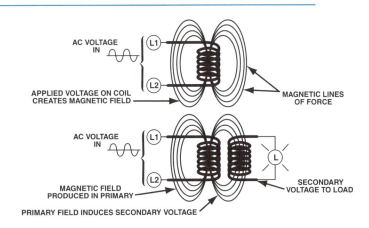

Turns Ratio

The ratio between the number of actual turns of wire in each coil is the key in determining the type of transformer and what the output voltage will be. The ratio between output voltage and input voltage is the

$$\frac{\text{Voltage (input)}}{\text{Voltage (output)}} = \frac{\text{Number of Primary Turns}}{\text{Number of Secondary Turns}}$$

same as the ratio of the number of turns between the two windings.

The relationship between the number of turns in the secondary and the number of turns in the primary is commonly called the turns ratio or voltage ratio. It is common practice to write the turns ratio with the primary (input) number first, followed by the secondary (output) number. The two numbers are often separated by a colon.

Consider this example:

Transformer Primary Voltage: 480 volts

Transformer Secondary Voltage: 120 volts

$$\frac{480 \text{ volts}}{120 \text{ volts}} = \frac{4 \text{ Primary Turns}}{1 \text{ Secondary Turn}} = \frac{4}{1}$$

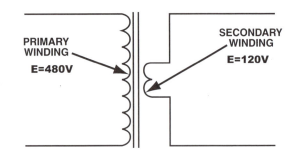

This transformer has four primary turns for every one secondary turn. Turns ratio is written as 4 to 1, or 4:1.

Step-Up vs. Step-Down

Step-Up Transformer:

The primary winding of a step-up transformer has fewer turns than the secondary winding, with the resultant secondary voltage being higher than the primary. Where a step-up transformer has a 1 to 2 turns ratio, the output voltage is double the input voltage. At first, this might seem like we are gaining or multiplying voltage without sacrificing anything. Of course, this is not the case. Ignoring small losses, the amount of power transferred in the transformer is equal on both the primary and secondary sides. Power is equal to Voltage multiplied by Current.

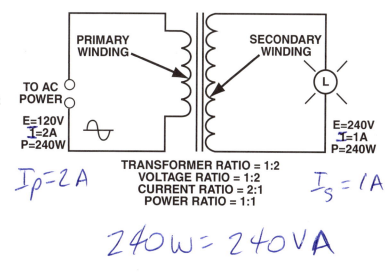

Handwritten annotations: $I_p = 2A$, $I_s = 1A$, 240W = 240VA

This is expressed by the formula:

P = E x I

Power is always equal on both sides of the transformer, meaning both sides of the equation must have the same value. This means we cannot change the voltage without changing the current also.

One big advantage of increasing the voltage and reducing the current is that power can be transmitted through smaller gauge wire. Think about how much wire is used by a utility company to get electricity to where it is used. For this reason, the generated voltages are stepped up very high for transmission across large distances. The voltage is then stepped back down to meet consumer needs.

Step-Down Transformer:

The primary winding of a step-down transformer has more turns than the secondary winding, so the secondary voltage is lower than the primary.

We can see that when voltage is *stepped down* from 240 to 120 V in a 2 to 1 ratio, the current is increased from 1 to 2 amps, keeping the power equal on each side of the transformer. In contrast, when the voltage is *stepped up* from 120 V to 240 V in a 1 to 2 ratio, the current is reduced from 2 to 1 amp to maintain the power balance.

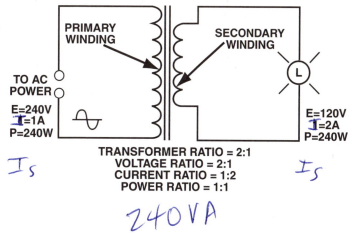

Handwritten annotations: I_s, I_s, 240VA

In other words, voltage and current may be changed for particular reasons, but power is merely transferred from one point to another.

Voltage Taps

As you know, the turns ratio determines the voltage transformation that takes place. There are times when the actual incoming voltage is different than the expected normal incoming voltage. When this happens, it could be advantageous to be able to change the turns ratio in order to get the desired (rated) output voltage. You might view this as fine tuning the input voltage to get the desired output. Voltage taps, designed into the transformer's primary winding, deliver this desired flexibility.

Suppose a transformer has a 4 to 1 turns ratio. Remember, this means the primary has 4 times as many turns as the secondary, which tells you the transformer is a step-down transformer. If the input voltage is 480 volts, the output would be 120 volts.

What if the input delivered to the transformer primary is less than the expected normal of 480 volts, say 456 volts for this example? Since the turns ratio is 4:1 the secondary voltage would be 114 volts. This could be significant if getting 120 volts from the secondary is critical.

Tapping the primary in a number of different spots helps to eliminate the problem by providing a means to adjust the turns ratio and fine-tune the secondary output voltage.

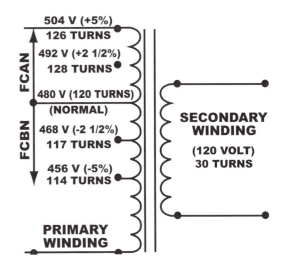

This transformer has taps at 2 1/2% and 5% below the normal voltage of 480 volts. In the industry, this would be referred to as having two 2 1/2% Full Capacity Below Normal Taps (FCBN). These two taps provide a 5% voltage range below the normal 480 volts. When taps are provided above the normal as illustrated, they are called Full Capacity Above Normal Taps (FCAN).

For standardization purposes, taps are in 2 1/2% or 5% steps. The tap arrangement used on many transformers is two @ 2 1/2% FCAN and four @ 2 1/2% FCBN, which provides a 15% total range of tap voltage adjustments. Refer to Connection Diagram 17, in Yr-2 Annex B, which provides this range of taps.

215.2 Principles of Operation

This is a good time to introduce several additional principles and common terms associated with transformers. This material will be especially helpful from a practical standpoint by helping to clarify concepts. In addition, a number of the terms are commonly used in the electrical industry. Being familiar with the nomenclature will simplify transformer discussions and selections at your work location. Remember the *Glossary of Transformer Terms* is located in Yr-2 Annex A at the back of this manual.

Single-Phase Transformers Part I

Transformer Ratings

Transformers are frequently rated in kilovolt-amperes (kVA or KVA), although there are other rating designations. Very large transformers are often rated in megavolt-amperes (MVA), and very small transformers can be rated in volt-amperes (VA). The rating enables you to calculate the maximum current that a transformer can deliver to a load without overheating. Knowing the ratings (power and voltage), the current can be calculated.

$$\text{Nameplate Current Rating (in A)} = \frac{\text{Nameplate Power Rating (in VA)}}{\text{Nameplate Voltage Rating (in V)}}$$

The nameplate current rating, or rated current, is the maximum current that can flow through the transformer windings without overheating the windings. The power rating of a transformer is the same for both the primary and the secondary.

Frequency

A transformer cannot change the frequency of the power supply. If the supply is 60 Hz, the output will also be 60 Hz. Because transformers are designed for operation at a particular frequency, frequency requirements must be known to make the proper transformer selection. The frequency of the source should be determined, and the frequency of the load should match. Transformers used in the United States and Canada are usually designed for 60 Hz. A great deal of the rest of the world uses 50 Hz.

Basic Impulse Level (BIL)

Outdoor electrical distribution systems are subject to lightning surges. Even if the lightning strikes the line some distance from the transformer, voltage surges can travel down the line and into the transformer. Other electrical equipment in the system can also cause voltage surges when they are opened and closed. These surges can be very damaging to transformers and other electrical equipment. The Basic Impulse Level (BIL) is a measure of the ability of the transformer's insulation system to withstand very high-voltage, short-time surges. The typical BIL level for 600 volt class transformers is 10 kV.

Sound

Although transformers are reliable static devices with no moving parts, they do produce a humming sound. The sound originates in the core. When the magnetic flux passes through the laminated core, the laminations expand and contract, generating a hum. Transformers are designed and constructed in such a manner that the noise is minimized, but not eliminated. The sound level of a transformer is measured in decibels (dB), and determined by tests conducted in accordance with NEMA standards.

Altitude

Air is thinner at higher altitudes, which impacts transformer cooling. Transformers are designed to operate with a normal temperature rise, at a specific height in feet above sea level. If the operation is to be at a higher altitude, the transformer's nameplate rating must be reduced. This reduction is called derating. The amount of the reduction depends on how much the standard altitude has been exceeded.

Efficiency

Although there are very small losses associated with transformers they are essentially very, very efficient machines. Efficiency is an indicator of how much you get out versus how much you put in.

Most of the energy provided to the primary of a transformer is transferred to the secondary. However, some energy is lost in the form of heat. These heat losses occur in the windings or the core. Losses and efficiency are very important concerns in the selection of a transformer. For example, a transformer with a lower initial cost may not be the best purchasing choice. Another transformer with a higher initial cost, but which is more efficient, could prove to be the best purchasing decision in the long run. Mathematically it can be expressed as:

Efficiency = Output Power / Input Power or
% Efficiency = [output power ÷ input power] × 100

The nameplate power rating on transformers is the amount of power it is rated to deliver (or output). As an illustration: if we had a transformer that was rated for 50 KVA but required an input of 51 KVA then this transformer's efficiency would be:

% Efficiency = [50,000 ÷ 51,000] × 100 = 98.04% efficient.

Put another way, the output power equals the input power, less the internal losses of the transformer.

Output Power = Input Power − Internal Losses

Transformer efficiency varies among manufacturers. Size (rating) and type also affect efficiency. Transformers with higher KVA (power) ratings are more efficient than smaller transformers. This is because the larger ones have larger cross-sectional area cores and use larger gauge wire for the windings. Both of these factors affect power losses. Typically, large transformers have an efficiency somewhere around 97%. A 20 MVA power transformer, for example, might have an efficiency of 99.4%, and a small 5 kVA transformer might be only 94% efficient. Compare this efficiency to a motor that might be only 50% efficient. You can see how efficiency is an important consideration when applying a transformer.

Since transformers are so efficient, we will generally ignore any losses in future studies and work with them as if they were 100% efficient (or $P_{output} = P_{input}$).

Copper Loss

One type of loss in transformers is copper loss. The copper windings, although a good conductor of electricity, are not perfect conductors. Copper has a certain resistance to current flow, as do all materials. One of the factors influencing copper loss is heat. Resistance increases with an increase in temperature. To minimize this problem, large electrical power distribution transformers are often cooled by circulation of water, forced air, or oil. Cooling also helps to prevent heat damage to winding insulation.

Eddy Currents

As a magnetic field expands and collapses about the windings of the iron-core transformer, its flux lines cut across both of the turns of the winding and the core. As a result, voltages are induced in the core itself. These voltages in the core create eddy currents. These currents move through the core in circular paths. Because eddy currents create heat in the core and do not aid the induction process, they are a waste of energy, referred to as eddy-current losses.

Core designs have been created in an attempt to minimize these losses. For example, basic transformers use a laminated core made up of insulated layers, rather than a solid core. Because the sheets are insulated from one another, the resistance across the core is high. Eddy currents are thereby reduced.

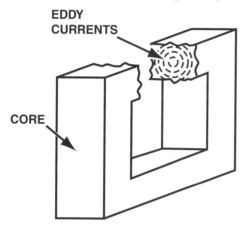
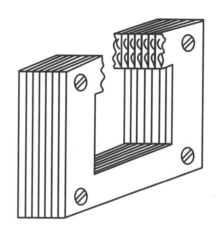

Heat Dissipation (Cooling)

Heat generated by losses must be removed to prevent deterioration of the transformer's insulation system, and the actual magnetic properties of the core. The insulation system is made up of the materials wound around the primary and secondary winding coils. A transformer's Insulation System Temperature Classification states the maximum temperature permitted in the hottest spot in the winding, at a specified ambient temperature, usually 40°C.

Methods of Dissipating Heat

Transformers can use a number of methods to dissipate heat. The method used depends primarily on the amount of heat that needs to be dissipated, and the application surroundings. This includes indoor vs. outdoor and hazardous vs. non-hazardous installations. There are two main transformer design types to deal with the problem: *liquid filled* and *dry*.

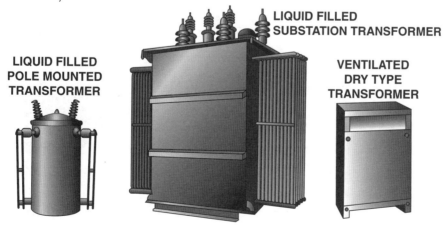

Liquid-Filled Transformers

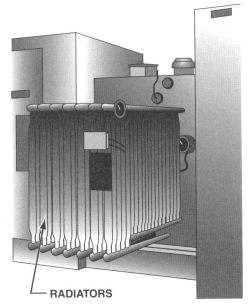

RADIATORS

Many transformers are contained in a tightly fitted sheet-metal case or tank of oil. Oil provides good electrical insulation, and carries heat away from the core and windings by convection. This type of transformer is referred to as a liquid-filled transformer.

Small transformers, such as pole-mounted trans-formers, might be cooled sufficiently by just allowing the oil to circulate inside the tank. Larger transformers, such as those shown at the right, might use fans or radiators to cool the oil.

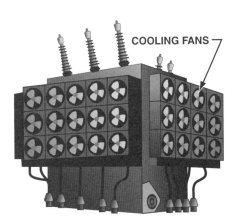

COOLING FANS

Fans for additional cooling and oil pumps for circulation could be required on even larger transformers. In addition to oil, other cooling liquids are used, such as silicone. Silicone might be used in an application where oil is not suitable, such as where flammability is an issue. Sealing the transformer case or tank is important, especially in the case of an oil-filled unit. Any penetrating moisture can reduce the insulating quality of the oil. Also, oxygen can cause oil decomposition, resulting in sludge.

Dry-Type Ventilated Transformers

A transformer designed to operate in air is called a dry type transformer. The design does not require the assistance of a liquid to dissipate excess heat. Natural or fan-assisted circulation through ventilation openings is all that is required to meet temperature classification requirements. Because a liquid is not used, a tank is not required. However, dry type transformers are contained in some type of an enclosure.

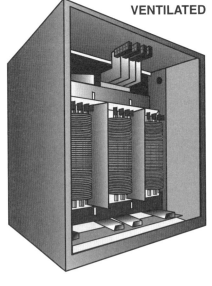

VENTILATED

A variety of specialty dry types exist. Typically not ventilated, this type of transformer usually has a small rating, and is capable of moving excessive heat away from the core and coils naturally, without the need for ventilation openings or other heat dissipation means. In most designs, this is accomplished by surrounding the core and coils with special material mixtures which absorb the heat and provide a solid seal. This type of transformer is ideal for hazardous locations, and is usually referred to as an encapsulated transformer.

ENCAPSULATED

215.3 Types of Transformers

To get a general idea of where transformers are used, let's look at a simple electrical utility system. Once electricity is generated, the voltage is increased by transformers. It is then transported to substations, where transformers decrease the voltage to usable levels for industrial plants, shopping centers and homes.

Some transformers can be several stories high. This type might be found at a generating station. Other transformers are small enough to hold in your hand. This type might be used with the charging cradle for a video camera. No matter what the shape or size, the purpose remains the same: transforming electrical power from one type to another, such as stepping up or stepping down the power. Let's take a closer look at several transformer types in use today.

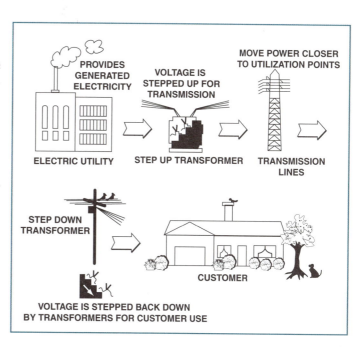

Power Transformer

Power transformers are used at generating stations to step up the generated voltage to high levels (115 kV to 765 kV) for transmission. Higher voltages are used for transmission because the current levels are reduced. This results in lower I2R losses in the transmission lines and greater efficiency thereby reducing transmission costs. Remember, the I2R power lost in transmission lines is simply an application of Ohm's Law: $P = I2R$. Resistance is a function of length and circular mil area (cma). Transmission voltages are stepped down (34 kV or 69 kV) by transformers at substations for local distribution. A power transformer's rating is given in terms of the secondaries' maximum voltage and current-delivering capacity.

Distribution Transformer

A distribution transformer is used to supply power to customers. Let's trace the path of this power from the generating station to a customer. From the generating station, the electrical power is fed to a distribution substation. The transmission voltage is stepped down in a number of steps, with the last step handled by the local pole-type or pad-mounted (mounted on the ground) distribution transformer. After all of these steps, the customer has access to power at voltages needed for lighting and equipment.

Autotransformer

The autotransformer is a special type of power transformer. It consists of a single, continuous winding that is tapped on one side to provide either a step-up or step-down function. Multi-tap ballasts are autotransformers.

Unlike a conventional two-winding transformer, which has the primary and secondary completely insulated from each other, the autotransformer's windings are both electrically and mechanically interconnected.

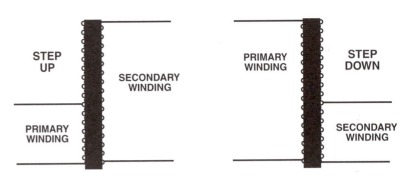

AUTOTRANSFORMERS

Instrument Transformer

For measuring high values of current or voltage, it is desirable to use standard low-range measuring instruments together with specially-constructed instrument transformers, also called accurate ratio transformers. An accurate ratio transformer does just as the name suggests. It transforms at an accurate ratio to allow an attached instrument to gauge the current or voltage without actually running full power through the instrument. It is required to transform relatively small amounts of power because its only load, called a burden, is the delicate moving elements of an ammeter, voltmeter or wattmeter.

There are two types of instrument transformers:
    Current - Used with an ammeter to measure current in AC voltages.
    Potential - Used with a voltmeter to measure voltage (potential difference) in AC voltages.

A current transformer has a primary coil of one or more turns of heavy wire while the secondary coil is very fine wire. The secondary coil is always connected in series in the circuit in which current is to be measured. The terminals of an in-line ammeter are connected in series with the secondary circuit.

A clamp-on ammeter works in a similar way. By opening the clamp and placing it around a single, current-carrying circuit conductor, the conductor itself acts as a single-turn primary and induces voltage into the secondary. The meter clamp serves as the secondary and the meter components complete the secondary circuit. If the single circuit conductor is wrapped twice around the clamp the turns ratio is doubled and twice the actual current appears on the clamp-on meter display. Note that the in-line ammeter indicates actual current. If a clamp-on meter is placed around both conductors that feed a single-phase load, the magnetic fields cancel each other and 0 A is indicated on the meter.

A potential transformer is a carefully designed, extremely accurate step-down transformer. These are frequently used for metering purposes. Common transformation ratios are 10:1, 20:1, 40:1, 80:1, 100:1, 120:1, and even higher. In general, a potential transformer is very similar to a standard two-winding transformer, except that it handles a very small amount of power. For safety, the secondary circuit is extremely well-insulated from the high-voltage primary. It is also grounded.

For metering purposes, larger services may require the use of current transformers (CT) and potential transformers (PT). These are frequently installed inside CT and PT cans (enclosures).

Isolation Transformer

Sometimes a transformer is used to supply "clean power" to sensitive electronic equipment, such as computers. These are referred to as isolation transformers, because a system neutral is created on the secondary. Occasionally, transformers with a 1:1 voltage ratio are used as isolation transformers. Refer to Connection Diagram 13, in Yr-2 Annex B. This diagram shows a transformer with a 1-to-1 voltage ratio.

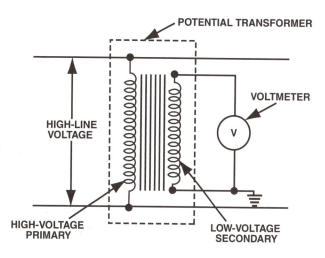

215.4 Dry-Type Transformer Typical Features, Special Considerations, and Enclosures

Dry-Type Distribution Transformer Typical Features

Typically, 10 KVA and smaller transformers:
- can be wall mounted
- have UL-3R enclosures for installation either indoors or outdoors
- are totally-enclosed encapsulated (resin filled)
  - moisture and air are sealed out
  - eliminates corrosion
  - protects insulation from deterioration
- two or more can be connected into a bank for 3-phase circuits
- have wire-type leads for field terminations
- have ground studs for enclosure bonding
- have knockouts for field connections

Typically, transformers larger than 10 KVA:
- are floor mounted
- are shielded
- are more efficient
- require a weathershield for outdoor installations
- operate quietly due to noise and vibration isolation pads
- can be installed on wall brackets (generally, up to 50 KVA)
- punched enclosure bottoms help ventilate the windings and core

Small Transformer with Leads

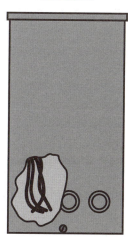

Dry-Type Distribution Transformer Special Considerations

The terms insulating and isolating are used to describe the isolation of the primary from the secondary windings. These are insulated from each other. All two, three and four winding transformers are of the insulating or isolating types. Only autotransformers, whose primary and secondary are connected to each other electrically, are not of the insulating or isolating variety.

A shielded transformer is designed with a metallic shield between the primary and secondary windings to reduce transient noise. This is especially important in critical applications such as computers, process controllers and many other microprocessor controlled devices.

Any single-phase transformer can be supplied by connecting the primary leads to any 2-wire circuit. The 2-wire circuit could be supplied from a single- or three-phase system. The transformer output will be single-phase. Transformers can not be used to convert single-phase to three-phase.

Only transformers with primary taps can be operated at voltages significantly below the nameplate rated voltage. In **NO** case should a transformer be operated at a voltage in excess of its nameplate rating, unless taps are provided for this purpose. When operating below the rated voltage, the KVA capacity is reduced correspondingly. For example: a 10 KVA transformer with a 480 volt primary and a 240 volt secondary has a rated secondary current of 41.7 amps. If the supply circuit voltage is 240 volts the secondary voltage will now be 120 volts. However, the rated secondary current still can't exceed 41.7 amps without overloading the windings. The reduced power rating of the transformer is:

$$P_{sec} = I \times E = 41.7 \times 120 = 5 \text{ KVA}$$

The result of overloading a transformer beyond its rating is excessive temperature within the windings. This excessive temperature causes overheating which will result in rapid deterioration of the insulation and cause complete failure of the transformer coils. Always select a transformer so that its power rating is not exceeded.

The expansion and contraction of the various parts of a transformer in response to temperature changes can exert mechanical forces that stretch, compress or abrade the insulation system. As insulation ages it can become physically brittle and be subject to damage, even when moving old transformers from one location to another. Also, transformers that are subjected to high current surges, such as during short-circuits, can cause enough movement to have long term effects on life expectancy.

Where transformers are used within their ratings and not subjected to excessive temperatures, they can have a life expectancy of 20 to 25 years.

Maximum Ambient Temperature	Maximum Percentage of Loading
40°C (104°F)	100%
50°C (122°F)	92%
60°C (140°F)	84%

Temperature rise in a transformer is the temperature of the windings and insulation above the existing ambient or surrounding temperature.

Transformers must be derated when the ambient exceeds 40°C, such as above a lay-in type ceiling or on the roof of a building. Use the following chart for de-rating standard transformers. For example, a 100 KVA transformer installed in a location where the temperature is 140°F has a derated capacity of 84 KVA.

Enclosures

The necessity to dissipate the losses within a transformer to a sufficient degree to avoid exceeding the insulation system rating is just one design goal for enclosures. Another is to provide adequate protection from external ambient conditions where the transformer will be used. Finally, vulnerability to tampering can be important where the public has ready access to the transformer, such as in parks, playgrounds or schools.

Transformers are used in all kinds of environments, ranging from clean, dry, air-conditioned environments, to dockside locations exposed to storms and salt spray. Unlike many other kinds of electrical equipment, it is often difficult to balance the requirements for loss dissipation and ambient protection. So, although NEMA has defined various classifications for electrical equipment enclosures, some of them are impractical for transformers and not generally available.

Standard, ventilated indoor transformers are supplied with NEMA Type 2 enclosures. This automatically qualifies them for any NEMA 1 specification. Most can be converted to NEMA 3R with the addition of weathershield accessories. The exception is a large, substation style enclosed transformer that requires a special enclosure design for a NEMA 3R rating.

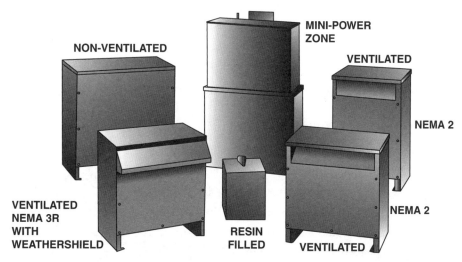

All ventilated dry type transformers, regardless of type, can be seriously de-rated if airflow is blocked or restricted. Most ventilated transformers bear a nameplate label indicating that 6 inches is required between any ventilation opening and an obstruction or wall. The exception is large, substation style transformers that are labeled for 12-inch clearance requirement.

Resin filled transformers can be placed directly on walls, but require 12 inches on both sides and top for adequate cooling. Non-ventilated transformers have no intentional ventilation openings, but do depend on their external surface to be in contact with adequate airflow to transfer heat away. It is recommended that a minimum of 3 inches be allowed on all sides of these products.

Enclosure Definitions

Type 1 Enclosures — are intended for indoor use, primarily to provide a degree of protection against contact with the enclosed equipment.

Type 2 Enclosures — are intended for indoor use, primarily to provide a degree of protection against limited amounts of falling water and dirt.

Type 3R Enclosures — are intended for outdoor use, primarily to provide a degree of protection against falling rain, sleet and external ice formation.

Refer to NEC Table 430.91 for additional enclosure types not listed above.

Definitions Pertaining to Enclosures

Ventilated — means constructed to provide for circulation of external air through the enclosure to remove excess heat, fumes or vapors.

Non-Ventilated — means constructed to provide no intentional circulation of external air through the enclosure.

Indoor Locations — are those areas protected from exposure to the weather.

Outdoor Locations — are those areas exposed to the weather.

215.5 Transformers with Single-Voltage Primaries

Single-Voltage Primary with Single-Voltage Secondary

At the right is an illustration of a transformer with one primary winding and one secondary winding. These are wound together around a core. The transformers we most routinely use are assembled at the factory in a metal enclosure. The primary winding would have a single voltage rating and the secondary winding would have a single voltage rating. The "H" leads (or lugs) on any transformer would be where the higher voltage connections are made with the "X" leads being for the connection of lower voltage. The drawings that appear below are equal to the one above. They are simply rotated on the page.

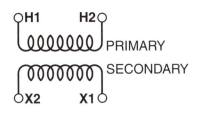

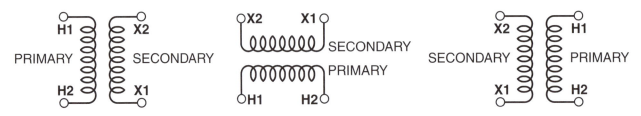

Single-Phase Transformers Part I

Typically a primary winding consists of a certain number of turns of wire and the secondary consists of a certain number of turns of wire. As electricians we are not concerned with the actual number of turns. Instead we need to know the voltage ratio. That is, if we apply a certain input voltage what will be our output voltage? The ratio of the number of turns on the primary to the number of turns on the secondary actually determines the voltage ratio.

Mathematically this can be expressed as:

$$N_p \div N_s = E_p \div E_s \quad \text{or} \quad T_p \div T_s = E_p \div E_s$$

$$\text{or} \quad N_p \div N_s = V_p \div V_s \quad \text{or} \quad T_p \div T_s = V_p \div V_s$$

Each of these formulas is equivalent. You will see in various textbooks where E or V is used to represent voltage and N or T will represent the number of turns. It makes no difference which you choose to use.

An enclosed transformer will have a nameplate attached that gives us the information we need. The transformer will have a rated primary voltage. This is the maximum voltage that can be applied to this winding without damage. Similarly, the transformer will have a rated secondary voltage. This is the maximum voltage that can be induced in this winding without damage. A voltage that is lower than the rating can be applied to the input but the result is that the output will have a proportionately lower voltage. The output voltage is determined by the voltage ratio. This voltage ratio is determined from the nameplate ratings:

$$E_{\text{rated primary}} \div E_{\text{rated secondary}} = \text{voltage ratio} = \text{V.R.}$$

Now we use this ratio to determine the actual output voltage on the secondary when something less than the rated voltage is applied to the primary. For example: a particular transformer has a nameplate that reads: 480 V primary − 120 volt secondary.

We can now calculate the Voltage Ratio: V.R. = 480/120 = 4

We need to know what the output (secondary) voltage would be if we had 450 volts available from our supply for the input (primary) voltage.

$$E_{\text{sec}} = E_{\text{pri}} \div \text{V.R.} = 450 : 4 = 112.5 \text{ volts}$$

On occasion you might need to know what minimum voltage you need to apply to get a certain output voltage. For example, you know that you will need at least 115 volts to operate a load that will be connected to the secondary. The formulas above can be rearranged to find the input voltage:

$$E_{\text{pri}} = E_{\text{sec}} \times \text{V.R.} = 115 \times 4 = 460 \text{ volts}$$

You have now determined that if you apply 460 volts you can get out 115 volts.

The *input* voltage is the voltage that is available from a supply circuit. Generally, a supply circuit originates at a circuit breaker in a panelboard or a fused safety switch. The *output* voltage is the voltage that is applied to the load. The load could be one specific piece of equipment, a control circuit, a panelboard that serves a number of loads, etc.

Year Two (Student Manual)

In rare instances the input voltage is applied to the secondary windings. When would this be required? Perhaps a piece of air-conditioning equipment arrived on the jobsite but the nameplate voltage rating was 480 volts. The plans required you to run a 240-volt circuit for the equipment. The mechanical (air-conditioning) contractor says he can get a replacement in 7 weeks but the store opens in 2 weeks. What can we do as electrical contractors? Most large transformers can be reverse-connected as step-up transformers. This could solve the mechanical contractor's problem although the cost may be prohibitive. You would install a transformer that could raise the voltage of your 240-volt circuit to 480. In this instance the input voltage would be applied to the secondary (X leads) and the output voltage (H leads) would be connected to the load. You would use a transformer with a rated 480-volt primary and a rated 240-volt secondary but instead of using it as a step-down transformer you will, instead, use it as a step-up transformer. This would be a 2:1 (or 2) voltage ratio. Using the same formula from above:

$$E_{pri} = E_{sec} \times V.R. = 240 \times 2 = 480 \text{ volts}$$

Note for Small Transformers:

NO LOAD output voltages are slightly higher than nameplate rating because the secondary windings of small transformers are "compensated". For this reason small transformers cannot be reverse-connected. Once a significant load is connected (as would be the case in the field) the secondary voltage will more closely match the nameplate rating. Without this compensation, the voltage would be less than the nameplate rating when loaded.

Let's go back to our transformer nameplate because we need to discuss power and current as it relates to the voltage ratio. We now move to a discussion of current. The ratio of primary current to secondary current is inversely proportional to the voltage ratio. Again, assuming that $P_{primary} = P_{secondary}$, we can rewrite this using Ohm's Law equivalents:

Where: $P_{pri} = I_{pri} \times E_{pri}$ and $P_{sec} = I_{sec} \times E_{sec}$ we can rewrite this as:

$$I_{pri} \times E_{pri} = I_{sec} \times E_{sec} \quad \text{or} \quad E_{pri} \div E_{sec} = I_{sec} \div I_{pri}$$

The rated power (P_{pri}) is a set nameplate value, when the primary voltage (E_{pri}) increases then the primary current (I_{pri}) decreases. Likewise, when the secondary voltage (E_{sec}) increases then the secondary current (I_{sec}) decreases since P_{sec} is a constant value and is equal to P_{pri}. This shows that primary voltage and primary current are inversely proportional and that secondary voltage and secondary current are also inversely proportional.

We can do an example with actual values to illustrate that the formulas above are correct. For this example we will select a 37.5 KVA transformer with a 480 V primary and a 120 V secondary. First we can find the **rated primary current** ($I_{rated\ pri}$). This is the maximum current that can flow through the primary winding without damage.

$$I_{rated\ pri} = P_{rated\ pri} \div E_{rated\ pri} = 37,500 \text{ VA} \div 480 \text{ V} = 78.125 \text{ A}$$

Then we can find the **rated secondary current** ($I_{rated\ sec}$). This is the maximum current that can flow through the secondary winding without damage.

$$I_{rated\ sec} = P_{rated\ sec} \div E_{rated\ sec} = 37,500 \text{ VA} \div 120 \text{ V} = 312.5 \text{ A}$$

Single-Phase Transformers Part I

Now let's insert the rated voltage and currents into this formula to make sure that the values on both sides of the equal sign are equal:

or $\dfrac{E_{rated\ pri}}{E_{rated\ sec}} = \dfrac{E_{rated\ sec}}{E_{rated\ pri}}$ $\dfrac{480}{120} = \dfrac{312.5}{78.125}$ or $4 = 4$

So now we have proven transformer currents are inversely proportional to the voltages.

We can re-configure the formula above into four different forms that set one unknown on one side of the equal sign:

$$E_{pri} = [I_{sec} \times E_{sec}] \div I_{pri} \quad \text{or} \quad E_{sec} = [I_{pri} \times E_{pri}] \div I_{sec}$$

$$\text{or} \quad I_{pri} = [I_{sec} \times E_{sec}] \div E_{pri} \quad \text{or} \quad I_{sec} = [E_{pri} \times I_{pri}] \div E_{sec}$$

One of these could be used where you know the three values on the right side of the equal sign. Continuing with our example of a 37.5 KVA transformer, how would we proceed if we wanted to know how much current would flow in the primary conductors (or I_{pri}) and knew the following:

the supply voltage is only 445 volts, instead of the rated 480

the secondary load will draw only 225 amps, instead of the rated 312.5

Here we have been given E_{pri} and I_{sec} so if we select $I_{pri} = [I_{sec} \times E_{sec}] \div E_{pri}$ then we need to find a value for E_{sec}. This is easy enough because we know the V.R. equals 4 and we can use $E_{sec} = E_{pri} \div V.R. = 445 \div 4 = 111.25$ volts. Now we know our 3 values for our formula and can simply plug them in and crunch the numbers:

$I_{pri} = [I_{sec} \times E_{sec}] \div E_{pri} = [225 \times 111.25] \div 445 = 56.25$ amps. Now we need to ask ourselves if this makes sense and does our primary power still equal our secondary power?

Let's check: First if 56.25 amps is the actual current then this is a little more than 2/3 of the rated primary current (72% to be exact). Looking at the actual secondary current (225 amps) it too is a little more than 2/3 the rated secondary current (again, 72% to be exact). So that part makes sense. Now let's check our power calculations:

$$P_{pri} = I_{pri} \times E_{pri} = (56.25) \times (445) = 25.031\ KVA$$

and $P_{sec} = I_{sec} \times E_{sec} = (225) \times (111.25)\ 25.031\ KVA$

Looks like we are good to go!

Single-Voltage Primary with Dual-Voltage Secondary

This is an illustration of a transformer that has a single primary winding and **two** secondary windings. The primary winding would have a single voltage rating. The secondary would have a *dual-voltage* rating. Meaning that EACH secondary winding has a voltage rating equal to the lower of the two voltages. For example, refer to the

Group V table shown in Yr-2 Annex B. Yr-2 Annex B is located at the back of this manual. Group V transformers have a primary rating of 208 volts and a secondary rating of 120/240 volts. *The 120/240 volt secondary rating means that EACH of the windings is rated to deliver 120 volts with the rated primary voltage applied.* Let's look at a 10 KVA transformer (Catalog No. 1507) as our example. Getting all the information from the Group V table for this particular transformer we now know that this transformer:

    a. is 15.19 inches tall.
    b. is 13.50 inches wide.
    c. is 10.84 inches deep (front-to-back).
    d. has a shipping weight of 125 pounds (note: shipping weight includes the packaging material but the actual weight of the equipment remains significant).
    e. designed to be wall-mounted – has built-in or is shipped with wall-mounting hardware.
    f. has ¾″ to 1¼″ KOs.
    g. does not require a weather shield because it is NEMA 3R.
    h. uses a Type D enclosure. Enclosure illustrations appear in Yr-2 Annex B.
    i. has wire leads instead of lug type connections.
    j. has a connection diagram as shown in *Diagram 6*.

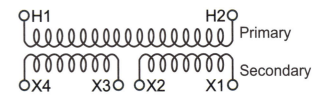

This connection diagram is shown at the right. The dashed line between the windings signifies that this is a shielded transformer. With four primary leads our transformer seems a little more complicated than the original drawing but it isn't really. It still has only one winding. This is between H1 and H4. H2 and H3 are taps that we can use to vary the turns (voltage) ratio between the primary and the secondary. Look at the chart shown below connection *Diagram 6*. Under *Primary Volts* we can see that we use the entire winding when our available supply voltage is 208. So we would connect one "hot" to H1 and the other "hot" to H4. We would need to "**cap off**" (with a wire nut or other insulating device) the leads marked H2 and H3 because we aren't going to use these if we have 208 volts available. DO NOT put H2 and H3 leads under one wire nut because you would be "shorting" across part of the winding. These unused leads must be insulated ("capped off") so that they don't come into contact with grounded surfaces or other energized parts.

Maybe you are asking yourself "when would we use H2 or H3 as connection points?" Think about voltage drop. Let's consider this situation: the length of the circuit from the 2-pole breaker in a 120/208 volt, 3Ø, 4-W panelboard to our transformer was excessive and the wire wasn't sized correctly. Or, it could be that the load had been increased so that now the wire isn't large enough to keep the voltage drop low. At any rate the voltage at the end of the "208 volt" circuit is now 189 volts. In this case we would use the H1 and H2 connections (because 189 is closer to 187 than it is to 198) to get our required secondary voltage. So, we would connect one "hot" from the breaker to H1 and the other to H2, capping off H3 and H4. In doing this you are actually decreasing the turns ratio because you are utilizing fewer turns ($187 \div 208 = 90\%$) on the primary while still using the same number of turns on the secondary.

Now we can discuss the secondary connections considering both voltage and currents. Again looking at the following chart the connection diagram:

Single-Phase Transformers Part I

1. If we need only 120 volts, and don't need 240 volts at all, we would connect our two 120 volt windings in **parallel**. It might be easier for you to see when drawn as shown at the right. Just as the chart tells us to do: we connect X2 to X4. Then we connect X1 to X3. Finally we connect our load to X4 (or X2) and X1 (or X3). Remember that we MUST use both windings to utilize the total available power of the transformer. The transformer we have selected is rated at 10 KVA. This means that EACH of our secondary windings is rated at 5 KVA. Using Ohm's Law we calculate that EACH winding can deliver 41.6666 amps to our load.

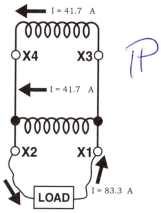

$$I_{X1 \text{ to } X2} = P \div E = [5000 \text{ VA}] \div [120 \text{ V}] = 41.6666 \text{ A}$$

and $$I_{X3 \text{ to } X4} = P \div E = [5000 \text{ VA}] \div [120 \text{ V}] = 41.6666 \text{ A}$$

Using both the windings we can get 41.6666 + 41.6666 = 83.3333 A delivered to our load. If this isn't clear think back to your study of parallel circuits. When batteries were parallel-connected the currents through each battery added together to get total current in the circuit. The same thing is happening here because a transformer winding is simply another type of power source.

2. If we want straight 240 volts for our load (no neutral) we would connect our two 120 volt secondary windings in **series**. That is, we would connect X2 and X3 together. Then we would connect our load to leads X1 and X4. What is the maximum current can we deliver to our load? Using Ohm's Law:

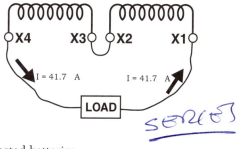

$$I_{X1 \text{ to } X4} = P_{rated} \div E_{sec} = [10,000 \text{ VA}] \div [240 \text{ V}] = 41.6666 \text{ A}$$

As a refresher, refer to your 1st Year lessons that addressed series-connected batteries.

3. If we need to connect a load (remember this could be a 120/240 volt 1Ø, 3-W panelboard) that requires 120 volts and 240 volts we would still connect our two 120 volt secondary windings in **series** to get the 240 volts by connecting X2 to X3. But we need to do something else to get 120 volts. We will "**center tap**" the two winding set at point X2 (or X3 because now they are electrically the same point since we have connected them together). This means that our neutral for our load is now connected to X2 (or X3).

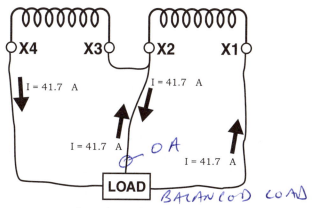

If we energized the transformer and put voltmeter leads between X1 and X2 (or X3) we should read about 120 volts. Between X2 (or X3) and X4 we should also read about 120 volts. Then across both windings (between X1 and X4) we should read about 240 volts. How much current can we deliver to our load? Using Ohm's Law:

$$I_{X1 \text{ to } X4} = P_{rated} \div E_{sec} = [10{,}000 \text{ VA}] \div [240 \text{ V}] = 41.6666 \text{ A}$$

So, as shown in the illustration above, we could have 240 volt loads connected that draw a maximum of 41.7 amps through each "hot". Or we could have 120 volt loads connected between X4 and X2 that draw a maximum of 41.7 amps $[(5000 \text{ VA}) \div (120 \text{ V}) = 41.7]$ and also additional 120 volt loads connected between X2 and X1 that draw a maximum of 41.7 amps $[(5000 \text{ VA}) \div (120 \text{ V}) = 41.7]$. Or we could have any combination of 120 v and 240 v loads.

Voltage Ratio = Turns Ratio

Let's revisit "voltage ratio" or "turns ratio". Remember, these are equivalent terms. The basic connection diagram is shown again at the right. The nameplate voltage ratings of our 10 KVA transformer indicate that When **exactly** 208 volts is applied between H1 and H4 then exactly 120 volts would be induced across winding X1-X2 and exactly 120 volt across X3-X4. The voltage ratio between the primary (H1-H4) to X1-X2 is:

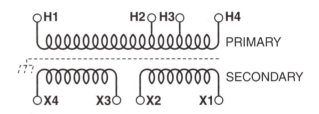

$$E_{rated\ primary} \div E_{rated\ secondary} = 208 \div 120 = 1.7333$$

Of course, it is the same for X3-X4. If we **did not** use the FCBN taps available to us, as shown in the table below Connection Diagram 6 in Yr-2 Annex B, and had a supply (or input) voltage of only 187 volts then the output voltage across X1-X2 (same for X3-X4) would be:

$$E_{output} = E_{input} \div V.R. = (187) \div (1.7333) = 107.8866 \text{ or } 107.9 \text{ volts}$$

This is too low and might damage equipment, especially motor-operated equipment. Now let's find the voltage ratio using the primary tap at H2. The voltage ratio between the primary (H1-H2) to X1-X2 is:

$$E_{rated\ primary} \div E_{rated\ secondary} = 187 \div 120 = 1.5583$$

Consider this. If you mistakenly connected your supply to H1 and H2 (again, capping off H3 and H4) BUT instead of the 187 volts you expected you actually have 208. You still have the 1.5583 voltage ratio! This means your output voltage would be:

$$E_{output} = E_{input} \div V.R. = (208) \div (1.5583) = 133.4787 \text{ or } 133.5 \text{ volts!!!!!}$$

Be extremely careful when using tap connections in transformers! If you apply too much voltage to the primary winding you will damage it. Similarly, if you connect loads to a voltage that is too high you will damage those. Not good at all. Always check your voltage before connecting any loads. This applies not just to transformer secondaries but to any equipment you are getting ready to energize.

On larger transformers the primary and secondary conductor leads are welded to "L" shaped brackets that are attached to an insulated rack. Each of these L-brackets has a hole drilled into the face. Mechanical or compression lugs are installed on the L-brackets using fasteners. As shown below, two flat bars (with a hole in each end) are shipped with the transformer and are used to jumper between the L-brackets to attain the desired output voltage.

Single-Phase Transformers Part I

Series-Connected

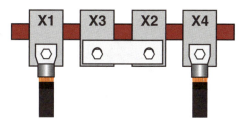

240-Volt Output

Parallel-Connected

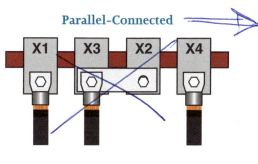

120-Volt Output

Series-Connected/Center-Tapped

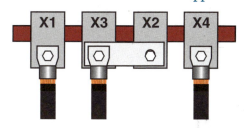

120/240-Volt Output

Single-Phase Transformers Part I

Objective 215.1 Worksheet

_____ 1. A(n) ___ is an electric device that uses electromagnetism to change voltage from one level to another.

    a. actuator   b. laminator   c. solenoid   **d. transformer**

_____ 2. Transformers operate on the principle of ___.

    a. opposites attract
    b. hysteresis
    c. electromagnetic induction
    d. opposites repel
    e. none of these

_____ 3. The ___ coil (input side) of a transformer is the coil to which the supply voltage is connected.

    a. primary   b. reactionary   c. impedance   d. secondary

_____ 4. The ___ coil (output side) of a transformer is the coil in which the voltage is induced.

    a. primary   b. reactionary   c. impedance   **d. secondary**

_____ 5. If the nameplate of a transformer shows the primary rated at 120 volts and the secondary rated at 12 volts then the ___ is 10 to 1.
    I. voltage ratio   II. turns ratio

    a. II only   b. neither I nor II   **c. both I & II**   d. I only

_____ 6. The voltage ratio for a transformer with a primary rated voltage of 120 and a secondary rated voltage of 24 is ___:___.

    a. 4:1   b. 1:5   c. 1:4   **d. 5:1**   e. none of these

_____ 7. A transformer has a primary rated voltage of 120 and a secondary rated voltage of 24 volts. If only 100 volts is applied to the primary then ___ volts is induced in the secondary.

    a. 120   **b. 20**   c. 500   d. 24   e. none of these

$$\frac{120}{24} = \frac{5}{1} = \frac{100}{x}$$

$$x = \frac{100}{5} = 20V$$

Year Two (Student Manual)

215-380　　　　　　　　　　　　　　　　　　　　　　　　　　　　　　　Single-Phase Transformers Part I

$\frac{120}{24} = \frac{5}{1} = \frac{x}{16}$ $x = 80$

_____ 8. A transformer has a primary rated voltage of 120 and a secondary rated voltage of 24. If you measure an output on the secondary of 16 volts then ___ volts has been applied to the primary.

   a. 30   b. 180   c. 120   (d. 80)   e. none of these

_____ 9. The current potential of a transformer is stepped down any time a transformer steps up the voltage.

   (a. True)   b. False

_____ 10. A single-phase 240 volt load is drawing 43 amps and connected to the secondary of a 25 KVA 480 primary to 240 secondary volt transformer. How much current is flowing in the primary winding?

   a. 43   b. 20   c. 86   d. 12.5   (e. 21.5)

_____ 11. A transformer has a primary rated voltage of 120 and a secondary rated voltage of 24. If a load connected to the secondary requires 3.5 amps to operate then the primary must be capable of delivering ___ mA.

   $120/24 = 5/1$ $3.5/5 = 0.7A$

   a. 1.43   (b. 700)   c. 1429   d. 70   e. none of these

_____ 12. A transformer has a primary rated voltage of 277 and a secondary rated voltage of 120. If the current in the circuit conductors connected to the primary is 12.6 amps then the load that is connected to the secondary must be drawing ___ amps. $277V \times 12.6A$

$277/120$

   a. 29.1   b. 5.46   c. 25.2   d. 21.98   e. none of these

_____ 13. Some transformers are provided with taps that are utilized when the ___ voltage is more or less than the rated voltage.

   (a. primary)   b. secondary

_____ 14. A transformer consists of a minimum of ___ coils (or windings).

   a. 1 — Autotransformer
   b. 2 — pri/sec.
   c. 3
   d. 4

Year Two (Student Manual)

Single-Phase Transformers Part I 215-381

____ 15. A transformer can be used to change (transform) an input voltage to an output voltage that is ____.

   a. higher   b. lower   c. the same   (d. either a or b or c)

____ 16. A transformer's primary winding and secondary winding ____ actually connected together.   ELECTRICALLY

   a. are   (b. are not)

____ 17. Power is transferred from the primary winding to the secondary winding of a transformer using magnetic coupling.

   a. False   (b. True)

____ 18. A coil in a transformer is sometimes called a ____.

   a. core   b. lead   (c. winding)   d. none of thes

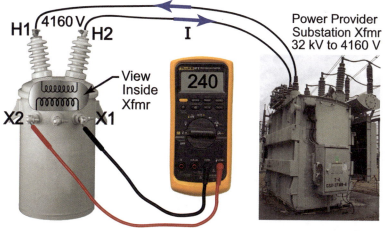

Figure 215.11

4160 / 120

____ 19. Refer to Figure 215.11, shown above. The turns-ratio for the substation transformer is ____.

   a. 52:3   b. 100:13   (c. 3:52)   d. 13:100   e. none of these

____ 20. A transformer with a primary winding that has 4 turns for every 1 on the secondary can be written as ____.

   (a. T.R. = 4 ÷ 1)   (b. T.R. = 4 to 1)   (c. T.R. = 4/1)   [d. either a or b or c]

_____ 21. When a transformer is used as a "step-up" transformer the output voltage is ___ than the input voltage.

   a. *higher*   b. lower

_____ 22. Electric utility providers step up the voltage for transmission across long distances because ___.

   a. the current is reduced
   b. smaller wire can be used
   c. the output power is increased
   d. a and b only
   e. a, b, and c

_____ 23. FCBN is an abbreviation for ___ taps.

   a. **F**ixed **C**athode **B**ackbone **N**etwork
   b. **F**requency **C**harged **B**efore **N**ew
   c. **F**erro-**C**onductive **B**asic **N**umbering
   d. **F**ull **C**apacity **B**elow **N**ormal

Single-Phase Transformers Part I

Objective 215.2 Worksheet

_____ 1. Transformers are normally rated in ___ and this indicates the maximum rated ___ power.

   a. W input   b. VA output   c. W output   (d. VA input)

_____ 2. A load that uses 6.2 KW of power is using ___ watts.   6,200

   a. 62   b. .062   c. 620   d. 6.2   (e. none of these)

_____ 3. A load that uses 56.1 VA of power is using ___ kVA.   ~~0.561~~ 0.0561

   a. 561   b. .0561   c. 56,100   d. 5610   e. none of these

_____ 4. Generally, transformers have an efficiency of more than 90%.

   (a. True)   b. False

   32 A × 480 =

_____ 5. A transformer requires 32 amps at 480 volts on the primary to deliver 120 volt power to a 14.6 kW load. The efficiency of this transformer is ___ %.

   $\frac{14.6 kW}{120 V} =$

   a. 22   b. 58.4   c. 95   d. 81

_____ 6. A transformer has a primary rating of 480 volts and supplies 240 volts to a 7.4 kVA load. The circuit that is connected to the primary originates from a 2-pole circuit breaker in a 277/480 volt 3Ø, 4-wire panelboard. The current measured in one of the circuit conductors connected to the breaker is 15.9 amps. The efficiency of this transformer is ___ %.

   a. 46.54   b. 59.52   c. 53.72   d. 96.96

_____ 7. ___ loss is loss caused by the induced currents that are produced in metal parts that are being magnetized.

   a. Conductivity   b. Hysteresis   c. Copper   (d. Eddy current)

_____ 8. Copper loss is a power loss (reduction of efficiency) caused by the voltage dropped in the (resistive) windings in a transformer.

   (a. True)   b. False

9. Typical "pole-mounted" transformers are filled with ___.

   a. oil   b. water   c. non-conductive cellulose   d. fiberglass insulation

10. When a 37.5 kVA 480 volt pri/240 volt 1Ø, 3-wire sec transformer is connected to a 31.5 kW 240 volt load the transformer is supplying ___ amps to the load.

    a. 131.25   b. 156.25   c. 65.625   d. 75.777   e. none of these

11. When loaded at 100% of its rating a 240 volt pri/24 volt sec 1.5 KVA transformer can deliver ___ amps to a connected load.

    a. 150   b. 36   c. 6.25   d. 15   e. 62.5

12. The NEMA standard BIL rating for transformers 600 volts and below is ___ KV.

    a. 10   b. 15   c. 20   d. 25

13. A dry type transformer uses ___ to dissipate excess heat that is generated by current passing through the windings.

    a. oil   b. deflection   c. air   d. cores

14. A 1Ø transformer that has a power rating of 37.5 KVA can deliver ___ amps of current to a 240 volt load without overheating the windings.

    a. 64   b. 15.6   c. 156   d. 312.5   e. 154

15. Transformers ___ change an input frequency of 60 Hz to an output frequency that is 10% higher or lower than the input frequency.

    a. can   b. cannot

16. Voltage surges can damage electrical equipment (including transformers). Voltage surges can be caused by ___.

    a. lightning strikes
    b. electrical equipment in the system opening and closing
    c. either a or b
    d. neither a nor b

Single-Phase Transformers Part I   215-385

_____ 17. ___ is an abbreviation for BIL.

    a. **B**usbar **I**nsulation **L**imiters
    b. **B**uilt-**I**n **L**aminations
    (c.) **B**asic **I**mpulse **L**evel
    d. **B**asic **I**ntensity (current) **L**evel

_____ 18. If a transformer "hums" after it is energized, ___.

    (a.) the transformer is operating normally
    b. the transformer is defective and should be de-energized immediately

_____ 19. A 112.5 KVA transformer that was installed in Denver (known as the mile-high city) could deliver ___ power if it had been installed in Miami, Fl.

    a. less
    b. the same
    (c. more) DEPENDING ON LOADING

_____ 20. Transformers are ___ efficient when the windings are cooled.

    (a. more)
    b. equally
    c. less

_____ 21. Statement ___, shown below, is true.
    I. efficiency = output power ÷ input power   ✓ pg. 362
    II. output power = input power - losses

    a. neither I nor II
    (b. both I & II)
    c. I only
    d. II only

Single-Phase Transformers Part I

Objective 215.3 Worksheet

_____ 1. The electric utility (power provider) generates power at a certain voltage. This voltage level is then ___ for transmission to substations.

   a. increased   b. decreased

_____ 2. Because ___ electric utility providers step up the voltage for distribution across long distances.
   I. I²R losses are reduced
   II. smaller wire can be used
   III. the current is reduced

   a. II only   b. I, II, & III   c. II & III only   d. I & III only   e. I & II only

_____ 3. When the cma size of transmission lines is doubled the power lost (wasted) in the conductors is ___.

   a. halved   b. quadrupled   c. squared   d. doubled

_____ 4. If 800 kW of power is produced at 2160 volts but will be transmitted across a long distance at 13.8 kV then the current flowing in the transmission wires is ___ amps.

   a. 57.97   b. 370.37   c. 37.037   d. 579.7   e. none of these

_____ 5. Normally 44 amps of current are flowing through a transmission line and 19.36 KW of power is lost in the conductors. However, during a peak period a current of 100 amps is pushed through the conductors. The I²R losses (power lost) in the conductors is ___ kW.

   a. 19.36   b. 100   c. 193.6   d. 44   e. none of these

_____ 6. ___ transformers are utilized for neon lighting and long distance transmission.

   a. Step-up   b. Step-down

_____ 7. When a large service is built CT cans are utilized. These enclosures contain ___ transformers.

   a. current   b. distribution   c. control   d. isolation

Year Two (Student Manual)

_____ 8. If 4160 volts is supplied by the utility but a customer requires 480 volts then a transformer is installed that will ___ the supplied voltage.

   a. step up   b. step down

_____ 9. A ___ is an example of an autotransformer.
   I. ballast in a fluorescent luminaire   II. ballast in an HID luminaire

   a. neither I nor II   b. II only   c. I only   d. both I & II

_____ 10. A clamp-on ammeter  is placed around a single circuit conductor.  If current is flowing in the conductor, a voltage is induced in the clamp, which is connected to the electronics portion of the meter, and current flow is indicated on the meter display.

   a. true   b. false

_____ 11. An ammeter is placed around the "hot" and neutral conductors that are connected to a lamp as shown below. The lamp draws 1.2 amps from the circuit. How much current will be indicated on the meter display?

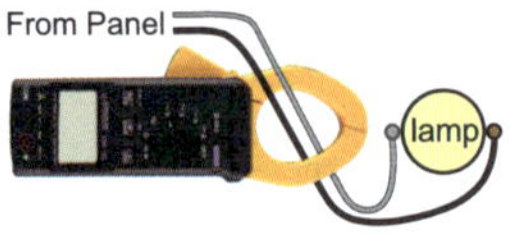

   a. 1.2   b. 0.6   c. 2.4   d. 0

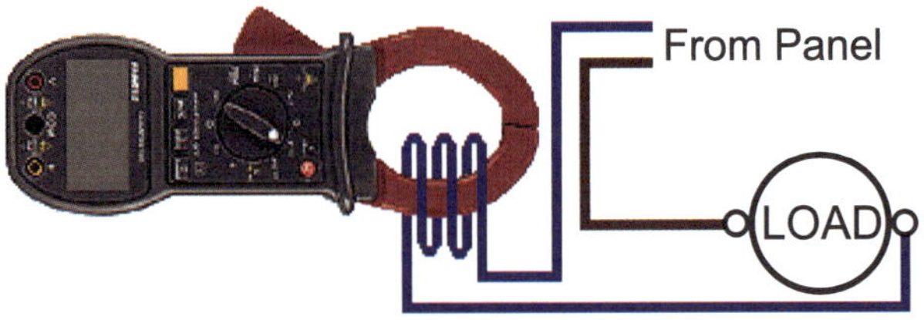

Figure 215.31

_____ 12. An 8 AWG circuit conductor to a load is passed though an ammeter clamp three times, as shown above in Figure 215.31. The ammeter indicates a current of 30 amps. The load is actually using ___ amps of current.

   a. 15   b. 90   c. 30   d. 10   e. none of these

_____ 13. An *isolation transformer* is routinely used where "clean power" is required to operate delicate electronic equipment such as computers.

   a. true   b. false

Objective 215.4 Worksheet

_____ 1. Temperature is the limiting factor in transformer loading.

   a. False   b. True

_____ 2. Which of the following situations would produce destructive **heat** within a 250 VA transformer that has a primary rating of 240 volts and a secondary rating of 120 volts?
   I. applying 480 volts to the primary (remember I = E ÷ Z)
   II. connecting a .3 kW load to the secondary

   a. neither I nor II
   b. either I or II
   c. I only
   d. II only

_____ 3. Which of the following conditions would damage a transformer that has a 1.5 kVA 480 volt pri. /24 volt sec. rating?
   I.   connecting a load of 150 watts
   II. connecting the primary to a circuit from a 480 volt panel
   III. installing the transformer inside an enclosure without adequate ventilation

   a. I & III only
   b. I & II only
   c. I, II, & III
   d. neither I nor II nor III
   e. III only

_____ 4. Temperature rise is the difference in temperature between the transformer windings and the ambient temperature surrounding the transformer.

   a. True   b. False

_____ 5. A transformer must be derated if the ambient temperature exceeds ___° F.

   a. 104
   b. 90
   c. 167
   d. 140

Single-Phase Transformers Part I

Maximum Ambient Temperature	Maximum Percentage of Loading
40°C (104°F)	100%
50°C (122°F)	92%
60°C (140°F)	84%

_____ 6. A transformer that is hung above a lay-in ceiling where the temperature is 122° F must be rated for at least ___ kVA (calculated value) where the connected load is 77 KW. Refer to Table 215.4, shown above.

    a. 100   b. 83.7   c. 77   d. 70.84

_____ 7. A 100 KVA rated transformer installed in an ambient temperature of 50°C has a derated value of ___ kVA. Refer to Table 215.4, shown above.

    a. 100   b. 104.167   c. 108.696   d. 92   e. none of these

_____ 8. If the temperature rise of a transformer is 50° C this equals _____° F.

    a. 50   b. 90   c. 104   d. 122   e. 58

_____ 9. An *insulating* transformer and an *isolating* transformer are ___.

    a. two different types of transformers   b. terms used to describe the same transformer

_____ 10. Refer to the "Connection Diagrams" that appear in Yr-2 Annex B. The transformer shown in ___ is a *shielded* transformer.
I. Diagram 13
II. Diagram 23
III. Diagram 58

    a. II only   b. II & III only   c. I, II, & III   d. I only   e. III only

_____ 11. Shielded transformers should be used where ___ loads are connected to the transformer secondary.

    a. computer   b. HID lighting   c. motor   d. all of these

_____ 12. A transformer has a nameplate rating of 480 volt primary – 120/240 volt secondary. If instead of rated voltage the applied primary voltage is 440 volts a 5 KVA transformer could deliver a maximum output of ___ KVA without overloading.

   a. 4   b. 5.45   c. 6.25   d. 4.58

_____ 13. A circuit from which of the following can be connected to a 1-ph transformer?

   a. a 2-pole breaker in a 120/208 V 3Ø, 4-W panelboard
   b. a 1-pole breaker (with a neutral) in a 120/208 V 3Ø, 4-W panelboard
   c. AØ and BØ in a 3-pole 480 volt safety switch
   d. any of the above
   e. none of the above

_____ 14. If a 1Ø transformer is connected to a 2-pole breaker in a 3Ø panelboard the output of the transformer will be ___ -phase.

   a. single   b. two   c. three

_____ 15. A well-designed transformer, when applied within its temperature limits, will have a life expectancy of ___ years.

   a. 1 to 5   b. 5 to 10   c. 10 to 15   d. 15 to 20   e. 20 to 25

_____ 16. NEMA Type 3R enclosures can be used ___.

   a. indoors   b. outdoors   c. either a or b   d. neither a nor b

_____ 17. When a transformer is loaded beyond its rating the effects in the transformer are ___.
   I. excessive temperature
   II. deterioration of insulation
   III. coil failure

   a. III only
   b. II & III only
   c. I & II only
   d. I, II, & III
   e. I & III only

Single-Phase Transformers Part I

_____ 18. A transformer enclosure that is ventilated ___ allows air from outside the enclosure to circulate through it.

   a. does   b. does not

_____ 19. 1Ø Group I (Re: Yr-2 Annex B) 150 VA transformers can be arranged in a bank for three-phase service.

   a. true   b. false

_____ 20. Which of the following 1Ø Group I (Re: Yr-2 Annex B) transformers are encapsulated?
   I. 7.5 KVA
   II. 15 KVA

   a. II only   b. neither I nor II   c. I only   d. both I & II

_____ 21. Large dry-type distribution type transformer enclosures have steel bottoms that are punched for ___.

   a. ventilation
   b. weight reduction
   c. conduit entries
   d. attachment of grounding lugs

_____ 22. The coil and core assembly of large dry-type distribution type transformers are mounted on vibration isolation pads so that ___.
   I. the transformer operates more quietly
   II. minimize the damage of shocks received during shipping

   a. II only   b. neither I nor II   c. I only   d. both I & II

_____ 23. A 7.5 KVA Group I (Re: Yr-2 Annex B) transformer can be installed outdoors if ___.
   I. weather shields are installed on the transformer
   II. the transformer is installed in a 18″ × 18″ × 12″ NEMA 3R enclosure

   a. II only   b. neither I nor II   c. I only   d. both I & II

_____ 24. Refer to Yr-2 Annex B. Which of these Enclosure Types is <u>ventilated</u>?
   Hint: If a weathershield is required then **air** and water can enter the enclosure.

   a. E & G   b. F & G   c. D & E   d. H & I

Single-Phase Transformers Part I

Objective 215.5 Worksheet

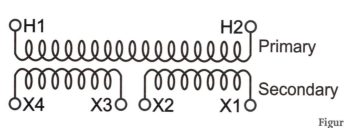

Figure 215.51

_____ 1. Some transformers have two separate secondary windings, as shown above in Figure 215.51. If a transformer's secondary windings are rated 120/240 these can be connected in ____ to deliver power to 120-volt loads only.

   a. series   b. parallel

_____ 2. The transformer shown above in Figure 215.51 has two separate secondary windings. The transformer is rated at 10 KVA and its secondary windings are rated 120/240 volts. The means that **each** secondary winding is rated at ____ KVA.

   a. 10   b. 5   c. 20

_____ 3. The transformer shown above in Figure 215.51 below has two separate secondary windings. The transformer is rated at 25 KVA and its secondary windings are rated 120/240 volts. If the secondary windings are connected to supply power to a 120-volt load then the transformer can deliver a maximum of ____ amps.

   a. 104   b. 52   c. 208   d. 240   e. 10.4

_____ 4. The transformer shown above in Figure 215.51 has two separate secondary windings. The transformer is rated at 15 KVA and its secondary windings are rated 120/240 volts. The secondary windings are connected for 120/240 volt operation. If several 120-volt loads are connected to X4 and X3 this winding can supply a maximum of ____ amps to those 120-volt loads.

   a. 62.5   b. 125   c. 31.25   d. 8   e. 78.125

_____ 5. You need to change from 240 volts to 480 volts to operate a piece of equipment that will require 25 KVA at 480 volts. You must use a transformer that is built specifically for step-up applications.

   a. False   b. True

_____ 6. Which of the following is NOT correct?

    a. $I_S \div I_P = E_S \div E_P$
    b. $I_S \div I_P = E_P \div E_S$
    c. $N_P \div N_S = E_P \div E_S$
    d. $N_P \div N_S = I_S \div I_P$

_____ 7. Ignoring small losses, if the primary current on a 240 to 120 volt (step-down) transformer is 24 amps then the secondary current would be ___ amps.

    a. 24
    b. .2
    c. 12
    d. 48
    e. 120

_____ 8. A .25 KVA transformer has a nameplate rating of 240 x 480 primary volts – 120/240 secondary volts. If rated voltage is applied to the primary the secondary voltage will be ___ than the nameplate secondary voltage because these small transformers have compensation built into their secondaries.

    a. higher
    b. lower

_____ 9. A 480 primary to 120/240-volt secondary single-phase transformer is actually supplied by 496 volts. If no tap changes are made what line-to-neutral output voltage will be available?

    a. 120   b. 124   c. 116   d. 240   e. 248

_____ 10. The unit of measurement for the answer to the previous question is represented by the letter or symbol ___.

    a. E   b. VA   c. A   d. V   e. W

_____ 11. A 2 KVA Group I (Yr-2 Annex B) 480 x 240 primary - 240/120 secondary volt single-phase transformer has ___ **secondary** ("X") field connections that are brought out of the windings as leads or terminal lugs.

    a. 2   b. 4   c. 7   d. 8

Single-Phase Transformers Part I

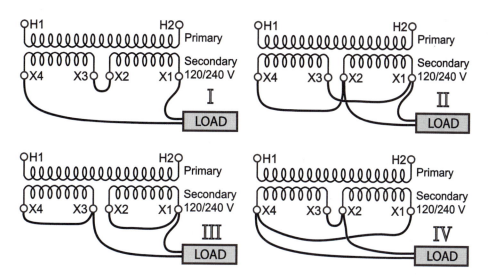

Figure 215.52

_____ 12. Which of the secondary transformer connections, shown above in Figure 215.52, is correct to supply the proper voltage to the 240-volt load.

a. IV
b. I
c. III
d. II
e. none of these

_____ 13. A transformer with a 120/240-volt secondary can deliver power at the proper voltage to ____.

a. 120-volt loads
b. 240-volt loads
c. both a and b
d. neither a nor b

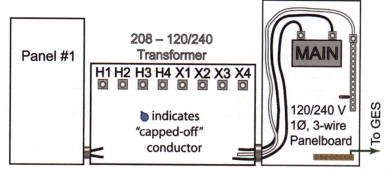

Figure 215.53

_____ 14. Refer to Figure 215.53 (above) and the information in Yr-2 Annex B for a 25 KVA transformer, catalog number 1509 from GROUP V. Which of the connections, shown below, is correct where a two-pole breaker in Panel #1 feeds the transformer? Panel #1 is a 120/208 volt, 3Ø, 4-wire system but the ACTUAL voltage at the transformer primary is **209** volts.

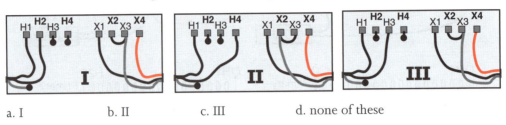

a. I   b. II   c. III   d. none of these

_____ 15. Refer to Figure 215.53 (above) and the information in Yr-2 Annex B for a 25 KVA transformer, catalog number 1509 from GROUP V. Which of the connections, shown below, is correct where a two-pole breaker in Panel #1 feeds the transformer? Panel #1 is a 120/208 volt, 3Ø, 4-wire system but the ACTUAL voltage at the transformer primary is **186** volts.

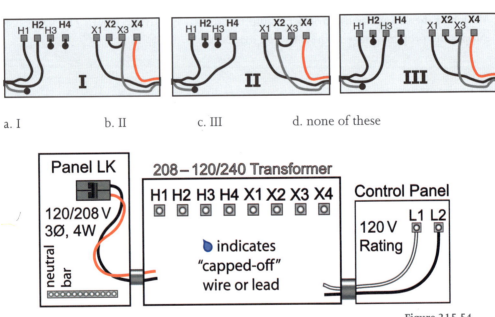

a. I   b. II   c. III   d. none of these

Figure 215.54

_____ 16. Refer to Figure 215.54 (above) and the information in Yr-2 Annex B for a 25 KVA GROUP V transformer (cat # 1509). Which of the connections, shown below, is correct? The ACTUAL voltage at the transformer primary is **196** volts.

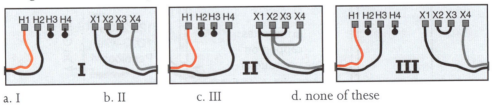

a. I   b. II   c. III   d. none of these

Single-Phase Transformers Part I

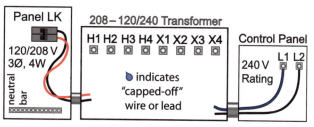

Figure 215.55

_____ 17. Refer to elevation drawing, shown above in Figure 215.55, and the information in Yr-2 Annex B for a 25 KVA GROUP V transformer (cat # 1509). The supply circuit voltage is ACTUALLY 188 volts at the point where it enters the transformer enclosure. Which connection, shown below is correct?

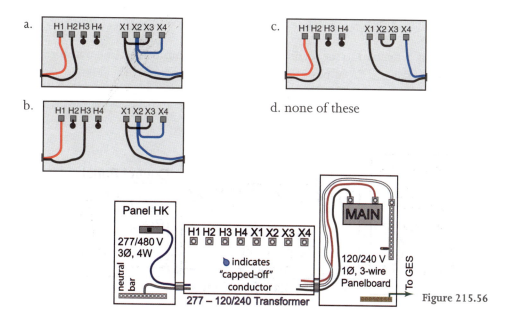

d. none of these

Figure 215.56

_____ 18. Refer to Figure 215.56 (above) and the information in Yr-2 Annex B for a 25 KVA transformer, catalog number 1609 from GROUP VI. The supply circuit voltage is ACTUALLY 279 volts at the point where it enters the transformer enclosure. Which of the connections, shown below, is correct?

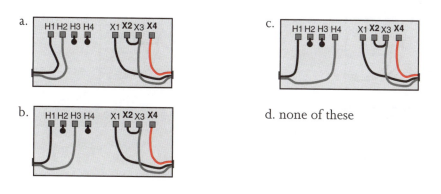

d. none of these

_____ 19. Refer to Figure 215.56 (above) and the information in Yr-2 Annex B for a 25 KVA transformer, catalog number 1609 from GROUP VI. The supply circuit voltage is ACTUALLY 251 volts at the point where it enters the transformer enclosure. Which of the connections, shown below, is correct?

a.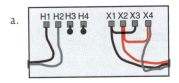

c.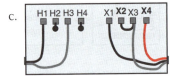

b.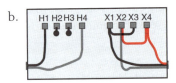

d. none of these

_____ 20. A 120 VA load is connected to the secondary of a 250 VA transformer. The transformer must be supplied with at least ____ VA of power on the primary.

a. 120   b. 57.6   c. 250   d. 2.08   e. 125

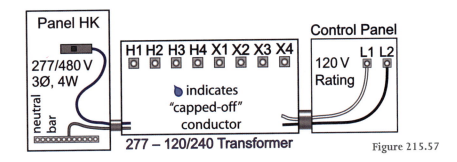

Figure 215.57

_____ 21. A GROUP VI, 5 KVA transformer (catalog number 1605) is shown above in Figure 215.57. Also refer to the information in Yr-2 Annex B to determine which of the connections, shown below, is correct where the supply circuit voltage is ACTUALLY 264 volts at the point where it enters the transformer enclosure.

a.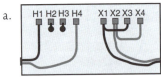

c.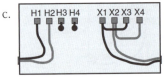

b.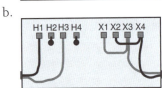

d. none of these

Single-Phase Transformers Part I   215-401

_____ 22. A GROUP VI, 5 KVA transformer (catalog number 1605) is shown above in Figure 215.57. Also refer to the information in Yr-2 Annex B to determine which of the connections, shown below, is correct where the supply circuit voltage is ACTUALLY 252 volts at the point where it enters the transformer enclosure.

a.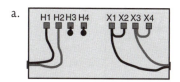

c.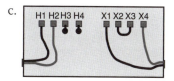

b.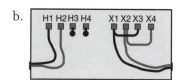

d. none of these

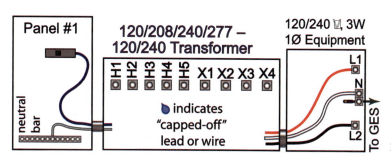

Figure 215.59

_____ 23. Refer to Figure 215.59, shown above, and the information in Yr-2 Annex B. The transformer is a Group VII – 5 KVA – Cat. No. 1705. Where Panel #1 is 120/208 V, 3Ø, 4-wire, connection diagram ____ (shown below) is correct.

a.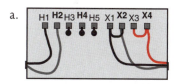

c.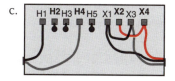

b.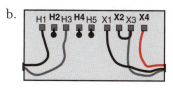

d. none of these

_____ 24. A catalog #1510 Group V (Yr-2 Annex B) 37.5 KVA transformer can be supplied from a ____ and deliver 120 volts on the secondary.

a. 2-pole breaker in a 120/208 v 3Ø, 4-W panelboard
b. 1-pole breaker installed on BØ (with a neutral) in a 120/240 V 3Ø, 4-W panelboard
c. 1-pole breaker installed on BØ (with a neutral) in a 277/480 V 3Ø, 4-W panelboard
d. 2-pole breaker in a 120/240 V 3Ø, 4-W panelboard
e. 2-pole breaker in a 277/480 V 3Ø, 4-W panelboard

_____ 25. You will feed a 3 KVA transformer (Cat. # 1704 in Yr-2 Annex B) with a circuit that is connected to a single-pole breaker in a 277/480 volt, 3Ø, 4-wire panel. You will connect your supply circuit (L1 and L2) as shown in drawing ___ below.

a.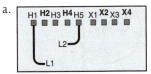

c.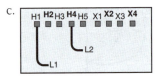

b.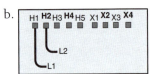

d. none of these

_____ 26. You will connect your load conductors to the secondary as shown in drawing ___ below to feed a single 240-volt load from a 3 KVA transformer (Cat. # 1704 in Yr-2 Annex B).

a.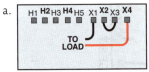

c.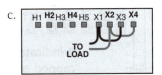

b.

d. none of these

_____ 27. A catalog #1704 Group VII (Yr-2 Annex B) 3 KVA transformer can be supplied from a ___ and deliver 120 volts on the secondary.

   a. 2-pole breaker in a 120/208 V 3Ø, 4-W panelboard
   b. 2-pole breaker in a 120/240 V 3Ø, 4-W panelboard
   c. 1-pole breaker installed on BØ (with a neutral) in a 277/480 V 3Ø, 4-W panelboard
   d. any of these
   e. none of these

Lesson 215 NEC Worksheet

_____ 1. In general, exposed metal transformer enclosures are required to be grounded.

      a. False   b. True

_____ 2. The minimum clearance from the bottom of a pole-mounted transformer to the ground is ___ feet.

      a. 14   b. 10   c. 8   d. no minimum required

_____ 3. A liquid-filled transformer has a primary voltage of 12,470 volts. The transformer is surrounded by a 6-foot high fence. The overall height of the fence must be extended 1-foot with at least ___ strands of barbed wire on top of the fence.

      a. 2   b. 3   c. 1   d. no minimum required

_____ 4. To allow for adequate ventilation or dissipation of heat, transformers are required to be marked with a minimum distance or clearance from walls or other obstructions.

      a. False   b. True

_____ 5. Individual dry-type transformers rated over 112½ kVA shall be enclosed in a fire-resistant room with materials having a ___-hour minimum fire rating.

      a. 1   b. 2   c. 3   d. 6

Single-Phase Transformers Part I   215-405

Lesson 215 Safety Worksheet

TRANSFORMER SAFETY

Transformers are a very common piece of electrical equipment. They are available in many forms and styles, from small open type control transformers to very large power distribution transformers.

There are numerous safety considerations that must be taken seriously regarding transformers. Below is a danger warning from an instruction sheet packaged with a transformer. You will notice that the last line of the instructions clearly states that failure to follow the instructions will result in death or serious injury.

HAZARD OF ELECTRIC SHOCK, EXPLOSION, OR ARC FLASH

- Apply appropriate personal protective equipment (PPE) and follow safe electrical work practices. See NFPA© 70E.
- This equipment must only be installed and service by qualified electrical personnel.
- Turn off all power supplying this equipment before working on or inside equipment.
- Always use a properly rated voltage sensing device to confirm power is off.
- Do not apply transformers to circuits where system voltage exceeds nameplate rated voltage by more than 5%.
- Always apply over-current protection on ungrounded input and output conductors as detailed in Article 450 of NFPA 70 (National Electrical Code© (NEC®)).
- Replace all devices, doors, and covers before turning on power to this equipment.

Failure to follow this instruction will result in death or serious injury.

Transformers will have several places that voltage will be present. There will be several voltages present in the transformer. You need to understand the locations where these voltages are present and where to test to verify the power is off.

Follow any and all of the manufacturers instructions that are included with every transformer pertaining to installation and any additional information they provide.

Questions

_____ 1. Before working on a transformer you must follow the Safe Work Practices in ___.

    a. OSHA   b. NEC Article 700   c. NFPA 70E   d. NEC Article 430

_____ 2. Before turning on the power to the transformer, replace all ___ and devices.
    I. doors
    II. fenders
    III. covers

    a. III only   b. I & II only   c. I only   d. II only   e. I & III only

_____ 3. All power must be turned off before working on or ___ a transformer.

    a. under   b. over   c. around   d. inside

_____ 4. A properly rated ___ must be used to confirm that the power is off before work is started on a transformer.

    a. power supply   b. PPE   c. voltage sensing device   d. none of these

_____ 5. Transformers must be installed only by ___.

    a. journeymen
    b. qualified persons
    c. quality persons
    d. none of these

Lesson 216
Single-Phase Transformers Part II

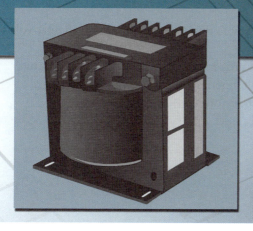

Electrical Curriculum

Year Two
Student Manual

IEC
PRIDE
NATIONAL

Purpose

This lesson will discuss the connections, loading, grounding, testing, troubleshooting and fault current calculation for single phase transformers with dual voltage primaries and dual voltage secondaries.

Homework

(Due at the beginning of this class)

For this lesson, you should:

- Thoroughly read the material contained within Lesson 216.
- Complete Objective 216.1 Worksheet.
- Complete Objective 216.2 Worksheet.
- Complete Objective 216.3 Worksheet.
- Complete Objective 216.4 Worksheet.
- Complete Lesson 216 NEC Worksheet.
- Read and complete Lesson 216 Safety Worksheet.

- Complete additional worksheets, if available, as directed by your instructor.

Objectives

By the end of this lesson, you should:

216.1 ....Understand the connection of transformers with dual-voltage primaries and dual-voltage secondaries.
216.2 ....Understand how to avoid overloading and properly select a single-phase transformer.
216.3 ....Understand grounding, testing, and troubleshooting of single-phase transformers.
216.4 ....Be able to calculate available single-phase transformer secondary fault current.

216.1 Connection of Dual-Voltage Primary/Dual-Voltage Secondary Transformers

Dual-Voltage Primary with Dual-Voltage Secondary

This is an illustration of a transformer that has two primary windings and two secondary windings. The primary has a "dual-voltage" rating. The secondary also has a "dual-voltage" rating. This means that EACH winding has a voltage rating equal to the lower of the two voltages given on the nameplate. A typical transformer, as shown here, would have **two** separate single-winding transformers in a single enclosure. That is, H1-H2 windings are wound on the same core as X3-X4 and form a set of windings. Similarly, H3-H4 windings are wound on the same core with X1-X2 and form the second set.

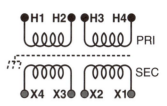

The voltage ratio is the ratio of the rated voltage across H1-H2 compared to the rated voltage across X3-X4. The voltage ratio of one set of windings will be the same as the second set of windings.

The secondary windings are connected in series or in parallel as discussed in Lesson 215 at *Single-Voltage Primary with Dual-Voltage Secondary*. The primary windings will be connected in series or in parallel depending upon the available supply voltage. For example, if we had a transformer with the primary rated at 120/240 volts then **each** primary winding (H1-H2 **or** H3-H4) would be rated for a maximum of 120 volts. If the available supply voltage were 120 then the primary windings would be connected in parallel so that 120 volts is applied to each primary winding. If the supply voltage were 240 then the windings would be connected in series so that half of the 240 (240 ÷ 2 = 120 V) would be applied to each primary winding. Remember, the lower of the two voltages in the voltage rating is the maximum voltage that can be applied to **each** winding.

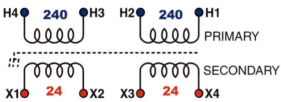

Let's look at some specific examples. If we had a transformer with a 240/480 volt primary and a 24/48 volt secondary then the voltage ratings would be as shown at the right. As you can see, if we applied 240 volts to the primary of the set on the left (H4-H3) then we would get 24 volts as the output voltage on the secondary (X1-X2) of this set. **Each set** has a voltage ratio of 240 to 24 or 240/24 or 10 to 1. Similarly if we applied 240 volts to the set on the right we would get the same 24-volt output on X3-X4. We can connect the primary windings in series or in parallel dependent upon the available supply voltage. As was done in Lesson 215 the secondary windings can also be connected in series or in parallel depending upon what output voltage is required for the load. This gives us 4 different combinations.

1) primary in **series**/secondary in **series**
   *This would be used where we have a supply voltage of 480 and wanted a secondary voltage of 48 volts. Recalling your study of series connections you can see that half (or 240) of the 480 volts is applied to **each** winding.*

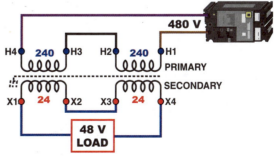

2) primary in **series**/secondary in **parallel**

This would be used where we have a supply voltage of 480 and wanted a secondary voltage of 24 volts. Following the connection wires from the load back to the windings you can see that only 24 volts is applied to the load.

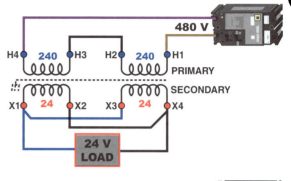

3) primary in **parallel**/secondary in **series**

*This would be used where we have a supply voltage of 240 and wanted a secondary voltage of 48 volts. This is shown at the left. Following the primary connection wires from the 240-volt breaker you can see that 240 volts is applied to **each** winding.*

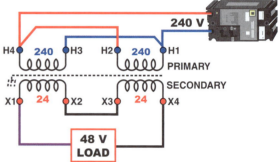

4) primary in **parallel**/secondary in **parallel**

This would be used where we have a supply voltage of 240 and wanted a secondary voltage of 24 volts. This is shown at the right.

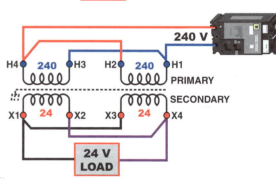

A pictorial elevation view of a large dual-voltage primary/dual-voltage secondary transformer is shown at the right. Although small transformers have lead wires, when larger transformers are used lugs must be installed on "tabs". These "tabs" have mounting holes for the installation of these lugs. Typically the lugs and the mounting hardware are not a part of the "gear package" for the job and the journeyman must order these. Of course, the job specifications will dictate if these are mechanical or compression-type lugs.

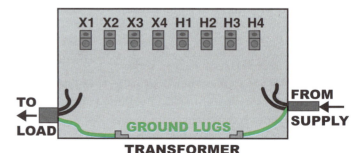

Lug kits are available from the manufacturer and are shipped with nuts, bolts (or machine screws), flat washers, and lock washers that are appropriate for the transformer rating.

Year Two (Student Manual)

Lugs, made of copper, or aluminum, are available. Additionally, lugs with multiple mounting holes and/or accommodation for multiple conductors are available. Lug manufacturers include Thomas & Betts, Burndy, and Panduit.

4-hole, 3-Conductor Aluminum Mechanical Lug	1-hole, Single Conductor Tin Plated Copper Lug	1-hole Copper Mechanical Lug

The primary connections have been made on the two drawings shown below. Note that the layout of the terminals is not the same as shown on the drawing above. These terminal arrangements differ among manufacturers as well as transformer types and ratings.

Series-Connected
480-Volt Input

Parallel-Connected
240-Volt Output

216.2 Loading and Selection of Single-Phase Transformers

When sizing single voltage transformers, we simply have to add the current of each load, and multiply the total by the voltage to get VA. If the loads are given in VA, we would add the loads to determine the total VA for the transformer.

When the windings are connected to produce two voltages, sizing the transformer becomes more complicated.

In the drawing at the right, we have an example of a two winding secondary that has been connected to produce two voltages. The winding voltages are equal, and the voltage across X1 and X4 will be the sum of the voltages of the two windings. The most common example seen in the field is the 120/240 volt single-phase, 3-wire system. Windings A and B (X4 to X3 and X2 to X1) each have a voltage of 120. The voltage from X4 to X1 will be 240.

The transformer power rating is the total power the entire secondary can supply. Each winding will supply 1/2 the total rating. For example, if the rating of the transformer is 25 KVA, then winding A is capable of supplying a load of 12.5 KVA and winding B can supply a 12.5 KVA load. When connecting loads to a transformer, it is important to balance loads between windings in order to avoid overloading either winding. Each winding will have to carry the **current** to any loads connected across it, plus the current from any load connected across both. For our 25 kVA transformer that has a 120/240 volt secondary rating, this means that EACH of the two windings can carry a **maximum** of:

$I_{\text{sec winding A}} = I_{\text{sec winding B}} = [25{,}000 \text{ VA} \div 2] \div 120 \text{ V} = 104.1667 \text{ A}$ *without overloading*

Refer to the previous drawing. We will have connected our 25 KVA transformer secondary to a 120/240 volt, 1Ø, 3-wire panel. If 120 volt Circuit 3 is turned "ON" and draws 25 amps and 120 volt Circuit 4 is also turned "ON" and pulls 30 amps then the total current on Winding B will be:

$I_{\text{sec winding B}} = 25 + 30 = 55 \text{ amps}$

We aren't overloading Winding B, yet, but we're more than half-way there with just 2 circuits!

Let's now add a 240 volt circuit. This we will connect to a 2-pole breaker in space 6-8 in our panel and the load will draw 40 amps. As shown in the drawing at the right this circuit is completed through both transformer windings so that the 40 amps are flowing through EACH winding. Now that we have added this load (Circuit 6-8) we have:

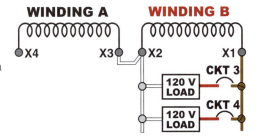

$I_{\text{sec winding B}} = 25 + 30 + 40 = 95 \text{ amps}$

of current flowing through Winding B, and

$I_{\text{sec winding A}} = 0 + 40 = 40 \text{ amps}$

of current flowing in Winding A.

Now we are really close to overloading Winding B! Remember we determined the maximum to be 104 amps and we have already loaded it to 95 amps.

We still have a lot of capacity remaining on Winding A so let's add a couple of loads to it, as shown at the right. Our customer has purchased two new pieces of equipment and your only option is to add these to Winding A. You connect these new loads to Circuits 1 and 2. The load connected to Circuit 1 has a nameplate rating of 4.2 kVA @ 120 volts and load connected to Circuit 2 has a name-plate rating of 5.16 KVA @ 120 volts. Remember, the limitation on a transformer winding is the amount of current it can carry without damage so we will have to convert these nameplate ratings into the amount of current required by each.

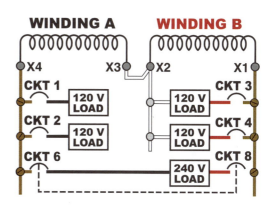

$I_{Ckt\ 1}$ = 4200 VA ÷ 120 V = 35 amps   and $I_{Ckt\ 2}$ = 5160 VA ÷ 120 V = 43 amps

Now, the amount of current flowing through Winding A is:

$I_{sec\ winding\ A}$ = Circuit 1 + Circuit 2 + Circuit 6-8 = 35 + 43 + 40 = 118 amps

Now we've done it! We have **overloaded** Winding A! At this point we have only two options.

- Option 1: Tell the customer he can't run whatever is connected to Circuits 1, 2, or 6-8. This will NOT make your customer happy.

- Option 2: Advise the customer that he needs a larger transformer and a new panel because the panel he has now is only rated at 100 amps. This won't make the customer happy either but, after all, he bought the new equipment to expand his capacity to make more products so he could make more money. This option WILL make your contractor happy.

The customer decides on Option 2 and now your job is to determine what size transformer and panel to install. Let's worry about the transformer first.

Doing the math you can see that the greater load is on Winding A so you will use this as your basis for sizing your transformer.

$P_{winding\ A}$ = $I_{winding\ A}$ × $E_{winding\ A}$ = 118 amps × 120 volts = 14,160 VA = 14.160 kVA

Since this is only half of the transformer you will **double** this to determine the minimum rating of the transformer required to serve these loads.

14.160 × **2** = 28.320 KVA will be the minimum rating of the new transformer

Referring to the Group I transformer table shown in Yr-2 Annex B you find that you can get a transformer with a rating of 37½ KVA. There is no other size available between the one you have now (25 kVA) and this one. This 37½ KVA transformer will work plus it will give him some additional capacity for future expansion.

Year Two (Student Manual)

Now we need to look at what size panel you should provide. It wouldn't make sense to get a panel larger than the maximum rating of the transformer so we will determine the maximum current that can be delivered by the 37.5 kva transformer.

$$I_{sec\ winding\ A} = I_{sec\ winding\ B} = I_{A\emptyset} = I_{B\emptyset} = [37,500\ VA \div 2] \div 120\ V = 156.25\ A \quad \text{or}$$

$$I_{sec\ winding\ A} = I_{sec\ winding\ B} = I_{A\emptyset} = I_{B\emptyset} = [37,500\ VA] \div \mathbf{240}\ V = 156.25\ A$$

Looks like you should get a panel with a 150 amp main breaker so that the transformer will not be overloaded.

How much more load, in amps, could the customer add if you install this new equipment?

New A∅ capacity − existing A∅ load = load that can be added to A∅ =

156.25 − 118 = 38.25 amps that could be added to A∅ (winding A)

New B∅ capacity − existing B∅ load = load that can be added to B∅ =

156.25 − 95 = 61.25 amps that could be added to B∅ (winding B)

Summary of Steps Required to Size Single-Phase Transformers

1. From the loads to be supplied, determine the following from the nameplates or the manufacturer's information:
a. voltage required
b. amperes or KVA required – Note that if the load is a motor that is to be started more than once an hour, you should select a transformer with a minimum rating of 20% greater. Additionally, for motor loads, select a transformer based on the nameplate FLA x SF.
c. Verify that the load can operate on single-phase.
2. Determine the actual voltage of the available circuit you will use to supply the transformer. Primary taps are available on many transformers to compensate for line voltages that are above or below the primary rating.
3. Select a transformer with a rating that is greater than or equal to the capacity required by the load.

216.3 Grounding, Testing, and Troubleshooting Single-Phase Transformers

Grounding

Grounding procedures for low voltage transformers are described in the National Electrical Code. Since the secondary of an isolation transformer is a separately derived system, many of the grounding rules associated with service entrance equipment also apply to transformers. In addition, equipment ground integrity must be maintained between the line side and the load side systems.

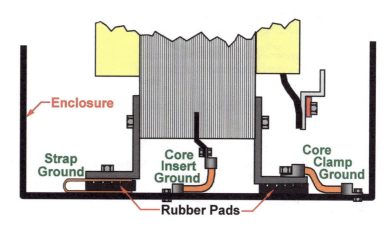

The transformer manufacturer's responsibility is to safety ground the magnetic core of the transformer to the enclosure. Underwriters Laboratories requires this safety measure, since the core is usually floated on rubber vibration isolating pads, electrically isolating it from the enclosure. Without this ground, it may be possible for the core to come in contact with live voltage without activating overcurrent protection on the primary, which would create a hazard for maintenance personnel. This ground strap is typically implemented as a strap bolted to the core clamping angles and the enclosure. Sometimes a tab is brought out from between the core laminations with an enclosure ground strap bolted to it. In other transformers, a simple flat conductor is used to bridge around the rubber vibration pads on one mounting foot. These three examples of core grounding are illustrated above.

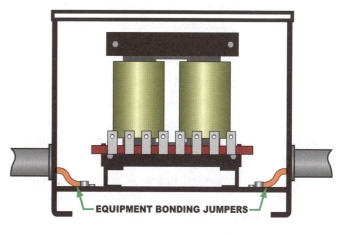

When installers wire transformers into electrical systems, they are responsible for connection of equipment grounding to the transformer enclosure. This provides a continuous path for ground between the primary and secondary distribution systems. This can be done with ground wires carried within the conduits, or conduit hub grounding lugs if the conduit itself is being used to carry equipment ground. Installers may use the same ground point that the transformer manufacturer used for core grounding, if available, or they may choose to install ground lugs in the transformer enclosure at convenient locations. None of the major US transformer manufacturers provide ground lugs in the enclosure for this purpose.

Finally, in the event that there is the need to ground the neutral on the secondary circuit in the transformer, that ground must be carried from the neutral terminal to equipment ground. In addition, a connection must be made to a legitimate building ground, such as a cold water pipe, structural steel or driven ground rod.

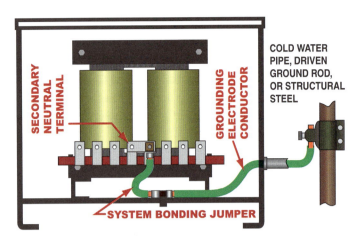

Contractors are sometimes required to create an isolated ground, or IG system, at the transformer secondary. Many installers have conceptual problems with IG systems. Some think that it means that the secondary neutral is not grounded, others interpret that a separate ground bar is required in the transformer. Neither of these is correct. To originate an IG ground wire at the transformer, simply carry a separate, IG ground conductor from the equipment ground point in the transformer to an isolated ground bar in the secondary panel. The isolated ground and the equipment ground originate from the same ground point within the transformer, where they are bonded to building ground, but remain separated in the panel and in all branch circuits. The purpose of isolated ground is to provide a single point ground for all load point receptacles, rather than the multiple grounds and potential ground current loops associated with a traditional, daisy chained equipment ground schemes. It's thought that IG systems provide improved resistance to load equipment disturbances caused by electrical noise. The minimum requirements for isolated ground systems are outlined in the National Electrical Code.

Grounding Summary

It's far too common for contractors and facility maintenance personnel to install transformers with improper or missing ground connections. Failure to ground secondary distribution systems has resulted in catastrophic load equipment damage when high level impulses or lightning hits occurred on the primary service. It takes very little energy to drive an ungrounded system to very high voltage levels above ground, easily breaking down wiring and equipment insulation. Although intentionally ungrounded systems exist in certain industrial installations, those companies accept the attendant risks.

Symptoms of Ungrounded Systems

Those AC systems that are required to be grounded are addressed in NEC® 250.20. One system that is required to be grounded is the 120/240 Volt, 1-phase, 3-wire secondary from a transformer. The neutral conductor, the equipment grounding conductors, and the grounding electrode system should be electrically connected together so that there is no difference of potential among them. Failure to install a system bonding jumper between the neutral and the equipment grounding conductors would be indicated by voltage readings taken between the neutral and any grounded surface or conductor. Since there should be no potential difference, the voltmeter should indicate 0 volts, or very close to zero.

Another example of an improperly grounded system is for a 120/208 V, 3-ph, 4-W system. Of course, line-to-neutral and line-to-ground voltage measurements for this system should be 120 volts. If the system were not properly grounded the following voltage measurements might be made: any of the three secondary lines to ground are 75V, 147V and 80V respectively, and the neutral to ground voltage is 50 volts.

Testing a "Good" Transformer

Transformer windings consist of many coils of wire wrapped around a core. Each end of a winding is labeled or marked, H1 & H2 for example. This winding is *electrically continuous* from one end (H1) to the other (H2). An **electrically continuous** winding (or conductor) has very little resistance between each end. An ohmmeter reading should indicate somewhere between 1 and maybe 20 ohms. This is commonly referred to as *continuity**. The actual resistance varies with the size of the transformer because larger transformers have to carry more current and so larger coil wire is used. And, of course, larger wire has less resistance than smaller wire. For the figure shown below the meter should indicate between 1 and 20 ohms on this digital meter if the leads are placed between:

H1 and H2 or
H3 and H4 or
X1 and X2 or
X3 and X4

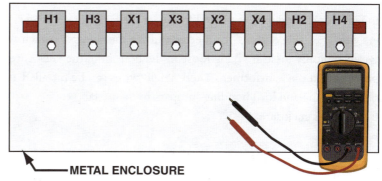
METAL ENCLOSURE

The core is insulated from the windings of wire and the coils do not touch each other electrically because the wire itself is insulated. In larger transformers the core is mounted on rubber isolation mounts and must be bonded to the metal enclosure. Typically a flat metal strip is used and this is installed at the factory. Smaller transformer cores are attached to the metal enclosure and are directly bonded through the attachment. Examples of small transformer cores with mounting "feet" are shown in Figure 12-4 in your Electrical Systems textbook. As previously discussed each transformer winding (X1-X2 for example) is insulated from the other windings (X3-X4 for example) as well as the metal enclosure surrounding the windings. This can be tested using an ohmmeter. For the figure shown above, the meter should indicate OL** on this digital meter if the leads are placed between:

H1	and	H3	or	H4	or	X1	or	X2	or	X3	or	X4			
H2	and	H3	or	H4	or	X1	or	X2	or	X3	or	X4			
H3	and	H1	or	H2	or	X1	or	X2	or	X3	or	X4			
H4	and	H1	or	H2	or	X1	or	X2	or	X3	or	X4			
X1	and	X3	or	X4	or	H1	or	H2	or	H3	or	H4			
X2	and	X3	or	X4	or	H1	or	H2	or	H3	or	H4			
X3	and	X1	or	X2	or	H1	or	H2	or	H3	or	H4			
X4	and	X1	or	X2	or	H1	or	H2	or	H3	or	H4			
Enclosure	and	X1	or	X2	or	X3	or	X4	or	H1	or	H2	or	H3	or	H4

Some transformers have shields that are inserted between the primary and secondary windings. The 3Ø transformer at the right is in the process of assembly. Note the **three** separate 1Ø transformers that are contained within the enclosure for this **three-phase** transformer. A dual-voltage **single-phase** transformer would have **two** separate 1Ø transformers within the enclosure. Note also in the picture that the secondary and primary windings for each of the 1Ø transformers are wound around the same core. This is a typical method of construction. The shields for this particular 3Ø transformer have been connected together and bonded to the metal frame, at the far right. However, many manufacturers ship their transformers with the "shield wires" connected only to the shields on each winding set. The other end of each of these wires, typically these are lying loose in the bottom of the enclosure, must be field connected to the grounding point within the transformer. These shield wires can be installed at the same "grounding" point used to connect your line and load bonding jumpers to the metal transformer enclosure.

Yr 2 – Annex B
Connection Diagram
2 Shielded Transformer

The connection diagrams shown in Annex B indicate if the transformer is shielded. Note that Connection Diagram #1 is for non-shielded transformers. Connection diagram #2 is for shielded transformers. This is indicated by the dashed line between the primary and secondary windings.

Checking a Transformer that Feeds a 120/240 volt, 1Ø Panelboard

Where a dual-voltage transformer is correctly connected to supply power to a 120/240 V, 1Ø, 3-wire panel, resistance measurements can be misleading. When this separately-derived system is properly connected, grounded and bonded, there should be an electrically-continuous path between all conductors, raceways and metal enclosures that are required by the NEC to be grounded and bonded. Remember, electrically continuous means that there should be 0 Ω resistance between these points.

Is the transformer bad?

Typically, when overloaded beyond the current carrying capacity of a winding, the coil of wire used for the winding will burn in two just like any other conductor. This, of course, could be indicated by a meter reading. You should expect to see "continuity" between X1 and X2 on the order of 1 to 20 ohms. However, if your ohmmeter indicates OL or something close to it such as 9.3 MΩ then the winding has opened (or nearly opened) and the transformer should be replaced.

One type of failure is for the insulated winding to "leak"*** to the core, or the enclosure, or to another winding. In any of these situations, if an ohmmeter reading between a winding lead and the metal case (to which the core is connected) shows something other than perfect insulation then the transformer is suspect and may need replacement****. This could have been caused by a severe overcurrent condition, such as a short circuit or ground fault having occurred, that damaged the insulation. Remember that high currents cause large

Single-Phase Transformers, Part II

magnetic fields that actually move the conductors, sometimes enough to damage them. The damage could also have been caused during shipment or installation if the transformer was dropped or otherwise roughly treated.

Another type of failure is where one winding is no longer perfectly insulated from another. This could be apparent if the transformer is rated to deliver 120 volts on the secondary when supplied on the primary with 240 volts but is only putting out 90 volts with a 240 volt input. This could be caused by several coils of one secondary winding having shorted together thus actually lowering the number of turns on the secondary. Remember $E_p \div E_s = N_p \div N_s$. The reverse could (theoretically) also happen. That is some of the primary coils short together which would cause the output on the secondary to be too high. However, when this happens and normal primary voltage is applied to the shorter primary winding it will most likely burn in two because the primary impedance was changed. $I = E \div Z$ and with a lower Z the primary current would be too high.

* **Continuity** can mean 0.2 Ω between a transformer core and the metal enclosure, 5 Ω for a good transformer winding, 220 Ω for a good incandescent lamp, or 4000 Ω for a good coil in an ice cube relay. Basically, any type of expected resistance within these ranges is referred to as having continuity because the conductor (or lamp filament) is continuous. Remember that if a meter is used in the continuity setting (audible) that beyond a certain resistance (90 Ω on some meters and 1000 Ω on others) the meter will indicate OL (or ∞) because it is out of range although the resistance isn't actually infinity.

** **OL** on a digital meter indicates perfect insulation when the meter is set in the resistance (Ω) position. If using an analog meter, an indication of perfect insulation would be ∞. Infinite resistance (infinity) is ∞ on an analog meter or OL on a digital meter.

*** **Leak** or leakage is less than perfect insulation where perfect insulation is expected.

**** **Caution**: to avoid misinterpretation of your measurements these tests should only be performed on transformers that have not had any leads (or lugs) interconnected or grounded, as you would with X2-X3 on a 120/240 volt secondary for example.

216.4 Single-Phase Fault Currents

Impedance (Z) is the current-limiting characteristic of a transformer and is expressed in percentage. This information is available on the nameplate of a transformer. Z can be used to calculate the available fault current on the secondary of a transformer should a bolted fault occur. A bolted fault is where accidental contact occurs between two secondary ungrounded conductors. Recall from 1st Year that EWR Unit 28 addressed these calculations to determine the interrupting ratings required for OCPDs. EWR Unit 28 uses a 2-step method to calculate bolted SECONDARY fault current. However, a single formula, combining both steps into one, can be used instead. This method requires that Z in % be changed to decimal form.

$$I_{sc} = VA \div [E_L \times \text{decimal } Z]$$

Single-Phase Transformers, Part II   216-421

Objective 216.1 Worksheet

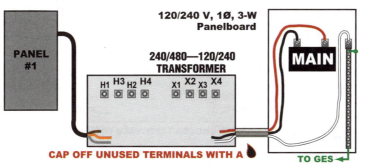

Figure 216.1

_____ 1. Refer to Figure 216.1, shown above, and the information in Yr-2 Annex B for a catalog number 1112, GROUP I, 5 KVA transformer. Which of the transformer connections, shown below, is correct? The transformer is fed from a 2-pole breaker in Panel #1 which is a 277/480 volt 3Ø, 4-wire panel.

a. 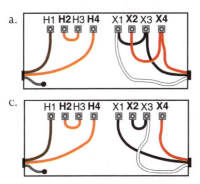 b.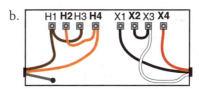

c. d. none of these

_____ 2. In Yr-2 Annex B, ___ KVA or larger Group I (240 x 480 primary volts – 120/240 secondary volts) transformers can NOT be wall mounted.

a. 5   b. 25   c. 15   d. 75

_____ 3. You have a 480 volt circuit available to supply power to a 500 VA Group I transformer Cat. #1105 (Yr-2 Annex B). You will connect your supply circuit (L1 and L2) as shown in drawing ___ below.

a. 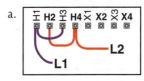 b.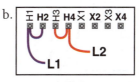

c. 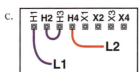 d.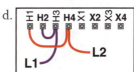

Year Two (Student Manual)

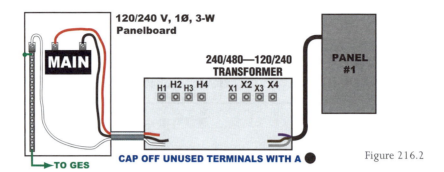

Figure 216.2

_____ 4. Refer to Figure 216.2, shown above, and the information in Yr-2 Annex B for a catalog number 1112, GROUP I, 5 KVA transformer. Which of the transformer connections, shown below, is correct? The transformer is fed from a 2-pole breaker in Panel #1 which is a 277/480 volt 3Ø, 4-wire panel.

a. 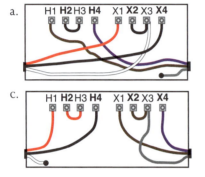 b.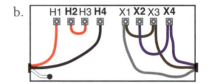

c. 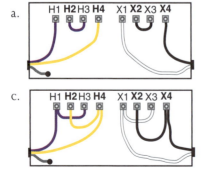 d. none of these

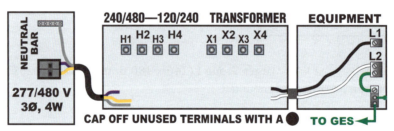

Figure 216.3

_____ 5. Refer to Figure 216.3, shown above, and the information in Yr-2 Annex B for a catalog number 1112, GROUP I, 5 KVA transformer. Which of the transformer connections, shown below, is correct? The equipment is rated to operate at 120 volts and includes a disconnecting means.

a. 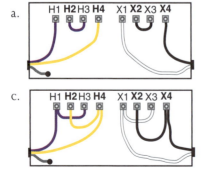 b.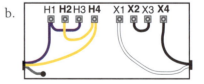

c. 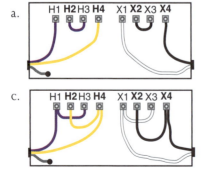 d. none of these

Single-Phase Transformers, Part II

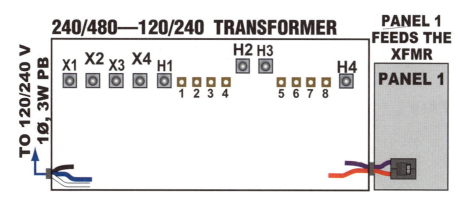

Figure 216.4

Terminal lugs are used in this transformer and any lugs not used are NOT required to be "capped off".

_____ 6. Refer to Figure 216.4, shown above, and the information in Yr-2 Annex B for a GROUP I – 37.5 KVA transformer, catalog number 1118. Panel #1 is 277/480 V, 3Ø, 4-W. The voltage where the primary circuit enters the transformer enclosure **has dropped to** 468 volts. Which of the transformer connections, shown below, is correct?

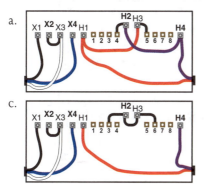
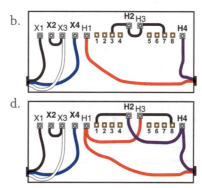

_____ 7. Refer to Figure 216.4, shown above, and the information in Yr-2 Annex B for a GROUP I – 37.5 KVA transformer, catalog number 1118. Panel #1 is 240 V, 3Ø, 3W. The voltage where the primary circuit enters the transformer enclosure **has dropped to** 216 volts. Which of the transformer connections, shown below, is correct?

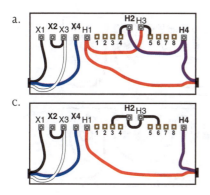
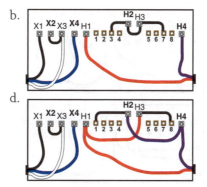

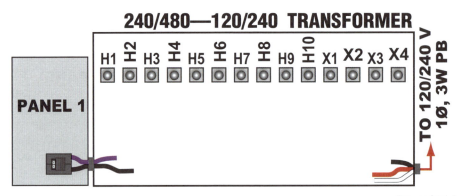

Figure 216.5

This transformer has leads and any unused leads must be "capped off".

_____ 8. Refer to Figure 216.5, shown above, and the information in Yr-2 Annex B for a GROUP I – 25 KVA transformer, catalog number 1117. Panel #1 (277/480 V, 3Ø, 4-W) feeds the transformer. The **actual** voltage where the primary circuit enters the transformer enclosure is 504 volts. Which of the transformer connections, shown below, is correct?

a. 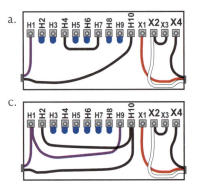 b.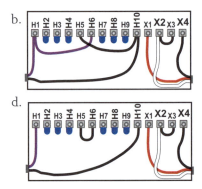

c. 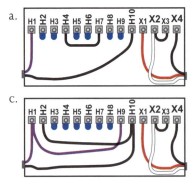 d.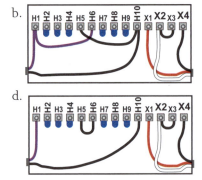

_____ 9. Refer to Figure 216.5, shown above, and the information in Yr-2 Annex B for a GROUP I – 25 KVA transformer, catalog number 1117. Panel #1 (240 V, 3Ø, 3-W) feeds the transformer. The **actual** voltage where the primary circuit enters the transformer enclosure is 252 volts. This transformer has **leads** and any unused leads have been "capped off". Which of the transformer connections, shown below, is correct?

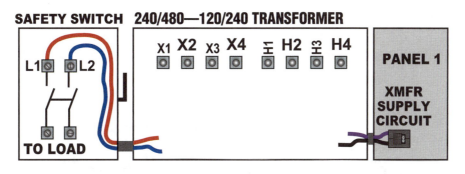

Figure 216.6

_____ 10. Refer to Figure 216.6, shown above, and the information in Yr-2 Annex B for a 3 KVA GROUP I transformer - catalog #1110. The safety switch supplies a 240 volt load. Panel 1 is 277/480 V, 3Ø, 4-W. Which of the transformer connections, shown below, is correct?

a. 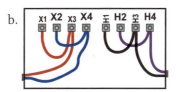 b. (diagram)

c. 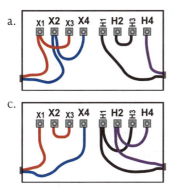 d. none of these

_____ 11. Refer to Figure 216.6, shown above, and the information in Yr-2 Annex B for a 3 KVA GROUP I transformer - catalog #1110. The safety switch supplies a 240 volt load. Panel 1 is 120/240 V, 3Ø, 4-W. Which of the transformer connections, shown below, is correct?

a. 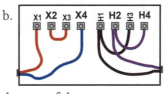 b. (diagram)

c. 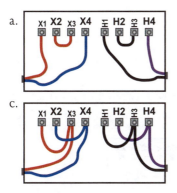 d. none of these

_____ 12. A .25 KVA Group I (Yr-2 Annex B) 480 x 240 primary - 240/120 secondary volt single-phase transformer has ___ **primary** ("H") field connections that are brought out of the windings as leads or terminal lugs.

   a. 2   b. 4   c. 7   d. 8

_____ 13. A Square D transformer has a nameplate that shows 240/480 V pri – 120/240 V sec. This transformer can be connected to deliver rated voltage where the 2-wire supply circuit is from a 277/480 volt 3Ø, 4-wire panelboard. The circuit conductors are connected to a 1-pole breaker in space 5 and the neutral bar.

   a. True   b. False

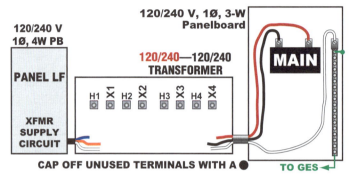

Figure 216.9

_____ 14. A *single-pole* breaker in Panel LF feeds the transformer primary. Refer to Figure 216.9 and Yr-2 Annex B for a 25 KVA GROUP III transformer Cat #1309. The panel on the right will be used for computer equipment. Which of the transformer connections, shown below, is correct?

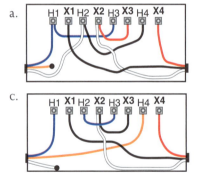

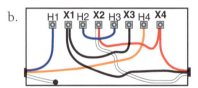

d. none of these

_____ 15. You have a 240 volt circuit available to supply power to a .5 KVA Group I transformer Cat. # 1105 (Yr-2 Annex B). You will connect your supply circuit (L1 and L2) as shown in drawing ___ below.

a.
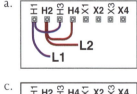

b.
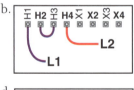

c.
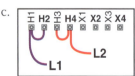

d.
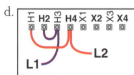

16. A Square D transformer with a nameplate rating of 240/480 V pri — 120/240 V sec can be connected to a 2-pole breaker in a ___ panelboard and deliver nameplate rated voltage to a load.

   I. 120/240 V 3Ø, 4-W
   II. 277/480 V 3Ø, 4-W

   a. either I or II
   b. II only
   c. neither I nor II
   d. I only

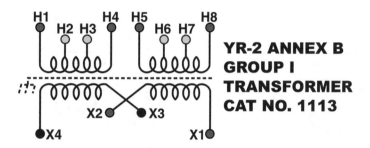

Figure 216.8

17. Refer to Figure 216.8, shown above. 468 volts is the actual supply voltage but 240 volts output is required. The portions of the primary windings between ___ are used in this situation.

   a. H1 & H3 and H6 & H8
   b. H1 & H2 and H5 & H7
   c. H1 & H3 and H5 & H8
   d. H1 & H2 and H5 & H6
   e. H1 & H4 and H5 & H8

18. Refer to Figure 216.8, shown above. 504 volts is the actual supply voltage but 240 volts output is required. The portions of the primary windings between ___ are used in this situation.

   a. H1 & H3 and H6 & H8
   b. H1 & H2 and H5 & H7
   c. H1 & H3 and H5 & H8
   d. H1 & H2 and H5 & H6
   e. H1 & H4 and H5 & H8

_____ 19. You have a 500 VA Group I transformer Cat. #1105 (Yr-2 Annex B). You need to have 240 volts for a piece of equipment. You will connect your load circuit (L1 and L2) as shown in drawing ___ below.

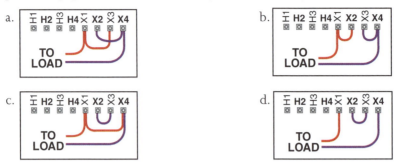

_____ 20. You have a 500 VA Group I transformer Cat. #1105 (Yr-2 Annex B). You should connect your load circuit as shown in drawing ___ below to supply 120 volts to a piece of equipment.

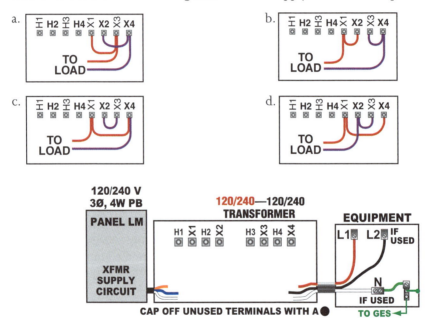

Figure 216.10

_____ 21. A 2-pole breaker in Panel LM feeds the transformer primary. The EQUIPMENT load is rated to operate at 120 volts. Refer to Figure 216.10 and Yr-2 Annex B for a 15 KVA GROUP III transformer Cat #1308.

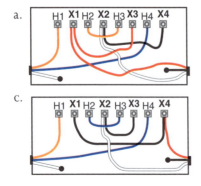

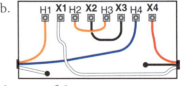

d. none of these

Single-Phase Transformers, Part II
216-429

_____ 22. For a 500 VA Group I transformer Cat. #1105 (Yr-2 Annex B) you should connect a 120/240 volt 1Ø, 3-wire load circuit as shown in drawing ___ below.

a.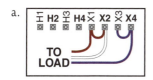

b.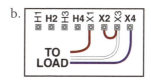

c.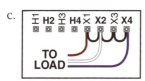

d.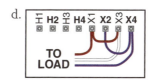

_____ 23. A Square D transformer has a nameplate that shows 240/480 V pri – 120/240 V sec. This transformer can be connected to deliver rated voltage where the supply circuit is from a 120/240 volt 1Ø, 3-wire panelboard. The circuit conductors are connected to a 2-pole breaker that is located in spaces 2-4.

   a. False   b. True

_____ 24. You have a 7.5 KVA Group III transformer Cat. # 1306 (Yr-2 Annex B). You will connect your available 240 volt supply circuit (L1 and L2) as shown in drawing ___ below.

a.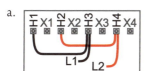

b.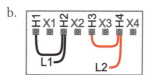

c.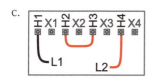

d.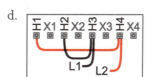

_____ 25. You have a 240-volt circuit available to supply power to a 7.5 KVA Group III transformer Cat. # 1306 (Yr-2 Annex B). You need to have 120/240 volts 3-W for a piece of equipment. Assuming that you correctly connected your supply circuit you will connect your 120/240 volt 1Ø, 3-wire load circuit as shown in drawing ___ below.

a.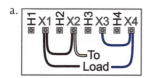

b.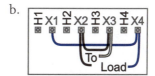

c.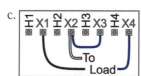

d.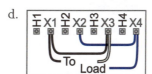

Year Two (Student Manual)

Single-Phase Transformers, Part II

Objective 216.2 Worksheet

_____ 1. The load to be connected to a new transformer is 62.5 KVA. The specifications require that the transformer be sized at 115% of the connected load to allow for future expansion. The minimum calculated size of the transformer is ___ kVA.

    a. 71.875   b. 62.5   c. 718.75   d. 625   e. none of these

_____ 2. To allow for future expansion the specifications limit transformer loading to 85%. The maximum load that can be connected to a 150 KVA transformer is ___ KVA.

    a. 176.47   b. 127.5   c. 162.75   d. 150   e. 137.25

_____ 3. An existing transformer is to be replaced with a new one. The specifications require that a new transformer be sized at 125% of the existing load to allow for additional loads in the future. The minimum calculated size of the transformer is ___ kVA. The existing load connected to the existing transformer is 79.2 KVA.

    a. 63.36   b. 125   c. 99   d. 19.8   e. 59.4

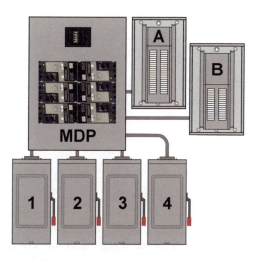

Figure 216.21

_____ 4. A transformer feeds the MDP shown above in Figure 216.21. The transformer must have a minimum (calculated) rating of ___ KVA to serve the following loads:
Panel A is 63 A at 120/240 V, 1Ø, 3W   Panel B is 10 kW at 240 V 1Ø
Switch 1 is 12 A at 240 V 1Ø   Switch 2 is 1200 W at 240 V 1Ø
Switch 3 is 21 A at 240 V 1Ø   Switch 4 is 16 A at 240 V 1Ø

    a. 123.2   b. 38.08   c. 158.667   d. 30.52   e. none of these

_____ 5. A .75 KVA, 120 Volt, 1Ø transformer loaded at 90% is delivering ___ amps to the connected loads.

    a. 5.625   b. .625   c. 8.1   d. 6.25   e. none of these

_____ 6. When fully loaded a 480 volt pri/240 volt sec 37.5 KVA 1Ø, 3-wire transformer can deliver ___ amps to a 240 volt load.

    a. 37.5   b. 78.125   c. 156.25   d. 90.211   e. none of these

_____ 7. A 120/240 volt 1Ø, 3-wire panelboard is to be fed from a transformer. The minimum calculated (NOT STD) size of the transformer is ___ kVA where the following loads are connected:
Circuit 1:   120 volt lighting that draws 14 amps
Circuit 3-5:   12.1 amp 240 volt swimming pool motor
Circuit 2-4:   14 kW 240 volt heater
Circuit 6:   5.2 amp 120 volt motor
Circuit 8:   11 amps of 120 volt lighting

    a. 21.512
    b. 24.152
    c. 20.528
    d. 18.630

_____ 8. Some transformers are provided with two secondary windings (coils). When the current through each of these two windings is equal the transformer is ___.

    a. overloaded
    b. in resonance
    c. electrically balanced
    d. tapped out

_____ 9. **Standard** 1Ø Group I transformer sizes are as shown in the KVA column in Yr-2 Annex B. If you had to select a transformer to serve a 240 volt load, that would draw 52 amps, you would select a ___ KVA transformer as the most economical choice.

    a. 10
    b. 25
    c. 15
    d. 7.5
    e. 5

Single-Phase Transformers, Part II

_____ 10. Which of the following is a standard size single-phase (240 x 480 volt primary – 120/240 volt secondary) dry-type distribution transformer? Refer to Yr-2 Annex B, Group I.
   I. 30 KVA   II. 45 KVA   III. 50 KVA

   a. I, II, & III
   b. II & III only
   c. II only
   d. I & III only
   e. III only

_____ 11. Refer to Yr-2 Annex B. A transformer will be supplied from a 240 volt 1-ph circuit that originates in a 240 V, 3-ph, 3-w panelboard. The transformer will supply power to a 120/240 V, 1-ph, 3-w panelboard. A Cat #____ is the smallest Group I transformer that can be used given these loads: There will be 132 amps of 120 volt loads connected to L1 in the panel and 122 amps of 120 volt loads connected to L2 in the panel. There will also be 52 amps of 240 volt loads connected in the panel.

   a. 1116   b. 1117   c. 1115   d. 1119   e. 1118

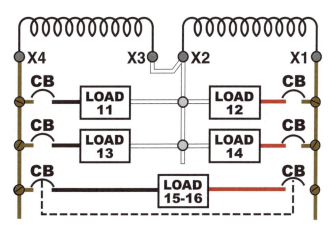

Figure 216.22

_____ 12. For clarity, only the secondary windings of a transformer are shown above in Figure 216.22. The transformer nameplate rating is 25 KVA with these voltage ratings: 240 x 480 primary volts – 120/240 secondary volts. Which load distribution (a, b, or c) offers the best (most balanced) distribution without exceeding the ratings of the transformer?

   a. Load 11 = 5.5 KVA   b. Load 11 = 2.4 KVA   c. Load 11 = 5.5 KVA
      Load 12 = 2.4 KVA   Load 12 = 5.5 KVA   Load 12 = 6.2 KVA
      Load 13 = 6.2 KVA   Load 13 = 6.2 KVA   Load 13 = 2.4 KVA
      Load 14 = 4.7 KVA   Load 14 = 4.7 KVA   Load 14 = 4.7 KVA
      Load 15-16 = 2.6 KVA   Load 15-16 = 2.6 KVA   Load 15-16 = 2.6 KVA

_____ 13. Refer to Figure 216.22, shown above. The transformer nameplate rating is 37.5 KVA with these voltage ratings: 240 x 480 primary volts – 120/240 secondary volts. Where the loads are as listed below the transformer is ___.

Load 11 = 8.4 KVA Load 12 = 9.4 KVA Load 14 = 4.3 KVA
Load 15-16 = 8.2 KVA Load 13 = 7.2 KVA

a. operating within its ratings
b. overloaded

_____ 14. Refer to Yr-2 Annex B. A transformer will be supplied from a 208 volt 1-ph circuit that originates in a 120/208 V, 3-ph, 4-w panelboard. The transformer will supply 240 volt power to a 1-ph load. The nameplate information on the load is "64 amps at 240 volts". A Cat #___ is the smallest Group V transformer that can be used.

a. 1506
b. 1509
c. 1507
d. 1505
e. 1508

_____ 15. A 50 KVA 1Ø transformer has a 120/240-volt secondary. The existing load is 143 amps on L1 and 172 amps on L2. Which of the following loads could be added without overloading the transformer?
I. a resistance heating unit rated for 8.6 KW at 240 volts; **plus** a 15 amp load consisting of 120 volt lighting (connected to L1)
II. 120 volt lighting with 30 amps connected to L1 and 35 amps connected to L2

a. Only I can be added
b. Only II can be added
c. Both I & II can be added
d. Either I or II but NOT both can be added
e. Neither I nor II can be added

_____ 16. A single-phase 240 volt load is drawing 43 amps and connected to the secondary of a 25 KVA 480 primary-to-240 secondary volt transformer. How much power, in kVA, is supplied to the load by the secondary winding?

a. 12.5   b. 25   c. 30.96   d. 10.32   e. 5.58

Objective 216.3 Worksheet

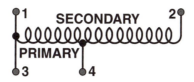

Figure 216.31

_____ 1. Refer to Figure 216.31, shown above. Using an ohmmeter on the autotransformer, ___ ohms would be read between points 1 and 3.

   a. 0   b. ∞ (or OL)   c. 2 to 20

_____ 2. Refer to Figure 216.31, shown above. Using an ohmmeter on the autotransformer, ___ ohms would be read between points 2 and 3.

   a. 0   b. ∞ (or OL)   c. 2 to 20

_____ 3. Refer to Figure 216.31, shown above. Using an ohmmeter on the autotransformer, ___ ohms would be read between points 3 and 4.

   a. 0   b. ∞ (or OL)   c. 2 to 20

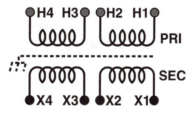

Figure 216.37

The transformer has NOT been connected or energized -- it has just been removed from the shipping container.

_____ 4. Using an ohmmeter on the transformer winding connection points, ___ Ω would be read between X1 and X2. Refer to Figure 216.37, shown above.

   a. 2 to 20   b. ∞ (or OL)   c. 0

_____ 5. Using an ohmmeter on the transformer winding connection points, ___ Ω would be read between X2 and X3. Refer to Figure 216.37, shown above.

   a. ∞ (or OL)   b. 0   c. 2 to 20

_____ 6. Using an ohmmeter on the transformer winding connection points, ___ Ω would be read between H1 and H4. Refer to Figure 216.37, shown above.

   a. ∞ (or OL)   b. 0   c. 2 to 20

_____ 7. Using an ohmmeter on the transformer winding connection points, ___ Ω would be read between H3 and H4. Refer to Figure 216.37, shown above.

   a. ∞ (or OL)   b. 0   c. 2 to 20

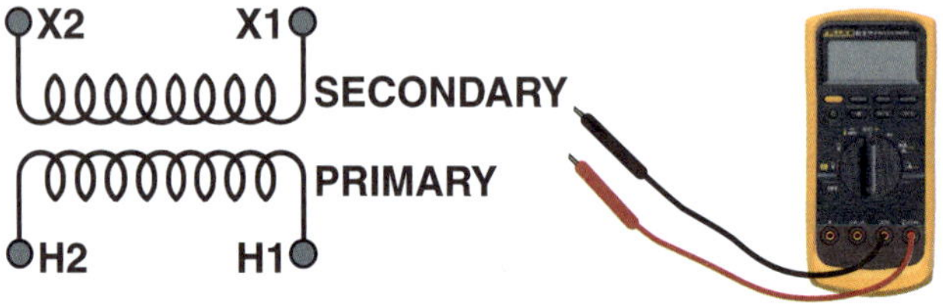

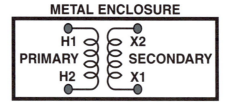

Figure 216.32

_____ 8. Refer to Figure 216.32, shown above. Using an ohmmeter on the transformer winding connection points, ___ Ω would be read between X1 and X2.

   a. 0   b. ∞ (or OL)   c. 2 to 20

_____ 9. Refer to Figure 216.32, shown above. Using an ohmmeter on the transformer winding connection points, ___ Ω would be read between X1 and H2.

   a. 0   b. ∞ (or OL)   c. 2 to 20

Figure 216.34

The transformer has NOT been connected or energized -- it has just been removed from the shipping container.

_____ 10. Using an ohmmeter on the transformer winding connection points, shown above in Figure 216.34, ___ Ω would be read between X1 or X2 or H1 or H2 and the metal transformer enclosure.

   a. 0   b. ∞ (or OL)   c. 2 to 20

Single-Phase Transformers, Part II 216-439

_____ 11. Using an ohmmeter on the transformer winding connection points, shown above in Figure 216.34, ___ Ω would be read between X1 or X2 or H1 or H2 and the metal core.

    a. 0   b. ∞ (or OL)   c. 2 to 20

_____ 12. A transformer has a dual-voltage primary and a dual-voltage secondary with four "X" leads and four "H" leads. No field connections have been made yet. Using an ohmmeter you should get a reading of ___ ohms between the metal enclosure and X1.

    a. 0.2   b. between 2 & 20   c. OL (or ∞)

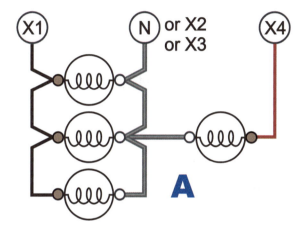
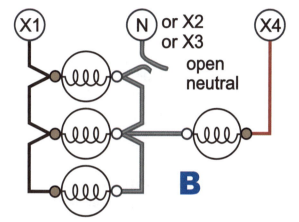

Drawing A shows a lamp load connected so that 3 lamps are on one circuit and 1 lamp is connected to a second circuit. These two circuits are on opposite phases in the 120/240 volt 1Ø, 3-wire panel, so you shared a neutral for the lamp circuits.

Drawing B shows these circuits energized when the neutral was disconnected, in a junction box, between the lamps and the panel.

Figure 216.39

Each lamp has a rating of 130 volts and 100 watts.

_____ 13. Refer to Figure 216.39A, shown above. The single lamp is ___.

    a. brighter than any lamp in the "3 Lamp Set"
    b. dimmer than any lamp in the "3 Lamp Set"
    c. OFF
    d. the same intensity as any lamp in the "3 Lamp Set"

_____ 14. Refer to Figure 216.39A, shown above. The rated current of each lamp is ___ A.

    a. 1.07   b. .36   c. .77   d. 1.3   e. 2.82

Year Two (Student Manual)

_____ 15. Refer to OPEN NEUTRAL - Figure 216.39B, shown above. The lamps in the "3-lamp set" are ___.

   a. any lamp in the "3-lamp set"
   b. dimmer than any lamp in the "3-lamp set"
   c. OFF
   d. the same intensity as any lamp in the "3-lamp set"

_____ 16. Refer to OPEN NEUTRAL - Figure 216.39B, shown above. The single lamp is ___.

   a. brighter than the single lamp
   b. dimmer than the single lamp
   c. OFF
   d. the same intensity as the others

_____ 17. Refer to OPEN NEUTRAL - Figure 216.39B, shown above. About ___ A of current is now flowing between X4 and the single lamp.

   a. 1.07   b. .36   c. .77   d. 1.3   e. 2.82

_____ 18. Refer to OPEN NEUTRAL - Figure 216.39B, shown above. The current through each lamp in the "3-lamp set" is ___ than the rated current.

   a. higher   b. lower

_____ 19. Refer to OPEN NEUTRAL - Figure 216.39B, shown on the previous page. Each lamp in the "3-lamp set" has ___ volts dropped across it.

   a. 120   b. 60   c. 240   d. 180   e. 144

_____ 20. Refer to OPEN NEUTRAL - Figure 216.39B, shown on the previous page. The single lamp is using about ___ watts.

   a. 192   b. 60   c. 180   d. 100   e. 21

_____ 21. Refer to OPEN NEUTRAL - Figure 216.39B, shown on the previous page. The single lamp is delivering ___ than its rated power output.

   a. more   b. less

Single-Phase Transformers, Part II

The transformer, shown below, is NOT energized but HAS been properly connected, grounded, and bonded. The secondary (L1, L2, and N) has been connected to a 120/240 volt, 1-ph, 3-wire main breaker panelboard where the main bonding jumper was installed. The iron core was bonded to the metal enclosure at the factory. You connected the shield wires to the enclosure.

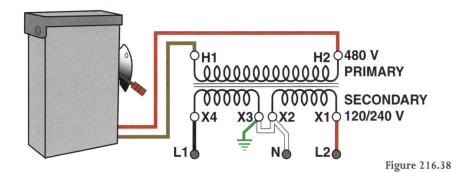

Figure 216.38

_____ 22. Refer to Figure 216.38, shown above. If the leads of an ohmmeter are placed between L2 & the shielding wire connection the meter should indicate a reading in the range of ___.

  a. 0.2 to 20 Ω   b. ∞ Ω (or OL)

_____ 23. Refer to Figure 216.38, shown above. If the leads of an ohmmeter are placed between N & the shielding wire connection the meter should indicate a reading in the range of ___.

  a. 0.2 to 20 Ω   b. ∞ Ω (or OL)

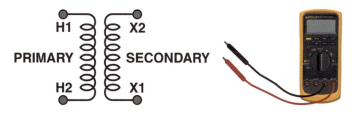

Figure 216.33

_____ 24. Refer to Figure 216.33, shown above. Using an ohmmeter on the transformer winding connection points, shown below, ___ Ω would be read between H1 and H2.

  a. 0   b. ∞ (or OL)   c. 2 to 20

_____ 25. Refer to Figure 216.33, shown above. Using an ohmmeter on the transformer winding connection points, ___ Ω would be read between X2 and H1.

  a. 0   b. ∞ (or OL)   c. 2 to 20

Year Two (Student Manual)

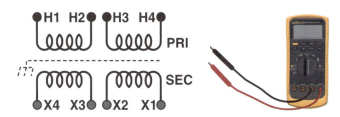

Figure 216.35

The transformer, shown above, is NOT energized but HAS been properly CONNECTED, grounded, and bonded. It has a primary rated for 240/480 and a secondary rated for 120/240. The secondary has been connected to a 120/240 volt, 1-ph, 3-wire main breaker panelboard. The shielding wires have been bonded to the metal enclosure.

_____ 26. Refer to Figure 216.35, shown above. If the ohmmeter probes (leads) are placed between X1 and the metal core (not shown) ___ Ω would be indicated on the meter.

    a. 0   b. ∞ (or OL)   c. 2 to 20

_____ 27. Refer to Figure 216.35, shown above. If the ohmmeter probes (leads) are placed between X3 and the metal transformer enclosure (not shown) ___ Ω would be indicated on the meter.

    a. 0   b. ∞ (or OL)   c. 2 to 20

Objective 216.4 Worksheet

_____ 1. A transformer is located in the rear corner of a residential lot. The transformer supplies four homes. It is marked 167 kVA, 120/240 volts, 2% impedance. The available line-to-line fault current at the load terminals of the transformers is ___ amps.

    a. 3479   b. 696   c. 1392   d. 34,792   e. 13,917

_____ 2. A 120/240 volt 1-ph, 3-wire panel is fed from a 50 KVA 1Ø transformer that has Z = 2.3% on its nameplate. If the two ungrounded conductors are connected together using a 4/0 AWG THHN/THWN copper jumper ___ amps of current will flow through the 4/0.

    a. 5217   b. 417   c. 104   d. 208   e. 9058

_____ 3. A 167 KVA 120/240 volt 1Ø, 3-wire transformer has Z = 1.83% on its nameplate. The available line-to-line (bolted) fault current on the secondary is ___ amps. The main breaker in the panelboard that this transformer feeds should have an AIC rating of at least this value.

    a. 21,902   b. 43,873   c. 76,047   d. 38,024   e. 91,257

_____ 4. A 120/240 volt 1-ph, 3-wire panel is fed from a 50 KVA 1Ø transformer that has Z = 2.3% on its nameplate. The transformer is supplied power from a 2-pole breaker in a 277/480 volt 3-ph, 4-wire panel. If X1 and X4 are connected together using a 4/0 AWG THHN/THWN copper jumper ___ amps of current will flow through the 4/0.

    a. 5217   b. 417   c. 104   d. 208   e. 9058

Lesson 216 NEC Worksheet

_____ 1. The requirements for protection of transformer secondaries in Article 450 will provide adequate protection of secondary conductors.

   a. True   b. False

_____ 2. Article 240 applies to the protection of secondary conductors and the protection of transformers.

   a. True   b. False

_____ 3. 1/0 AWG copper THHN/THWN conductors are installed from a 120/240 V, 1-ph, 3-wire panel to the secondary terminals on a 50 KVA transformer. The maximum rating of the main breaker in the panel is ___ amps. The breaker terminals are rated at 75°C.

   a. 200   b. 250   c. 175   d. 225   e. 150

_____ 4. Open-type transformers (one of these is shown below) are permitted to be installed in a ___.
   I. stainless steel control cabinet
   II. fiberglass combination starter enclosure
   III. sheetrock-covered furdown (soffit)

   a. I only
   b. I, II, or III
   c. I or II only
   d. I or III only
   e. none of these

_____ 5. Transformers with ventilating openings shall be installed so that the ventilating openings are not blocked by walls or other obstructions. The required clearances shall be ___.
   a. clearly marked on the transformer
   b. as listed in Article 450
   c. maintained on the top, bottom, and all sides
   d. as given in the job specifications

Lesson 216 Safety Worksheet

Safe Work Practices – Energized Circuits

Safety-related work practices shall be employed to prevent electric shock or other injuries resulting from either direct or indirect electrical contacts, when work is performed near or on equipment or circuits which are or may be energized. The specific safety-related work practices shall be consistent with the nature and extent of the associated electrical hazards.

Live parts to which an employee may be exposed shall be de-energized before the employee works on or near them, unless the employer can demonstrate that de-energizing introduces additional or increased hazards or is infeasible due to equipment design or operational limitations. Live parts that operate at less than 50 volts to ground need not be de-energized if there will be no increased exposure to electrical burns or to explosion due to electric arcs.

Conductive articles of jewelry and clothing (such a watch bands, bracelets, rings, key chains, necklaces, metalized aprons, cloth with conductive thread, or metal headgear) may not be worn if they might contact exposed energized parts. However, such articles may be worn if they are rendered nonconductive by covering, wrapping, or other insulating means.

Employees working in areas where there are potential electrical hazards shall be provided with, and shall use, electrical personal protective equipment that is appropriate for the specific parts of the body to be protected and for the work to be performed.

Employees shall wear nonconductive head protection wherever there is a danger of head injury from electric shock or burns due to contact with exposed energized parts.

Employees shall wear protective equipment for the eyes or face wherever there is danger of injury to the eyes or face from electric arcs or flashes or from flying objects resulting from electrical explosion.

When working near exposed energized conductors or circuit parts, each employee shall use insulated tools or handling equipment if the tools or handling equipment might make contact with such conductors or parts. If the insulating capability of insulated tools or handling equipment is subject to damage, the insulating material shall be protected.

Safety signs and tags. Safety signs or accident prevention tags shall be used where necessary to warn employees about electrical hazards which may endanger them.

Barricades shall be used in conjunction with safety signs where it is necessary to prevent or limit employee access to work areas exposing employees to un-insulated energized conductors or circuit parts. Conductive barricades may not be used where they might cause an electrical contact hazard.

The above paragraphs are taken from OSHA Part 1910 Occupational Safety and Health Standards.

QUESTIONS

_____ 1. Safety related work practices, with energized circuits shall be employed to prevent ___.

   a. electric shock
   b. electrical fault currents
   c. electrical overcurrent
   d. electrical overload

_____ 2. Live parts to which an employee may be exposed shall be ___.

   a. decommissioned   b. destroyed   c. Energized   d. deenergized

_____ 3. Employees working in areas with potential electrical hazards shall use ___ to protect specific parts of the body.

   a. NFPA   b. PPE   c. EPA   d. NBA

_____ 4. When working near exposed energized conductors, each employee shall use ___.

   a. hand tools   b. power tools   c. insulated tools   d. none of these

_____ 5. ___ or accident prevention tags shall be used to warn employees about electrical hazards which may endanger them.

   a. Safety slogans   b. Safety signs   c. both a and b   d. neither a or b

Lesson 217
Mid-Term Review

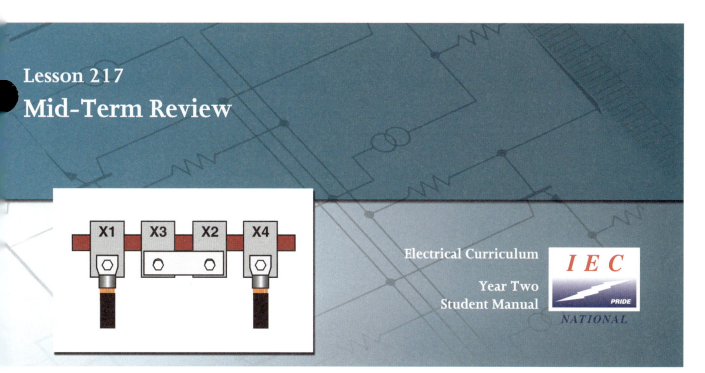

Electrical Curriculum
Year Two
Student Manual

Purpose
The first semester mid-term exam review is scheduled for this lesson.

Homework
(Due at the beginning of this class)
For this lesson, you should:

- Review all your worksheets, homework, and quizzes.

- Complete your "Practice Exam."

- Be prepared to clarify any questions or problem areas you encountered during your review.

Objectives
This lesson will determine your proficiency in the subject matter from the previous lessons. The instructor should review next week's material beforehand to determine if the students will require additional instruction before assigning the worksheets for that lesson.

217-450 Mid-Term Review

Year Two (Student Manual)

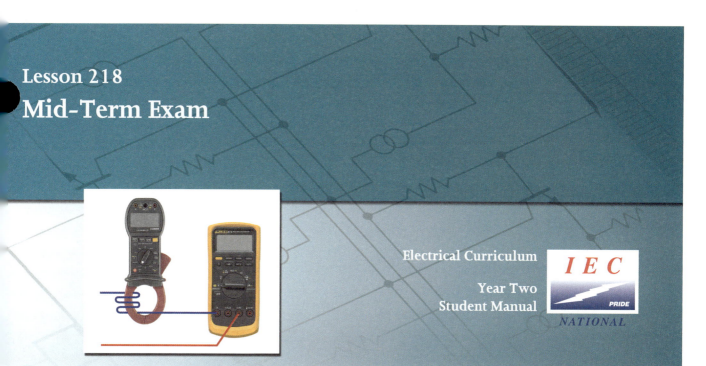

Lesson 218
Mid-Term Exam

Electrical Curriculum

Year Two
Student Manual

IEC
PRIDE
NATIONAL

Purpose
The first semester mid-term exam is scheduled for this lesson.

Homework
(Due at the beginning of this class)

For this lesson, you should:

- Review all your worksheets, homework, and quizzes.
- Complete your "Practice Exam."
- Be prepared to clarify any questions or problem areas you encountered during your review.

Objectives
This lesson will determine your proficiency in the subject matter from the previous lessons. The instructor should review next week's material beforehand to determine if the students will require additional instruction before assigning the worksheets for that lesson.

Lesson 219
Power Generation – Three-Phase Transformers, Circuits, and Calculations

Electrical Curriculum

Year Two
Student Manual

Purpose

This lesson will discuss sources of power and how power is transmitted and distributed using transformers. This lesson also introduces three-phase AC circuits and transformers with delta-connected primaries.

Homework

(Due at the beginning of this class)

For this lesson, you should:

- Thoroughly read the material contained within Lesson 219.
- Complete Objective 219.1 Worksheet.
- Complete Objective 219.2 Worksheet.
- Complete Objective 219.3 Worksheet.
- Read and complete Lesson 219 Safety Worksheet.
- Complete additional assignments or worksheets if available and directed by your instructor.

Objectives

By the end of this lesson, you should:

219.1 ....Recognize delta and wye voltage systems and be able to utilize 3-phase formulas to calculate values for various circuits.

219.2 ....Understand how to determine the rating of a three-phase bank of transformers, be able to determine supply voltage requirements, and perform calculations.

219.3 ....Recognize methods and components utilized in the transmission and distribution of power from various types of power plants.

219.1 Power Generation and Calculations

Refer to Lesson 201 for a review on Common Voltage Systems.

A single-phase (1Ø) generator consists of a coil of wire (or winding) attached to the inside of a cylinder. (Think of a spool of wire on a reel – *it has two ends.*) This "winding" doesn't rotate and is called the *stator* – meaning stationary. The two ends of the stator winding are extended (or brought out) and connected to the output terminals. There is also a set of magnets mounted on a shaft that runs through the middle of the stator winding – again think of the spool of wire. This shaft is called the *rotor* – because it rotates. Common generators have a gasoline or diesel engine and this rotor is coupled to the shaft of the engine. When the engine is engaged it turns the rotor shaft inside the stator. This rotating magnetic field induces a voltage in the stator winding. A load (lights or equipment) can be connected to the output terminals of the stator. The 1Ø voltage output is a sine wave as shown above.

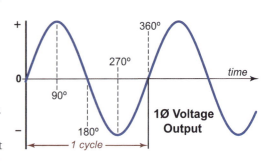

A three-phase (3Ø) generator can be thought of as three single-phase generators in one. A 3Ø generator has three sets of stator windings and three sets of magnets on the rotor. These are mounted so that they are offset 120° from each other. This produces a voltage output where the three voltages are 120° out of phase with each other as shown at the right.

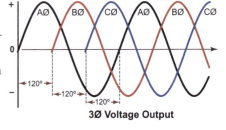

Hoover Dam Hydroelectric Plant

Generators are available in sizes ranging from hundreds of megawatts to a few kilowatts. Generally, the source of three-phase power is a utility generating plant. Power plants, such as the hydroelectric plant at Hoover Dam, require several huge generators. However, it could be an on-site permanently mounted generator, or a skid or trailer-mounted portable generator, or a very-small portable generator.

View from above

Hoover Dam Hydroelectric Plant

View from the floor level

Power Generation — Three-Phase Transformers, Circuits, and Calculations

| Skid-Mounted | Trailer-Mounted | Portable |

In the case of the generating plant, the output voltage is relatively small but is stepped up to many thousands of volts for transmission to a substation where transformers step the voltage down to a lower level. A customer can purchase a generator (genset) with voltage outputs that suit their needs. Many 3-phase generators can be connected to deliver various output voltages. These outputs can be connected in either a delta (Δ) or wye (Y) configuration. These are so called, Δ and Y, because they are frequently drawn schematically in these shapes, as shown below.

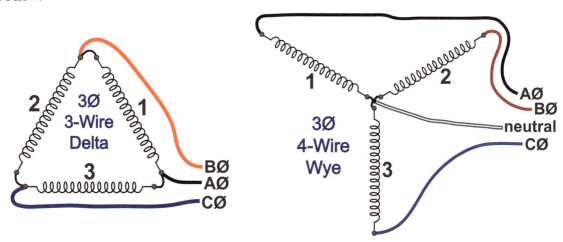

If the output is 3Ø, 3-W, the output consists of three ungrounded conductors and no neutral is available. Conversely, if the output is 3Ø, 4-W, there are three ungrounded conductors plus a neutral conductor. For example, a generator that has a 2400 volt 3Ø, 3-W output does not have an available neutral. However, 277/480V 3Ø, 4-W; 120/208 V 3Ø, 4-W; and 120/240 V 3Ø, 4-W systems all have neutrals in addition to ungrounded conductors. These systems can deliver power to 3Ø loads and line-to-neutral loads.

Whatever the source, 3Ø power consists of three outputs (AØ, BØ, and CØ) that are made available to the user at output lugs, leads, or terminals. The voltage between each ungrounded output (line) conductor is equal. This is the line-to-line voltage (E_{line} or E_L). Previously, when considering single-phase systems, Ohm's Law and the power formulas were used to determine various quantities. With three-phase systems, these formulas are modified. Two generator systems will be used to illustrate the differences between these formulas. Both generators, a 1Ø and a 3Ø, have ratings of 15 KW. The maximum line current (I_{line} or I_L) that can be delivered (or output) is based on this nameplate power rating and will be determined using the appropriate formulas.

DETERMINING LINE CURRENT - 1Ø versus 3Ø	
1-phase, 240-Volt 15 KW Generator	**3-phase, 240-Volt 15 KW Generator**
For 1Ø:	For 3Ø:
$I_L = P_{1\emptyset} \div E_L$	☞ $I_L = P_{3\emptyset} \div (E_L \times \sqrt{3})$
$I_L = 15{,}000 \div 240$	$I_L = 15{,}000 \div (240 \times 1.732)$
$I_L = 62.5\ A$	$I_L = 36.1\ A$

If a load is to be supplied by a generator, the generator must have a rating that is large enough to supply the line current required by the load. Again, a single-phase and a three-phase load will be compared.

DETERMINING POWER - 1Ø versus 3Ø	
1-phase, 240-Volt 62.5 A Load	**3-phase, 240-Volt 136.1 A Load**
For 1Ø:	For 3Ø:
$P_{1\emptyset} = I_L \times E_L$	☞ $P_{3\emptyset} = I_L \times (E_L \times \sqrt{3})$
$P_{1\emptyset} = 62.5 \times 240$	$P_{3\emptyset} = 36.1 \times (240 \times 1.732)$
$P_{1\emptyset} = 15{,}000\ W = 15\ KW$	$P_{3\emptyset} = 15{,}000\ W = 15\ KW$

The power supply (source) that supplies power to a 3Ø load can be a generator or a transformer. The capacity (rated output) of the power source must be at least as large as the load it is to supply. The rated power output (capacity) that can be delivered by a generator is given on the nameplate in watts or KW. The rated power output (capacity) of a transformer is given on the nameplate in volt-amperes or KVA.

Basic three-phase loads contain three elements. For 3Ø motors these three elements are windings. For duct heaters, boilers, or other resistive loads, the elements are heating elements (resistors). The elements of a 3Ø load are connected internally in either a delta or a wye configuration. If the nameplate power rating, in W or VA, is given and rated nameplate voltage is applied to a 3Ø load, the line current can be determined using the same formula given previously for determining 3Ø line current.

☞ $IL = P3\emptyset \div (EL \times \sqrt{3})$

When rated current and rated voltage of a 3Ø load are provided on the nameplate, but not the power rating, the power used (or delivered in the form of heat or motion) can be determined using the same formula given previously for determining 3Ø power.

☞ $P3\emptyset = IL \times (EL \times \sqrt{3})$

A voltage system with only one voltage output is referred to as having a *straight* voltage rating. Systems with straight voltage ratings include 2400 volt 3Ø, 3-W and 480 volt 3Ø, 3-W. Straight rated voltage systems are typically derived from a delta-connected system.

Power Generation — Three-Phase Transformers, Circuits, and Calculations

A voltage system with two voltage outputs is referred to as having a *slash* voltage rating. Systems with slash voltage ratings include 2400/4160 volt 3Ø, 4-W; 120/208 V 3Ø, 4-W; 120/240 V 3Ø, 4-W and 277/480 volt 3Ø, 4-W. Systems with slash ratings can be either wye or delta connected. If the higher voltage rating of a slash-rated system equals the lower voltage rating times 1.732, then the slash-rated system is wye connected. An example of a slash-rated, wye-connected system is a 2400/4160 Volt 3Ø, 4-W where:

4160 = 2400 × 1.732

If the higher voltage of a slash-rated voltage system equals twice the lower voltage, the system is delta connected. An example of a slash-rated, delta-connected system is a 120/240 Volt 3Ø, 4-W where:

240 = 120 × 2

Lesson 219.2   Three-Phase Transformers – Delta-Connected Primaries

If the generated voltage available to a customer is not suitable for the equipment he needs to energize, a transformer (Xfmr) must be installed to step up or step down the voltage. Generally, the voltage is stepped down. Our discussions will be directed at stepping down voltage.

To provide power to a load, the circuit conductors to the load must be sized to carry the load current. These load circuit conductors must also be protected at their ampacities. Since the voltage is to be stepped down to the load's rated voltage, a transformer must be installed. The load circuit conductors are connected to this step-down transformer. The load circuit conductors are connected to the transformer with the OCPD between the load and the transformer. The circuit conductors between the OCPD and the transformer are referred to as secondary conductors because they are connected to the transformer secondary windings. The OCPD in this secondary circuit is referred to as the secondary OCPD because it protects the transformer secondary windings against overload. This OCPD must also protect these secondary conductors at their ampacities. The transformer must also be sized to carry the load connected to its secondary. The size, or rating, of the transformer is determined once the amount of current that will flow through its secondary windings is determined. To transform power to the secondary, the transformer must be provided with power to its primary windings through a supply circuit. These supply circuit, or primary, conductors must also be sized based on the amount of current that will flow through them. The amount of current flowing through the supply circuit conductors is determined by knowing the amount of current flowing through the transformer primary windings. Then, of course, the circuit conductors supplying the primary are required to be protected at their ampacities. The OCPD protecting these primary conductors also protects the primary windings in the transformer against overload. To this point only 1Ø transformers have been discussed. The rating of a 1Ø transformer is the maximum power that it can deliver (or supply) to a connected load without damage to the transformer windings. This rating is expressed in either KVA or in VA. The power that the transformer supplies to a load is exactly the same as the power required by the load. The transformer should be selected so that the current supplied to the load(s) is less than the rating of the transformer. A review of single-phase is provided in Example 219.21, below.

Example 219.21: Using: $P = I \times E$ or $I = P \div E$ or $E = P \div I$

It has been determined that, if a particular 1Ø transformer is rated at 25 KVA and has a secondary a voltage rating of 240 volts, it can supply a 240 volt load with a maximum of 104.17 amps to the load.

$I = P \div E$

$I = 25000 \div 240$

$I = 104.17$ amps

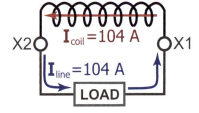

Figure 219.21

As shown in **Figure 219.21**, the maximum current that this transformer can supply to the load is 104 amps where the voltage between X1 and X2 is 240 volts. Note that the current flowing through the coil (or winding) is 104 amps. If the load has a nameplate rating of 104.1666 amps at 240 volts then the power required from (delivered by) the transformer is 25 KVA.

$P = I \times E$

$P = 104.1666 \times 240$

$P = 24,999.98$ VA

$P = 25$ KVA

Before beginning the discussion of 3Ø, it is necessary to make certain that distinctions are made among some very similar terms.

Line Current versus Coil (or winding) Current

Refer to **Figure 219.21**. The current flowing in the circuit conductors to the load is designated as *line current* and represented by I_{line} or I_L. The current flowing through the *winding* of a transformer is called *coil current* or *winding current* and is represented as I_{coil} or $I_{winding}$.

Line Voltage versus Coil (or winding) Voltage

Refer to **Figure 219.22**. The voltage between the circuit conductors is designated as *line voltage* and represented by E_{line} or E_L. The voltage across a winding of a transformer is called *coil voltage* or *winding voltage* and is represented as E_{coil} or $E_{winding}$.

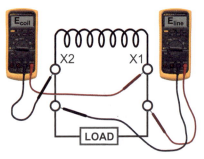

Figure 219.22

Power Generation — Three-Phase Transformers, Circuits, and Calculations

Three-Phase Transformers

Dry-Type 3Ø Transformer

Dry-type transformers are the most commonly seen on jobsites. Three-phase transformers are simply a configuration using three <u>individual</u> single-phase transformers. A dry-type transformer contains three transformers and each of these transformers is the same size (rating). **Figure 219.23** shows <u>three separate</u> 1Ø transformers contained within one enclosure. At the factory the three *primary* windings are interconnected in a wye or delta configuration. The factory also interconnects the *secondary* windings in a wye or delta configuration. Wye (**Y**) and delta (Δ) is a "short-hand" designation for how connections are generally drawn. If a 3Ø transformer is connected Δ-Y then the *primary* is connected in Δ and the *secondary* is connected in Y. The remaining three connections for 3Ø transformers are: Δ-Δ, Y-Δ, and Y-Y. For example, a 480Δ to 208Y/120 transformer is, of course, a delta-wye. This is the

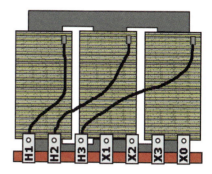

Figure 219.23

most common configuration seen in the field with a Δ-Δ running a close second. The job of the electrician is to connect a 3Ø, 3-wire supply circuit to the three "H" leads and the load to the four "X" leads. For example, a 3Ø, 4-wire panelboard could be the load connected to the secondary. Three ungrounded conductors from the panel would be connected to X1, X2, and X3. Then, the grounded conductor would be connected to X0.

Single-Voltage Delta-Connected Primaries

As mentioned before, most banks of transformers have Δ-connected primaries. Although available with dual-voltage primaries, most transformers seen on a pole or rack have primaries that are rated to operate at a single voltage. A pictorial illustration of a transformer bank with a Δ-connected primary is shown in **Figure 219.24A**. A schematic representation of these same connections is shown in **Figure 219.24B**.

Single-Voltage Delta-Connected Primaries

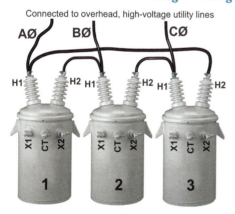

Pictorial - Figure 219.24A

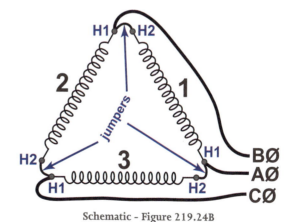

Schematic - Figure 219.24B

Dual-Voltage Delta-Connected Primaries

In previous lessons where single-phase transformers were discussed, connections were shown using a transformer with a dual-voltage primary rated at 240/480 volts and a dual-voltage secondary rated at 120/240 volts. These transformers have two sets of windings for the primary and two for the secondary.

Year Two (Student Manual)

The supply voltage to the primary could be 240 volts. In this case, since each winding is rated to be operated at 240 volts, the two primary windings would be connected in parallel. Or, the supply voltage could be 480 volts. For a 480 volt supply, the two primary windings would be connected in series so that half of the 480 (240 volts) would be applied to each of the two windings. Remember, each winding **wants** to have 240 volts applied to it because that is its rated voltage!

A bank of three transformers can be built where the supply circuit is 480 volts, 3Ø. For now, primary connections will be the focus. The secondary windings are not shown for this discussion.

Three individual 1Ø transformers, with 240/480V primaries, will be connected into a bank. The primary windings will be connected so that 480 volts can be applied to each primary winding set, as shown at the right. As shown in the drawing, AØ and BØ are connected to Xfmr 1 so that 480 volts is applied. BØ & CØ are connected to Xfmr 2. CØ & AØ are connected to Xfmr 3. Note that transformer is often seen abbreviated as xfmr.

Instead of running a wire from the AØ power supply circuit to H1 on Xfmr 1 **plus** another wire from AØ to H4 on Xfmr 3, only one wire could be run to H1 on Xfmr 1. Then a jumper could be run from here over to H4 on Xfmr 3. The same could be done for the BØ and the CØ wires. The drawing at the left shows these jumpers and is equivalent to the drawing shown above. The connection drawing at the left more clearly shows this Δ connection but the drawing above is also delta-connected.

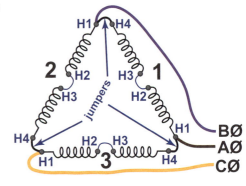

Transformer Primary Data Info for the above scenario:
- E_{line}: The voltage between AØ and BØ of the supply circuit is **480** volts.
  The voltage between AØ and CØ of the supply circuit is **480** volts.
  The voltage between BØ and CØ of the supply circuit is **480** volts.

- Coil or winding voltage measurements on each transformer (1, 2, and 3):
  480 volts would be measured between H1 & H4.
  240 volts would be measured between H1 & H2.
  240 volts would be measured between H3 & H4.

- The two primary windings on each transformer are connected in **series**.

Typically, the conductors connected to the primary windings of a transformer (or transformer bank) are sized to not exceed the rating of the overcurrent protective device (OCPD) protecting these conductors. Although not always the case, a typical installation consists of a primary OCPD and conductors that are sized according to the rating of the transformer (or bank rating). Before determining the rating of a transformer, or bank, required to supply a given load, discussions will begin with a given transformer bank. The total bank rating will be determined first. Then the amount of load that can be connected will be determined. The 3Ø formulas required to determine the total rating, or capacity, of a bank of 1Ø transformers will be introduced in the following discussions.

Three-Phase Transformer BANKS

The transformers banks most routinely encountered are installed on a pole or a rack. Banks, where each of the three individual transformers is the same size, will be considered first. Three 1Ø transformers are shown at the right. Note that each transformer has an X1 and an X2 secondary connection point. These transformers also have a point designated as CT. This is the midpoint (or center-tap) of the secondary winding.

Dry-Type 3Ø Transformer

Three-Phase POWER

The power rating of single-phase transformers was discussed previously. The power rating or capacity of a bank of three individual 1Ø transformers is simply the sum of the ratings of each of the single transformers. Where the three transformers have ratings that are not equal, the following formula is used to determine the capacity of the bank.

☞ $P_{3\emptyset} = P_{1\emptyset} + P_{1\emptyset} + P_{1\emptyset}$

For example, if each of the 1Ø transformers shown above is rated at 25 KVA, the rating of the 3Ø bank equals:

$P_{3\emptyset} = P_{1\emptyset} + P_{1\emptyset} + P_{1\emptyset}$

$P_{3\emptyset} = 25 + 25 + 25$

$P_{3\emptyset} = 75$ KVA

Where all three 1Ø transformers are the **same size,** the following formula is used to determine the capacity of the bank.

☞ $P_{3\emptyset} = 3 \times P_{1\emptyset}$

Again, if each of the 1Ø transformers shown above is rated at 25 KVA, the rating of the 3Ø bank equals:

$P_{3\emptyset} = 3 \times 25$

$P_{3\emptyset} = 75$ KVA

In *Example* 219.21 it was found that a 25 KVA 1Ø transformer had a rated current of 104.17 amps on its 240 volt secondary **coil**. This rated current is actually the rated **coil** current (I_{coil}). Although a transformer is rated in VA, the windings are limited by the amount of current that can flow through them without damage. If the current through the coil (winding) of this transformer exceeds 104.17 amps, it will be overloaded and will suffer damage. Remember, a winding is simply a wire and you can't exceed its ampacity!

For delta-connected banks of three transformers, coil currents do not equal line currents. Consequently, the simple power formulas used to this point must be better defined using subscripts to distinguish between coil and line values used in the power formulas.

- ☞ $P_{1\emptyset} = I_{coil} \times E_{coil}$
- ☞ $I_{coil} = P_{1\emptyset} \div E_{coil}$
- ☞ $E_{coil} = P_{1\emptyset} \div I_{coil}$

These subscripted formulas will be used for the following discussions.

Balanced loads are connected loads that result in equal coil currents through each transformer. When loads are **balanced**, you can use the following formula to determine the power required from the 3Ø transformer bank.

- ☞ $P_{3\emptyset} = 3 \times P_{1\emptyset}$
- ☞ $P_{3\emptyset} = 3 \times I_{coil} \times E_{coil}$

Once again, using the rated current and rated voltage of 25 KVA, 1Ø transformers, the rating of the 3Ø bank equals:

$P_{3\emptyset} = 3 \times I_{coil} \times E_{coil}$
$P_{3\emptyset} = 3 \times (104.17 \text{ Amps}) \times (240 \text{ Volts})$
$P_{3\emptyset} = 75 \text{ KVA}$

Example 219.22: A bank of three individual 1Ø transformers is shown at the right. Each of these has a 240 V secondary and is rated at 25 KVA. It was found in *Example* 219.21 that the rated secondary winding current is 104.17 amps because:

$I_{coil} = P_{1\emptyset} \div E_{coil}$
$I_{coil} = 25000 \div 240$
$I_{coil} = 104.17 \text{ amps}$

This is the same current as found in Example 219.21. Nothing has changed!

To find the capacity of the bank:

$P_{3\emptyset} = 3 \times P_{1\emptyset}$
$P_{3\emptyset} = 3 \times 25$
$P_{3\emptyset} = 75 \text{ KVA}$

The formula: P3Ø = 3 × P1Ø can be applied to any bank of three equally-sized transformers where the loads are balanced. This formula applies to both delta and wye-connected secondaries.

219.3 Power Generation, Transmission, and Distribution

Many steps are required to deliver power from the point of generation to an end user.

Step 1 is to generate the power.

Step 2 is to step the voltage up for transmission across a great distance.

Step 3 is to step the voltage down for distribution.

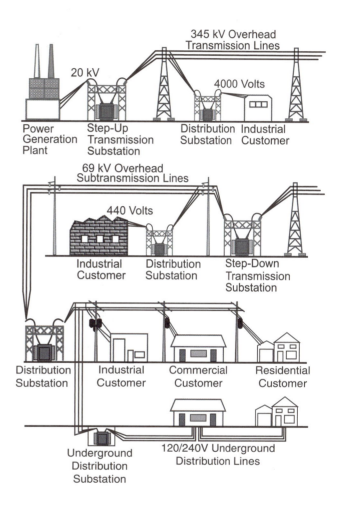

Power Generation Plant

A power generation plant is a facility designed to produce electric energy from another form of energy, such as:

- Heat (thermal) energy generated from:
    - fossil fuels, such as coal, petroleum, or natural gas
    - solar thermal energy
    - geothermal energy
    - nuclear energy

- Potential energy from falling water in a hydroelectric facility

- Wind energy

- Solar electric from solar (photovoltaic) cells

- Chemical energy from fuel cells or batteries

There are many different types of electric power generating plants. The major types generating electric power today are shown below.

Hoover Dam - Hydroelectric Plant

Nuclear Power Plant

Solar – Thermal Power Plant

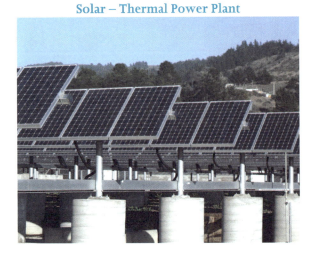

Navajo Power Plant – Coal Fired

Geothermal Power Plant

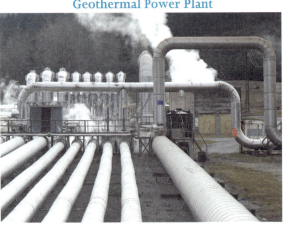

Wind Turbine Generating Farm

SUBSTATIONS

A substation is a high-voltage electric system facility. It is used to switch generators, equipment, and circuits or lines in and out of a system. It also is used to change alternating current (AC) voltages from one level to another, and/or change alternating current to direct current (DC) or direct current to alternating current. Some substations are small with little more than a transformer and associated switches. Others are very large with several transformers and dozens of switches and other equipment.

Substation Functions

Substations are designed to accomplish the following functions (although not all substations have all these functions):

- Change voltage from one level to another
- Regulate voltage to compensate for system voltage changes
- Switch transmission and distribution circuits into and out of the grid system
- Measure electric power qualities flowing in the circuits
- Connect communication signals to the circuits
- Eliminate lightning and other electrical surges from the system
- Connect electric generation plants to the system
- Make interconnections between the electric systems of more than one utility
- Control reactive kilovolt-amperes supplied to and the flow of reactive kilovolt-amperes in the circuits

Typical High-Voltage Substation

Step-up Transmission Substation

A step-up transmission substation receives electric power from a nearby generating facility and uses a large power transformer to increase the voltage for transmission to distant locations. A transmission bus is used to distribute electric power to one or more transmission lines. There can also be a tap on the incoming power feed from the generation plant to provide electric power to operate equipment in the generation plant.

A substation can have circuit breakers that are used to switch generation and transmission circuits in and out of service as needed or for emergencies requiring shut-down of power to a circuit or redirection of power.

The specific voltages leaving a step-up transmission substation are determined by the customer needs of the utility supplying power and to the requirements of any connections to regional grids. Typical voltages are:

High voltage (HV) ac:	69 kV, 115 kV, 138 kV, 161 kV, 230 kV
Extra-high voltage (EHV) ac:	345 kV, 500 kV, 765 kV
Ultra-high voltage (UHV) ac:	1100 kV, 1500 kV
Direct-current high voltage (dc HV):	±250 kV, ±400 kV, ±500 kV

Direct current voltage is either positive or negative polarity. A DC line has two conductors, so one would be positive and the other negative.

Distribution Substation

These are located near the end-users. Distribution substation transformers change the transmission or sub trans-mission voltage to lower levels for use by end-users. Typical distribution voltages vary from 34,500Y/19,920 volts to 4,160Y/2400 volts. Both are 3Ø circuits with a grounded neutral source for a total of four wires. For the 4160 system, the voltage between the three phase conductors or wires would be 4,160 volts and the voltage between one phase conductor and the neutral ground would be 2400 volts. From here the power is distributed to industrial, commercial, and residential customers.

Step-down Transmission Substation

Step-down transmission substations are located at switching points in an electrical grid. They connect different parts of a grid and are a source for subtransmission lines or distribution lines. The step-down substation can change the transmission voltage to a subtransmission voltage, usually 69 kV. The subtransmission voltage lines can then serve as a source to distribution substations. Sometimes, power is tapped from the subtransmission line for use in an industrial facility along the way. Otherwise, the power goes to a distribution substation.

Underground Distribution Substation

Underground distribution substations are also located near the end-users. Distribution substation transformers change the subtransmission voltage to lower levels for use by end users. Typical distribution voltages vary from 34,500Y/19,920 volts to 4,160Y/2400 volts. An underground system may include duct runs, conduits, manholes, high voltage underground cables, risers, pad-mounted transformers, and transformer vaults. From the substation the power is distributed to industrial, commercial, and residential customers.

Underground distribution lines frequently enter vaults that contain underground distribution transformers. A transformer vault is a structure or room in which power transformers, network protectors, voltage regulators, circuit breakers, meters, and other types of equipment are housed.

Risers are used on a power line pole to connect an overhead system to an underground system. A riser has a conduit from the ground up the pole where potheads are used to connect conductors from an underground system to the overhead lines.

Underground Vault

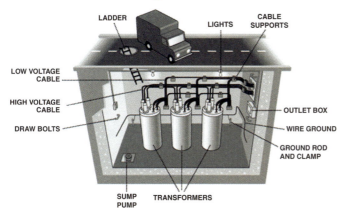

Power Pole with Risers

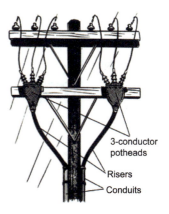

Shielded cables are used for distribution voltages as required by the NEC® at 310.6. These cables are used for aerial, direct burial, conduit, and underground duct installations.

TRANSMISSION LINES

Transmission lines carry electric energy from one point to another in an electric power system. These systems can carry alternating current or direct current or a combination of both. Also, electric current can be carried by either overhead or underground lines. Transmission lines are operated at relatively high voltages, transmit large quantities of power, and transmit power over large distances.

35 kV Type XLP Underground Distribution Cable

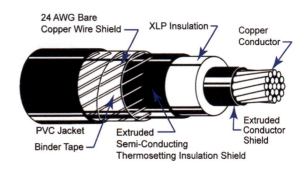

Overhead Transmission Lines

Overhead alternating current (AC) transmission lines share one characteristic: they carry 3-phase (3Ø) current. The voltages vary according to the particular grid system they belong to. Transmission voltages vary from 69 kV up to 765 kV. The direct current (DC) voltage transmission tower has lines in pairs rather than in threes (for 3Ø current), as in AC voltage lines. One line is the positive current line and the other is the negative current line.

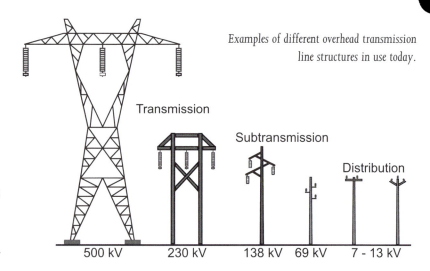

Examples of different overhead transmission line structures in use today.

Subtransmission Lines

Subtransmission lines carry voltages reduced from the major transmission line system. Typically, 34.5 kV to 69 kV, this power is sent to regional distribution substations. Sometimes the subtransmission voltage is tapped along the way for use in industrial or large commercial operations. Some utilities categorize these as transmission lines.

Frequently, overhead subtransmission lines are installed above with distribution lines installed below.

DISTRIBUTION SYSTEM

A distribution system consists of all the facilities and equipment connecting a transmission system to the customer's equipment. A typical distribution system can consist of substations, distribution feeder circuits, switches, protective equipment, primary circuits, distribution transformers, secondaries, and services.

Protective Equipment

Protective equipment in a distribution system consists of protective relays, cutout switches, disconnect switches, lightning arresters, and fuses. These work individually or may work in concert to open circuits whenever a short circuit, lightning strikes or other disruptive event occurs. When a circuit breaker opens, the entire distribution circuit is de-energized. Since this can disrupt power to many customers, the distribution system is often designed with many layers of redundancy. Through redundancy, power can be shut off in portions of the system only, but not the entire system, or can be redirected to continue to serve customers. Only in extreme events, or failure of redundant systems, does an entire system become de-energized, shutting off power to large numbers of customers. The redundancy consists of the many fuses and fused cutouts throughout the system that can disable parts of the system but not the entire system. Lightning arresters also act locally to drain off electrical energy from a lightning strike so that the larger circuit breakers are not actuated.

Distribution Primaries and Secondaries on a Subtransmission Pole

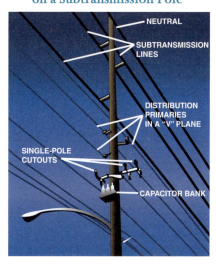

Pole-Mounted Type Lightning Arrester

Bus-Mounted Lightning Arresters

Fused Cutout

A single-pole disconnect (cutout) combined with a fuse is a fused cutout.

Distribution systems have switches installed at strategic locations to redirect or cut-off power flows for load balancing or sectionalizing. Also, this permits repairing of damaged lines or equipment or upgrading work on the system.

Air-Break Isolator Switch

Air Circuit Breaker

Oil-Filled Circuit Breakers

Air Circuit Breaker Inside	Air Circuit Breaker Outside

SERVICE FROM DISTRIBUTION TRANSFORMERS

Most **industrial** customers need 2,400 to 4,160 volts to run heavy machinery. They usually require 3-phase lines to power 3-phase motors. These customers usually own their substation or substations. Substations (distribution transformers) reduce the voltage from the transmission line to the desired level for distribution throughout the plant area.

Commercial customers are usually served at distribution voltages, ranging from 14.4 kV to 7.2 kV. Power is supplied through a service drop line which leads from a transformer on or near the distribution pole to the customer's end use structure. They may require 3-phase lines to power 3Ø motors.

The distribution electricity is reduced to the end use voltage (120/240 volts single phase) via a pole mounted or pad-mounted transformer. Power is delivered to the **residential** customer through a service drop line which leads from the distribution pole transformer to the customer's structure, for overhead lines, or underground.

Power Generation — Three-Phase Transformers, Circuits, and Calculations

Objective 219.1 Worksheet

_____ 1. If the output of a 3Ø generator is 240 volts, line-to-line, and has a neutral connection, this is a ___ system.

   a. 3Ø, 3-W   b. 3Ø, 4-W

_____ 2. A 3Ø 240 volt load is supplied power by a 100 KW generator. The nameplate rating on the load is 59.2 KVA. The line (circuit) conductors between the generator and the load will be carrying ___ amps.

   a. 164   b. 247   c. 427   d. 493   e. 285   f. 142

_____ 3. The abbreviation or symbol used for the unit of measurement for the previous question is ___.

   a. E   b. Ω   c. I   d. V   e. A

_____ 4. A 3Ø, 480-volt, 45 KVA, wye-connected load draws ___ a 3Ø, 480-volt, 45 KVA, delta-connected load.

   a. more line current than
   b. less line current than
   c. the same line current as

_____ 5. When a 3Ø, 4-W supply system (typically a transformer) delivers 277/480 volts, the system has line-to-neutral voltages that are ___ volts.

   a. 120   b. 277   c. 240   d. 480   e. 208

1	38.7		8.4 kVA		2	Schedule
3	KW		load		4	
5		load	10.1 kVA		6	120/240 V
7	18.2		load		8	3Ø 4W
9	kVA	12.6			10	
11		load	kW		12	Panel LB
13	1.7 KVA load		load		14	

Figure 219.102

_____ 6. Refer to Figure 219.102, shown above. Each circuit conductor feeding the load connected to Circuit LA-7,9,11 will carry ___ amps.

   a. 43.8   b. 151.6   c. 50.5   d. 75.8   e. 87.5

7. Refer to Figure 219.102, shown on the previous page. Each circuit conductor feeding the load connected to Circuit LA-2,4 will carry ___ amps.

   a. 23.3   b. 70.0   c. 40.4   d. 20.2   e. 35.0

8. Power can be measured in ___.
   I. volt-amperes   II. watts

   a. neither I nor II
   b. either I or II
   c. I only
   d. II only

9. A single-phase, 240 volt, 9600 watt water heater is accidentally connected to a 120-volt line. The water heater will deliver about ___ of its 9600 W rating.

   a. 50%   b. 33%   c. 25%   d. 100%   e. 300%   f. 200%

1	38.7	8.4 kVA	2	
3	KW	load	4	Schedule
5	load	10.1 kVA	6	
7	18.2	load	8	Panel LA
9	kVA	12.6	10	
11	load	kW	12	120/208 V
13	1.7 KVA load	load	14	3Ø 4W

Figure 219.101

10. Refer to Figure 219.101, shown above. Each circuit conductor feeding the load connected to Circuit LA-10,12,14 will carry ___ amps.

    a. 90.9   b. 52.5   c. 60.6   d. 30.3   e. 35.0

11. Refer to Figure 219.101, shown above. The circuit conductors between the load and the breaker in space LA-6,8 will carry ___ amps.

    a. 48.6   b. 42.1   c. 72.9   d. 24.3   e. 28.0

12. When a 3Ø, 4-W supply system delivers 120/240 volts, the system is ___ connected.

    a. delta   b. wye

Power Generation — Three-Phase Transformers, Circuits, and Calculations

_____ 13. If the output of a 3Ø generator is 480 volts, line-to-line, and has a neutral connection, this is a ___ system.

a. 3Ø, 4-W
b. 3Ø, 3-W

_____ 14. A single-phase, 480-volt duct heater is rated for 10 KVA. It is being fed from a 277/480 volt three-phase panel. The duct heater will draw ___ amps.

a. 12.03   b. 26.04   c. 27.71   d. 20.83   e. 36.08   f. 48.0

_____ 15. A 480-volt, 3-phase panel has nine 2-pole breakers that supply a total parking lot lighting load of 252 amps. The lighting load is equally divided among the 2-pole breakers so that each phase is carrying 84 amps. The panel also has three 3-pole breakers that supply three-phase loads of 18, 34, and 62 amps respectively. What is the total load in KVA?

a. 175.7   b. 164.6   c. 241.1   d. 366.0   e. 304.3   f. 209.5

_____ 16. A three-phase motor draws 28 amps at 208 volts. How much power, in KVA, is being used by the motor?

a. 6.720   b. 10.087   c. 7.280   d. 5.824   e. 12.609   f. 11.639

_____ 17. A three-phase motor draws 28 amps at 230 volts. The power used by the motor is ___ KVA.

a. 8.050   b. 6.440   c. 10.411   d. 13.943   e. 11.154   f. 8.214

_____ 18. A three-phase 69 KVA load will be connected to a circuit that is fed from a 3-pole breaker in a 120/240 volt 3Ø, 4-wire panel. This load will draw ___ amps of current through each circuit conductor.

a. 110.6   b. 166   c. 191.5   d. 287.5   e. 331.7

_____ 19. A distribution substation near a residential neighborhood is delivering 4160/2400 volts to the transformers for the homes. 4160/2400 is a ___ system.

a. delta   b. wye

Year Two (Student Manual)

Power Generation — Three-Phase Transformers, Circuits, and Calculations

Objective 219.2 Worksheet

_____ 1. A bank of transformers will require that three transformers with ratings of at least ___ KVA each be used to supply enough power at 120/240 volts to a 69 KVA 3Ø load.

    a. 40   b. 23   c. 34.5   d. 69   e. 120

_____ 2. 480-volt 3-phase circuit is connected to the delta primary of a 480 to 240 volt 3-phase transformer. The voltage across any one primary winding is ___ volts.

    a. 208   b. 240   c. 120   d. 480   e. 277

_____ 3. A 3Ø, 45 KVA load will be supplied power by a 3Ø, 480 Δ to 120/208 volt "Y" transformer bank. This transformer bank must have a 3Ø rating of at least ___ KVA.

    a. 15   b. 135   c. 26   d. 45

_____ 4. A "bank" of three individual 1Ø transformers is to be connected to deliver 277/480 volts. Each of the transformers is rated at 150 KVA. The total rating of the bank is ___ KVA.

    a. 300   b. 450   c. 50   d. 150

_____ 5. A "bank" of three 1Ø transformers has a total rating of 60 KVA. This means that each transformer is rated to deliver ___ KVA.

    a. 30   b. 15   c. 20   d. 60

Objective 219.3 Worksheet

_____ 1. A(An) ___ is the typical OCPD on the primary of an overhead pole-mounted distribution transformer.

   a. air breaker
   b. fused cutout
   c. molded case circuit breaker
   d. any of these
   e. none of these

_____ 2. A typical CenterPoint Energy overhead distribution system operates at 34,500 volts. This would be equal to ___ kV.

   a. 345   b. 34.5   c. 3.45   d. .345   e. none of these

_____ 3. Power is distributed in grid systems so that if one substation goes "down" another substation can be connected to power the grid.

   a. True   b. False

_____ 4. Overhead distribution lines that have a neutral run with them are configured so that the neutral conductor is above the ungrounded conductors.

   a. False   b. True

_____ 5. 500 kV = ___ volts

   a. 5,000   b. .500   c. 50,000   d. 500   e. none of these

_____ 6. Which of these are typical Ultra-High transmission voltages?

   a. 230 kV   b. 1100 kV   c. 765 kV   d. all of these   e. none of these

_____ 7. Underground transmission or distribution lines can be installed in a trench without conduit.

   a. True   b. False

Lesson 219 Safety Worksheet

Lockout/Tagout

OSHA Fact Sheet

What is the OSHA standard for control of hazardous energy sources?

The OSHA standard for *The Control of Hazardous Energy (Lockout/Tagout)*, Title 29 Code of Federal Regulations (CFR) Part 1910.147, addresses the practices and procedures necessary to disable machinery or equipment, thereby preventing the release of hazardous energy while employees perform servicing and maintenance activities. The standard outlines measures for controlling hazardous energies—electrical, mechanical, hydraulic, pneumatic, chemical, thermal, and other energy sources.

In addition, 29 CFR 1910.333 sets forth requirements to protect employees working on electric circuits and equipment. This section requires workers to use safe work practices, including lockout and tagging procedures. These provisions apply when employees are exposed to electrical hazards while working on, near, or with conductors or systems that use electric energy.

Why is controlling hazardous energy sources important?

Employees servicing or maintaining machines or equipment may be exposed to serious physical harm or death if hazardous energy is not properly controlled. Craft workers, machine operators, and laborers are among the 3 million workers who service equipment and face the greatest risk. Compliance with the lockout/tagout standard prevents an estimated 120 fatalities and 50,000 injuries each year. Workers injured on the job from exposure to hazardous energy lose an average of 24 workdays for recuperation.

How can you protect workers?

The lockout/tagout standard establishes the employer's responsibility to protect employees from hazardous energy sources on machines and equipment during service and maintenance.

The standard gives each employer the flexibility to develop an energy control program suited to the needs of the particular workplace and the types of machines and equipment being maintained or serviced. This is generally done by affixing the appropriate lockout or tagout devices to energy-isolating devices and by deenergizing machines and equipment. The standard outlines the steps required to do this.

What do employees need to know.

Employees need to be trained to ensure that they know, understand, and follow the applicable provisions of the hazardous energy control procedures. The training must cover at least three areas: aspects of the employer's energy control program; elements of the energy control procedure relevant to the employee's duties or assignment; and the various requirements of the OSHA standards related to lockout/tagout.

What must employers do to protect employees?

The standards establish requirements that employers must follow when employees are exposed to hazardous energy while servicing and maintaining equipment and machinery. Some of the most critical requirements from these standards are outlined below:

- Develop, implement, and enforce an energy control program.

- Use lockout devices for equipment that can be locked out. Tagout devices may be used in lieu of lockout devices only if the tagout program provides employee protection equivalent to that provided through a lockout program.

- Ensure that new or overhauled equipment is capable of being locked out.

- Develop, implement, and enforce an effective tagout program if machines or equipment are not capable of being locked out.

- Develop, document, implement, and enforce energy control procedures. [See the note to 29 CFR 1910.147(c)(4)(i) for an exception to the documentation requirements.]

- Use only lockout/tagout devices authorized for the particular equipment or machinery and ensure that they are durable, standardized, and substantial.

- Ensure that lockout/tagout devices identify the individual users.

- Establish a policy that permits only the employee who applied a lockout/tagout device to remove it. [See 29 CFR 1910.17(e)(3) for exception.]

- Inspect energy control procedures at least annually.

- Provide effective training as mandated for all employees covered by the standard.

- Comply with the additional energy control provisions in OSHA standards when machines or equipment must be tested or repositioned, when outside contractors work at the site, in group lockout situations, and during shift or personnel changes.

How can you get more information?

OSHA has various publications, standards, technical assistance, and compliance tools to help you, and offers extensive assistance through its many safety and health programs: workplace consultation, voluntary protection programs, grants, strategic partnerships, state plans, training, and education. Guidance such as OSHA's *Safety and Health Management Program Guidelines* identify elements that are critical to the development of a successful safety and health management system. This and other information are available on OSHA's website at www.osha.gov.

- For a free copy of OSHA publications, send a self-addressed mailing label to this address: OSHA Publications Office, P.O. Box 37535, Washington, DC 20013-7535; or send a request via fax to (202) 693-2498, or call (202) 693-1888.

- To file a complaint by phone, report an emergency, or get OSHA advice, assistance, or products, contact your nearest OSHA office under the "U.S. Department of Labor" listing in your phone book, or call toll-free at (800) 321-OSHA (6742). The teletypewriter (TTY) number is (877) 889-5627.

- To file a complaint online or obtain more information on OSHA federal and state programs, visit OSHA's website.

Questions

_____ 1. Insure that Lockout/Tagout devices identify the ___.

    a. owner   b. foreman   c. company   d. individual

_____ 2. Use only Lockout/Tagout devices authorized for the particular equipment. These devices are to be ___.
       I. standardized   II. substantial   III. durable

    a. I, II, & III
    b. I only
    c. III only
    d. I & III only
    e. II & III only

_____ 3. Compliance with Lockout/Tagout prevents an estimated ___ fatalities each year.

    a. 160   b. 120   c. 150   d. 180

_____ 4. A lockout can be removed by ___.

    a. only the person who applied it
    b. a helper
    c. a foreman
    d. any qualified person

_____ 5. OSHA's Lockout/Tagout standards outline measures required for controlling hazardous ___ energies.

    a. thermal
    b. pneumatic
    c. hydraulic
    d. electrical
    e. all of these

Lesson 220
3Ø Transformers – Delta-Delta

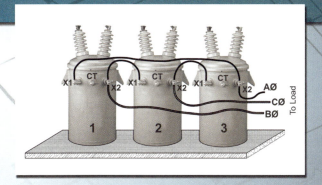

Electrical Curriculum

Year Two
Student Manual

Purpose
This lesson will discuss the wiring configurations, applications, operations, and calculations for three-phase transformers with delta-connected secondaries.

Homework
(Due at the beginning of this class)

For this lesson, you should:

- Thoroughly read the material contained within Lesson 220.
- Complete Objective 220.1 Worksheet.
- Complete Objective 220.2 Worksheet.
- Complete Objective 220.3 Worksheet.
- Complete Objective 220.4 Worksheet.
- Complete Objective 220.5 Worksheet.
- Read and complete Lesson 220 Safety Worksheet.

- Complete additional assignments or worksheets if available and directed by your instructor.

Objectives

By the end of this lesson, you should:

220.1 ....Understand single-voltage 3Ø, 3-wire, closed-delta secondary outputs, capacities, and connections, and be able to perform calculations and select proper primary tap connections for the transformers shown in Yr-2 Annex C.

220.2 ....Understand single-voltage 3Ø, 4-wire, closed-delta secondary outputs, capacities, and connections, and be able to perform calculations and select proper primary tap connections for the transformers shown in Yr-2 Annex C.

220.3 ....Understand single-voltage 3Ø, 3-wire and 4-wire, open-delta secondary outputs, capacities, and connections.

220.4 ....Understand dual-voltage 3Ø, 3-wire and 4-wire, closed-delta secondary outputs and connections.

220.5 ....Understand delta-delta voltage and current transformations, and be able to perform calculations and select proper primary and secondary conductors and OCPDs based on the requirements of NEC Article 240.

Objective 220.1 Single-Voltage Delta Connected Secondaries

When equipment, a system, or a circuit has a Δ (delta) or Y (wye) designation, the equipment, system, or circuit is 3Ø. The discussion here is limited to the secondary windings of 3Ø transformers and transformer banks. It is assumed that the proper (rated) voltage has been applied to the primaries to produce the desired (rated) secondary voltages. As with primaries, a transformer secondary can be connected in Δ or in Y. Here, delta connections are addressed. One method to draw the three windings of a bank is illustrated in Figure 220.11. This connection provides a 3Ø, 3-wire secondary. Note that the three windings are drawn to look like a Δ. A delta connection requires that the 1Ø transformers be interconnected so that X1 of one transformer is connected to X2 of the next. Extreme care is required when these connections are made in the field or in the lab. If misconnected, an explosion could occur when the bank is energized!

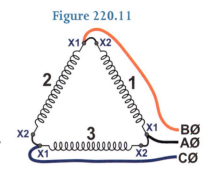

Figure 220.11

Two other methods of drawing the transformers are shown in **Figures 220.12** and **220.13**. These are equivalent to **Figure 220.11**. Notice that in **Figures 220.11, 220.12, and 220.13** the transformers are numbered 1-2-3 for easier reference.

Winding-to-Winding Transformation and the Use of Taps on Delta-Delta Transformers

Figure 220.12

Figure 220.12 *is a pictorial drawing of how the connections are actually made in the field. These transformers are mounted on a rack, although they could just as easily been "pole-mounted". The connections would have been the same.*

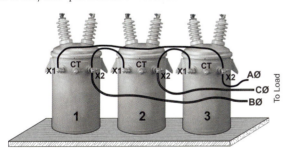

Figure 220.13

Figure 220.13 *shows a line-type connection drawing. Note that the primary windings are delta-connected. Again, this is the more common method of primary connection.*

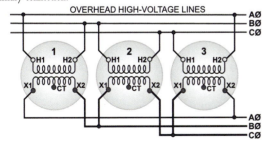

On occasion, electricians are required to make adjustments for supply voltages that are other than the rated voltage required by the transformer. On dry-type transformer installations this is accomplished by moving taps provided on the primary windings. Be aware that not all transformers are provided with taps.

Before discussing taps, locate a Group H, Δ-Δ, 75 KVA transformer in Yr-2 Annex C, and become familiar with all the information available in the transformer tables. The Catalog number for this transformer is 3303. Note all the information provided in the row on this table for this transformer. This transformer does not have wire leads. Instead it has provisions for the installation of terminal lugs. The transformer is a little over 41½" in height (top-to-bottom), is almost 33" wide (left-to-right, facing the front), and is almost 30" deep (front-to-back). This 75 KVA transformer weighs 1125 pounds. Although 1125 pounds is the shipping weight, the shipping material typically includes a light-weight, pine pallet with cardboard over the transformer enclosure. For all practical purposes, the transformer weighs 1125 pounds and a mechanical means will have to be utilized to move it into place for connection. Also note that the weight prohibits it from being mounded on wall brackets. Of course, it can be lifted onto a rack instead of setting it on a concrete pad or the floor. No knockouts are provided in this enclosure and no weathershield is required. This transformer is totally-enclosed and is suitable for outdoor locations. The table heading indicates that Group H transformers are non-ventilated. When a transformer is non-ventilated some means must be provided to allow dissipation of the heat generated from the windings. In the case of the Group H transformers, the windings are encapsulated. This is the principal reason a Group H, 75 KVA transformer weighs 625 pounds more than a Group G, 75 KVA transformer. In the last column of the table, note that the enclosure type is H. Enclosure types are shown in Yr-2 Annex B. Looking at the enclosure type illustration will give a reference for the dimensions since the nameplate is shown on the front of the transformer. Finally, also in the last column, the connection diagram for this transformer is Diagram 26.

The winding diagram from Diagram 26 is shown at the right. Note that this is a shielded type transformer and has three "high-voltage" primary inputs. These are labeled H1, H2, and H3. A 480-volt supply

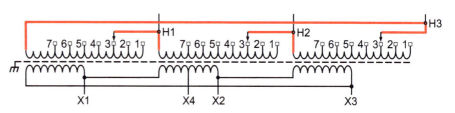

circuit is connected to these three points. Although there are many taps on each of the three primary windings, each winding is single voltage. At the factory the unlabeled end of each winding is connected to the "H" terminal connection point. Also, a lead is connected between each "H" terminal point and the "3" tap connection tab on each winding. These factory connections are shown in red on the diagram above. Referring to the table from Diagram 26, it can be seen that tap "3" is used when the input voltage is exactly 480 volts. With the taps connected at "3" the winding-to-winding voltage ratio is 480:240 or 2:1

because the rated voltage of each primary winding is 480 and the rated voltage of each secondary winding is 240. Winding-to-winding voltage transformation is critical to understanding tap connections. If the supply circuit is very close to its source, the line-to-line voltage could be closer to 504 volts instead of 480. If the taps remain set at 3 the voltage ratio is still 2:1 and would result in a secondary winding voltage of:

$E_{sec} = E_{pri} \times (N_{sec} \div N_{pri}) = 504 \times (240 \div 480) = 252$ volts

For this higher-than-rated input voltage the table indicates that the taps should be moved to position "1" on each of the three windings. Moving the taps actually changes the number of turns in the primary winding. Moving the tap from 3 to 1 adds more turns. In position 1 the voltage ratio (turns ratio) is now 504 to 240. Using this adjusted ratio the output voltage on each secondary winding is now:

$E_{sec} = E_{pri} \times (N_{sec} \div N_{pri}) = 504 \times (240 \div 504) = 240$ volts

If the input voltage is lower than rated, the taps are moved in the opposite direction. If, for example, the supply circuit is run a long distance from its source the voltage may have dropped to 435 volts. If the taps are left in the factory-set "3" position, the secondary winding voltage would be:

$E_{sec} = E_{pri} \times (N_{sec} \div N_{pri}) = 435 \times (240 \div 480) = 217.5$ volts

This is not acceptable and the taps should be moved to position "7" as shown in Diagram 26's table. "7" was selected because the 435-volt input voltage is closer to the 432 volt table value than it is to the 444 volt table value. Moving the taps to tab 7 reduces the number of turns in the primary. In position 7 the turns ratio (voltage ratio) is 432:240. With the taps set at 7, the output voltage is:

$$E_{sec} = E_{pri} \times (N_{sec} \div N_{pri}) = 435 \times (240 \div 432) = 241.7 \text{ volts}$$

The "low-voltage" secondary windings are also single voltage. The secondary is 3-phase, 4-wire delta and has three outputs for the phase conductors. These are labeled X1, X2, and X3. Additionally, the neutral point on this secondary is labeled X4. The midpoint of the middle winding is center-tapped and connected to the output labeled X4.

Ungrounded 3Ø, 3-Wire Delta Secondaries

Line-to-line voltage measurements on a 3Ø, 3-wire delta secondary are equal to the system voltage. Examples include: for a 480-volt system, $E_{Line} = 480$ volts and for a 240-volt system, $E_{Line} = 240$ volts.

Under very specific conditions, NEC Sections 250.20 and 250.21 permit an ungrounded 3Ø, 3-wire delta secondary. That is, the windings are not grounded. However, NEC Section 250.4(B) requires that the metal enclosures and raceways for any system or circuit are to be grounded. Line-to-ground voltages for an ungrounded system are *theoretically* zero because the windings are isolated from ground. However, in reality there may be some voltage (difference of potential) but this is due to the insulation leaks (to ground) in the connected loads.

Although the windings are ungrounded, the last paragraph in NEC Section 250.21 requires that ground detectors be installed on ungrounded systems that have met all the provisions of 250.21. Ground detectors offer capacitive coupling (high impedance) between a faulted conductor and ground so that the fault current is greatly reduced. This reduced fault current prevents the OCPDs from opening and a continuous process can continue. When a ground detector detects a fault, the maintenance electricians should have been alerted so that the fault can be corrected. If a second fault occurs before the first is corrected, the OCPDs will "see" the full fault current and operate, opening the circuit. The second fault will cause fault current to flow through only one pole of a 3-pole breaker. If the system is 480 volts, 480 volts (faulted line-to-ground voltage) is applied to the one pole of the breaker. For these 3Ø, 3-wire systems, slash rated circuit breakers are not permitted. For example, a breaker listed for 277/480 volt operation is not permitted on a 480-volt, 3Ø, 3-wire delta system.
A breaker with a straight 480-volt rating would be required. NEC Section 240.85 provides these requirements.

Corner-Grounded 3Ø, 3-Wire Delta Secondaries

These 3Ø, 3-wire corner-grounded systems have one leg (phase conductor) grounded. This affects the voltage to ground on all phases. The grounded leg is referred to as the "dead man's leg" because checking voltage from this grounded leg to ground will yield a zero reading. This "0 volt" reading leads the "dead man" to think that the system was de-energized. Section 250.26 in the NEC does not specify which phase conductor to ground. If CØ is selected as the grounded conductor, the voltage measurements on a 240-volt 3Ø, 3-W system would be:

$E_{AØ}$ to ground = 240 volts $E_{AØ}$ to $E_{BØ}$ = 240 volts

$E_{BØ}$ to ground = 240 volts $E_{AØ}$ to $E_{CØ}$ = 240 volts

$E_{CØ}$ to ground = 0 volts $E_{BØ}$ to $E_{CØ}$ = 240 volts

For any voltage system, ALWAYS check both line-to-line and line-to-ground voltages before servicing any equipment. Remember that where individual breakers are handle-tied, for disconnection purposes **only**, all poles may not trip together and might leave one, or more, "legs" energized.

More Three-phase power

As discussed in the previous lesson, distinctions were made between coil and line currents as well as coil and line voltages. Coil currents must be considered to prevent overloading any one of the 1Ø transformer windings. However, the loads connected to the secondaries determine line currents in the conductors. These line currents are then used to determine coil currents.

Example 220.11: Refer to **Figure 220.14** and the E (voltage) and I (current) values shown at the various points in the circuit. A 3Ø load is connected to the 3Ø transformer bank. The nameplate *on the load* indicates that it will draw 180.422 amps at 240 volts 3Ø. 180.422 amps is the amount of current flowing in the circuit conductors between the bank and the load. ⇒ Iline = 180.422 amps ⇒ 180.4 A (as shown on the ammeters in Figure 220.14). The I_{line} (line current) in the AØ conductor equals that in BØ and in CØ (180.4 amps).

3Ø Transformers – Delta-Delta

The line-to-line (or line) voltage supplied by the transformer bank is 240 volts $\Rightarrow E_{line} = 240$ volts. As shown in Figure 220.14, line-to-line voltage measurements between L1 & L2, or between L1 & L3, or between L2 & L3 are 240 volts.

Use the following formula to find 3Ø power delivered to, or used by, a 3Ø load when line voltage and line current are known.

☞ $P_{3\emptyset} = \sqrt{3} \times I_{line} \times E_{line}$

☞ $P_{3\emptyset} = (1.732) \times I_{line} \times E_{line}$

For this example:

$P_{3\emptyset} = \sqrt{3} \times I_{line} \times E_{line}$

$P_{3\emptyset} = (1.732) \times I_{line} \times E_{line}$

$P_{3\emptyset} = \sqrt{3} \times (180.4) \times (240)$

$P_{3\emptyset} = 75$ KVA

Each of the transformers in the bank must supply an equal amount of power to the 3Ø load. The amount of power supplied by each transformer to a 3Ø load can be determined using the following formula.

☞ $P_{1\emptyset} = P_{3\emptyset} \div 3$

For this example:

$P_{1\emptyset} = P_{3\emptyset} \div 3$

$P_{1\emptyset} = 75$ KVA $\div 3$

$P_{1\emptyset} = 25$ KVA

$\Rightarrow$ each of these three 1Ø transformers must be rated for at least 25 KVA to supply sufficient power to the load.

These are the same 25 KVA 1Ø transformers used in all the examples, so far. That is, the voltage between X1 and X2 is 240 volts. A coil voltage measurement between X1 & X2 on Transformer 2 is shown in **Figure 220.14**. If a voltage measurement were made across Transformer 1 or 3, 240 volts would also be indicated. Knowing this, the current flowing in each winding can now be determined using the formula for coil current.

$I_{coil} = P_{1\emptyset} \div E_{coil}$

$I_{coil} = 25000 \div 240$

$I_{coil} = 104.17$

$I_{coil} \approx 104.2$ amps

Again, this is the amount of current flowing though the secondary winding (or coil) of each 1Ø transformer in the bank. Although an ammeter is not shown around the winding of Transformer 1, it equals the current flowing in the windings of Transformer 2 and Transformer 3.

True or False: $P_{3\emptyset} = \sqrt{3} \times I_{line} \times E_{line}$ and $P_{3\emptyset} = 3 \times I_{coil} \times E_{coil}$

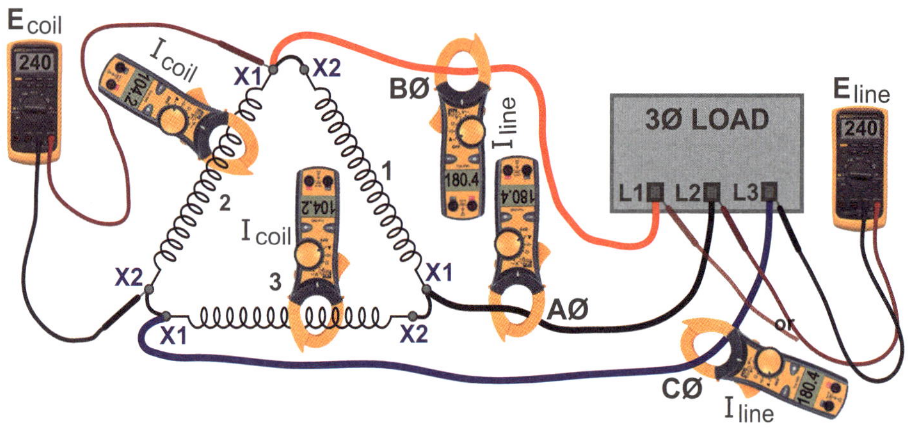

The value for line current selected in *Example* 220.11 was deliberate. This value was selected to illustrate the power relationships between coil values and line values. Since *Example* 220.11 used a 3Ø load, $I_{line\ A\emptyset} = I_{line\ B\emptyset} = I_{line\ C\emptyset}$. A 3Ø load is an example of a balanced load because all line currents are equal. When **balanced loads** are connected to the bank the following formulas can be used.

Figure 220.14

☞ $P_{3\emptyset} = \sqrt{3} \times I_{line} \times E_{line}$

☞ $P_{3\emptyset} = 3 \times I_{coil} \times E_{coil}$

These can be proven using the known values from *Example* 220.11, where:

$I_{line} = 180.422\ A$ $E_{line} = 240\ V$ and $I_{coil} = 104.167\ A$ $E_{coil} = 240\ V$

$P_{3\emptyset} = \sqrt{3} \times I_{line} \times E_{line}$ $P_{3\emptyset} = 3 \times P_{1\emptyset}$

$P_{3\emptyset} = (1.732) \times I_{line} \times E_{line}$ $P_{3\emptyset} = 3 \times I_{coil} \times E_{coil}$

$P_{3\emptyset} = \sqrt{3} \times (180.422) \times (240)$ $P_{3\emptyset} = 3 \times (104.167) \times (240)$

$P_{3\emptyset} = 75\ KVA$ $P_{3\emptyset} = 75\ KVA$

☞ **True!** ☜

For the **delta**-connected secondary in **Figure 220.14**, notice that the value for

$I_{line} \ne I_{coil}$ but $E_{line} = E_{coil}$. This is a characteristic of a delta connection.

Year Two (Student Manual)

3Ø Transformers – Delta-Delta

Now that it has been proven that $P_{3\emptyset} = \sqrt{3} \times I_{line} \times E_{line}$, and $P_{3\emptyset} = 3 \times I_{coil} \times E_{coil}$ is true, these formulas can be used to find the current (I) and voltage (E) relationships in a delta. The coil and line relationships for a **delta** configuration are as follows.

☞ $E_{line \Delta} = E_{coil \Delta}$

☞ $I_{line \Delta} = \sqrt{3} \times I_{coil \Delta}$

☞ $I_{coil \Delta} = I_{line \Delta} \div \sqrt{3}$

Proof: $E_{line \Delta} = E_{coil \Delta}$ $I_{line \Delta} = \sqrt{3} \times I_{coil \Delta}$ $I_{coil \Delta} = I_{line \Delta} \div \sqrt{3}$

  $240 = 240$ $180.422 = (1.732)(104.167)$ $104.167 = (180.422) \div (1.732)$

Finally, a few more necessary 3Ø formulas that apply to delta-connected secondaries with *balanced* loads can be derived. These are shown in the table below.

Balanced loads are connected loads that cause exactly the same amount of current to flow in each line (phase conductor). These loads can consist of one or more 3Ø loads. Or the loads can be several 1Ø loads that require the same current in each line. The loads could also be a combination of both 1Ø and 3Ø loads – but the current in the line conductors must be the same to be a balanced load! For example:

Coil and Line Values for BALANCED Loads Connected to Delta Secondaries

☞ $I_{line} = P_{3\emptyset} \div (\sqrt{3} \times E_{line})$

☞ $E_{line} = P_{3\emptyset} \div (\sqrt{3} \times I_{line})$

☞ $I_{coil} = P_{3\emptyset} \div (3 \times E_{coil})$

☞ $E_{coil} = P_{3\emptyset} \div (3 \times I_{coil})$

Load #1 has a nameplate rating of: 3Ø   20 A @ 240 volts Ckt 1-3-5 (AØ-BØ-CØ)

Load #2 has a nameplate rating of: 1Ø   10 A @ 240 volts Ckt 2-4 (AØ-BØ)

Load #3 has a nameplate rating of: 1Ø   10 A @ 240 volts Ckt 6-8 (AØ-CØ)

Load #4 has a nameplate rating of: 1Ø   10 A @ 240 volts Ckt 10-12 (BØ-CØ)

The current that will flow in each line conductor is shown in the table below.

Load	AØ Conductor	BØ Conductor	CØ Conductor
#1	20 amps	20 amps	20 amps
#2	10 amps	10 amps	
#3	10 amps		10 amps
#4		10 amps	10 amps
TOTALS	40 AMPS	40 AMPS	40 AMPS

This is a classic example of a **balanced load** because the current in each line conductor is the **same**.

Year Two (Student Manual)

Objective 220.2   120/240 V 3Ø, 4-Wire Closed-Delta Secondary

Another commonly available system voltage is 120/240 volt, 3Ø, 4-wire. This is a system used where some 3Ø 240-volt loads will be connected, however, the majority of the loads require 120 volts or 1Ø, 240 volts to operate. Typically, these systems serve *unbalanced* loads.

This 3Ø, 4-wire grounded system has a grounded conductor as required by Section 250.20(B)(3) in the NEC. The grounded conductor is the midpoint (center tap in this case) of any one of the windings. The nominal voltage measurements on a 240-volt 3Ø, 4-W system are:

$E_{AØ}$ to ground = 120 volts $E_{AØ}$ to $E_{BØ}$ = 240 volts

$E_{BØ}$ to ground = 208 volts $E_{AØ}$ to $E_{CØ}$ = 240 volts

$E_{CØ}$ to ground = 120 volts $E_{BØ}$ to $E_{CØ}$ = 240 volts

With a 120/240 volt, 3Ø, 4-wire system, care must be exercised when connected to a 3Ø, 4-wire panel because of the presence of the "high-leg". The "high-leg", or BØ, conductor is connected to the middle busbar in the panel. Single-pole breakers can NOT be connected to any space in the panel where the breakers are connected to the BØ bus. Single-pole circuit breakers are not listed for 208 volts. Additionally, any 1Ø 120/240 volt load that requires a neutral (a clothes dryer, blast chiller, cappuccino machine, or range are typical examples because they contain 120 volt components) should not be connected to a 2-pole breaker unless the breaker is connected to CØ & AØ. If this 2-pole breaker were connected to BØ then 208 volts, instead of 120 volts, could be applied to the 120 volt components of the load.

One method to draw these three secondary windings is illustrated in **Figure 220.21**. Note that, as before with a 3-wire delta (Figure 220.11), the three windings are drawn to look like a Δ with X1 of one transformer connected to X2 of the next. The only difference between Figure 220.11 and Figure 220.21 is the addition of a connection to the center tap (CT) connection point on one of the transformers. This connection could have been made to any one of the three transformers. For convention, we selected Transformer 3.

Two other methods of drawing the transformers are shown below in **Figures 220.22** and **220.23**. These are equivalent to **Figure 220.21**. Note that in **Figures 220.21, 220.22, & 220.23** the transformers are numbered 1-2-3 for easier reference. The connections would have been the same. **Figure 220.22** shows a line-type connection drawing. **Figure 220.23** is a pictorial drawing of how the field connections are actually made. These transformers are pole-mounted, although they could just as easily been set on a pad. Note that the primary windings are delta-connected in all three. Again, this is the more common method of primary connection.

Figure 220.21

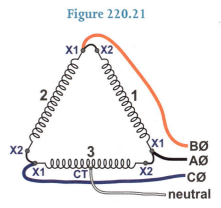

Figure 220.22

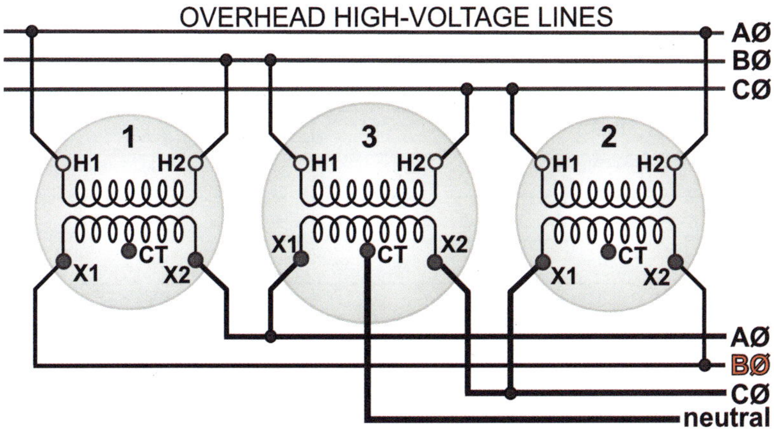

Figure 220.23

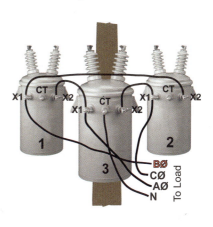

Although a wye system (such as 120/208 V) requires three equally sized transformers, a delta system can have one transformer that has a larger rating. This is shown in **Figure 220.23** where Transformer 3 (center transformer) is larger than the other two transformers. This installation is typical for a 4-wire, closed-delta system. For a 4-wire, closed-delta system, each of the three transformers supplies 1/3 of the power required for the 3Ø loads. Additionally, when 1Ø loads are connected between AØ and CØ, the two smaller transformers (1 and 2) supply 1/3 of this 1Ø power and the larger transformer (3) supplies 2/3 of this 1Ø power. An example of this system is illustrated in the following example.

Example 220.21: This is an example of an unbalanced closed-delta secondary. A bank of three individual transformers is used. One is a larger size than the other two. Generally, the larger transformer supplies most of the power to 120-volt loads. This transformer is called the *lighting* transformer. The smaller transformers are called the *power* transformers because these are required when 3Ø loads are to be served. This example is fairly simple because the 1Ø loads connected to AØ will be equal to the 1Ø loads connected to CØ. The following loads are connected in a 120/240 volt 3Ø, 4-wire panelboard as shown in Figure 220.24.

Figure 220.24

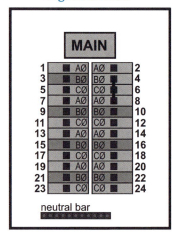

Circuit #1 supplies 120-volt loads that draw 28 amps

Circuit #5 supplies 120-volt loads that draw 28 amps

Circuit #7 supplies 120-volt loads that draw 20 amps

Circuit #11 supplies 120-volt loads that draw 20 amps

Circuit #13 supplies 120-volt loads that draw 39 amps

Circuit #17 supplies 120-volt loads that draw 39 amps

Circuit #4-6-8 supplies a **3Ø** 240 V load that draws 48 A

The line currents flowing in each secondary conductor are shown in the table below.

Circuit	AØ	BØ	CØ
#1	28		
#5			28
#7	20		
#11			20
#13	39		
#17			39
1Ø Subtotals	87 AMPS	0 AMPS	87 AMPS
#4-6-8	48	48	48
3Ø Subtotals	48 AMPS	48 AMPS	48 AMPS
TOTAL I_{line} (1Ø +3Ø)	135 AMPS	48 AMPS	135 AMPS

The approximate size of 120/240 delta lighting and power transformers for unbalanced loads can be determined using the following method. The calculated size for each power transformer must be determined first. Use the **3Ø Subtotals** plus the larger of the two **1Ø Subtotals**, shown in the table above. Since the 1Ø loads are balanced, AØ and CØ are both 87 amps. Use 87 amps in the formula shown below.

☞ $P_{each\ power\ xfmr} = [(I_{line(3Ø\ subtotal)} \div 1.732) + (1/3 \times I_{line(1Ø\ subtotal)})] \times [240]$

For this example:

$P_{each\ power\ xfmr} = [(48 \div 1.732) + (1/3 \times 87)] \times [240]$

$P_{each\ power\ xfmr} = [(27.713) + (29)] \times [240]$

$P_{each\ power\ xfmr} = [56.713] \times [240] = 13{,}611$ VA

$P_{each\ power\ xfmr} = \mathbf{13.611\ KVA}$

3Ø Transformers – Delta-Delta

Next, determine the calculated size of the lighting transformer using the formula shown below.

☞ $P_{lighting\ xfmr} = [(I_{line(3\emptyset\ subtotal)} \div 1.732) + (^2/_3 \times I_{line(1\emptyset\ subtotal)})] \times [240]$

For this example:

$P_{lighting\ xfmr} = [(48 \div 1.732) + (^2/_3 \times 87)] \times [240]$

$P_{lighting\ xfmr} = [(27.713) + (58)] \times [240]$

$P_{lighting\ xfmr} = 20,571\ VA$

$P_{lighting\ xfmr} = \mathbf{20.571\ KVA}$

For these approximations it isn't necessary to find coil currents. But, remember, this gives only an *approximate* size for the transformers.

In Example 220.21, AØ and CØ could be sized at a 1/0 AWG to carry the 135-amp line currents, while BØ could be sized at 8 AWG to carry the 48-amp line current. The secondary conductors can be protected at their ampacities with 150 amp fuses on AØ and CØ and a 50 amp fuse on BØ. These secondary conductors could NOT be terminated on a 150 amp main breaker in a panelboard, because the smaller BØ conductor would not be properly protected at its ampacity. 8 AWG can be protected at no more than its ampacity of 50 amps!

Reminders:

For a 120/240 3Ø, 4-W closed-delta secondary:

E_{line}:   The voltage between AØ and BØ is **240** volts.

           The voltage between AØ and CØ is **240** volts.

           The voltage between BØ and CØ is **240** volts.

$E_{line-to-neutral}$: The voltage between AØ and neutral is **120** volts.

           The voltage between BØ **(high-leg)** and neutral is **208** volts.

           The voltage between CØ and neutral is **120** volts.

E_{coil}:   **240** volts would be measured between X1 & X2 *on any transformer.*

Objective 220.3 OPEN-Delta Secondaries

240 V, 3Ø, 3-wire OPEN-Delta Secondary

A variation of a 3Ø, 3-wire system utilizes only two transformers. Refer to **Figure 220.31** at the right. This is an economical method used to provide a 3Ø power system because *only two* transformers are used. As before, 240 V secondary windings will be used to illustrate the concepts of this system.

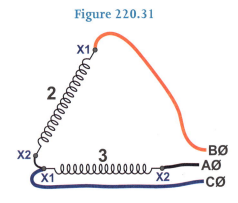

Figure 220.31

For an open-delta, 240-volt, 3Ø, 3-W secondary:

E_{line}: The voltage between AØ and BØ is **240** volts.

The voltage between AØ and CØ is **240** volts.

The voltage between BØ and CØ is **240** volts.

With this system the voltage values are the same as with a 3Ø, 3-W closed delta. The difference between an open delta and a closed delta is the amount of power available. A 45 KVA 3Ø load would require a 3Ø bank with a capacity of at least 45 KVA. For a **closed** delta system, three transformers that have a capacity (or rating) of at least 15 KVA **each** would be required.

$1/3 \times 45$ KVA = 15 KVA EACH for a closed-delta system of three transformers

Opening the Delta of a Closed-Delta System

Another advantage of a closed-delta system is that, if one of the transformers is lost, 3Ø power can still be provided. However, the available power has been reduced because now only two transformers are providing the power. If, initially, all three transformers of a closed-delta system were equally sized and operational but one of them went out (opening the delta), the following formula is used to determine the reduced 3Ø power available. (OD = *open-delta*)

☞ $P_{3Ø\ OD} = [P_{single\ xfmr} \times 3] \times [\sqrt{3} \div 3]$

or

☞ $P_{3Ø\ OD} = [P_{single\ xfmr} \times 3] \times .577$

*3 is used in this formula because this system **initially** had 3 transformers*

Using this formula, can the 45 KVA load still be supplied by the bank if one of the transformers goes out?

$P_{3Ø\ OD} = [15\ KVA \times 3] \times [\sqrt{3} \div 3]$

$P_{3Ø\ OD} = 25.981$ KVA **NO!**

The two remaining transformers would be overloaded because they would still try to supply enough current to the load. The affected facility can continue to function temporarily if some of the loads are disconnected. If some loads are not disconnected, the two remaining transformers will be overloaded and the secondary voltage

3Ø Transformers – Delta-Delta

will drop severely. This reduced voltage could damage equipment that is operating at the lower voltage. If the remaining transformers remain in an overload condition, the windings will be damaged.

For an open-delta system, the bank must have a capacity large enough to supply the load without overloading (45 KVA in this example). This will have to be accomplished using only two transformers. If only two transformers are installed, the KVA ratings can't simply be added together as was done with a closed delta. Instead, the following formula is used to determine the 3Ø rating of an open-delta bank.

☞ $P_{3Ø\ OD} = [P_{single\ xfmr} \times 2] \times [\sqrt{3} \div 2]$

or

☞ $P_{3Ø\ OD} = [P_{single\ xfmr} \times 2] \times .866$

2 is used in this formula because this system **initially** had **2** transformers

Routinely, an open-delta bank is designed for the expected load using only two transformers. The size of each of these transformers is determined using the following formula.

☞ $P_{single\ xfmr} = [P_{3Ø\ Load} \div (.866)] \times ½$

½ is used in this formula because this transformer is **half** of the bank of **2** transformers

Using this formula for the 45 KVA load, determine the minimum size of each of the two transformers for an open-delta system.

$P_{single\ xfmr} = [45\ KVA \div (.866)] \times ½$

$P_{single\ xfmr} = 25.981\ KVA$

Although three 15 KVA transformers were used for a closed-delta, now two 25.981 KVA transformers will be required for an open-delta. If transformers with ratings of at least 25.981 KVA are used, these won't be overloaded by the 45 KVA load.

Using the formula, the capacity of the open delta (OD) bank using two 25.981 KVA transformers is:

$P_{3Ø\ OD} = [P_{single\ xfmr} \times 2] \times [\sqrt{3} \div 2]$

$P_{3Ø\ OD} = [25.981 \times 2] \times .866$

$P_{3Ø\ OD} = 45\ KVA$

As with closed-deltas, the following 3Ø formulas for open-deltas apply only to balanced loads.

Open-Delta Coil and Line Values for BALANCED Loads	
☞	$I_{line} = P_{3Ø\ OD} \div (\sqrt{3} \times E_{line})$
☞	$E_{line} = P_{3Ø\ OD} \div (\sqrt{3} \times I_{line})$
☞	$I_{coil\ OD} = I_{line\ OD}$ This is because, in an open-delta, the line current can't go anywhere but through the winding connected to the line conductor.
☞	$E_{line} = E_{coil}$ This is the same as for a closed-delta.

120/240 V, 3Ø, 4-wire OPEN-Delta Secondary

A variation of a 120/240 volt, 3Ø, 4-wire system utilizes only two transformers. As with the closed delta 120/240 volt 3Ø system, the center-tapped transformer is generally larger than the smaller one that will be used to supply 3Ø power. This is an economical method used to provide a small amount of 3Ø power along with 120/240 volt 1Ø power. The secondary voltages available are exactly the same as for a closed-delta. Refer to Figure 220.32.

Reminders:

For a 120/240 3Ø, 4-W open-delta secondary:

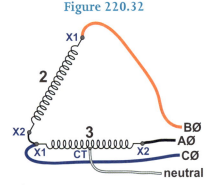

Figure 220.32

E_{line}: The voltage between AØ and BØ is **240** volts.

The voltage between AØ and CØ is **240** volts.

The voltage between BØ and CØ is **240** volts.

$E_{line-to-neutral}$: The voltage between AØ and neutral is **120** volts.

The voltage between BØ **(high-leg)** and neutral is **208** volts.

The voltage between CØ and neutral is **120** volts.

E_{coil}: 240 volts would be measured between X1 & X2 on either transformer.

$I_{coil\ OD} = I_{line\ OD}$

Again, as with the 3-wire open-delta system, the design difference between open and closed delta is the capacity (or power available) of the bank.

$P_{3Ø\ OD} = [P_{single\ xfmr} \times 3] \times .577$

Use this formula if there were 3 transformers initially

$P_{3Ø\ OD} = [P_{single\ xfmr} \times 2] \times .866$

Use this formula if there were 2 transformers initially

$P_{single\ xfmr} = [P_{3Ø\ Load} \div (.866)] \times \frac{1}{2}$

Use this formula to size EACH of the two transformers in an open-delta bank

Objective 220.4 Dual-Voltage Delta-Connected Secondaries

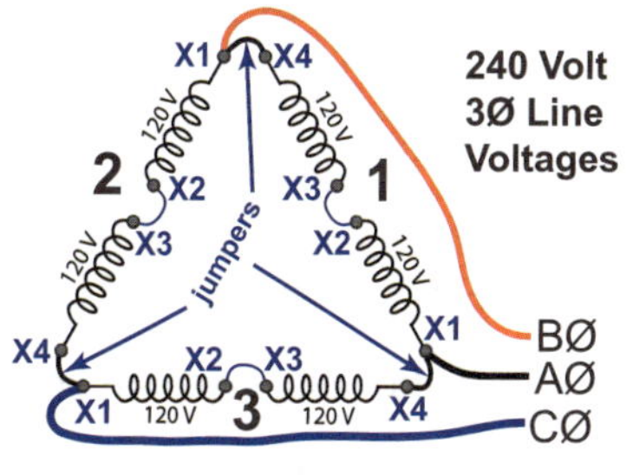

As used previously, many single-phase transformers have dual-voltage secondaries (120/240 V). These single-phase transformers have two sets of windings for the secondary. When rated voltage is applied to the primary winding(s), 120 volts is induced in **each** of the two secondary windings. When connecting three of these 1Ø transformers to form a 3Ø delta bank, the two 120-volt secondary windings on each transformer are connected in series to produce 240-volt line voltages. These connections are shown in **Figure 220.41** where the voltage between secondary conductors AØ and BØ is 240 volts, between AØ and CØ is 240 volts, and between BØ and CØ is 240 volts. This output could supply 240-volt 3Ø, 3-W loads or 240-volt 1Ø, 2-W loads.

If 120-volt loads are also required to be connected, one of the three transformers is "center-tapped" so that two 120-volt outputs are available. As before, Transformer 3 is used for convenience and is shown below in **Figure 220.42**. The line-to-line voltages remain the same (240 volts) but now two additional voltages are available. Between AØ and neutral the voltage is 120 and between CØ and neutral the voltage is 120 volts.

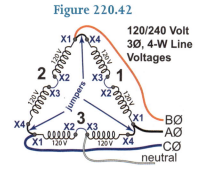

Figure 220.41

Figure 220.42

Objective 220.5
Delta-Delta Transformer Voltage and Current Ratios

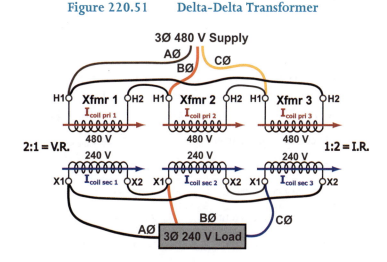

Figure 220.51   Delta-Delta Transformer

The connected load on the secondary of a transformer causes current to flow in the primary. The amount of current flowing in the primary windings is proportional to the amount of current flowing in the secondary windings. The proportion is dictated by the voltage (turns) ratio. As with single-phase transformers, the current ratio is inversely proportional to the voltage ratio. However, with 3-phase transformers, the current in each of the three windings must be considered independently. A 3Ø transformer with single windings on both the primary and secondary, as shown in **Figure 220.51**, will be used to illustrate how to determine primary coil current when secondary coil current is known. The primary windings are delta connected and the secondary windings are delta connected. This transformer is referred to as delta-delta or Δ-Δ. The primary of each of the three transformer windings is rated for 480 volts and the secondary is rated for 240 volts. The winding-to-winding voltage ratio (V.R.) is 2:1 and the winding-to-winding current ratio (I.R.) is 1:2.

Recall that the basic ratio formula is: **V.R. = $E_{pri} \div E_{sec} = I_{sec} \div I_{pri}$ = I.R.**

Don't forget that these are *winding-to-winding* and NOT line-to-line ratios. If the load is balanced, the coil currents are equal in each of the three windings.

When the rated winding voltages are known, use the following formula to find primary coil current:

☞ $I_{coil\ pri} = (I_{coil\ sec} \times E_{coil\ sec}) \div E_{coil\ pri}$

Example 220.51: A 4160 – 480 Volt delta-delta transformer supplies power to a 3Ø load. The line current in the secondary conductors is 162 Amps. The line current in the primary conductors is ___ Amps.

Example 220.51 Solution:

First the coil current in the secondary windings must be determined. The current in each of the three transformer windings will be equal since this is a balanced load. Additionally, the coil voltage and line voltage are equal since this is a delta secondary.

$I_{coil\ sec} = I_{line\ sec} \div \sqrt{3} = 162 \div 1.732 = 93.53$ Amps

Now that the secondary coil current is known, the primary coil current can be determined.

$I_{coil\ pri} = (I_{coil\ sec} \times E_{coil\ sec}) \div E_{coil\ pri} = (93.53 \times 480) \div 4160 = 10.79$ Amps

Now that the primary coil current is known, the primary line current can be determined.

$I_{line\ sec} = I_{coil\ pri} \times \sqrt{3} = 10.79 \times 1.732 = \mathbf{18.69}$ Amps

Example 220.52: Refer to the 480 to 240 V, Δ-Δ transformer shown in Figure 220.51. As is standard in 2nd year U.N.O., THHN/THWN copper conductors with 75°C terminations are used. Recall from printreading assignments that U.N.O. is frequently used as an abbreviation for Unless Noted Otherwise. The 3Ø, 3-W, 240-volt load is rated at 33.2 kVA. Determine the following:

    A. Secondary line current = _____ Amps = $I_{line\ sec}$

    B. Secondary coil current = _____ Amps = $I_{coil\ sec}$

    C. Primary coil current = _____ Amps = $I_{coil\ pri}$

    D. Primary line current = _____ Amps = $I_{line\ pri}$

    E. Minimum secondary line conductor size to load = _____ AWG

    F. Maximum OCP rating to protect secondary line conductors = _____ Amps

    G. Minimum primary line conductor size for given load = _____ AWG

3Ø Transformers – Delta-Delta

H. Maximum OCP rating to protect primary line conductors = _____ Amps

I. The minimum standard size 3-phase transformer required to supply the load has a rating of ___ KVA.
   a. 100   b. 45   c. 75   d. 112½   e. 150   f. 30

Example 220.52 Solutions:

A. $E_{line\,\Delta} = E_{coil\,\Delta} = 240$ volts on the Δ secondary

   $I_{sec\,line} = P_{3\emptyset} \div (E_{sec\,line} \times 1.732) = 33200 \div (240 \times 1.732) =$ **79.87** Amps

B. $I_{coil\,sec} = I_{line\,sec} \div \sqrt{3} = 79.87 \div 1.732 =$ **46.11** Amps in the Δ secondary windings

C. $I_{coil\,pri} = (I_{coil\,sec} \times E_{coil\,sec}) \div E_{coil\,pri} = (46.11 \times 240) \div 480 =$ **23.06** Amps in the Δ primary windings

D. $I_{line\,sec} = I_{coil\,pri} \times \sqrt{3} = 23.06 \times 1.732 =$ **39.93** Amps on the primary

E. 79.87 Amps on the secondary line conductors $\Rightarrow$ **4** AWG

F. 4 AWG has an ampacity of 85 Amps at the 75°C terminals $\Rightarrow$ **80*** A OCP

   * NEC 240.21(C) does not permit utilizing 240.4(B)

G. 39.93 Amps on the primary line conductors $\Rightarrow$ **8** AWG

H. 8 AWG has an ampacity of 50 Amps at the 75°C terminals $\Rightarrow$ **50** A OCP

I. 33.2 kVA load $\Rightarrow$ **45** KVA transformer $\Rightarrow$ answer **b**

   or: $P_{3\emptyset} = I_{line} \times E_{line} \times 1.732 = 79.87 \times 240 \times 1.732 = 32,200\,W \Rightarrow$ **45** KVA

3Ø Transformers – Delta-Delta   220-503

Objective 220.1 Worksheet

_____ 1. A 480-volt to 240-volt three-phase transformer is rated for 112.5 KVA. What is the maximum available secondary line current on B phase?

   a. 468.8   b. 46.8   c. 234.4   d. 140.3   e. 156.3   f. 270.6

_____ 2. In a 240 volt ungrounded delta, what is the voltage between B phase and ground?

   a. 240 volts   b. 208 volts   c. 120 volts   d. 360 volts   e. 0 volts

_____ 3. Three-phase loads of 16, 42, and 75 amps are fed from a 240 volt panel. This total 3Ø load is ___ KVA.

   a. 55.3   b. 47.9   c. 133   d. 399   e. 230.4   f. 27.7

_____ 4. A Group G, Catalog #3204, transformer is to be connected to supply a 3Ø, 4-Wire panelboard. The secondary conductors feeding the panelboard should be connected at terminal points ___.

   a. H1, H2, H3, & X4
   b. H1, H2, & H3
   c. X1, X2, X3, & X4
   d. X1, X2, & X3
   e. None of these options is correct.

_____ 5. The taps are set at position 2 on a 30 KVA, Group H transformer. The line-line secondary voltage will be ___ volts if the primary is connected to a 442 volt line.

   a. 221   b. 267   c. 255   d. 261   e. 216   f. 208

_____ 6. A 480Δ — 240 volt 3-wire Δ transformer is delivering power to a 3Ø load and has 75 amps of coil current flowing through each winding. The current flowing in each circuit conductor to the load is ___ amps.

   a. 25   b. 129.9   c. 75   d. 43.3

_____ 7. A 480 – 240 volt, Δ - Δ transformer is connected to a line with a voltage of 518. If compensating taps are not used, the secondary line voltage will be ___ volts.

   a. 240   b. 259   c. 299   d. 480   e. 222   f. 518

Year Two (Student Manual)

8. A 150 KVA transformer is feeding 240-volt, 3-phase loads of 125 amps and 225 amps. How much current is available to supply additional 3-phase loads?

   a. 120.3 A   b. 10.8 A   c. 275 A   d. 91.7 A   e. 17.3 A   f. 244.2 A

9. A 3Ø load will consume 21 KVA. The circuit is fed from breakers in a 120/240 volt 3-ph, 4-W panelboard. The line current drawn by the load is ___ A.

   a. 87.5   b. 100.9   c. 58.3   d. 50.5   e. 19.8

10. A 480-volt to 240-volt three-phase transformer is rated for 112.5 KVA. The transformer serves an existing 3Ø load of 190 amps. No more than ___ kVA of 3-phase load can be added.

    a. 8.4   b. 49.2   c. 2.8   d. 33.5   e. 44.4   f. 80.6

11. The delta-connected secondary of a transformer is delivering a line current of 54 amps. What is the phase (coil or winding) current? The secondary voltage is 240 volts line-to-line.

    a. 31.2 amps   b. 54 amps   c. 93.5 amps   d. 162 amps   e. 18 amps

12. A Group H 112.5 KVA 480 volt Δ primary to 240 volt delta secondary transformer has a supply voltage of 447 volts at its primary terminals. To get close to the rated output voltage you must move the tap inter-connection from its factory shipped position 3 to position ___ on each winding.

    a. 3   b. 5   c. 1   d. 2   e. 4   f. 6

13. Which of the formulas below is correct? Power factor is 1.0 or 100%.

    a. $P_{3\emptyset} = I_L \times E_L \times 1.732$ b. $P_{3\emptyset} = I_{coil} \times E_{coil} \times 3$ c. both a and b

14. When a 120/240 volts 3Ø, 4-W supply system (typically a transformer) delivers power to a load, the coil (or winding) current equals ___.

    a. $I_L \times 1.732$ b. I_L c. $I_L \div 1.732$

15. Refer to Yr 2-Annex C. A 480-volt to 240-volt three-phase transformer serves a 190 amp, 3Ø load. What is the minimum standard size Group G, 3-phase transformer required to supply the load?

    a. 75   b. 112.5   c. 45   d. 150   e. 225   f. 300

3Ø Transformers – Delta-Delta

_____ 16. A 3Ø load is connected to the delta-connected secondary of a transformer. The transformer has 102 amps of current flowing in each secondary winding. ____ amps of line current is flowing in the conductors between the load and the transformer.

a. 177   b. 102   c. 34   d. 198   e. 59   f. 306

_____ 17. A Group G, Catalog #3204, transformer is to be connected to supply a 3Ø, 4-Wire panelboard. The supply circuit conductors should be connected at primary terminal points ____.

a. X1, X2, & X3
b. X1, X2, X3, & X4
c. H1, H2, H3, & X4
d. H1, H2, & H3
e. None of these options is correct.

_____ 18. A 480-volt to 240-volt three-phase transformer is rated for 112.5 KVA. The transformer serves an existing load of 3Ø 190 amps. Would a three-phase load of 54 amps cause an overload?

a. Yes   b. No

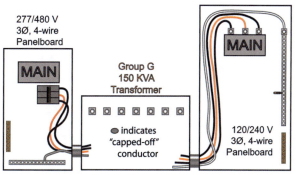

Figure 220.103

_____ 19. Refer to Figure 220.103 (above). Drawing ____ (below) shows the transformer correctly connected.

a.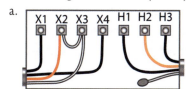

b.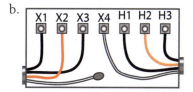

c.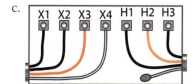

d.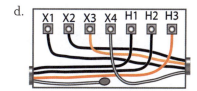

3Ø Transformers – Delta-Delta

Objective 220.2 Worksheet

_____ 1. A 15 kVA 1Ø load has a rated voltage of 240 volts. This load will pull ___ amps when supplied from a circuit in a 120/240 volt 3-ph, 4-W panelboard.

   a. 41.6   b. 62.5   c. 72.1   d. 36.1   e. 108.3

Scenario 220.23
A closed-delta bank of transformers delivers 120/240 volt power to several loads. The 1-phase loads are drawing 120 amps on AØ and 120 amps on CØ. The bank is also supplying power to a 3Ø load that draws 46 amps.

_____ 2. Refer to Scenario 220.23, shown above. The approximate calculated size required for the "power" transformer(s) is ___ KVA.

   a. 16   b. 40   c. 26   d. 69   e. 48   f. 29

_____ 3. Refer to Scenario 220.23, shown above. The approximate calculated size required for the "lighting" transformer(s) has a rating of ___ KVA.

   a. 69   b. 16   c. 40   d. 29   e. 48   f. 26

Scenario 220.21
A 3Ø transformer bank supplies 120/240 volts to a 3Ø, 4-wire panelboard. There is a 30 KVA 3Ø load connected to a 3-pole breaker in spaces 1-3-5. There is a 240-volt 15 KW load connected to a 2-pole breaker in spaces 2-4. There is a 120-volt 10 KW load connected to a 1-pole breaker in space 6.

_____ 4. Refer to Scenario 220.21, shown above. If all the loads are operating and a clamp-on ammeter were placed around the AØ feeder conductor to the panel, there should be ___ amps of current flowing in the feeder conductor.

   a. 229.2   b. 113.8   c. 132.3   d. 134.7   e. 187.5

_____ 5. Refer to Scenario 220.21, shown above. If all the loads are operating and a clamp-on ammeter were placed around the feeder neutral conductor to the panel, there should be ___ amps of current flowing in the neutral.

   a. 155.5   b. 83.3   c. 72.2   d. 134.7   e. 104.2

_____ 6. Three 3-phase loads are fed from a 240 volt panel. Load 1 draws 16 amps, Load 2 draws 42 amps, and Load 3 draws 75 amps. What minimum size standard size Group G, three-phase transformer will be required to supply the panel?

   a. 150   b. 45   c. 112.5   d. 75   e. 30   f. none of these

220-508   3Ø Transformers – Delta-Delta

_____ 7. A Group G, 300 KVA 480 volt Δ primary to 240 volt delta secondary transformer has a supply voltage of 494 volts at its primary terminals. To get close to the rated output voltage, you must relocate tap connections to Position ___ on each winding.

   a. 4   b. 7   c. 2   d. 1   e. 5   f. 6

Scenario 220.22
A 3Ø transformer bank supplies 120/208 volts to a 3Ø, 4-wire panelboard. There is a 30 KVA 3Ø load connected to a 3-pole breaker in spaces 1-3-5. There is a 208-volt 15 KW load connected to a 2-pole breaker in spaces 2-4. There is a 120-volt 10 KW load connected to a 1-pole breaker in space 6.

_____ 8. Refer to Scenario 220.22, shown above. If all the loads are operating, there should be ___ amps of current flowing in the neutral feeder conductor where it enters the panel.

   a. 72.2   b. 155.5   c. 83.3   d. 104.2   e. 134.7

_____ 9. Refer to Scenario 220.22, shown above. If all the loads are operating, there should be ___ amps of current flowing in the BØ feeder conductor where it enters the panel.

   a. 134.7   b. 124.9   c. 155.4   d. 229.2   e. 216.3

_____ 10. Refer to Scenario 220.22, shown above. If all the loads are operating, there should be ___ amps of current flowing in the CØ feeder conductor where it enters the panel.

   a. 152.7   b. 83.3   c. 120.2   d. 208.3   e. 166.6

_____ 11. What minimum standard size Group H, three-phase transformer would be required where the following loads are supplied from the 3-phase 240 volt panel circuits shown:

   Circuit 2-4 feeds a 15 amp single-phase load
   Circuit 6-8 feeds a 10 amp single-phase load
   Circuit 10-12 feeds a 15 amp single-phase load
   Circuit 1-3 feeds a 10 amp single-phase load
   Circuit 5-7 feeds a 15 amp single-phase load
   Circuit 9-11 feeds a 10 amp single-phase load
   Circuit 14-16-18 feeds a 55 amp three-phase load

   a. 30   b. 150   c. 45   d. 112.5   e. 75

3Ø Transformers – Delta-Delta   220-509

_____ 12. Where a delta bank of three transformers is installed to serve a load that is comprised mostly of 120 volt loads (balanced), each power transformer must supply ___ required by the 120 volt loads.

   a. about a third of the current
   b. about half the current
   c. about three times the current
   d. about twice the current
   e. the total current

_____ 13. A 3Ø transformer supplies 120/240 volts to a 3Ø, 4-wire panelboard. From the breaker ispace ___ there will be 208 volts to ground (or neutral).

   a. 2   b. 13   c. 5   d. 23   e. 10

_____ 14. A Group G, 500 KVA 480 volt Δ primary to 240 volt delta secondary transformer has a supply voltage of 494 volts at its primary terminals. To get close to the rated output voltage you must install jumpers between ___ on each winding.

   a. 1&6
   b. 3&4
   c. 3&6
   d. 1&2
   e. 1&4
   f. 2&3

_____ 15. A 240-volt panel has the following 3Ø loads connected: 40 amps, 60 amps, and 50 amps. Group I, 1Ø transformers will be used to make a closed-delta bank. The minimum standard size transformers required to serve these loads have a minimum rating of ___ KVA.

   a. 15
   b. 45
   c. 37½
   d. 50
   e. 25
   f. 75

_____ 16. A 480 delta to 240 delta transformer, without taps, is connected to a 510-volt supply circuit. The line-to-line voltage on the secondary will be ___ volts.

   a. 120   b. 294   c. 208   d. 135   e. 255   f. 240

Year Two (Student Manual)

Objective 220.3 Worksheet

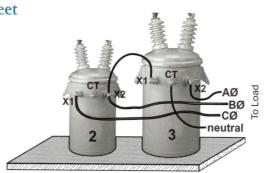

Figure 220.301

_____ 1. Refer to Figure 220.301, shown above, where the secondary voltage is 240 volts. This is a(an) ___ transformer bank.

   a. closed delta   b. open delta   c. wye   d. none of these

_____ 2. Refer to Figure 220.301, shown above, where the secondary voltage is 240 volts. ___ supplies more power to 120 volt loads.

   a. The larger transformer   b. The smaller transformer

_____ 3. Refer to Figure 220.301, shown above, where the secondary voltage is 240 volts. If this bank feeds a fused safety switch, instead of a panelboard, the switch can be a ___.
   I. 250 volt 3Ø, 4-wire
   II. 600 volt 3Ø, 4-wire

   a. either I or II   b. I only   c. II only   d. neither I nor II

_____ 4. Refer to Figure 220.301, shown above, where the secondary voltage is 240 volts. This bank can supply the proper voltages to a ___ panelboard(s).
   I.   120/240 volt, 3Ø, 4-wire
   II. 120/208 volt, 3Ø, 4-wire
   III. 120/240 volt 1Ø, 3-wire

   a. I & II only   b. I only   c. I & III only   d. III only   e. I, II, & III

_____ 5. When a 120/240 volts 3Ø, 4-W **open-delta** bank of transformers delivers power to a load, the AØ-to-BØ voltage is ___ volts.

   a. 0   b. 240   c. 120   d. 208

6. When a 120/240 volts 3Ø, 4-W **open-delta** bank of transformers delivers power to a load, the BØ-to-CØ voltage is ___ volts.

   a. 208   b. 0   c. 120   d. 240

7. An **open-delta** bank of transformers delivers ___ power.

   a. 2Ø   b. 3Ø   c. 1Ø

8. To deliver 3Ø power the minimum number of 1Ø transformers required is ___.

   a. 3   b. 1   c. 2

9. A 240-volt panel has the following 3Ø loads connected: 40 amps, 60 amps, and 50 amps. Group I, 1Ø transformers will be used to make an open-delta bank. The minimum standard size transformers required to serve these loads are rated at ___ KVA.

   a. 75   b. 37½   c. 50   d. 45   e. 15   f. 25

10. When a 120/240 volts 3Ø, 4-W **open-delta** bank of transformers delivers power to a load, the line current equals ___.

    a. $I_{coil\ (or\ winding)}$
    b. $I_{coil\ (or\ winding)} \times 1.732$
    c. $I_{coil\ (or\ winding)} \div 1.732$

11. When a 120/240 volts 3Ø, 4-W **open-delta** bank of transformers delivers power to a load, the BØ-to-neutral voltage is ___ volts.

    a. 240   b. 0   c. 208   d. 120

12. When a 120/240 volts 3Ø, 4-W **open-delta** bank of transformers delivers power to a load, the line voltage equals ___.

    a. $E_{coil\ (or\ winding)} \times 1.732$
    b. $E_{coil\ (or\ winding)}$
    c. $E_{coil\ (or\ winding)} \div 1.732$

3Ø Transformers – Delta-Delta

13. A bank of three 100 KVA transformers is delivering 120/240 volts to a building service. This bank can deliver ___ amps of current to the service if one of the transformers goes out.

    a. 833   b. 481   c. 722   d. 417   e. 200

14. An **open-delta** bank of transformers uses 100 KVA transformers and supplies 120/240 volts. This bank can deliver a maximum of ___ amps of current to loads.

    a. 481   b. 200   c. 417   d. 833   e. 722

15. When a 120/240 volts 3Ø, 4-W **open-delta** bank of transformers delivers power to a load, the AØ-to-neutral voltage is ___ volts.

    a. 0   b. 240   c. 120   d. 208

16. An **open-delta** bank of transformers will require that two transformers with ratings of at least ___ KVA each be used to supply enough power at 120/240 volts to a 69 KVA 3Ø load.

    a. 69   b. 34.5   c. 120   d. 23   e. 40

Objective 220.4 Worksheet

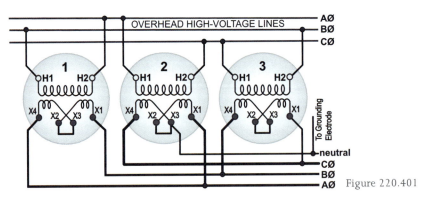

Figure 220.401

_____ 1. Refer to Figure 220.401, shown above. The transformer bank supplies 120/240 volts on its secondary ("X") windings. This means that each winding (X1 to X2 or X3 to X4) is supplying ___ volts.

    a. 240   b. 208   c. 120   d. 277   e. 480

3Ø Transformers – Delta-Delta

_____ 2. Refer to Figure 220.401, shown above. The transformer bank has a ___-connected secondary ("X" leads).

   a. delta   b. wye

_____ 3. Refer to Figure 220.401, shown above. The transformer bank is a ___ connected bank.

   a. delta-wye   b. wye-delta   c. delta-delta   d. wye-wye

_____ 4. Refer to Figure 220.401, shown above. The transformer bank has each of its secondary ("X") windings connected in ___.

   a. parallel   b. series

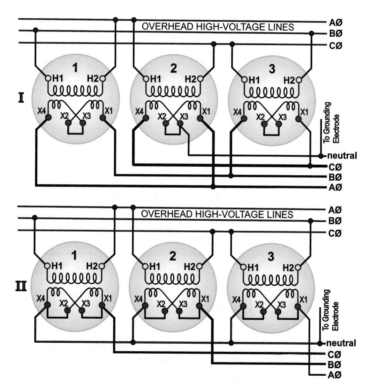

Figure 220.402

_____ 5. Each transformer in the bank, shown above in Figure 220.402, is rated to deliver 120/240 when rated primary voltage is applied. Which diagram shows the correct connections for a 120/240 volt 3-ph, 4-wire secondary?

   a. II only   b. I only   c. neither I nor II

220-516　　　　　　　　　　　　　　　　　　　　　　　　　　　3Ø Transformers – Delta-Delta

Objective 220.5 Worksheet

Scenario 220.53
The secondary of a 300 KVA transformer delivers power to a 180-amp, 3Ø load. This is a 480 — 240 volt Δ - Δ transformer.

_____ 1. Refer to Scenario 220.53, shown above, the smallest primary line conductors permitted are ___ AWG.

　　a. 4　　　　b. 2　　　　c. 3　　　　d. 1/0　　　　e. 4/0　　　　f. none of these

_____ 2. Refer to Scenario 220.53, shown above, the primary line conductors must be sized to carry at least ___ amps.

　　a. 180　　　b. 144　　　c. 104　　　d. 113　　　e. 90　　　　f. 52

_____ 3. Refer to Scenario 220.53, shown above. The secondary OCPD can have a rating no greater than ___ A, where the smallest permitted secondary conductors are installed.

　　a. 200　　　b. 125　　　c. 100　　　d. 150　　　e. 175　　　f. none of these

_____ 4. Refer to Scenario 220.53, shown above. The primary OCPD can have a rating no greater than ___ A, where the smallest permitted primary line conductors are installed.

　　a. 125　　　b. 150　　　c. 175　　　d. 100　　　e. 200　　　f. none of these

_____ 5. Refer to Scenario 220.53, shown above, the smallest secondary conductors permitted are ___ AWG.

　　a. 1/0　　　b. 2　　　　c. 4　　　　d. 4/0　　　e. 3　　　　f. none of these

3Ø Transformers – Delta-Delta   220-517

_____ 6. A 3Ø load is connected to the secondary of a Δ-Δ transformer that has a 12,470-volt primary and a 480-volt secondary. When the secondary conductors to the load are carrying 456 Amps, the primary line conductors are carrying ___ Amps.

    a. 10.13   b. 17.55   c. 52.65   d. 87.76   e. 30.4   f. 152

Year Two (Student Manual)

Lesson 220 Safety Worksheet

HAZCOM PCB TRANSFORMER OIL

POLYCHLORINATED BIPHENYLS (PCB) FACTS

Synonyms: Chlorodiphenyls; Aroclors®; Kanechlors®
CAS Number: 1336-36-3
DOT Numbers: UN2315
DOT Designation: N/A

NFPA Hazard Rating:

Health--2
Flammability--1
Reactivity—0

Carcinogen

Poisonous gasses are produced in fire

Exposure Levels

Inhalation may produce irritation to the nose, throat and lungs. Levels above 10 mg/m3* are reported to be unbearable. Inhalation may contribute significantly to all symptoms of long-term exposure. Skin absorption is moderate, and contributes significantly to all symptoms of long-term exposure. Sensitized individuals may develop a rash after two days of exposure by contact or inhalation.

PCBs may produce eye irritation. Levels of 10 mg/m3 are severely irritating. Ingestion contributes significantly to all symptoms of long-term exposure. There are no reported deaths of humans due to a single ingestion; however, experiments in animals suggest that ingestion of 6 to 10 fluid ounces would cause death to a healthy 150 pound adult.

PCBs are readily absorbed into the body by all routes of exposure and may persist in tissues for years after exposure. Symptoms may be felt immediately, or they may be delayed for weeks or months, and may last for months. High levels of PCB vapor (1 to 10 mg/m3) may produce a burning feeling in the eyes, nose and face; lung and throat irritation; nausea; dizziness; and chemical acne.

Characteristics and Potential Exposures

PCBs are mixtures of chemicals that form clear to yellow, oily liquids, or mixtures that form white, crystalline (sand-like) solids and hard resins. They are used in insulation for electric cables and wires in the production of electric transformers and condensers, as additives for extreme pressure lubricants, and as a coating in foundry use.

Health Hazard Information

Short-term exposure may cause irritation to the skin, nose, throat, eyes and lungs. Long-term exposure may cause a burning feeling in the eyes, nose and face; lung and throat irritation; nausea; dizziness; and chemical acne. Liver damage and digestive disturbance have been reported in some individuals. PCBs may impair the function of the immune system.

Personal Protective Equipment Guidelines

Avoid skin contact with PCBs. Wear protective gloves and clothing. Wear splash proof chemical goggles and a faceshield when working with PCB liquid, unless full facepiece respiratory protection is worn. Wear dust proof goggles and face shield when working with crystalline solids or dust, unless full facepiece respiratory protection is worn. When the potential exists for exposures over 0.001 mg/m3, use a National Institute for Occupational Safety and Health (NIOSH) approved supplied-air respirator with a full facepiece operated in a pressure-demand or other positive-pressure mode.

Handling and Storage

A regulated, marked area should be established where PCBs are handled, used or stored. Store PCBs in tightly closed containers in a cool, well-ventilated area away from strong oxidizers (such as chlorine, bromine and fluorine).

Spills and Emergencies

Most environmental emergencies involve spills of hazardous materials that must be reported to the emergency responders, call 9-1-1. When reporting a spill, callers can also obtain technical assistance regarding response, containment and cleanup of hazardous materials. Evacuate and isolate the area of the spill or leak, and restrict persons not wearing protective equipment from areas of spills or leaks until cleanup is complete. Ventilate the area of the spill or leak. Absorb liquid spills in vermiculate, dry sand, earth, or a similar material, and deposit in sealed containers.

Disposal Methods

Incineration (3000° F) with scrubbing to remove any chlorine-containing products is recommended. In addition, some chemical waste landfills have been approved for PCB disposal.

Fire Extinguishing

PCBs may burn, but do not readily ignite. Use dry chemical, CO2, water spray, or foam extinguishers. Poisonous gasses are produced in fires, including dioxin and chlorinated dibenzofurans.

Emergency First Aid Measures

Eye Contact

Immediately flush with large amounts of water. Continue for at least 15 minutes, occasionally lifting upper and lower lids.

Skin Contact

Quickly remove contaminated clothing. Immediately wash skin with large amounts of soap and water.

Respiratory

Remove the victim from the site of the release.
Begin rescue breathing if breathing has stopped, and CPR if heart activity has stopped. Transfer the victim promptly to a medical facility.

Ingestion

If PCBs are swallowed, give large quantities of saltwater and induce vomiting. Seek medical attention immediately.

THIS FACT SHEET DOES NOT REPLACE THE MATERIAL SAFETY DATA SHEET (MSDS) REQUIRED FOR A HAZARDOUS CHEMICAL UNDER THE OCCUPATIONAL HEALTH AND SAFETY ACT OF 1970 (OSHA) AND REGULATIONS PROMULGATED UNDER THIS ACT.

Questions

_____ 1. When handling PCB transformer oil, you must wear ___.

    a. protective gloves   b. goggles and face shield   c. protective clothing   d. all of these

_____ 2. PCB's were used as insulation in ___.

    a. cables   b. transformers   c. wires   d. all of these

_____ 3. In case of a spill, do the following ___.

    a. call 9-1-1
    b. evacuate and isolate the area
    c. restrict persons not wearing PPE
    d. all of these

_____ 4. PCB's contained in oil do not contribute to airborne health hazards.

    a. False   b. True

_____ 5. PCB oils are ___.
    I. non-hazardous   II. carcinogenic   III. poison gas producing

    a. II only
    b. II & III only
    c. III only
    d. I only
    e. neither I nor II nor III
    f. I, II, & III

Lesson 221
3Ø Transformers – Delta-Wye

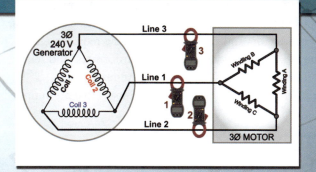

Electrical Curriculum

I E C

Year Two
Student Manual

NATIONAL

Purpose

This lesson will discuss the wiring configurations, applications, operations, and calculations for three-phase transformers with wye-connected secondaries.

Homework

(Due at the beginning of this class)

For this lesson, you should:

- Thoroughly read the material contained within Lesson 221.
- Complete Objective 221.1 Worksheet.
- Complete Objective 221.2 Worksheet.
- Complete Objective 221.3 Worksheet.
- Complete Objective 221.4 Worksheet.
- Read and complete Lesson 221 Safety Worksheet.

- Complete additional worksheets, if available, as directed by your instructor.

Objectives

By the end of this lesson, you should:

221.1 ....Understand single-voltage 3Ø, 4-wire, wye secondary outputs, capacities, and connections.

221.2 ....Understand dual-voltage 3Ø, 4-wire, wye secondary outputs, capacities, and connections.

221.3 ....Understand delta-wye voltage and current transformations, be able to perform calculations, select proper primary and secondary conductors and OCPDs, based on the requirements of NEC Article 240, for single-voltage 3Ø, 4-wire, wye secondary transformers, and select proper primary tap connections for the transformers shown in Yr-2 Annex C.

221.4 ....Be able to perform calculations required for delta and wye-connected loads that are supplied from either delta or wye power sources.

221.1 Single-Voltage Wye Connected Secondaries

As with the Δ secondary connections, it is assumed that the proper (rated) voltage has been applied to the primaries to produce the desired (rated) secondary voltages. Wye connections will now be addressed. One method to draw these three windings (Remember, these are three individual 1Ø transformers.) is illustrated in **Figure 221.11**. Note that the three windings are drawn to look like a **Y**. A Y (wye) connection requires that the 1Ø transformers be interconnected so that X1s (or the X2s – one or the other) of all three transformers are connected together.

Figure 221.11

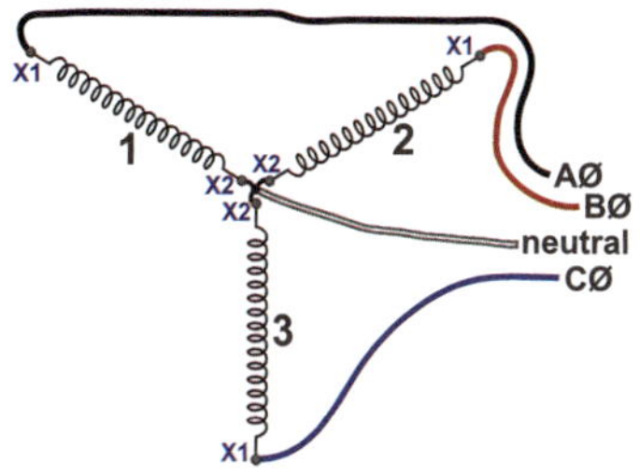

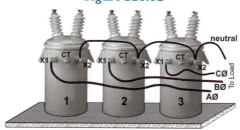

Two other methods of drawing the transformers are shown in **Figures 221.12** and **221.13**. These are equivalent to **Figure 221.11**. Note that in **Figures 221.11, 221.12, & 221.13** the transformers are numbered 1-2-3 for easier reference.

Figure 221.12

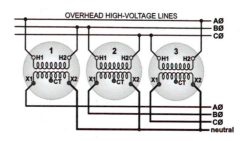

Figure 221.12 shows a pictorial drawing of how the connections are actually made in the field. Although they could just as easily been "pole-mounted", these transformers are mounted on a rack. The connections would have been the same, regardless of mounting type.

Figure 221.13

Figure 221.13 shows a line type connection drawing. Note that the primary windings are still shown to be Δ connected.

Three-phase power

Three-phase transformer banks with wye-connected secondaries consist of three equally rated 1Ø transformers. As with delta-connected secondaries, the power rating or capacity of a bank of three individual 1Ø transformers (all the same size) is simply the three times the rating of any one of the singles. For example, if each of the 1Ø transformers shown in Figure 221.11 is rated at 25 KVA, the rating of the 3Ø bank would equal:

$P_{3\emptyset} = 3 \times P_{1\emptyset}$
$P_{3\emptyset} = 3 \times 25$
$P_{3\emptyset} = 75$ KVA.

For the following discussion, recall (once again) that:

$\mathbf{P}_{1\emptyset} = I_{coil} \times E_{coil}$ and $\mathbf{I}_{coil} = P_{1\emptyset} \div E_{coil}$ and $\mathbf{E}_{coil} = P_{1\emptyset} \div I_{coil}$

If the transformers in **Figures 221.11**, **221.12**, and **221.13** had 277 volt secondaries, a bank could be built that was rated to deliver 277 volts line-to-neutral and 480 volts line-to-line. This would, of course, be a 277/480 volt, 3Ø, 4-wire system. Alternately, if these transformers had 120 volt secondaries, a bank could be built that was rated to deliver 120 volts line-to-neutral and 208 volts line-to-line. This would, of course, be a 120/208 volt, 3Ø, 4-wire system. Although both of these low voltage systems are commonly used, the 120/208 system will be addressed in the examples. Other wye systems include 7200/12470 and 2400/4160 volts. Accepting that the coil voltage (X1 to X2) on each transformer is 120 volts, line-to-line voltages on the secondary of the **Y**-connected transformer bank can now be considered.

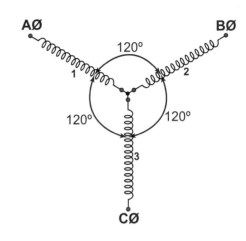

Since the 3Ø voltages that were applied to our primaries are 120° out of phase, the secondary 3Ø voltages will also be 120° out of phase, as shown at the right.

If voltage is measured between AØ and BØ, this voltage measurement is across the secondary of Transformer 1 **plus** the voltage across Transformer 2. **However**, these two voltages are 120° out of phase so **they can't simply be added together**.

For a **wye:** **120 + 120 ≠ 240**

If not 240 volts, what is the line-to-line voltage in a wye? On the drawing shown at the left, lines are drawn between AØ, BØ, and the neutral point. Similarly, lines are drawn from BØ to CØ to the neutral point; and from CØ to AØ to the neutral point. The result is three triangles with the neutral tie point as a corner for each triangle. Arbitrarily selecting the upper triangle, it is a triangle with AØ, BØ, and the neutral as its corners. If these lines are drawn to scale so that the length of the line from the neutral to AØ was 120 volts and the length of the line from the neutral to BØ was the same length, the length of the line from AØ to BØ could be measured. This measurement would show that the length of the line from AØ to BØ was 208 volts.

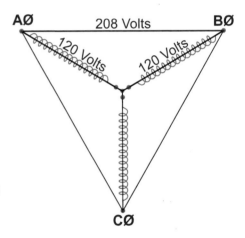

3Ø Transformers – Delta-Wye

An easier way to find the line-to-line voltage for a **Y**, and the <u>method used by electricians</u>, is shown in the following formula. You **will** be responsible for knowing this method!

☞ $E_{line} = E_{coil} \times \sqrt{3}$

☞ $E_{line} = E_{coil} \times 1.732$

For a system with a 120-volt coil voltage:

$E_{line} = 120 \times 1.732$

$E_{line} = 207.85$ or **208** volts

For a system with a 277-volt coil voltage:

$E_{line} = 277 \times 1.732$

$E_{line} = 479.78$ or **480** volts

Refer to **Figure 221.14**. Once again, 25 KVA 1Ø transformers will be used to illustrate wye calculations. These have a rated secondary voltage of 120 volts. Having worked previously with these 1Ø transformers, it was determined that:

$I = P \div E$

$I = 25000 \div 120$

$I = 208.33$ amps.

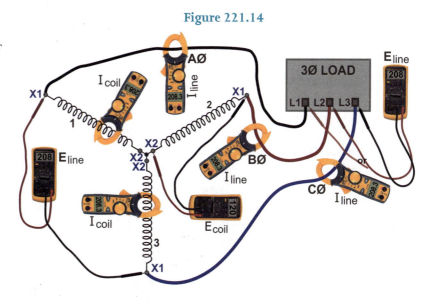

Figure 221.14

208.33 amps is the rated current of the 120 volt secondary winding of each of the 1Ø transformers. This rated current is actually the rated coil current (I_{coil}). Remember that, although a transformer is rated in VA, the windings are limited in the amount of current that can flow through the winding conductor. If the current through the secondary coil (winding) of this transformer exceeds 208.33 amps, the winding will be damaged because it is simply a wire and its ampacity can't be exceeded!

For a **Y** secondary, as with a Δ secondary, when the loads are **balanced** (the coil current in each transformer is the same), the following formula can be used to determine the total bank capacity.

☞ $P_{3\emptyset} = 3 \times P_{1\emptyset}$

☞ $P_{3\emptyset} = 3 \times I_{coil} \times E_{coil}$

For the 3Ø load shown in Figure 221.14:

$P_{3\emptyset} = 3 \times P_{1\emptyset}$

$P_{3\emptyset} = 3 \times I_{coil} \times E_{coil}$

$P_{3\emptyset} = 3 \times (208.33) \times (120)$

$P_{3\emptyset} = 75 \text{ KVA}$

A neutral connection is not shown in Figure 221.14 because a neutral is not required by the 75 KVA 3Ø load. For the **wye**-connected secondary, shown in **Figure 221.14,** the values for coil and line voltages are not equal ($E_{line} \neq E_{coil}$). However, the coil and line currents are equal ($I_{line} = I_{coil}$). These are characteristic of a wye connection.

It was previously determined that the 3Ø capacity of a wye-connected bank is stated mathematically as:

$P_{3\emptyset} = \sqrt{3} \times I_{line} \times E_{line}$, and $P_{3\emptyset} = 3 \times I_{coil} \times E_{coil}$

These formulas will be used to find the current (I) and voltage (E) relationships for a **wye** configuration. These relationships are shown in the following formulas.

☞ $I_{line\ Y} = I_{coil\ Y}$

☞ $E_{line\ Y} = \sqrt{3} \times E_{coil\ Y}$

☞ $E_{coil\ Y} = E_{line\ Y} \div \sqrt{3}$

Proof   208.33 A = 208.33 A
208 V = (1.732)(120 V)
120 V = (208 V) ÷ (1.732)

Example 4: Loads connected to Y secondaries are usually NOT balanced. This example uses the same bank of three individual 25 KVA transformers that have a total capacity of 75 KVA 3Ø. The following loads are connected in a 120/208 volt 3Ø, 4-wire panelboard. Refer to **Figure 221.15**, shown below.

Load #1 has a nameplate rating of: 3Ø 20 A @ 208 volts Ckt 1-3-5 (AØ-BØ-CØ)

Load #2 has a nameplate rating of: 1Ø 5 A @ 120 volts Ckt 2 (AØ)

Load #3 has a nameplate rating of: 1Ø 30 A @ 208 volts Ckt 4-6 (BØ-CØ)

Load #4 has a nameplate rating of: 1Ø 40 A @ 120 volts Ckt 8 (AØ)

The current flowing in each line conductor is shown in the table at the right.

Load	AØ	BØ	CØ
#1	20 amps	20 amps	20 amps
#2	5 amps		
#3		30 amps	30 amps
#4	40 amps		
TOTALS	65 AMPS	50 AMPS	50 AMPS

3Ø Transformers – Delta-Wye

This is a classic example of an unbalanced load because the current in each line conductor is not equal. As shown, the greatest current is flowing in the AØ conductor and through Transformer 1. This means that, if Transformer 1 doesn't have a rated current of at least 65 amps, it will be overloaded. Using the three individual 25 KVA transformers, the rated current of each is:

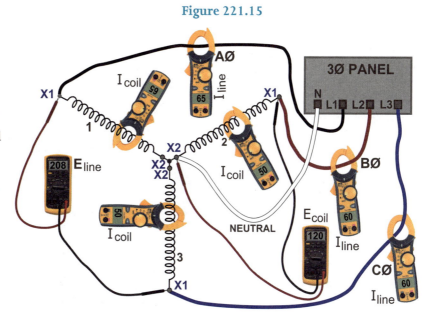

Figure 221.15

$I_{coil} = 25000 \div 120 \text{ volts}$

$I_{coil} = 208.33 \text{ amps.}$

These transformers are well within their ratings.

Notice, in this example, that a slightly better balance would have been achieved if Load #2 had been connected to BØ (or to CØ). This would have resulted in AØ = 60 amps; BØ = 55 amps; and CØ = 50 amps.

Example 5: This is another example of an unbalanced Y secondary. Once again, three individual, equally-sized transformers are connected into a 3Ø bank. For this example, the minimum size required for each individual transformer must be determined. Of course, since these are all the same size, the one that has the greatest load will dictate the minimum size. The following loads are connected in a 120/208 volt 3Ø, 4-wire panelboard. Refer to **Figure 221.16**.

Load #1 has a nameplate rating of: 3Ø   18 KVA @ 208 volts   Ckt 1-3-5 (AØ-BØ-CØ)
Load #2 has a nameplate rating of: 1Ø   26 A @ 208 volts   Ckt 7-9 (AØ-BØ)
Load #3 has a nameplate rating of: 1Ø   30 A @ 208 volts   Ckt 4-6 (BØ-CØ)
Load #4 has a nameplate rating of: 1Ø   45 A @ 120 volts   Ckt 11 (CØ)

For this example, Load #1 is given in KVA instead of amps, so the line current for this load will have to be calculated.

$I_{line} = P_{3\emptyset} \div [E_{line} \times (1.732)]$

$I_{line} = 18000 \div [208 \times 1.732]$

$I_{line} = 49.96 \text{ amps.}$

The total currents that will flow in each line conductor are shown in the table at the left.

Load	AØ	BØ	CØ
#1	49.96	49.96	49.96
#2	26 amps	26 amps	
#3		30 amps	30 amps
#4			45 amps
TOTALS	75.96 AMPS	105.96 AMPS	124.96 AMPS

Year Two (Student Manual)

For this example, the greatest current is flowing in the CØ conductor and through Transformer 3. This means that, if Xfmr 3 doesn't have a rated current of about 125 amps, it will be overloaded. The smallest permitted rating of Xfmr 3 can now be determined using the coil current through Xfmr 3 (124.96 amps) and the coil voltage across Xfmr 3 (120 volts). Using this information:

$$P_{1\emptyset} = I_{coil} \times E_{coil}$$

$$P_{1\emptyset} = 124.96 \times 120$$

$$P_{1\emptyset} = 14{,}995 \text{ VA}$$

$$P_{1\emptyset} \approx 15 \text{ KVA}.$$

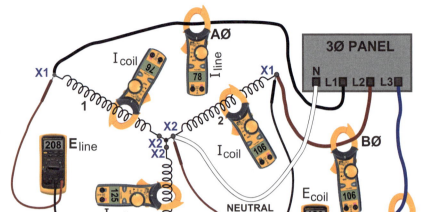

Figure 221.16

If three transformers that have a rating of at least 15 KVA are selected, the bank will have just enough capacity.

Again, an electrician's job is to keep a panel as closely balanced as possible. A better balance would have resulted if Load #2 had been placed on BØ-CØ (Ckt 9-11, for example) and Load #4 had been placed on AØ (Ckt 2, for example). This is shown in the table at the right. This arrangement will still require transformers that have minimum standard ratings of 15 KVA. However, the CØ winding will remain cooler because it does not have as much coil current flowing through it. Remember, heat losses are measured in watts and equal I²R. These losses are commonly referred to as "I²R" losses and should be minimized when possible. When these losses are minimized the transformer operates more efficiently.

Load	AØ	BØ	CØ
#1	49.96	49.96	49.96
#2		26 amps	26 amps
#3		30 amps	30 amps
#4	45 amps		
TOTALS	94.96 AMPS	105.96 AMPS	105.96 AMPS

221.2 Dual-Voltage Wye Connected Secondaries

120/208 V 3Ø, 4-wire Wye Secondary

If a 3Ø panel (or other load) requires supply power, where the majority of the loads require 120 volts to operate, the secondary will be wye connected. This system can also supply 208 volt, 3Ø power to any 3Ø loads. During Single-Phase Transformer Lab 1, one of the 1Ø transformers was connected to a 480 volt supply and the secondary was connected to deliver 120 volts to a load. In

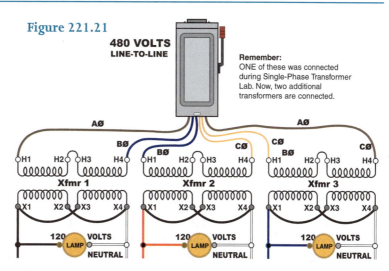

Figure 221.21

Figure 221.21, note that the only complication now is that three of these are connected. Here the 120 volt secondary of each of the transformers is connected to a 120 volt load. Note in the drawing that each secondary has a "hot" and a neutral and is operating independently from the other two. Arbitrarily, X2-X4 was selected to be the neutral in Figure 221.21. Consequently, X1-X3 is the "hot" or "phase" conductor for each secondary.

In order to be able to connect a 3Ø load to the bank, the secondaries of the transformers are interconnected. That is, all three are connected together. X1-X3 on Transformer 1 can be selected to be AØ; X1-X3 on Transformer 2 can be BØ; and X1-X3 on Transformer 3 can be CØ. Since each secondary winding has a neutral, all three of these neutrals are connected together. The result is a 3Ø, 4-wire system (3 "hots" and 1 neutral). This is shown in Figure 221.22.

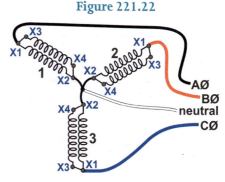

Figure 221.22

The connections in 221.22 have been drawn in the shape of a **Y** or wye. This is a 120/208 3Ø, 4-W secondary where:

- E_{line}:
    The voltage between AØ and BØ is **208** volts.
    The voltage between AØ and CØ is **208** volts.
    The voltage between BØ and CØ is **208** volts.

- E_{coil} for each of the three transformers (1, 2, and 3):
    120 volts would be measured between X1 & X2.
    120 volts would be measured between X3 & X4.
    120 volts would be measured between AØ, or BØ, or CØ and neutral.

- The two secondary windings on each transformer are connected in **parallel**.

Note that the capacity of each of the single-phase transformers used to form a wye bank should have an equal rating. Additionally, the rating of a dual-voltage, single-phase transformer is the sum of the rating of each individual winding set within the transformer. For example, if the rating of a dual-voltage, single-phase transformer is 50 KVA, the rating of each set (H1-H2 & X1-X2, or H3-H4 & X3-X4) is half of the total. For a 50 KVA, half the rating is 25 kVA.

221.3 Delta-Wye Transformer Voltage and Current Ratios

The connected load on the secondary of a transformer causes current to flow in the primary. The amount of current flowing in the primary windings is proportional to the amount of current flowing in the secondary windings. The proportion is dictated by the voltage (turns) ratio. As with single-phase transformers, the current ratio is inversely proportional to the voltage ratio. However, with 3-phase transformers, the current in each of the three windings must be considered independently. A 3Ø transformer with single windings on both the primary and secondary, as shown in **Figure 221.31**, will be used to illustrate how to determine primary coil current when secondary coil current is known. The primary windings are delta connected and the secondary windings are wye connected. This transformer is referred to as delta-wye or Δ-Y. The primary of each of the three transformer windings is rated for 480 volts and the secondary is rated for 120 volts. The winding-to-winding voltage ratio (V.R.) is 4:1 and the winding-to-winding current ratio (I.R.) is 1:4.

Figure 221.31 Delta-Wye Transformer

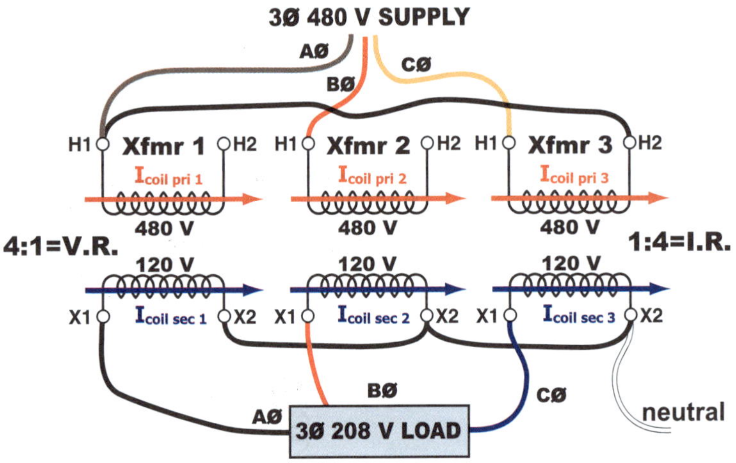

Recall that the basic ratio formula is:

$$\text{V.R.} = E_{pri} \div E_{sec} = I_{sec} \div I_{pri} = \text{I.R.}$$

Don't forget that these are *winding-to-winding* and NOT line-to-line ratios.

If the load is balanced, the coil currents are equal in each of the three windings.

When the rated winding voltages are known, use the following formula to find primary coil current:

3Ø Transformers – Delta-Wye

☞ $I_{coil\ pri} = (I_{coil\ sec} \times E_{coil\ sec}) \div E_{coil\ pri}$

Example 221.31: A 4160 – 277/480 Volt delta-wye transformer supplies power to a 3Ø load. The line current in the secondary conductors is 162 Amps. The line current in the primary conductors is ___ Amps.

Example 221.31 Solution:

First the coil current in the secondary windings must be determined. The current in each of the three transformer windings will be equal since this is a balanced load. Additionally, the coil current and line current are equal since this is a wye secondary.

$I_{coil\ sec} = I_{line\ sec} = 162$ Amps

Remember that this is a wye system, and as such, the line voltages and coil voltages are not equal. The coil voltages for this system are 277 volts since the line voltages are 480.

Now that the secondary coil current is known, the primary coil current can be determined.

$I_{coil\ pri} = (I_{coil\ sec} \times E_{coil\ sec}) \div E_{coil\ pri} = (162 \times 277) \div 4160 = 10.79$ Amps

Now that the primary coil current is known, the delta-connected primary line current can be determined.

$I_{line\ sec} = I_{coil\ pri} \times \sqrt{3} = 10.79 \times 1.732 = \mathbf{18.69}$ Amps

Example 221.32: Refer to the 480 to 240 V, Δ-Y transformer shown in Figure 221.31. As is standard in 2nd year U.N.O., THHN/THWN copper conductors with 75°C terminations are used. The 3Ø, 3-W, 208-volt load is rated at 33.2 kVA. Determine the following:

    A. Secondary line current = _____ Amps = $I_{line\ sec}$

    B. Secondary coil current = _____ Amps = $I_{coil\ sec}$

    C. Primary coil current = _____ Amps = $I_{coil\ pri}$

    D. Primary line current = _____ Amps = $I_{line\ pri}$

    E. Minimum secondary line conductor size to load = _____ AWG

    F. Maximum OCP rating to protect secondary line conductors = _____ Amps

    G. Minimum primary line conductor size for given load = _____ AWG

    H. Maximum OCP rating to protect primary line conductors = _____ Amps

    I. The minimum standard size 3-phase transformer required to supply the load has a rating of ___ KVA.

        a. 100   b. 45   c. 75   d. 112½   e. 150   f. 30

Example 221.32 Solutions:

A. $I_{sec\ line} = P_{3\emptyset} \div (E_{sec\ line} \times 1.732) = 33200 \div (208 \times 1.732) = $ **92.15** Amps

B. $I_{coil\ sec} = I_{line\ sec}$ **92.15** Amps on the Y secondary windings

C. $I_{coil\ pri} = (I_{coil\ sec} \times E_{coil\ sec}) \div E_{coil\ pri} = (92.15 \times 120) \div 480 = $ **23.04** Amps on the Y primary windings

D. $I_{line\ sec} = I_{coil\ pri} \times \sqrt{3} = 23.04 \times 1.732 = $ **39.9** Amps on the primary

E. 92.15 Amps on the secondary line conductors ⇒ **3** AWG

F. 3 AWG has an ampacity of 100 Amps at the 75°C terminals ⇒ **100** A OCP

G. 23.04 Amps on the primary line conductors ⇒ 12 AWG

However, the 23 Amp load will require a 25 A OCPD (minimum).
Additionally, NEC 240.4(D) requires that 10 AWG conductors be used with a 25 A OCPD. ⇒ **10** AWG

H. 10 AWG has an ampacity of 35 Amps at the 75°C terminals ⇒ 35 A OCP

However, NEC 240.4(D) requires that 10 AWG be protected by a 30 A OCP (maximum).
Since the load is 23 Amps, the smallest permitted is a **25** A OCPD.

I. 33.2 kVA load ⇒ **45** KVA transformer ⇒ answer **b**
   or $P_{3\emptyset} = I_{line} \times E_{line} \times 1.732 = 92.15 \times 208 \times 1.732 = 32{,}200$ W ⇒ **45** KVA

Primary Voltages Above or Below Rated Voltage

The nameplate voltage ratings of the primary and secondary windings are used to determine the output voltages for each secondary winding. If a voltage higher than the rated voltage is applied to the primary, a voltage higher than the rated secondary voltage will be delivered to a load. Conversely, if a voltage that is lower than the rated voltage is applied to the primary, a lower than rated voltage will be induced in the secondary windings.

To illustrate this concept, refer to Yr 2-Annex C. Locate a 50 KVA, Group D transformer in the table. Group D transformers have primary windings that are rated at 480 volts and secondary windings that are rated at 120 volts. This is a 4-to-1 voltage ratio. Applying this ratio, when 496 volts is the actual supply circuit line voltage, a secondary voltage of 124 volts would be induced in each secondary winding.

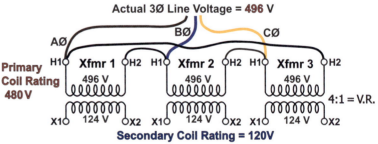

Supply Circuit Voltage Above Rated Primary Voltage

☞ $E_{actual\ sec} = (E_{rated\ sec} \div E_{rated\ pri}) \times E_{actual\ pri}$

$E_{actual\ sec} = (120 \div 480) \times 496$

$E_{actual\ sec} = 124$ volts

Where 124 volts is induced in the windings of this wye-connected secondary, shown above, the line voltage would be almost 215 volts:

$$E_{line} = E_{coil} \times \sqrt{3} = 124 \times \sqrt{3} = 214.8 \text{ volts}$$

If the supply circuit voltage is less than the rated primary voltage, a less-than-rated secondary voltage will be induced. This condition is shown at the right. Using the 4:1 voltage ratio:

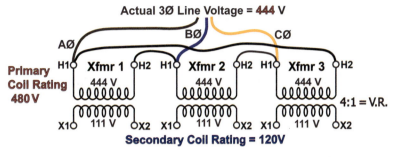

$E_{actual\ sec}$

$= (E_{rated\ sec} \div E_{rated\ pri}) \times E_{actual\ pri}$

$= (120 \div 480) \times 444$

$= 111 \text{ volts}$

Where 111 volts is induced in the windings of this wye-connected secondary, shown above, the line voltage only be about 192 volts:

$$E_{line} = E_{coil} \times \sqrt{3} = 111 \times \sqrt{3} = 192.3 \text{ volts}$$

It is imperative that the voltage applied to a load be as close as possible to the nameplate rated voltage of the load. To accomplish this, when the supply circuit voltage is above or below the transformer's rated primary voltage, a transformer with several primary taps would be selected. The 50 KVA, Group D transformer uses Wiring Diagram 22 for connections. The winding and tap connection points from Diagram 22 are shown above. As shown in the Annex C diagram, this particular transformer has two taps that are *above* the standard 480 volts.

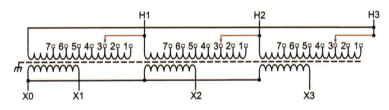

The #2 tap is 2½ percent above 480: $480 + (480 \times .025) = 480 + 12 = 492$ volts.
The #1 tap is 5 percent above 480: $480 + (480 \times .05) = 480 + 24 = 504$ volts.

When the taps are moved from the factory-set Position 3 (shown above in red) to either Position 2 or Position 1, windings are added to the primary winding. This changes the voltage ratio of the transformer.

When moved to Position 2, the voltage ratio is changed to: 492:120.
When moved to Position 1, the voltage ratio is changed to: 504:120.

As was found previously, a supply circuit with an actual voltage of 496 volts produced a secondary output of 124/215 volts. This would be the case if the taps remained in the factory-set position. However, with this particular transformer, the taps would be moved to Position 2. Position 2 is selected because 492 volts is closer to 496 than is 504. With a ratio of 492-to-120 the voltage induced in each secondary winding is:

$$E_{\text{actual sec}} = (E_{\text{rated sec}} \div E_{\text{rated pri}}) \times E_{\text{actual pri}}$$

$$E_{\text{actual sec}} = (120 \div 492) \times 496$$

$$E_{\text{actual sec}} = 120.98 \text{ volts} \approx 121 \text{ volts}$$

The resultant line voltage is:

$$E_{\text{line}} = E_{\text{coil}} \times \sqrt{3} = 121 \times \sqrt{3} = 209.54 \text{ volts} \approx 210 \text{ volts}$$

This will enable the 120 and 208-volt connected loads to operate at close to their rated voltages.

This 50 KVA, Group D transformer has four additional taps that are **below** the standard 480 volts.

The #4 tap is 2½ percent below 480: $480 - (480 \times .025) = 480 - 12 = 468$ volts.
The #5 tap is 5 percent below 480: $480 - (480 \times .05) = 480 - 24 = 456$ volts.
The #6 tap is 7½ percent below 480: $480 - (480 \times .075) = 480 - 36 = 444$ volts.
The #7 tap is 10 percent below 480: $480 - (480 \times .1) = 480 - 48 = 432$ volts.

When the taps are moved from the factory-set Position 3 to Positions 4, 5, 6, or 7, windings are removed from the primary winding. This changes the voltage ratio of the transformer.

When moved to Position 4, the voltage ratio is changed to: 468:120.
When moved to Position 5, the voltage ratio is changed to: 456:120.
When moved to Position 6, the voltage ratio is changed to: 444:120.
When moved to Position 7, the voltage ratio is changed to: 432:120.

Consider again, the condition where the supply circuit was 444 volts. If the taps are left at Position 3, the secondary would deliver 111/192 volts. This would not be acceptable. For the 50 KVA, Group D transformer, the taps should be moved to Position 6. Having done this, the resultant line-to-neutral voltage would be:

$$E_{\text{actual sec}} = (E_{\text{rated sec}} \div E_{\text{rated pri}}) \times E_{\text{actual pri}}$$

$$E_{\text{actual sec}} = (120 \div 444) \times 444$$

$$E_{\text{actual sec}} = 120 \text{ volts}$$

And, of course, the resultant line voltage is:

$$E_{\text{line}} = E_{\text{coil}} \times \sqrt{3} = 120 \times \sqrt{3} = 207.8 \text{ volts} \approx 208 \text{ volts}$$

This will enable any 120 and 208-volt connected loads to operate at their rated voltages.

221.4 Delta versus Wye Connected Three-Phase Loads

Any load, including a three-phase load, requires that nameplate rated voltage be applied to deliver nameplate rated power (or current). If rated voltage is applied, rated line current will be drawn or rated power will be delivered.

A 3-phase piece of equipment (a load) contains at least three individual elements. If the equipment contains more than three elements, these will be multiples of three and the equipment will have a dual-voltage rating. For example, a piece of medical equipment has six elements and is rated to operate at 240 or 480 volts. The equipment has two elements for each phase. The elements, for each phase, are connected in series when the higher voltage is applied, and are connected in parallel when the lower voltage is applied.

If a 3-phase piece of equipment contains three elements, the rated voltage of each element must be applied to that element to produce rated power. The three elements inside the equipment can be delta-connected or wye connected.

Delta-Connected Loads

If the elements in the equipment are delta-connected and each element has a rating of 480 volts, a 480-volt circuit would be connected to the equipment terminals. This 480-volt circuit could originate from a 480 V, 3Ø, 3W panelboard. The panelboard is connected to the delta secondary of a 480-volt transformer and the line-to-line voltage is 480 volts. Alternately, the equipment could be supplied from a 480 V, 3Ø, 4W panelboard. Although the panelboard is supplied from the wye secondary of a 277/480 volt transformer, the line-to-line voltage is still 480 volts. The equipment doesn't care where the circuit originates, as long as it gets a line voltage of 480 so that the applied voltage equals the rated voltage of each contained element. The line current required by the load equals the line current delivered by the transformer (circuit).

Example 221.41: Refer to Figure 221.41, shown below. As discussed previously, the line voltages output by the power supply are 480 and the coil voltages are 277. 480 volts is connected across (applied to) Heating Elements A, B, and C. The nameplate on the 3Ø EQUIPMENT is 150 kVA at 480 volts.

1. Determine the line currents in Line 1, Line 2, and Line 3.

2. Determine the currents flowing through Coil 1, Coil 2, and Coil 3 in the power supply.

3. Determine the currents flowing through Heating Elements A, B, and C.

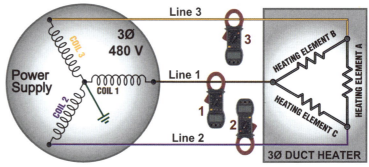

Figure 221.41 Wye-Connected Power Supply to Delta-Connected Load

Solutions for Example 221.41:

1. The line currents can be determined from the nameplate information. Since this is a 3-phase load, recall that 3Ø loads are balanced loads, the currents will be equal in each line conductor.

   $I_{line} = P_{3\emptyset} \div [E_{line} \times (1.732)]$
   $= 150{,}000 \div [480 \times (1.732)]$
   $= \mathbf{180.42}$ amps $= I_1 = I_2 = I_3$

   Each ammeter, shown in the above drawing, should indicate 180.42 amps.

2. Since the power supply is wye connected, the coil currents equal the line currents.

   $I_{coil} = I_{line} = \mathbf{180.42}$ Amps $= I_{Coil\ 1} = I_{Coil\ 2} = I_{Coil\ 3}$

3. The current in each delta-connected heating element could be determined using the delta formula:

   $I_{coil} = I_{line} \div \sqrt{3}$
   $= 180.42 \div 1.732$
   $= \mathbf{104.17}$ Amps $= I_{Element\ A} = I_{Element\ B} = I_{Element\ C}$

Example 221.42: Refer to Figure 221.42, shown below. As discussed previously, the line voltages output by the power supply are 480 and the coil voltages are 480. 480 volts is connected across (applied to) Heating Elements A, B, and C. The nameplate on the 3Ø EQUIPMENT is 150 kVA at 480 volts.

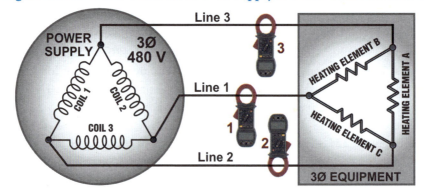

Figure 221.42 Delta-Connected Power Supply to Delta-Connected Load

3Ø Transformers – Delta-Wye

1. Determine the line currents in Line 1, Line 2, and Line 3.
2. Determine the currents flowing through Coil 1, Coil 2, and Coil 3 in the power supply.
3. Determine the currents flowing through Heating Elements A, B, and C.

Solutions for Example 221.42:

1. The line currents can be determined from the nameplate information. Since this is a 3-phase load, the currents will be equal in each line conductor.

$$I_{line} = P_{3\emptyset} \div [E_{line} \times (1.732)]$$
$$= 150{,}000 \div [480 \times (1.732)]$$
$$= \mathbf{180.42} \text{ amps} = I_1 = I_2 = I_3$$

Each ammeter, shown in the above drawing, should indicate 180.42 amps.

2. The current in each delta-connected power supply coil could be determined using the delta formula:

$$I_{coil} = I_{line} \div \sqrt{3}$$
$$= 180.42 \div 1.732$$
$$= \mathbf{104.17} \text{ Amps} = I_{Coil\ 1} = I_{Coil\ 2} = I_{Coil\ 3}$$

3. The current in each delta-connected heating element could be determined using the delta formula:

$$I_{coil} = I_{line} \div \sqrt{3}$$
$$= 180.42 \div 1.732$$
$$= \mathbf{104.17} \text{ Amps} = I_{Element\ A} = I_{Element\ B} = I_{Element\ C}$$

Wye-Connected Loads

If the elements in the 3Ø, 3W equipment are wye-connected and each element has a rating of 120 volts, a 208-volt circuit would be connected to the equipment terminals. This 208-volt, 3Ø, 3W circuit would, most likely, originate from a 120/208 V, 3Ø, 4W panelboard. The line-to-line voltage is, of course, 208 volts. Again, the equipment doesn't care where the circuit originates, as long as it gets a line voltage of 208 so that 120 volts is applied to each wye-connected element. The line current required by the load equals the line current delivered by the transformer (circuit).

Example 221.43: Refer to Figure 221.43, shown below. As discussed previously, the line voltages output by the power supply are 208 and the coil voltages are 120. However, the heating elements are wye-connected, so 120 volts is applied to (connected across) Heating Elements A, B, and C. The nameplate on the 3Ø EQUIPMENT is 150 kVA at 208 volts.

Figure 221.43 Wye-Connected Power Supply to Wye-Connected Load

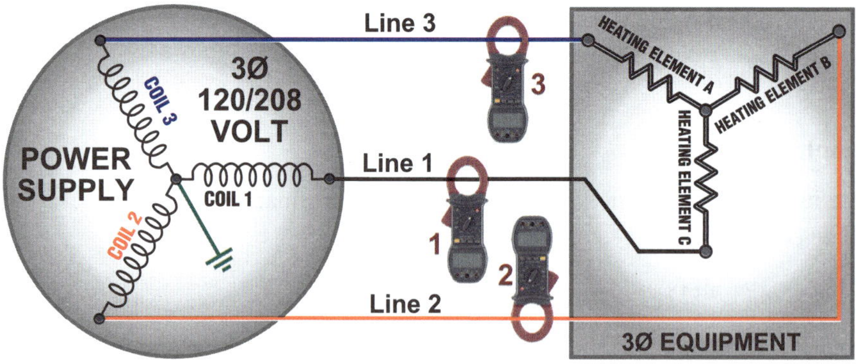

1. Determine the line currents in Line 1, Line 2, and Line 3.
2. Determine the currents flowing through Coil 1, Coil 2, and Coil 3 in the power supply.
3. Determine the currents flowing through Heating Elements A, B, and C.

Solutions for Example 221.43:

1. The line currents can be determined from the nameplate information. Since this is a 3-phase load (once again, 3Ø loads are balanced loads), the currents will be equal in each line conductor.

$$I_{line} = P_{3\emptyset} \div [E_{line} \times (1.732)]$$

$$= 150{,}000 \div [208 \times (1.732)]$$

$$= \mathbf{416.36} \text{ amps} = I_1 = I_2 = I_3$$

Each ammeter, shown in the above drawing, should indicate 416.36 amps.

2. Since the power supply is wye connected, the coil currents equal the line currents.

$$I_{coil} = I_{line} = \mathbf{416.36} \text{ Amps} = I_{Coil\ 1} = I_{Coil\ 2} = I_{Coil\ 3}$$

3. The current in each wye-connected heating element equals the line current:

$$I_{coil} = I_{line} = \mathbf{416.36} \text{ Amps} = I_{Element\ A} = I_{Element\ B} = I_{Element\ C}$$

Example 221.44: Refer to Figure 221.44, shown below. As discussed previously, the line voltages output by the power supply are 480 and the coil voltages are 480. However, the heating elements are wye-connected, so 277 volts is applied to (connected across) Heating Elements A, B, and C. The nameplate on the 3Ø EQUIPMENT is 150 kVA at 480 volts.

3Ø Transformers – Delta-Wye

Figure 221.44 Delta-Connected Power Supply to Wye-Connected Load

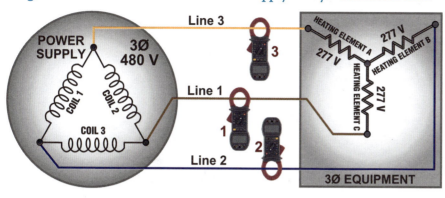

1. Determine the line currents in Line 1, Line 2, and Line 3.

2. Determine the currents flowing through Coil 1, Coil 2, and Coil 3 in the power supply.

3. Determine the currents flowing through Heating Elements A, B, and C.

Solutions for Example 221.44:

1. The line currents can be determined from the nameplate information. Since this is a 3-phase load, the currents will be equal in each line conductor.

    $I_{line} = P_{3\emptyset} \div [E_{line} \times (1.732)]$

    $= 150{,}000 \div [480 \times (1.732)]$

    $= \mathbf{180.42}$ amps $= I_1 = I_2 = I_3$

    Each ammeter, shown in the above drawing, should indicate 180.42 amps.

2. The current in each delta-connected power supply coil could be determined using the delta formula:

    $I_{coil} = I_{line} \div \sqrt{3}$

    $= 180.42 \div 1.732$

    $= \mathbf{104.17}$ Amps $= I_{Coil\ 1} = I_{Coil\ 2} = I_{Coil\ 3}$

3. The current in each wye-connected heating element equals the line current:

    $I_{coil} = I_{line} = \mathbf{180.42}$ Amps $= I_{Element\ A} = I_{Element\ B} = I_{Element\ C}$

3Ø Transformers – Delta-Wye

Objective 221.1 Worksheet

_____ 1. What is the maximum line current the secondary of a 500 KVA Group H1 transformer can deliver without exceeding its rating?

   a. 1203
   b. 601
   c. 801
   d. 463
   e. 1388
   f. 1042

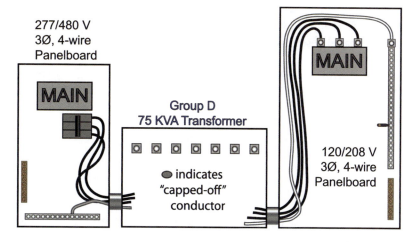

Figure 221.102

_____ 2. Refer to Figure 221.102 (above). Drawing ___ (below) shows the transformer correctly connected.

a.
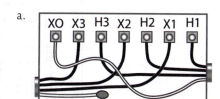

c.
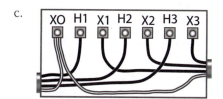

b.
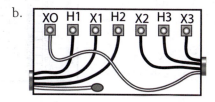

d.
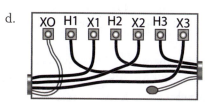

_____ 3. When a 120/208-volt, 3Ø, 4-W supply system (typically a transformer) delivers power to a load, the line current equals ___.

   a. $I_{coil \ (or \ winding)} \div 1.732$
   b. $I_{coil \ (or \ winding)}$
   c. $I_{coil \ (or \ winding)} \times 1.732$

Scenario 221.14

A 277/480 volt panel has eighteen, 1-pole breakers that supply a total 277 volt lighting load of 252 amps. The 277-volt lighting load is equally divided among the 1-pole breakers. Six of the 1-pole breakers are connected to A-phase, six are connected to B-phase, and six are connected to C-phase busbars. The panel also has three, 3-pole breakers that supply three phase loads of 18, 34, and 62 amps respectively.

_____ 4. A Group H1 transformer with a rating of ___ KVA is the smallest that can be used to supply the panel described above in Scenario 221.14.

   a. 300
   b. 112.5
   c. 500
   d. 225
   e. 150
   f. 75

_____ 5. The total load on the panel described in Scenario 221.14 (above) is ___ KVA.

   a. 198   b. 122   c. 366   d. 164.6   e. 304.3   f. 101.4

Scenario 221.15

A 42-circuit 120/208 volt 3-ph, 4-wire panel is shown at the right. Circuits 1, 2, 3, 7, 25, & 27 supply 120 volt loads that are drawing 19 amps on each circuit. Circuits 4 & 16 supply 120 volt circuits that are connected to loads that pull 5 amps each. Breakers 12, 15, 17, 18, & 22 are spares. Breakers 19 & 20 supply connected loads that have nameplates that read 16 amps at 120 volts. Circuits 14, 21, & 23 supply a 3Ø, 4-wire multi-wire branch circuit that supplies lighting loads that draw 7 amps on each circuit. Breakers 29 through 42 are also spares. Circuit 6-8-10 supplies a 17 amp, 3Ø load. Circuit 9-11-13 supplies a 41 amp three phase load. Circuit 24-26-28 supplies a 54 amp three phase load.

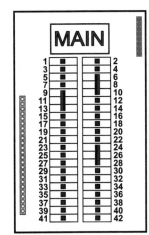

_____ 6. Refer to Scenario 221.15, shown above. The smallest standard size Group D, 480 Δ - 208/120 3Ø, 4-W wye transformer that can supply this panel is a ___ KVA.

   a. 75   b. 112.5   c. 45   d. 50   e. 150   f. 37.5

_____ 7. Refer to Scenario 221.15, shown above. With all the connected loads operating, the current flowing through CØ in the main breaker is ___ amps.

   a. 167   b. 112.5   c. 272.4   d. 81.72   e. 227   f. 119

3Ø Transformers – Delta-Wye   221-543

_____ 8. Refer to Scenario 221.15, shown on the previous page. With all the connected loads operating, the current flowing through AØ in the main breaker is ____ amps.

   a. 272.4   b. 167   c. 227   d. 81.72   e. 119   f. 112.5

_____ 9. Refer to Scenario 221.15, shown on the previous page. With all the connected loads operating, the current flowing through BØ in the main breaker is ____ amps.

   a. 119   b. 112.5   c. 272.4   d. 167   e. 81.72   f. 227

Scenario 221.12

A voltmeter is used to measure voltages between various points on a 120/208-volt, 3Ø, 4W system.

_____ 10. Refer to Scenario 221.12, shown above. ____ volts will be read between AØ and BØ.

   a. 480   b. 120   c. 277   d. 0   e. 208   f. 240

_____ 11. Refer to Scenario 221.12, shown above. ____ volts will be read between BØ and the neutral.

   a. 240   b. 208   c. 277   d. 0   e. 480   f. 120

_____ 12. Refer to Scenario 221.12, shown above. ____ volts will be read between neutral and ground.

   a. 277   b. 120   c. 0   d. 480   e. 240   f. 208

_____ 13. Refer to Scenario 221.12, shown above. ____ volts will be read between CØ and ground.

   a. 208   b. 240   c. 277   d. 120   e. 480   f. 0

_____ 14. A 480-volt, 3-phase supply circuit is connected to a Group D, Cat #3112, transformer. What is the voltage across any one primary winding?

   a. 480   b. 240   c. 0   d. 208   e. 277   f. 120

_____ 15. When a 120/208-volt, 3Ø, 4-W supply system (typically a transformer) delivers power to a load, the line voltage equals ____.

   a. $E_{coil\ (or\ winding)} \times 1.732$ b. $E_{coil\ (or\ winding)} \div 1.732$ c. $E_{coil\ (or\ winding)}$

Year Two (Student Manual)

_____ 16. If a 4-wire, wye-connected supply system has a line-to-line voltage of 13.5 kV, what is the line-to-neutral voltage?

   a. 0 volts   b. 40,500 volts   c. 7,800 volts   d. 23,400 volts   e. 13,500 volts

_____ 17. A 15 kVA 1Ø load has a rated voltage of 208 volts. This load will pull ___ amps when supplied from a circuit in a 120/208 volt 3-ph, 4-W panelboard.

   a. 108.3   b. 41.6   c. 72.1   d. 36.1   e. 62.5

_____ 18. A single-phase load is rated to deliver 17 KVA at 240 volts. When the load is connected to a 2-pole breaker in a 120/208 volt, 3Ø, 4-W panelboard, the load will draw ___ amps of current.

   a. 94.3   b. 47.2   c. 35.4   d. 61.4   e. 54.4   f. 81.7

Table 221.1					
Ckt #'s	Load	AØ	BØ	CØ	A 480Δ — 208Y/120 transformer will supply power to a panel that has these loads:
17-19	24 amps				
5-7-9	12.4 KVA				
1	13 amps				
13-15	7.2 KVA				
11	1.86 KVA				
					current totals

You can use this table to calculate the currents.

_____ 19. Refer to Table 221.1, shown above. Which transformer winding has the smallest load in amps?

   a. AØ   b. BØ   c. CØ

_____ 20. Refer to Table 221.1, shown above. The smallest standard size Group D 3Ø transformer you would select is rated at ___ KVA.

   a. 45   b. 37.5   c. 30   d. 75   e. 50

_____ 21. When a 277/480-volt, 3Ø, 4-W supply system (typically a transformer) delivers power to a load, the coil (or winding) current equals ___.

   a. I_L b. $I_L \div 1.732$ c. $I_L \times 1.732$

3Ø Transformers – Delta-Wye   221-545

_____ 22. A 277/480 volt panel has several 277-volt loads that total 240 amps. These 277-volt loads are equally balanced (or divided) among all three phases. The panel also supplies three-phase loads of 20, 60, and 80 amps. The total load is ___ kVA.

   a. 400   b. 240   c. 133.3   d. 115.2   e. 166.7   f. 199.5

_____ 23. A 277/480 volt 3Ø, 4W service is supplied by a 300 KVA transformer. What is the maximum rated line current the transformer can deliver to a load?

   a. 451   b. 361   c. 300   d. 208   e. 625

_____ 24. The unit of measurement for the answer to the previous question is represented or symbolized by ___.

   a. A   b. KVA   c. VA   d. V   e. KW

_____ 25. A 120/208 volt 3Ø, 4W panelboard is fed by a 45 KVA transformer. The existing loads are as follows: AØ is 46 amps, BØ is 79 amps, and CØ is 92 amps. How many more amps of load could you add to AØ without overloading the transformer?

   a. 62   b. 79   c. 33   d. 46   e. 56

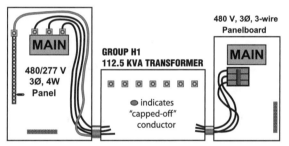

Figure 221.105

_____ 26. Drawing ___ (below) shows the transformer correctly connected for the equipment layout shown above in Figure 221.105.

a.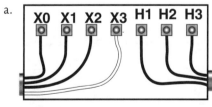

b.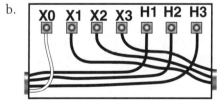

c.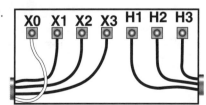

d. none of these is correct

_____ 27. A three-phase, 208-volt load of 84.3 kVA will require what standard-size, Group I, single-phase transformers to be connected in a delta bank?

    a. 25   b. 37.5   c. 50   d. 100   e. 15   f. 75

_____ 28. A 480 — 120/208 volt wye transformer is delivering power to a 3Ø load and has 75 amps of coil current flowing through each winding. The current flowing in each circuit conductor to the load is ___ amps.

    a. 75   b. 25   c. 43.3   d. 129.9

_____ 29. What minimum standard size Group H1, three-phase transformer would be required for the following loads?
    80 amps to 277 volt loads that are connected on AØ
    95 amps to 277 volt loads that are connected on BØ
    87 amps to 277 volt loads that are connected on CØ
    plus
    one 15 amp three-phase 480 volt load
    one 25 amp three-phase 480 volt load
    one 40 amp three-phase 480 volt load
    one 55 amp three-phase 480 volt load

    a. 150   b. 225   c. 45   d. 75   e. 300   f. 112.5

_____ 30. A 480 volt to 120/208 volt three-phase transformer is rated for 125 KVA. If the AØ secondary conductor from the transformer has a load of 140 amps, BØ has 150 amps, and CØ has 130 amps, how much load, in amps, could be added to CØ?

    a. 471   b. 0   c. 217   d. 171   e. 130   f. 391

_____ 31. If three, single-phase, 37.5 KVA transformers are connected together to form a three-phase wye bank, what will be the total rating in KVA?

    a. 65   b. 112.5   c. 150   d. 225   e. 195   f. 75

_____ 32. When a 277/480-volt, 3Ø, 4-W supply system (typically a transformer) delivers power to a load, the coil (or winding) voltage equals ___.

    a. $E_L \times 1.732$ b. E_L c. $E_L \div 1.732$

3Ø Transformers – Delta-Wye

_____ 33. A Group D transformer is to be supplied from a breaker in a 277/480 volt 3Ø, 4W panelboard. You will have to install ___ circuit conductors between the 480 volt panel and the transformer.

a. 3 ungrounded + 1 grounded
b. 1 grounded + 2 ungrounded
c. 3 ungrounded
d. 2 ungrounded

_____ 34. A 480 to 120/208 V transformer has a current in each wye-connected secondary winding of 102 amps. How much line current is the load drawing from the transformer?

a. 58.9   b. 25.5   c. 44.2   d. 185   e. 102   f. 73.6

_____ 35. A wye-connected transformer secondary is properly grounded and has a line-to-ground voltage of 277. The voltage across each winding is ___ volts.

a. 0   b. 277   c. 240   d. 208   e. none of these

_____ 36. A 277/480 volt 3Ø, 4W service is supplied by a utility transformer that is located far away from the service equipment and the line-to-line voltage has dropped to 456 volts at the service equipment. The actual line-to-neutral voltage available at the service equipment is ___ volts.

a. 152   b. 263   c. 228   d. 277   e. 208

Objective 221.2 Worksheet

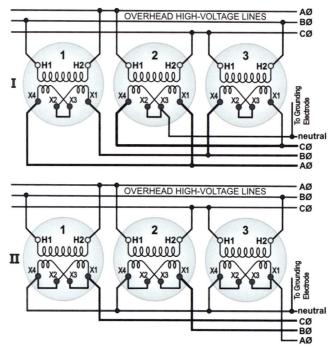

Figure 220.402

_____ 1. Each transformer in the bank, shown above in Figure 220.402, is rated to deliver 120/240 when rated primary voltage is applied. Which diagram shows the correct connections for a 120/208 volt 3-ph, 4-wire secondary?

a. I only   b. neither I nor II   c. II only

_____ 2. Each transformer in the bank, shown above in Figure 220.402, is rated to deliver 120/240 when rated primary voltage is applied. Which diagram shows the correct connections for a 120/240 volt 3-ph, 4-wire secondary?

a. I only   b. neither I nor II   c. II only

_____ 3. Three individual 240 @ 480 @ 120/240 volt single-phase transformers could be connected to produce a ___ output.
I.   240 volt 3-phase, 3-wire
II. 120/240 volt 3-phase, 4 wire
III. 120/208 volt 3-phase, 4 wire

a. II & III only   b. I & II only   c. III only   d. I & III only   e. I, II, & III

3Ø Transformers – Delta-Wye   221-551

Scenario 221.21

You are to connect a 3-ph battery charger system. The charger has 3 terminal strips: TA, TB, and TC; each strip is numbered 1 through 6. The nameplate indicates that the input power can be 208/230/460 volts 3-ph. The connection diagrams from the manufacturer require the following:

    460 volts: L1-L2-L3 to TA6-TB6-TC6 with shorting bars between 3&4 on each strip
        also interconnect TA1-TB1-TC1
    230 volts: L1-L2-L3 to TA6-TB6-TC6 with shorting bars between 3&6 and 1&4 on each strip
        also interconnect TA1-TB1-TC1
    208 volts: L1-L2-L3 to TA5-TB5-TC5 with shorting bars between 2&5 and 1&4 on each strip
        also interconnect TA1-TB1-TC1

_____ 4. Refer to Scenario 221.21, shown above. If your supply circuit is from a 3-pole breaker in a 120/240 volt 3Ø, 4-wire panelboard, you will connect the charger according to the instructions shown for ___ volts.

    a. 208   b. 230   c. 460

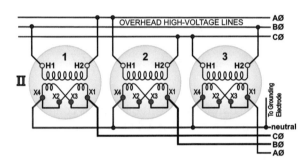

Figure 221.201

_____ 5. Refer to Figure 221.201, shown above. The transformer bank has a ___-connected secondary ("X" leads).

    a. delta   b. wye

_____ 6. Refer to Figure 221.201, shown above. The transformer supplies 120/208 volts on its secondary ("X") windings. This means that each winding (X1 to X2 or X3 to X4) is supplying ___ volts.

    a. 480   b. 208   c. 277   d. 120   e. 240

_____ 7. As shown above in Figure 221.201, each transformer has its secondary ("X") windings connected in ___.

    a. parallel   b. series

_____ 8. Refer to Figure 221.201, shown above. This is a ___ connected bank.

    a. wye-wye   b. delta-delta   c. wye-delta   d. delta-wye

Objective 221.3 Worksheet

_____ 1. When a Group D, Cat #3111, transformer secondary is delivering 100 amps to a three-phase load, ___ amps of current flows through each primary supply circuit conductor.

    a. 100   b. 14.4   c. 25   d. 57.7   e. 43.3   f. 75

_____ 2. A 480 delta to 120/208 wye transformer, without taps, is connected to a 510-volt supply circuit. The line-to-neutral voltage on the secondary will be ___ volts.

    a. 221   b. 120   c. 208   d. 255   e. 128   f. 113

_____ 3. A wye-connected generator has a line-line voltage of 380 volts. What is the voltage across each winding?

    a. 658   b. 240   c. 208   d. 127   e. 380   f. 219

_____ 4. A 75 KVA Group D 3Ø transformer is to be supplied from a 3-pole breaker in a 277/480 volt 3Ø, 4-wire panelboard. However, due to the distance of the transformer from the panel the voltage has dropped to 444 volts. The primary taps should be moved from "3" (factory set) to "___".

    a. 6   b. 7   c. 5   d. 4

_____ 5. A Group D, 112.5 KVA 480 volt Δ primary to 208/120 volt wye secondary transformer has a supply voltage of 494 volts at its primary terminals. To get close to the rated output voltage you must move the tap inter-connection from its factory shipped position 3 to position ___ on each winding.

    a. 1   b. 7   c. 6   d. 5   e. 2   f. 4

3Ø Transformers – Delta-Wye

_____ 6. A 500 KVA, Group D transformer is supplied from a circuit that has a line voltage of 508 at the input terminals. If compensating taps are not used, the secondary line-to-line voltage will be ___ volts.

    a. 126   b. 121   c. 209   d. 218   e. 254   f. 147

_____ 7. A Group D, Cat #3102, transformer is connected to a 450-volt supply circuit. If the taps remain in the factory-set position, the line-to-line voltage on the secondary will be ___ volts.

    a. 195   b. 225   c. 120   d. 128   e. 113   f. 208

_____ 8. A 150 KVA Group D transformer has a primary voltage of 493 volts applied because the utility supply is so close. To compensate for the higher than rated primary voltage you will move the primary taps from 3 to ___.

    a. 6   b. 5   c. 4   d. 1   e. 2

_____ 9. The taps are left at the factory-set position on a 50 KVA, Group D transformer. If the primary is connected to a 460 volt circuit, the line-to-neutral secondary voltage will be ___ volts.

    a. 230   b. 199   c. 133   d. 89   e. 115   f. 345

Objective 221.4 Worksheet

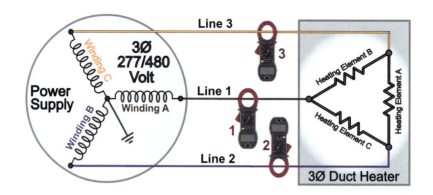

Figure 221.401

_____ 1. Refer to Figure 221.401, shown above. The voltage applied to Heating Element B is ___ volts.

   a. 120   b. 208   c. 240   d. 480   e. 277

_____ 2. Refer to Figure 221.401, shown above. The current flowing in Line 3 is ___ the current flowing in Winding C.

   a. equal to   b. less than   c. more than

_____ 3. Refer to Figure 221.401, shown above. Each of the Heating Elements (A, B, and C) is rated equally. If the total rating of the 3-phase heating load is 90 KW, what is the KW rating of each individual heating element?

   a. 90   b. 30   c. 156   d. 36   e. 52

_____ 4. Refer to Figure 221.401, shown above. The winding (coil) voltage produced by the power supply in Winding C is ___ volts.

   a. 120   b. 208   c. 480   d. 240   e. 277

_____ 5. Refer to Figure 221.401, shown above. If Ammeter 3 indicates 45.8 amps then the duct heater is delivering ___ KW of power.

   a. 38.007   b. 12.687   c. 26.443   d. 10.480   e. 21.984

Year Two (Student Manual)

3Ø Transformers – Delta-Wye   221-555

_____ 6. Refer to Figure 221.401, shown on the previous page. The nameplate rating of the 3Ø duct heater is 90 KW at 480 volts. Each circuit conductor to the duct heater will carry ___ amps at the applied voltage.

   a. 135   b. 325   c. 188   d. 108   e. 62.5

_____ 7. Refer to Figure 221.401, shown on the previous page. The line (line-to-line) voltage produced by the power supply is ___ volts.

   a. 480   b. 240   c. 208   d. 120   e. 277

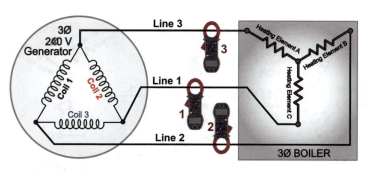

Figure 221.404

_____ 8. Refer to Figure 221.404, shown above. The voltage applied to Heating Element A is ___ volts.

   a. 138.6   b. 240   c. 415.7   d. 360.3   e. 120   f. 208

_____ 9. Refer to Figure 221.404, shown above. The total rating of the 3-phase boiler is 72 KW. Each of the Heating Elements (A, B, and C) has the same KW rating. The rating of each individual heating element in the boiler is ___ KW.

   a. 24   b. 124.7   c. 216   d. 17.32   e. 72   f. 41.57

_____ 10. Refer to Figure 221.404, shown above. If Ammeter 1 indicates 32 amps then the boiler is delivering ___ KW of power.

   a. 23.04   b. 7.68   c. 6.66   d. 15.53   e. 13.30

_____ 11. Refer to Figure 221.404, shown above. The nameplate rating of the 3Ø boiler is 48 KW at 240 volts. Each circuit conductor to the boiler will carry ___ amps at the applied voltage.

   a. 346.4   b. 133.2   c. 230.8   d. 115.5   e. 200

Year Two (Student Manual)

221-556   3Ø Transformers – Delta-Wye

_____ 12. Refer to Figure 221.404, shown on the previous page. The current flowing in Line 3 is ___ the current flowing in Winding 1.

   a. more than   b. less than   c. equal to

_____ 13. Refer to Figure 221.404, shown on the previous page. The line (line-to-line) voltage produced by the generator is ___ volts.

   a. 120   b. 360.3   c. 240   d. 138.6   e. 208   f. 415.7

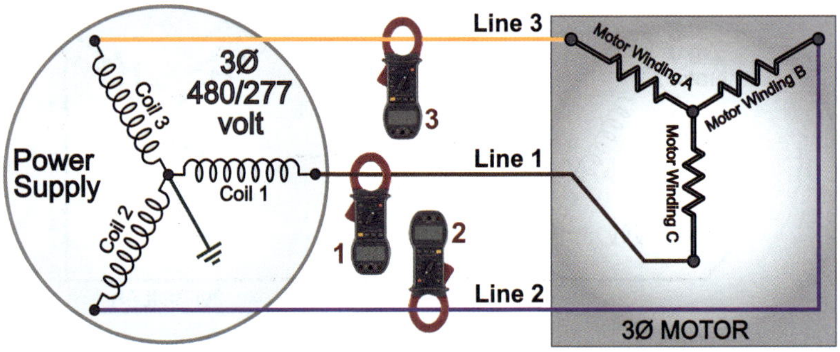

Figure 221.403

_____ 14. Refer to Figure 221.403, shown above. If Ammeter 2 indicates 38.5 amps then the motor is using ___ KW of power.

   a. 55.4   b. 10.7   c. 32   d. 18.5   e. 96

_____ 15. Refer to Figure 221.403, shown above. The voltage applied to Motor Winding A is ___ volts.

   a. 480   b. 240   c. 208   d. 277   e. 120

_____ 16. Refer to Figure 221.403, shown above. The line (line-to-line) voltage produced by the power supply is ___ volts.

   a. 240   b. 120   c. 480   d. 208   e. 277

_____ 17. Refer to Figure 221.403, shown above. If the nameplate rating of the 3Ø motor is 60 KW at 480 volts, Ammeter 2 will indicate ___ amps at the applied voltage.

   a. 125   b. 42   c. 375   d. 217   e. 72

3Ø Transformers – Delta-Wye   221-557

_____ 18. Refer to Figure 221.403, shown on the previous page. The voltage produced by the power supply in Coil 2 is ___ volts.

a. 480   b. 277   c. 208   d. 120   e. 240

_____ 19. Refer to Figure 221.403, shown on the previous page. Each of the Motor Windings (A, B, and C) is rated at 15 KVA. The total 3Ø rating of the motor must be ___ KVA.

a. 15   b. 45   c. 8.66   d. 5   e. 26

_____ 20. Refer to Figure 221.403, shown on the previous page. The current flowing in Line 2 is ___ the current flowing in Motor Winding B.

a. equal to   b. less than   c. more than

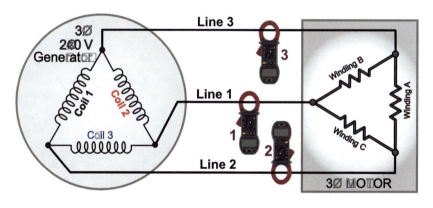

Figure 221.402

_____ 21. Refer to Figure 221.402, shown above. The voltage produced in Coil 2 in the generator is ___ volts.

a. 240   b. 120   c. 208   d. 480   e. 277

_____ 22. Refer to Figure 221.402, shown above. The voltage applied to Motor Winding C is ___ volts.

a. 480   b. 208   c. 277   d. 240   e. 120

_____ 23. Refer to Figure 221.402, shown above. If the nameplate rating of the 3Ø motor is 60 KVA at 240 volts, Ammeter 2 will indicate ___ amps at the applied voltage.

a. 180   b. 250   c. 167   d. 144   e. 288

Year Two (Student Manual)

_____ 24. Refer to Figure 221.402, shown on the previous page. Is the current flowing in generator Coil 1 equal to the current flowing in Motor Winding A?

    a. No   b. Yes

_____ 25. Refer to Figure 221.402, shown on the previous page. If Ammeter 2 indicates 38.5 amps then the generator is delivering ___ KW of power.

    a. 6.23   b. 13.87   c. 9.24   d. 8.01   e. 16.0

3Ø Transformers – Delta-Wye

Lesson 221 Safety Worksheet

Hoists – Hand Operated Chain and Cable Types

As electricians we are required to install many different types of equipment. Some of this equipment is large and/or heavy. We need some form of hoist equipment to help us move it or lift it into place. There are many different manufacturers that make hoists that we can use.

These can be of the type that use chains or they can use cables as the lifting source. If these tools are used as they are designed and intended, they can make our job significantly easier and safer than trying to lift or move things that are too heavy or too cumbersome to handle safely.

Some basic safety items must be followed to assure that you do not injure yourself or someone else. Let's review these basics:

- Make sure the hoist is in good working order.
- The rated capacity of the hoist is sufficient to lift the intended load.
- Be sure the chain or cable is in good condition, not stretched or broken.
- Do not allow anyone under a suspended load.
- Never leave a load that is suspended.
- Never attempt to lift more than the hoist's capacity.
- Never add a "cheater handle" to attempt to lift more load.
- Never use a hoist that is damaged or has been abused.
- Make sure the structure that you are attaching to is strong enough.
- Do not pull over any sharp edges or corners.
- Pull the load with even motion, do not jerk the load.
- Lift only with the load centered under the hoist.

If you follow these basic safety steps when lifting or pulling with a chain or cable hoist, you should be able to move your equipment safer and easier than without using the hoist.

Year Two (Student Manual)

Questions

_____ 1. Hoists can be of a type that use ___ as the lifting source.
   I. cable   II. chains

   a. either I or II   b. II only   c. neither I nor II   d. I only

_____ 2. When ___ a hoist to a structure, make sure the structure is strong enough.

   a. attaching   b. working   c. pulling   d. moving

_____ 3. Never use a hoist that is damaged or has been ___.

   a. repaired   b. abused   c. assigned to another job   d. any of these

_____ 4. When operating a hoist, do not allow anyone under a ___ load.

   a. supervised   b. suspended   c. suspect   d. superseded

_____ 5. The ___ capacity of a hoist is sufficient to lift the intended load.

   a. reflected   b. minimum   c. usual   d. rated

_____ 6. Never add a ___ to a hoist to attempt to lift more load.

   a. come-a-long   b. tugger   c. cheater handle   d. any of these

_____ 7. When using a hoist, move the load with ___ motion. Do not jerk the load.

   a. fast   b. even   c. varying   d. purposeful

Lesson 222
Transformers for Non-Linear Loads – 3Ø Fault Currents – Voltage Drop

Electrical Curriculum

Year Two
Student Manual

Purpose

This lesson will discuss the problems associated with non-linear loads, calculations for available fault currents at transformer secondaries, and voltage drop.

Homework

(Due at the beginning of this class)

For this lesson, you should:

- Thoroughly read the material contained within Lesson 222.
- Complete Objective 222.1 Worksheet.
- Complete Objective 222.2 Worksheet.
- Complete Objective 222.3 Worksheet.
- Read and complete Lesson 222 Safety Worksheet.

- Complete additional worksheets, if available, as directed by your instructor.

Objectives

By the end of this lesson, you should:

    222.1 ....Understand problematic symptoms and solutions, terms, and transformer construction associated with systems serving non-linear loads

    222.2 ....Be able to calculate three-phase and single-phase bolted-fault currents.

    222.3 ....Be able to calculate three-phase and single-phase voltage drop.

222.1 Transformers for Non-Linear Loads

Typical K-Rated Transformer Construction

- Iron Core
  - *windings wound around iron core*
- Tap Connections
- Isolation Pads for Vibration
- Faraday Shielding for Noise and Voltage Spikes
- Aluminum Windings Reduce Weight and Dissipate Heat to Reduce Losses
- Double-Sized Neutral Connections for Excessive Currents
- Rigid Steel Bottom is Punched for Ventilation

    Although most dry-type distribution transformers are shielded for cleaner power, special transformers are generally specified for non-linear loads. Non-linear loads such as computers, variable speed drives for motors, HID lighting, fluorescent lighting using electronic ballasts, inverters, welders, etc produce high levels of harmonic currents. Many devices and equipment that utilizes a DC power supply will generate these harmonic currents. These currents can far exceed the current in any ungrounded conductor and a standard transformer would be damaged.

When supplying power to non-linear loads special types of transformers should be used. K-rated transformers are commonly used for this purpose.

There are four basic K-factor classifications for transformers that supply non-linear loads:
- K-1 includes resistance heating, incandescent lighting, motors, and control and distribution transformers
- K-4 includes welders, induction heaters, HID lighting, fluorescent lighting, and solid state controls
- K-13 includes telecommunications equipment and branch circuits in health care facilities
- K-20 includes computers, variable speed drives, and data processing equipment branch circuits

Transformers and Non-Linear Load Terminology

Sinusoidal Current or Voltage: This is a term used to describe a periodic waveform that can be expressed as the sine of a linear function of time.

Linear Loads: Linear loads produce a current waveform that conforms to, or is in phase with, the waveform of the applied voltage. Restated, these are loads where a change in the current is directly proportional to a change in the applied voltage. These include resistance heating, incandescent lighting, water heaters, or any load that consists of resistance elements.

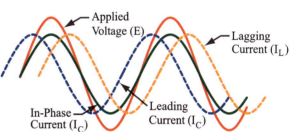

When inductive loads are present the current lags the applied voltage.

When capacitive loads are present the current leads the applied voltage.

60 Hz	Fundamental
120 Hz	2nd Harmonic
180 Hz	3rd Harmonic
240 Hz	4th Harmonic etc.

Harmonic: A sinusoidal waveform with a frequency that is an integral multiple of the fundamental 60 Hz frequency.

Current waveforms from non-linear loads appear distorted because the non-linear waveform is the result of adding harmonic components to the fundamental current.

Triplen Harmonics: These are odd multiples of the 3rd harmonic (3rd, 9th, 15th, 21st, etc).

Non-linear Loads: These loads produce current waveforms that do NOT conform to the waveform of the applied voltage. A change in current in a non-linear load does NOT result in proportional to the applied voltage. Non-linear loads include computer power supplies, motor drives (variable frequency drives), fluorescent lighting, and HID lighting. Frequently the electronics contained in these loads use current in pulses. Non-linear loads produce non-sinusoidal current or voltage waveforms. The resultant waveforms are distorted due to the presence of harmonics.

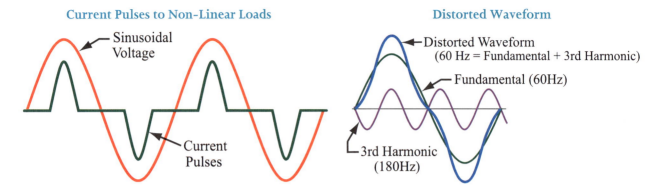

Voltage Harmonic Distortion (VHD): Voltage harmonic distortion is caused by harmonic currents flowing through the system impedance. The utility power system has relatively low system impedance, and the VHD is very low. But, VHD on the distribution power system can be significant due to its relatively high system impedance.

Harmonic Distortion: This is a non-linear distortion of a system characterized by the appearance in the output of harmonic currents (or voltages) when the input is sinusoidal.

Total Harmonic Distortion (THD): This is the sum of the squares of all harmonic currents present in the load excluding the 60 Hz fundamental. It is usually expressed as a percent of the fundamental.

Root Mean Squared Current (or Voltage) RMS: This can be either:
  a. the vector sum of the fundamental current and the THD, or
  b. the square root of the sum of the squared value of the fundamental current and the squared value of the THD.

Eddy Currents: These are currents that flow in a conducting material that is in the presence of a varying magnetic field. These currents are in addition to any current drawn by a connected load.

Eddy Current Losses: This is power dissipated due to the presence of eddy currents. This includes eddy current losses in the transformer core, the windings, the case, and any associated hardware.

I^2R Losses: These are losses, in the form of heat, due to the passage of current through resistance.

Harmonic Spectrum "K" Losses: The "K" factor for a linear load (without harmonics) is 1.

Symptoms & Solutions: Loads supplied from delta systems, such as 120/240 volt, do not produce harmonic currents. However, in wye systems, harmonics are a major consideration. A system that is plagued with the presence of harmonic producing elements is frequently apparent because of overloaded neutrals. This occurs in a 3Ø, 4-W wye system because the neutral currents from balanced multiwire branch circuits DO NOT cancel each other because the neutral currents are not proportional to the applied voltages. This current can be as high as twice the current in one of the ungrounded conductors. Because of this overloading the neutral conductors of both branch circuits and feeders are overheated which damages the insulation and the conductor. Increasing the size of the neutral conductors alleviates the conductor overheating but does not remove the presence of harmonics. This overheating is also present in the transformer and can overheat the windings and their insulation. Many products are available to help reduce the problems associated with harmonics. One such product is a "K" rated transformer. Typically these transformers have "double" or 200% neutrals and generally paralleled secondary neutral conductors are connected. Refer to BCES plans, Dwg E5.1, where paralleled (double) neutrals are required in *each* conduit to several of the low-voltage panelboards. These panels supply power to many non-linear (harmonic producing) loads.

> Note: Frequently, plans and specifications require that each circuit have its own neutral or super-neutrals (larger than the ungrounded conductor) be installed with multiwire circuits, including feeder circuits.

The K-Rated transformer was created to address the additional heat being produced by standard Delta-Wye transformers when feeding non-linear loads. The goal of the K-Rated transformer is to dissipate the heat produced over a larger area, thus providing the illusion of correction. The K-Rated transformer does nothing to reduce the root issue of harmonic currents or correct Power Quality issues. Because of their larger size, they also consume additional energy to 'do the same work' and are counterproductive to an energy efficient electrical distribution design.

Harmonic Mitigating Transformers (HMTs) are a better solution, than K-Rated Transformers, to alleviate problems caused by the presence of harmonics and as a means to reduce energy costs. The recent shift of our electrical loads from predominately AC loads (resistive heating elements, incandescent lighting, three-phase motor load) to DC loads (such as computers, fax machines, printers, cell phone chargers, etc.) requires that our electrical infrastructure changes as well. These new loads now introduce other currents and frequencies into our electrical power systems — commonly known as "harmonics". Harmonic currents can cause additional heating, which may cause transformers, generators and conductors to become overloaded. Excessive heat is one of the major reasons that standard transformers and conductors fail prematurely. These harmonic currents have various other effects (such as reduced lifespan and mysterious mis-operation of equipment) on the components and loads of an electrical distribution system. Eaton's Cutler-Hammer HMTs, when used properly within an electrical system, will help keep the loads operating the way the manufacturer designed them and keep the facility's electrical system free from voltage distortion.

222.2 Three-Phase Fault Currents

Short-circuit calculations for single-phase transformers were addressed in previous lessons. The calculation given was for determining the available short circuit current at the transformer secondary. The current limiting characteristic of a transformer is impedance (Z). Transformer impedance, (Z) is generally given on the nameplate in % form. This must be converted to a decimal value for use in fault-current formulas. Calculation of the impedance of a transformer will not be included in this lesson. However, the ability to calculate the available bolted-fault (short-circuit) current at the transformer secondary terminals for both 1Ø and 3Ø is necessary. Use the following formulas to calculate the maximum available short-circuit fault current (I_{SC} or I_{SCA}).

- **Single-phase** $I_{sc} = VA \div [E_L \times Z]$
- **Three-phase** $I_{sc} = VA \div [E_L \times (1.732) \times Z]$

A "rule of thumb" is that the available secondary fault current is about 50 to 20 times the rated secondary current. These numbers are the reciprocal of the impedance $[1 \div Z]$ of the transformer. That is: Z = 2% corresponds to 50 times or $[1 \div .02]$ and Z = 5% corresponds to 20 times or $[1 \div .05]$.

Key considerations in electrical design involve the current carrying capacities of conductors and components. Conductor sizes and overcurrent protection ratings are based on preventing excessive temperature rise under steady state conditions. The system, however, must be designed with equipment that meets or exceeds the short-circuit current rating available during a fault. The purpose of this objective is to understand the TREMENDOUS amount of current that flows through the windings and the conductive path created during a short-circuit, and that the closer this occurs to the power source (such as a transformer) the higher the current. Additionally, the KVA or KW rating of the power source determines the amount of current available at its output terminals. If the cover is removed from an energized transformer and the cover contacts the "hot" terminals in the process, a bolted-fault will occur. Bolted-faults are those faults that are very close to the power source. The interrupting rating of the first OCPD used on a transformer secondary (main breaker or fused switch) must be rated for the available fault current. If not, it could very well explode. If the designer of a system specifies particular (manufacturer-specific) OCPDs the installer (field electrician) MUST install these devices without substitution! These devices are most likely series-rated and coordinated to prevent complete system shutdown or hazardous conditions.

222.3 Three-Phase Voltage Drop

Single-phase voltage drop was addressed in previous studies. To calculate the approximate voltage dropped in the conductors of a single-phase, or three-phase circuit use the following formulas.

- $V_{d\,1\varnothing} = (2 \times K \times I \times L) \div cma$
- $V_{d\,3\varnothing} = (\sqrt{3} \times K \times I \times L) \div cma$

Where, as studied previously:

K = calculated or provided value
I = line current
L = circuit length (one way)
cma = circular mil area of the conductor

In any AC circuit the value for K varies with ambient temperature and circuit impedance. Voltage drop in AC circuits is the product of current and impedance ($E = I \times Z$). Impedance is affected by the conductor size, conductor material, the inductance produced by adjacent conductors, and the wiring method used. Occasionally, other factors are considered when engineering circuits. Because of the complexity, engineers and designers frequently use computer software to calculate voltage drop.

The formula, shown above, is used by electricians to determine *approximate* voltage drop. Typical K values used range from 10 to 13 for copper. Alternate 3Ø formulas for determining minimum conductor size (cma), maximum conductor length (L), and maximum load (I) are shown in the following formulas. When using these formulas, $V_{d\,3\varnothing}$ is the maximum limit.

- $cma = (\sqrt{3} \times K \times I \times L) \div (V_{d\,3\varnothing})$
- $L = (cma \times V_{d\,3\varnothing}) \div (\sqrt{3} \times K \times I)$
- $I = (cma \times V_{d\,3\varnothing}) \div (\sqrt{3} \times K \times L)$

Transformers for Non-Linear Loads – 3Ø Fault Currents – Voltage Drop

Objective 222.1 Worksheet

_____ 1. One symptom of problems in a system supplying power to non-linear loads is overheated neutral conductors.

   a. False   b. True

_____ 2. Which of the following are harmonic triplens?

   a. 3rd   b. 21st   c. 9th   d. 15th   e. all of these

_____ 3. Transformers with aluminum windings ___ than those with copper windings.

   a. more easily dissipate heat
   b. are lighter in weight
   c. have fewer losses
   d. all of these

_____ 4. Steel transformers enclosures have punched bottoms specifically ___.

   a. to install a GEC
   b. for circulation of air
   c. to reduce weight
   d. all of these

_____ 5. A transformer with a K rating of ___ can be used to supply a panelboard that serves only linear loads.

   a. 4   b. 13   c. 1   d. 20

_____ 6. Which of the following loads are considered to be non-linear?
   I. computers
   II. incandescent lamps
   III. variable frequency drives

   a. I & II only
   b. II only
   c. I & III only
   d. I, II, & III
   e. II & III only

_____ 7. Non-linear loads produce a current waveform that ___ with the applied voltage waveform.

   a. does not conform   b. is in phase

_____ 8. K-Rated non-linear load isolation transformers have Faraday shields that ___.

   a. eliminate voltage spikes
   b. eliminate electrical noise
   c. both a and b
   d. neither a nor b

_____ 9. Which of the following are linear loads?

   a. HID lighting
   b. a booster heater for a dishwasher
   c. neither a nor b
   d. both a and b

_____ 10. A transformer with a K rating of ___ should be used for computer rooms.

   a. 20   b. 1

_____ 11. Non-linear loads supplied from a ___ system would require the installation of a HMT or K-Rated transformer.
   I. 120/240 V, 3Ø 4W   II. 277/480 V, 3Ø 4W   III. 120/208 V, 3Ø 4W

   a. II & III only
   b. III only
   c. I only
   d. II only
   e. I, II, & III

_____ 12. "Non-Linear Load Isolation" (or "K-rated") transformers are manufactured with ___ to handle excessive currents due to harmonics.

   a. aluminum strip windings
   b. wound cores
   c. double-sized neutral conductors
   d. Faraday shields

Objective 222.2 Worksheet

_____ 1. A 150 KVA 480 to 120/240 volt 3Ø, 4-wire transformer has Z = 2.07% on its nameplate. The available line-to-line (bolted) fault current on the secondary is ___ amps. Any secondary OCP installed on the secondary of this transformer should have an AIC rating of at least this value.

    a. 17,432   b. 8,716   c. 1,119   d. 15,097   e. 20,114

_____ 2. An electrician installed a Group I, 250 KVA, 1Ø transformer and the secondary conductors to the main breaker in a 120/240 volt 1Ø, 3-wire panelboard. He left the transformer cover and deadfront off overnight. Someone sabotaged the transformer during the night by installing a jumper between the X1 and X4 transformer lugs. The following morning the electrician installed the covers and energized panel. About ___ amps of current flowed through the jumper. The transformer has Z = 1.0 % on its nameplate.

    a. 40,064   b. 12,028   c. 104,167   d. 69,393   e. 120,192   f. 60,140

_____ 3. A 225 KVA transformer has a 120/208 volt, 4-wire delta secondary. The nameplate shows 3.2% Z. The available secondary bolted (line-to-line) fault current is ___ amps.

    a. 5,413   b. 7,200   c. 19,517   d. 29,297   e. 16,915

_____ 4. You are installing a 150 KVA, 480 volt to 120/208 volt 3-phase transformer. If the %Z is 2.7, the available fault current on the secondary is ___ amps.

    a. 11,574   b. 26,709   c. 13,365   d. 23,148   e. 6,682   f. 15,42

Transformers for Non-Linear Loads – 3Ø Fault Currents – Voltage Drop 222-573

Objective 222.3 Worksheet

Voltage Drop Scenario #7

A load draws 12.9 amps and the distance between the load and the panel is 164 feet. The circuit to the load consists of three #10 THHN/THWN stranded uncoated copper conductors that are connected to a 3-pole breaker in a 120/240 volt 3Ø, 4-wire panel. Use K = 12, as required in any calculations.

_____ 1. Refer to Voltage Drop Scenario #7, shown above. The voltage applied to the load is ___ volts.

    a. 200.7   b. 236.3   c. 235.8   d. 231.5   e. 203.8

_____ 2. Refer to Voltage Drop Scenario #7, shown above. If the voltage drop is to be limited to 2.3% then this is ___ volts.

    a. 2.39   b. 5.22   c. 5.52   d. 4.78   e. 2.76

_____ 3. Refer to Voltage Drop Scenario #7, shown above. The voltage dropped in the circuit conductors is ___ %.

    a. 3.06   b. 3.53   c. 2.04   d. 1.77   e. 1.02

_____ 4. Refer to Voltage Drop Scenario #7, shown above. What is the voltage dropped in the circuit conductors?

    a. 8.47   b. 4.24   c. 7.34   d. 2.12   e. 3.67

Voltage Drop Scenario #1

For a load the conductors consist of two #12 THHN/THWN solid coated copper conductors connected to the neutral bar and a 1-pole breaker in a 120/208 volt 3Ø, 4-wire panelboard. The circuit serves a load that draws 9.8 amps. The distance between the panel and the load is 150 feet.

_____ 5. Refer to Voltage Drop Scenario #1, shown above. Use the formula $K = (R_{per\ foot} \times cma)$ to determine the actual K value. The actual K value = ___ using Table 8.

    a. 12.6   b. 12.93   c. 13.39   d. 13.13   e. 12.0

Year Two (Student Manual)

_____ 6. Refer to Voltage Drop Scenario #1, shown above. Use Table 8 resistance and Ohm's Law to find the voltage dropped in the conductors and then determine the voltage applied to the load. The voltage applied to the load is ___ volts.

   a. 202.6   b. 202.1   c. 114.1   d. 117.1   e. 119.6

_____ 7. Refer to Voltage Drop Scenario #1, shown above. If the voltage drop is to be limited to 2.2% then this is ___ volts.

   a. 4.58   b. 2.64   c. 5.45   d. 3.60   e. 1.83

_____ 8. A 3Ø circuit supplies power to a 23.1 A load. A 3-pole breaker in a 480/277 volt 3Ø, 4-wire panel provides this power through 196 feet of 12 AWG MC cable. The voltage applied to the load is ___ volts. The K value for this cable installation is 11.7.

   a. 472   b. 263   c. 466   d. 270   e. 249   f. 452

_____ 9. A 2-pole circuit breaker in a 120/208 volt panel feeds a circuit that draws 14 amps. 12 AWG copper conductors are used. To avoid exceeding the maximum voltage drop per the NEC the circuit length can't exceed ___ feet. Use K = 10.93

   a. 140   b. 89   c. 154   d. 133   e. 121

_____ 10. Two THHN/THWN copper circuit conductors are connected to the neutral bar and a 1-pole breaker in a 120/208 volt 3Ø, 4-wire panelboard. The circuit serves a load that draws 11.9 amps. The distance between the panel and the load is 176 feet. The smallest conductors you can install to the load are ___ AWG. The specifications limit the voltage drop to 2.1 %. Use K = 10.86 to solve this problem.

   a. 14   b. 10   c. 6   d. 12   e. 8

_____ 11. The manufacturer limits the voltage drop on a piece of 3Ø, 480-volt equipment to ± 4%. The equipment will draw 48 amps and will be connected to the panel with 8 AWG THHN/THWN circuit conductors. Using K = 11.2, the maximum distance permitted between the equipment and the panel is ___ feet.

   a. 550   b. 318   c. 590   d. 238   e. 340

Transformers for Non-Linear Loads – 3Ø Fault Currents – Voltage Drop

_____ 12. The heating element in an electric forced-air heating unit will produce nameplate rated heat output if connected to a voltage supply that is ___ the rated voltage.

    a. less than   b. greater than   c. equal to   d. any of these

_____ 13. The three THHN/THWN copper conductors of a circuit are connected a 3-pole breaker in a 208/120 volt 3Ø, 4-wire panelboard. The circuit serves a load that draws 23.1 amps. The distance between the panel and the load is 216 feet. The smallest conductors you can install to the load are ___. Using K = 11.2 and the 2.2 % voltage drop limitation imposed by the job specifications, solve this problem.

    a. 10   b. 8   c. 6   d. 14   e. 12

_____ 14. A 3Ø 480-volt, 41 KVA load is located a distance of 280 feet from the 480 volt panel. If K=11.3, and the wire is #8, calculate the voltage applied to the load.

    a. 447   b. 471   c. 460   d. 452   e. 464

_____ 15. A load was located so that the circuit length was 100 feet from the panel. However the load has been relocated so that the circuit length is now 150 feet from the panel. The total circuit resistance ___ when the load was relocated. The circuit was extended using the same size wire.

    a. decreased   b. did not change   c. increase

Lesson 222 Safety Worksheet

OSHA's Most Frequently Cited Construction Violations

The most frequently cited Federal OSHA Construction Safety and Health Violations appear to change little from year to year. We are listing them here to raise the level of awareness to these violations to provide some additional information to help you learn how to better protect yourself on the construction sites. We want to stimulate conversation about these violations and further discussion to see what can be done to help correct these things before you become a victim yourself.

The following is listed by OSHA Standard number with description. The number in brackets is the number of citations issued by Federal OSHA for these violations only for our Electrical Work job category, SIC1731.

1. 1926.405 Electrical – (320) Wiring methods, components and equipment for general use these violations are for items such as temporary wiring, box and fitting use, conductors, cords and cables and equipment.

2. 1926.403 Electrical – (221) General requirements deals with safety of equipment, listing and labeling, mounting, marking, identification, working space and guarding of live parts.

3. 1926.404 Electrical – (108) Wiring design and protection includes identification of conductors, GFCI protection, conductor clearances, disconnecting means, grounding and bonding of equipment.

4. 1926.501 Fall Protection – (103) Duty to have Fall Protection requires fall protection systems, for unprotected sides and edges, leading edges, hoist areas, holes, excavations, dangerous machinery, roof work, wall openings and protection from falling objects.

5. 1926.453 Aerial Lifts (99) covers numerous types of lifts and references ANSI standards and procedures to be followed when operating these lifts.

6. 1926.1053 (80) Stairways and Ladders – The requirements of capacity and design criteria for fixed ladders and portable ladders. It also states the correct usage of ladders.

7. 1926.416 Electrical – (74) Safety-related work practices states no work on energized circuits, locating underground electric lines and guarding of work spaces.

8. 1926.1052 Stairways and Ladders – (49) Stairways outlines the rules for stairways, landings, stairrails, midrails, and handrails. It also covers use of stair pans and temporary stairs.

9. 1926.020 General Safety and Health Provisions – (43) General safety and health provisions describes the requirements to provide frequent and regular inspections of the jobsite and use of any equipment or tool that is not in compliance.

10. 1926.021 General Safety and Health Provisions – (39) Safety training and education requirements are the employers responsibility to instruct to each employee.

11. 1926.451 Scaffolds – (39) General requirements for scaffolding details many rules for many different types of scaffolds. It outlines what must be provided to employees who must erect, dismantle or work on scaffolds. Falling object protection is also required.

12. 1926.100 Personal Protective Equipment – (38) Head Protection requirements are outlined in this section.

13. 1910.1200 Hazard Communication (33) covers all the requirements for HazCom, labeling, information, definitions, handling of chemicals and emergency procedures necessary.

14. 1910.333 Selection and use of work practices – (28) Sets the criteria for work on or near any energized circuits, safety procedures and training required, qualifications of employees, lock-out requirements to be followed and more.

15. 1926.95 Criteria for personal protective equipment (28) outlines the need for and use of PPE. Also the responsibility for maintaining the equipment.

16. 1910.67 Vehicle-mounted elevating and rotating platforms (24) shows the rules of use, fall protection requirements, training requirements and more.

17. 1926.503 Fall Protection – Training requirements (23) include recognition of hazards of falling, correct procedures of erecting, maintaining, disassembling and inspection of fall protection systems, and the retraining of employees who show a lack of knowledge retention.

18. 1910.335 Electrical – Safeguards for personnel protection (22) requires PPE, signs, insulated mats and barricades.

19. 1926.502 Fall protection systems criteria and practices (22) provides a detailed system for fall protection, PFAS personal fall arrest systems, safety nets, positioning systems, warning line systems, controlled access zones, safety monitoring systems and fall protection plans.

20. 1926.1060 Stairways and Ladders – Training requirements (22) outline how to recognize hazards, use of fall protection and correct use of ladders and stairs.

21. 5A1 General Duty clause (22) is a rule that requires each employer to provide a place of employment that is free of conditions that are causing or likely to cause death or serious harm to employees.

Talk openly in class about these violations and whether you or anyone in your class has seen any of these, and what was done to correct the situation. OSHA has a very easy to use website at www.osha.gov. Click on standards, then click on 1910 for General Industry or 1926 Construction standard as needed. You can see the complete standards in detail.

Questions

_____ 1. OSHA 1910.67 covers ___.

    a. aerial lifts   b. electrical safeguards   c. wiring methods   d. vehicle mounted platforms

_____ 2. OSHA ___ sets forth standards for fall protection systems.

    a. 1926.502   b. 5A1   c. 1910.335   d. 1926.404

_____ 3. The OSHA ___ standard is violated more than any other in our work category.

    a. Fall Protection   b. Electrical   c. Scaffolds   d. Aerial Lifts

_____ 4. OSHA 1910.333 requires the use of lock-out and sets criteria for work on or near ___ circuits.

    a. multi-wire   b. energized   c. branch   d. data

_____ 5. The HazCom standard covers ___ plus training and emergency procedures.
    I. labeling   II. information   III. definitions

    a. I & II only   b. II & III only   c. I, II, & III   d. III only   e. II only   f. none of these

_____ 6. The most frequently cited OSHA construction standards change ___ from year to year.

    a. often   b. more   c. much   d. little

Lesson 223
NEC® Requirements for Transformers

Electrical Curriculum

Year Two
Student Manual

IEC
PRIDE
NATIONAL

Purpose

This lesson will discuss various types of transformers and their construction. The lesson will also address NEC requirements for installation, including vaults. Transformer and conductor overcurrent protection requirements and calculations are also addressed in this lesson.

Homework
(Due at the beginning of this class)

For this lesson, you should:

- Thoroughly read the material contained within Lesson 223.
- Read ES Units 12 in its entirety.
- Complete Objective 223.1 Worksheet.
- Complete Objective 223.2 Worksheet.
- Complete Objective 223.3 Worksheet.
- Complete Objective 223.4 Worksheet.
- Complete Objective 223.5 Worksheet.
- Read and complete Lesson 223 Safety Worksheet.

- Complete additional worksheets, if available, as directed by your instructor.

Objectives

By the end of this lesson, you should:

223.1 ....Understand the various types of transformers used and their construction.
223.2 ....Understand transformer installation requirements of the NEC and those of the manufacturers.
223.3 ....Understand the NEC requirements for transformer vaults.
223.4 ....Understand the NEC requirements for the protection of transformers against overcurrent.
223.5 ....Understand transformer connections and the NEC requirements for the protection of transformer secondary conductors against overcurrent.

223.1 Transformer Construction and Types

223.2 Transformer Installation

Typically, a single layer of 5/8″ sheetrock attached to a wall provides a 30-minute fire-resistant wall, and a double layer provides a 1-hour fire rating. These ratings are modified by the type of structural or framing materials used. Fire-resistance ratings are also modified with the installation of insulating materials within the wall space.

Reprinted with permission from Square D, Schneider Electric
©2003 Schneider Electric All Rights Reserved

PROPER VENTILATION AND CLEARANCES

All ventilated dry type transformers, regardless of type, can be seriously de-rated if airflow is blocked or restricted. Most ventilated transformers bear a nameplate label indicating that 6 inches is required between any ventilation opening and an obstruction or wall. The exceptions are large, substation style transformers that are labeled for 12-inch clearance requirement. Resin filled transformers can be placed directly on walls, but require 12 inches on both sides and top for adequate cooling.

Non-ventilated transformers have no intentional ventilation openings, but do depend on their external surface to be in contact with adequate airflow to transfer heat away. It is recommended that a minimum of 3 inches be allowed on all sides of these products.

Occasionally, users will request recommended spacing between transformers that will be vertically stacked, either on walls or equipment racks. For transformers up to 500kVA the Square D recommendation historically has been 18 inches.

SYSTEM GROUNDING

Neutral grounding is sometimes better accomplished at the secondary panel neutral bar, where allowed by code. There is typically very limited wiring space and neutral terminal capacity in the transformer compared to the load side panel.

Contractors are sometimes required to create an isolated ground, or IG system, at the transformer secondary. Many installers have conceptual problems with IG systems. Some think that it means that the secondary neutral is not grounded, others interpret that a separate ground bar is required in the transformer. Neither of these is correct. To originate an IG ground wire at the transformer, simply carry a separate, IG ground conductor from the equipment ground point in the transformer to an isolated ground bar in the secondary panel. The isolated ground and the equipment ground originate from the same ground point within the transformer, where they are bonded to building ground, but remain separated in the panel and in all branch circuits. The purpose of isolated ground is to provide a single point ground for all load point receptacles, rather than the multiple grounds and potential ground current loops associated with traditional, daisy-chained equipment ground schemes. It's thought that IG systems provide improved resistance to load equipment disturbances.

Typical Grounding Problem:

A contractor suspects that a 480 Delta – 208Y/120 transformer he has installed is defective. Voltage measurements at the secondary lines to ground are 75V, 147V and 80V respectively, and the neutral to ground voltage is 50 volts.

Explanation and Solution:

The contractor should be advised that the transformer secondary is a separately derived service, and that the neutral in 208Y/120 systems is normally required to be grounded in accord with the National Electrical Code. The presence of voltage between neutral and ground indicates that the neutral is not bonded properly.

223.3 Transformer Vaults

223.4 Transformer Overcurrent Protection

223.5 Transformers - NEC Article 240 and Connections

Revisit Chapter 5 in ES and review the 25′ feeder tap rule, 240.21(B)(3), associated with Figure 5-19.

NEC Section 240.21 addresses where overcurrent protection is required in circuits. Section 240.21(B) addresses feeder taps and Section 240.21(C) addresses transformer secondary conductors, specifically. These secondary conductor "tap" rules are addressed in this objective.

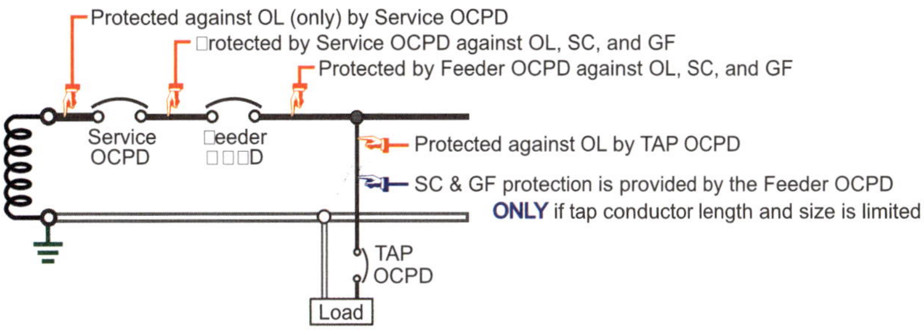

Figure 223.51   Only AØ of a 4-Wire Wye System is Shown

Feeder taps conductors CAN be protected against short-circuit and ground-fault (SC & GF) by the feeder OCPD. Refer to Figure 223.51. The time required for the feeder OCPD to operate, in the event of a SC or GF, varies with the length and size of the tap conductors. This is because the total circuit impedance is affected by the length and the size of the wire used for the tap conductor. Feeder taps are protected against overload (OL) by the OCPD at the termination of the tap conductor.

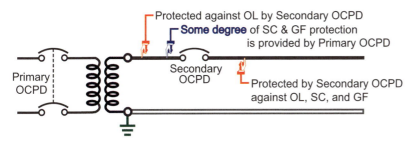

Figure 223.52   Only AØ of a 4-Wire Wye System is Shown

Transformer secondary conductors are protected against overload by the OCPD at the termination of the secondary conductor. Refer to Figure 223.52. To some degree, secondary conductors are protected against SC & GF by the primary OCPD. However, unlike feeder taps, there is no direct, wire-to-wire connection between the Primary OCPD and the secondary conductors. The Primary OCPD provides short-circuit and ground-fault protection **only** when the secondary conductors are of limited length and are sized large enough to lower the circuit impedance to a point where the Primary OCPD "sees" the fault current induced in the primary circuit.

The requirements contained in NEC 240.21 are the result of case studies of various circuit fault scenarios. If these requirements are met, a circuit should operate safely when faults (short circuits or ground faults) occur.

Among the secondary conductor "tap rules" listed at NEC Section 240.21(C), the requirements of NEC 240.21(C)(2) and 240.21(C)(6) are used most frequently. Although the secondary conductor lengths differ between (2) and (6) they are very similar. Both require physical protection and both require that the conductors terminate at a single OCPD. However, an unlimited number of sets of conductors can be connected to the secondary "lugs" with each set terminating in a single OCPD. Both NEC 240.21(C)(2) & (6) place additional restrictions on the size of the secondary conductors. Note that the voltage ratios used within NEC Article 240 are line voltages (secondary and primary) and NOT "winding-to-winding" voltages.

The Code language at NEC 240.21(C)(2)(1)(c) can be written as the formula:

☞ **[Sec. Cond. Ampacity] ≥ [OCP$_{Pri}$ ÷ 10] × [E$_p$ ÷ E$_s$]**

Where E$_p$ = Primary Line-to-Line Voltage and E$_s$ = Secondary Line-to-Line Voltage

The Code language at 240.21(C)(6)(1) can be written as the formula:

☞ **[Sec. Cond. Ampacity] ≥ [OCP$_{Pri}$ ÷ 3] × [E$_p$ ÷ E$_s$]**

Where E$_p$ = Primary Line-to-Line Voltage and E$_s$ = Secondary Line-to-Line Voltage

The term *secondary conductors* refer to the conductors that are connected to transformer secondaries. Occasionally, these conductors meet the NEC definition of a feeder, however, when used in discussions of transformers these conductors are referred to as *secondary conductors*.

Additional requirements for secondary conductors:
- Secondary conductors can NOT be tapped.
- Secondary conductors must be able to carry the connected load.
- Secondary conductors must have an ampacity ≥ the rating of the OCPD to which it is connected.

Student Notes

NEC® Requirements for Transformers   223-589

Objective 223.1 Worksheet

_____ 1. Liquid-filled transformers are commonly used in ___.

   a. 2400 V to 13,800 V voltages
   b. 150 kVA to 3000 kVA ratings
   c. both a and b
   d. neither a nor b

_____ 2. The NEMA insulation classifications for dry-type transformers are Classes ___.

   a. A, B, F, and H   b. A, B, C, and D   c. A, B, C, and F

_____ 3. Some types of transformers are liquid-filled. The liquid helps to dissipate the heat generated by the transformer windings.

   a. True   b. False

_____ 4. ___ is the process by which oxygen mixes with other elements and forms a type of rust-like material.

   a. Transference   b. Covalence   c. Induction   d. Oxidation

_____ 5. Askarel is a flammable liquid that contains PCBs.

   a. False   b. True

_____ 6. A ___ point is the temperature at which liquids give off vapor sufficient to form an ignitable mixture with the air near the surface of the liquid.

   a. fire   b. flash   c. either a or b   d. neither a nor b

_____ 7. Transformers are electrical devices that contain ___.

   a. motors   b. overcurrent devices   c. no moving parts   d. switches

_____ 8. Dry-type transformer insulation is based on an ambient temperature of ___°C.

   a. 75   b. 90   c. 40   d. 60

Year Two (Student Manual)

Objective 223.2 Worksheet

_____ 1. Askarel-insulated transformers are permitted to be installed outdoors only.

   a. False   b. True

_____ 2. Temperature ___ is the amount of heat that an electrical component produces above the ambient temperature.

   a. insulation   b. rise   c. limitation   d. none of these

_____ 3. Type ___ buildings have fire resistance ratings from 0 to 2 hours.

   a. IV   b. II   c. I   d. III

A Group H1, 150 KVA transformer will sit on the housekeeping pad shown at the right. The job specifications require that the transformer be centered on the pad and that the pad extend 6" beyond the transformer, on all sides. The transformer nameplate requires 6" of clearance on the front and the back for proper ventilation.

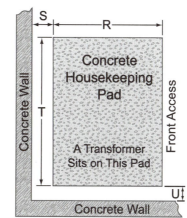

Figure 223.114

_____ 4. Refer to Figure 223.114, shown above. The minimum dimension for **T** is ___ inches.

   a. 41.87   b. 0   c. 53.52   d. 44.90   e. 38.90   f. none of these

_____ 5. Refer to Figure 223.114, shown above. The minimum dimension for **R** is ___ inches.

   a. 0   b. 41.87   c. 53.52   d. 44.90   e. 38.90   f. none of these

_____ 6. Refer to Figure 223.114, shown above. The shipping weight of this transformer is ___ lb.

   a. 970   b. 1125   c. 1155   d. none of these

_____ 7. Refer to Figure 223.114, shown on the previous page. The minimum dimension for **S** is ___ inches.

      a. 44.90   b. 0   c. 53.52   d. 41.87   e. 38.90   f. none of these

_____ 8. Refer to Figure 223.114, shown on the previous page. The minimum dimension for **U** is ___ inches.

      a. 44.90   b. 41.87   c. 53.52   d. 0   e. 38.90   f. none of these

_____ 9. An electrician has installed a 480 Delta – 208Y/120 transformer. Voltage measurements at the secondary lines to ground are 75V, 147V, and 80V respectively, and the neutral-to-ground voltage is 50 volts. This is most likely happening because ___.

      a. the MBJ has not been installed   b. the transformer is defective

_____ 10. What minimum size aluminum main bonding jumper is required for the secondary of a transformer with two 250 kcmil aluminum conductors per phase?

      a. 2   b. 1/0   c. 3/0   d. 2/0   e. 4

_____ 11. A liquid-filled transformer has a primary voltage of 12,470 volts. The transformer is surrounded by a 6-foot high fence. The overall height of the fence must be extended 1-foot with at least ___ strands of barbed wire on top of the fence.

      a. 3   b. 1   c. 2   d. no minimum required

_____ 12. The minimum size copper THHN/THWN conductor that can be installed between XO and the ground bar in a transformer is ___. Two parallel sets of 3/0 THHN/THWN copper are run from the transformer to the secondary panelboard. The grounding electrode conductor from the "I" beam terminates at the ground bar in the transformer.

      a. 1/0   b. 3/0   c. 2/0   d. 2   e. 4

_____ 13. A 150 KVA, 480 to 208Y/120 volt dry-type transformer is to be located in front of a plywood telephone equipment board that is attached to the wall over one layer of 5/8 inch sheetrock. The NEC will permit this installation if the transformer is installed at least 12 inches from the plywood.

      a. True   b. False

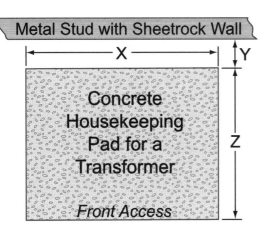

The nameplate on a Group D, 150 KVA transformer requires 6" of clearance on the front and the back for ventilation. It will sit on the housekeeping pad shown at the left. The specifications for this job require that the transformer be centered on the pad and extend 4" beyond the transformer, on all sides.

Figure 223.113

_____ 14. Refer to Figure 223.113, shown above. The minimum dimension for **Z** is ___ inches.

   a. 40.9   b. 37.87   c. 32.9   d. 49.52   e. 29.87

_____ 15. Refer to Figure 223.113, shown above. The minimum dimension for **Y** is ___ inches.

   a. 0   b. 6   c. 4   d. 2   e. 10

_____ 16. Refer to Figure 223.113, shown above. The minimum dimension for **X** is ___ inches.

   a. 37.87   b. 29.87   c. 40.9   d. 32.9   e. 49.52

_____ 17. For transformers up to 500 kVA, the Square D recommended spacing between transformers that will be vertically stacked (either on walls or equipment racks) historically has been ___ inches.

   a. 2   b. 18   c. 36   d. 48

_____ 18. An autotransformer may be used to buck (decrease) or boost (increase) the supply voltage.

   a. False   b. True

_____ 19. Transformers are required by the NEC to be marked with a minimum distance or clearance from walls or other obstructions to facilitate the dissipation of heat through the ventilation openings.

   a. False   b. True

_____ 20. A 45 KVA, 480 to 208Y/120 volt dry-type transformer is suspended below the ceiling of an equipment room where the ceiling is plywood attached to wood joists. The general rule is that the minimum clearance from the transformer to the ceiling is ___ inch(es).

    a. 12   b. 4   c. 3   d. 24   e. 6

_____ 21. In general, exposed metal transformer enclosures are required to be grounded.

    a. True   b. False

Objective 223.3 Worksheet

_____ 1. Doors for transformer vaults shall open ___ the vault.

   a. into   b. out from

_____ 2. ___ hardware is door hardware that is designed to open easily in an emergency situation.

   a. Escape   b. Magnetic release   c. Panic   d. None of these

_____ 3. The minimum size of transformer vault ventilation openings for transformers rated less than 50 kVA is ___ sq ft.

   a. 4   b. 2   c. 1   d. 3

_____ 4. Where a transformer vault is not protected by an automatic fire suppression system, the minimum fire rating for the roof is ___ hour(s).

   a. 2   b. 6   c. 1   d. 4   e. 3

_____ 5. Where a transformer is installed in a vault on the 3rd floor of a high-rise building and is not protected by an automatic fire suppression system, the minimum fire rating for the floor is ___ hour(s).

   a. 6   b. 4   c. 3   d. 1   e. 2

Objective 223.4 Worksheet

_____ 1. The maximum size circuit breaker permitted by Note 1, to NEC® Table 450.3(A), to protect the primary of a 4160 to 480 V, 3Ø, 75 KVA transformer is ___ amps. The transformer has an impedance of 1.7% and is installed in a nonsupervised location.

   a. 30   b. 35   c. 60   d. 70   e. 20

_____ 2. A 7200 to 4160 volt 3Ø, 150 kVA transformer is installed in a supervised location and is supplied from a feeder tap that is protected with a circuit breaker in an enclosure. The transformer supplies power to a 3Ø fused safety switch. As limited by NEC® 450.3, the fuses to be installed in the switch are permitted to be no larger than ___ amps. The transformer Z = 7%.

   a. 60   b. 125   c. 45   d. 70   e. 50

_____ 3. A supervised transformer installation has conditions of maintenance which ensure that only persons who are ___ to monitor and service the electrical equipment.

   a. authorized   b. qualified   c. management approved   d. all of these

_____ 4. A feeder is tapped and a circuit breaker in an enclosure is installed to protect the tap. The tap feeds a 4160 to 2400 volt 3Ø, 225 kVA transformer that is installed in a nonsupervised location. The transformer Z = 4%. The transformer supplies power to a 3Ø fused safety switch. As limited by NEC® 450.3, the fuses to be installed in the switch can have a rating of not more than ___ amps.

   a. 125   b. 150   c. 100   d. 175   e. 300

_____ 5. Note 1, to NEC® Tables 450.3(A) and (B), permits the next standard size OCPD to be used to protect a transformer. When protecting secondary conductors, Section 240.21(C) ___ the next standard size OCPD.

   a. does not permit   b. permits

_____ 6. Although NEC® 450.3 permits a certain OCPD rating for transformer protection, don't forget to check NEC® 240.4(F) for secondary conductor protection requirements. The maximum size fuse permitted, among those listed below, to protect the primary of a 480 to 240 V, 1Ø, 500 VA dry-type transformer is ___ amps. The secondary conductors are 14 AWG copper THHN/THWN and overcurrent protection is not provided on this secondary.

   a. 4   b. 1.25   c. 6   d. 3   e. 1

NEC® Requirements for Transformers

_____ 7. A 1.5 KVA, single-phase, 120 to 24 volt transformer is fed by a branch circuit from a 20 amp circuit breaker. The secondary conductors are 4 AWG copper THHN/THWN. Is secondary protection required for the transformer? Compare the requirements of NEC® 450.3 with 240.4(F).

   a. Yes   b. No

_____ 8. A fused safety switch feeds a 480 to 208Y/120 volt 3Ø, 30 KVA transformer. The transformer supplies power to a 3Ø main breaker panelboard. To provide adequate overload protection for the transformer, the maximum size main breaker permitted to be installed in the panel is ___ amps.

   a. 110   b. 200   c. 175   d. 100   e. 150   f. 125

_____ 9. The maximum size fuse permitted to protect the primary of a 480 to 240 V, 1Ø, 15 kVA dry-type transformer is ___ amps. The secondary conductors are 6 AWG copper THHN/THWN and overcurrent protection is not provided on this secondary. Don't forget to check 240.4(F) and 240.21(C) in the NEC® then pick the smaller permitted by Article 240 or Note 1 to T.450.3(B).

   a. 60   b. 35   c. 30   d. 40   e. 70

Objective 223.5 Worksheet

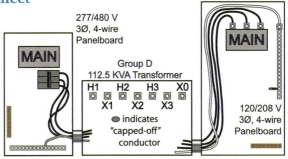

Figure 223.504

_____ 1. Refer to Figure 223.504, shown above. The maximum length of the conductors, between the transformer and the 120/208 volt panelboard, is ___ meters per the NEC®.

a. 25   b. unlimited   c. 10   d. 3

_____ 2. Refer to Figure 223.504, shown above. If the 277/480-volt panelboard and the transformer are not within sight of each other, a disconnect (safety switch) is required to be installed within sight of the transformer.

a. False   b. True

_____ 3. Refer to Figure 223.504, shown above. The maximum length of the conductors between the transformer and the 277/480 volt panelboard is ___ meters per the NEC®.

a. 25   b. 3   c. unlimited   d. 10

_____ 4. Refer to Figure 223.504, shown above. Are the conductors correctly connected in the 208-volt panelboard?

a. No   b. Yes

_____ 5. Refer to Figure 223.504, shown above. If the transformer connections are correctly made they would appear as shown in ___ below.

a.   b.   c.

d. none of these

NEC® Requirements for Transformers 223-601

_____ 6. The secondary conductors from a 45 KVA, Group H1 transformer are connected to a fusible safety switch, fused at 70 amps. The smallest secondary conductors permitted to be installed, *between the transformer and the switch*, are ___ AWG. For this particular load, the NEC® permits 8 AWG conductors to be installed between the disconnect and the load.

  a. 10 b. 8 c. 6 d. 3 e. 4

_____ 7. A 225 KVA 480 – 120/208 volt 3Ø, 4-W transformer feeds a panelboard with a 225 amp main breaker. The length of the secondary conductors between the transformer and the main is 9 feet. These conductors must be at least ___ in size where the primary breaker is 300 Amps.

  a. 1/0 AWG b. 1 AWG c. 4 AWG d. 4/0 AWG e. none of these

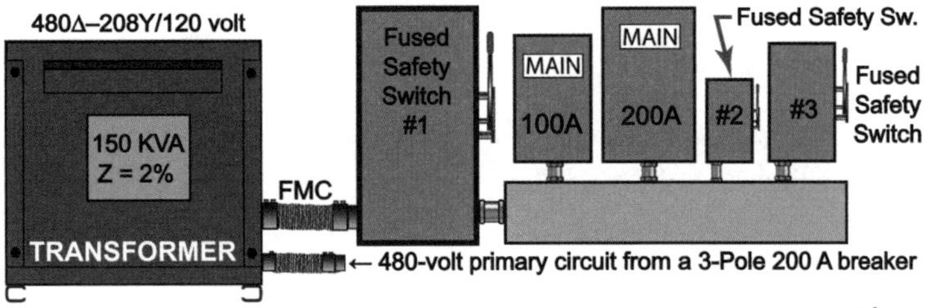

Figure 223.503

One set of secondary conductors is connected to Safety Switch #1. The feeder conductors in the wireway are protected by the fuses in Switch #1. Taps are made in the wireway (gutter) to supply power to the MBO panels and Safety Switches #2 & #3.

_____ 8. Refer to the indoor installation shown above in Figure 223.503. The conductors between the transformer and the line side of Switch #1 are 600 KCMIL. The largest fuses that can be installed in Switch #1 are ___ amp.

  a. 600 b. 350 c. 400 d. 450 e. 500

_____ 9. Refer to the indoor installation shown above in Figure 223.503. If properly fused in Switch #1, the minimum NEC® size requirement for the conductors between the transformer and the line side of Switch #1 is ___ AWG or kcmil. These conductors are 22 feet in length.

  a. 2/0 b. 350 c. 300 d. 250 e. 4/0

_____ 10. For the **outdoor** installation, shown above in Figure 223.503, the secondary conductors between the transformer and Switch #1 are 22 feet in length. If Switch #1 is properly fused, the minimum NEC® size requirement for the conductors between the transformer and the line side of Switch #1 is ___ AWG or kcmil.

  a. 300 b. 2/0 c. 4/0 d. 350 e. 250 f. no minimum

Year Two (Student Manual)

_____ 11. Refer to the indoor installation shown above in Figure 223.503. The conductors that are connected to the two main breaker panels and to Switches #1 & #2 from the wireway have NEC® requirements in ___.

   a. 240.21(B)(3)
   b. 240.21(C)(6)
   c. 240.21(B)(1)
   d. 240.21(C)(2)

_____ 12. Refer to the indoor installation shown above in Figure 223.503. The smallest wires that can be connected to the transformer primary are ___ kcmil or AWG .

   a. 2/0   b. 300   c. 3/0   d. 1/0   e. 250

_____ 13. Refer to the indoor installation shown above in Figure 223.503. If the conductors between the transformer and the line side of Switch #1 are 13 feet in length there are NEC® requirements in ___.

   a. 240.21(B)(3)
   b. 240.21(C)(6)
   c. 240.21(B)(1)
   d. 240.21(C)(2)

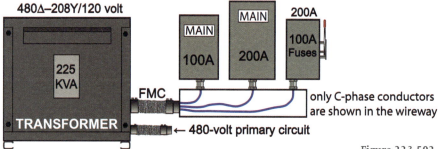

Figure 223.502

The transformer is supplied from a 300 Amp, 480-volt breaker. Three sets of secondary conductors are connected to triple lugs in the transformer and run through the gutter to feed: one panelboard with a 100 amp main breaker, **plus** another panelboard with a 200 amp main breaker, **plus** a safety switch fused at 100 amps.

_____ 14. Refer to the indoor installation shown above in Figure 223.502. From the transformer, 3 AWG conductors are run to the 100 amp main breaker, 3/0 AWG conductors are installed to the 200 amp main breaker, and 3 AWG conductors are run to the 100 amp main breaker. None of the conductor lengths, between the transformer lugs and the switch or MAIN breaker lugs, exceeds 10 feet. According to the NEC®, this installation is ___.

   a. permitted   b. a violation of 240.21(C)

_____ 15. Refer to the indoor installation shown above in Figure 223.502. From the transformer, 3 AWG conductors are run to the 100 amp main breaker, 3/0 AWG conductors are installed to the 200 amp main breaker, and 3 AWG conductors are run to the 100 amp main breaker. The conductor length, between the transformer lugs and the 100 A MAIN breaker lugs, is 14 feet. According to the NEC®, this installation is ___.

    a. permitted   b. a violation of 240.21(C)

_____ 16. Three sets of secondary conductors are run from a 225 KVA 480 – 120/208 volt 3Ø, 4-W transformer.
        Set #1 feeds a panelboard with a 100 amp main breaker.
        Set #2 feeds another panelboard with a 200 amp main breaker.
        Set #3 feeds a safety switch fused at 100 amps.
Is this permitted by the NEC where triple lugs are installed in the transformer?

    a. Yes   b. No

Scenario 223.51

The secondary of a 300 KVA, 480 – 240 volt Δ–Δ transformer is delivering 100 amps to 3Ø loads. Based on this load, the smallest permitted primary and secondary conductors are installed.

_____ 17. Refer to Scenario 223.51, shown above. The maximum rating of the primary OCPD is ___ A.

    a. 50   b. 80   c. 90   d. 70   e. 100   f. none of these

_____ 18. A transformer supplies power to a 240-volt, 3Ø, non-continuous 35.286 KVA load. The smallest secondary circuit conductors permitted for this load are used. The secondary OCPD can have a rating not greater than ___ Amps.

    a. 60   b. 90   c. 100   d. 80   e. 110   f. none of these

_____ 19. A 150 KVA, Group D transformer supplies power to a panel where AØ has a load of 300 amps, BØ has 312 amps, and CØ has 280 amps. How much load, in amps, could be added to AØ?

    a. 61
    b. 421
    c. 325
    d. 116
    e. none, the transformer is overloaded

_____ 20. A 90 amp breaker supplies power to a 75 KVA 480 – 120/208 volt 3Ø, 4-W transformer. The transformer feeds a panelboard with a 200 amp main breaker. The length of the secondary conductors between the transformer and the main is 19 feet. These conductors must be at least ___ AWG in size.

a. 3   b. 4   c. 2/0   d. 4/0   e. 3/0   f. no minimum

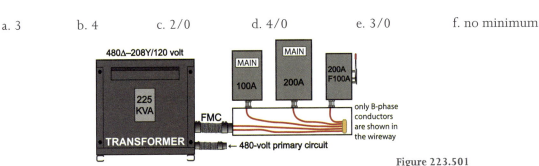

Figure 223.501

The transformer is supplied from a 300 Amp, 480-volt breaker. Two paralleled 600 kcmil copper secondary conductors are installed between the transformer and the gutter. These are connected to double lugs in the transformer and to terminal block lugs in the gutter. Smaller conductors are connected to the block in the gutter to feed: one panelboard with a 100 amp main breaker, **plus** another panelboard with a 200 amp main breaker, **plus** a safety switch fused at 100 amps.

_____ 21. Refer to the indoor installation shown above in Figure 223.501. From the gutter (wireway), 3 AWG conductors are run to the 100 amp main breaker and the fused switch. 3/0 AWG conductors are installed to the 200 amp main breaker. The conductor lengths, between the transformer lugs and the switch or MAIN breaker lugs, do not exceed 10 feet. According to the NEC®, this installation is ___.

a. permitted   b. a violation of 240.21(C)

_____ 22. Section 240.21(C)(2) in the NEC® is the "10 foot rule". This section requires that the secondary conductors be 10 feet or shorter in length, be installed in a raceway, and ___.
  I. be sized (have sufficient ampacity) to carry the load
  II. have an ampacity that is ≥ the rating of the circuit breaker or fuses at the end
  III. have an ampacity that is ≥ [(Pri OCPD ÷ 10) × (Epri line ÷ Esec line)]

a. II & III only   b. III only   c. I & II only   d. I, II, & III   e. I only

_____ 23. Section 240.21(C)(6) in the NEC® is the "25 foot rule". This section requires that the secondary conductors be 25 feet or shorter in length and ___.
  I. have an ampacity that is ≥ [(Pri OCPD ÷ 3) × (Epri line ÷ Esec line)]
  II. have an ampacity that is ≥ the rating of the circuit breaker or fuses at the end
  III. be installed in a raceway or otherwise protected from physical damage

a. I & II only   b. II & III only   c. III only   d. I only   e. I, II, & III

Lesson 223 Safety Worksheet

Fall Protection

Falls continue to be the number 1 cause of death on construction sites, year after year. NIOSH (The National Institute of Occupational Safety and Health) indicates that every year, on average, 400 or more workers fall to their death on construction sites. These deaths are needless and tragic. The workers leave spouses and children behind to live on without them. These deaths place a hardship on the employee's family, the employer and everyone on the jobsite. These unfortunate deaths are, for the most part, preventable. Some workers choose not to wear a (PFAS) personal fall arrest system, and then fall to their death. Some workers are exposed to work at heights without proper training of the hazards associated with this type of work, or provided with the correct safety equipment and training that could save their lives.

OSHA has a large standard covering Fall Protection (CFR1926.500). This subpart has many requirements that outline numerous methods to accomplish safe means to work at heights. Some excerpts of the requirements are following:

- The employer shall determine if the walking/working surface has strength and structural integrity to support employees safely.

- Each employee on a walking/working surface with an unprotected side or edge which is 6 feet or more above a lower level shall be protected from falling by the use of guardrail systems, safety net systems, or a personal fall arrest systems.

- Each employee in a hoist area shall be protected from falling 6 feet or more to lower levels, by guardrail systems or personal fall arrest systems.

- If an employee can fall through a hole, 6 feet or more to a lower level, they have to be protected by personal fall arrest systems, covers over the holes, or guardrail systems.

- Each employee on a ramp, runway or other walkway shall be protected from falling 6 feet or more by a guardrail system.

- Each employee at the edge of an excavation, well, pit, shaft or similar excavation 6 feet or more in depth shall be protected from falling by guardrail systems, fences, or barricades.

- Working over dangerous equipment that one could fall onto or into must be protected by guardrail systems, machine guards or personal fall arrest systems.

You will notice that ladders, scaffolds, aerial lifts and other elevated types of work are not mentioned here. This is because fall protection is covered in each of these subparts of OSHA as part of that subpart.

NEC® Requirements for Transformers

Questions

_____ 1. If an employee can fall through a ___, 6 feet or more to a lower level, the employee has to be protected by personal fall arrest systems, covers over the holes, or guardrail systems.

   a. skylight   b. roof   c. hole   d. ceiling

_____ 2. Falls are the number ___ cause of construction deaths.

   a. two
   b. three
   c. four
   d. one

_____ 3. Each employee at the edge of ___ a shaft or similar excavation 6 feet or more in depth shall be protected from falling by guardrail systems, fences, or barricades.
   I. a pit   II. a well   III. an excavation

   a. I or III only
   b. I only
   c. I or II or III
   d. II or III only
   e. II only
   f. I or II only

_____ 4. The ___ shall determine if the walking/working surface has strength and structural integrity to support employees safely.

   a. employer
   b. steel erector
   c. employee
   d. general contractor

_____ 5. Working over dangerous equipment that one could fall onto or into must be protected by guardrail systems, ___ or personal fall arrest systems.

   a. machine nets
   b. safety nets
   c. machine guards
   d. any of these

Year Two (Student Manual)

Lesson 224
Buck-Boost Transformers: Single- and Three-Phase

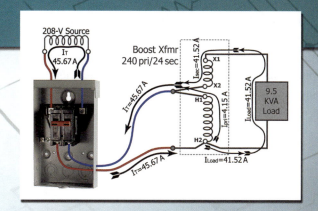

Electrical Curriculum

Year Two
Student Manual

IEC
PRIDE
NATIONAL

Purpose

This lesson will discuss how transformer polarity and connections are utilized to buck or boost circuit voltage. This lesson will also address compensated windings, back-feeding, and calculations for determining output capacities and voltages for various transformer connections.

Homework

(Due at the beginning of this class)

For this lesson, you should:

- Thoroughly read the material contained within Lesson 224.
- Locate Yr-2 Annex D at the back of this manual.
- Complete Objective 224.1 Worksheet.
- Complete Objective 224.2 Worksheet.
- Complete Objective 224.3 Worksheet.
- Read and complete Lesson 224 Safety Worksheet.

- Complete additional worksheets, if available, as directed by your instructor.

Objectives

By the end of this lesson, you should:

    1.1 ....Understand additive and subtractive polarity and the output voltages of buck-boost transformers.

    224.2 ....Understand applications involving buck-boost transformers and be able to perform the calculations for determining the output voltages and capacities of these transformers.

    224.3 ....Understand compensation and back-feeding of small transformers.

224.1 Polarity

Polarity is the instantaneous voltage obtained from the primary winding in relation to the secondary winding. Polarity is generally a term used with single-phase transformers.

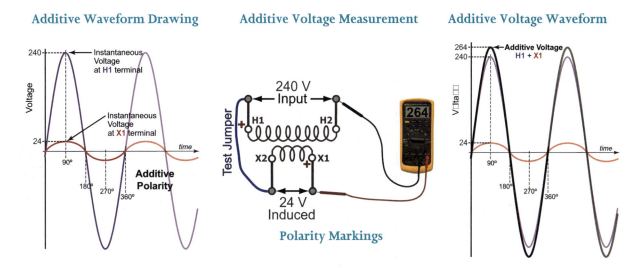

Additive Polarity

Smaller transformers, typically 50 KVA and smaller, operating at 600 volts and below are normally wound for additive polarity. When field connection terminals H1 and X1 are spaced as far apart as possible from each other, the transformer is wound for additive polarity. A typical arrangement for additive polarity, facing the transformer terminals is: X1-H1-X2-H2-X3-H3-X4-H4. The **Additive Waveform Drawing** for a 240 pri – 24 V sec transformer with additive polarity is shown on the previous page. This drawing shows that the voltage levels of H1 and X1 reach their positive peaks and zeros at the same instant in time. The instantaneous voltage measured at terminal H1 is *positive* 240 volts at a point in time. The instantaneous voltage induced in the secondary winding at terminal X1 is *positive* 24 volts if measured at that same point in time. Some reference material may indicate **Polarity Markings** with a "**+**", as shown on the previous page. Other reference material may indicate the polarity using a "**•**". To test a transformer for additive polarity, a test jumper is installed between X2 and H1, as shown on the previous page. The **Additive Voltage Measurement** between H2 and X1 is the sum of the voltage across the higher voltage winding and the lower voltage winding. If the voltage of the higher voltage waveform were added to the voltage of the lower voltage waveform, the resultant waveform would have a peak value of 240 + 24 = 264 volts. This resultant **Additive Voltage Waveform** is shown on the previous page.

Subtractive Polarity

Larger capacity and higher than 600-volt transformers are typically wound for subtractive polarity. When the field connection terminals H1 and X1 are adjacent to each other, the transformer is wound for subtractive polarity. A typical arrangement for subtractive polarity, facing the transformer terminals is: X1-X2-X3-X4-H1-H2-H3-H4. The **Subtractive Waveform Drawing** for a 240 pri – 24 V sec transformer with subtractive polarity is shown below. This drawing shows that the voltage level of H1 reaches its *positive* peak at the same instant that the voltage level at X1 reaches its *negative* peak. To test a transformer for subtractive polarity, a test jumper is installed between X2 and H2, as shown below. The **Subtractive Voltage Measurement** between H1 and X1 is the difference between the voltage across the higher voltage winding and the lower voltage winding. If the voltage of the higher voltage waveform were added to the *negative* voltage of the lower voltage waveform, the resultant waveform would have a peak value of $240 + (-24) = 216$ volts. This resultant **Subtractive Voltage Waveform** is shown below.

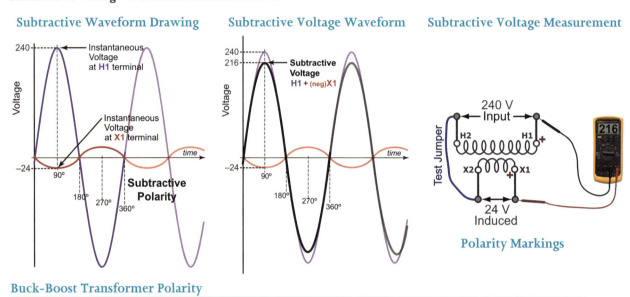

Buck-Boost Transformer Polarity

Buck-boost transformers are small capacity transformers and are wound for additive polarity. These are connected to either boost (increase) the input voltage to a higher output voltage, or buck (decrease) the input voltage to a lower output voltage.

When buck-boost transformers are connected to **boost** the input voltage, the primary and secondary voltages are *additive*.

An analogy used to illustrate the *additive* voltage of a boost connection is a string of batteries. As shown below, twenty 12-volt batteries are series-connected in a string to represent the 240-volt primary windings (H1 to H4), and two 12-volt batteries represent the 24-volt secondary windings of a 240 to 24-volt transformer. The voltage measurements are as indicated in the drawing.

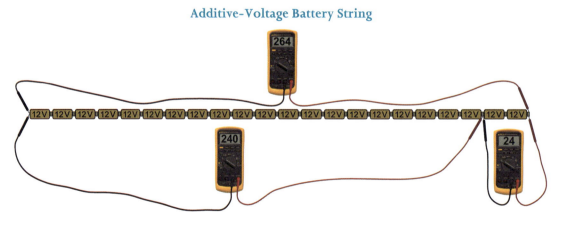

When buck-boost transformers are connected to **buck** the input voltage, the primary and secondary voltages are *subtractive*.

An analogy used to illustrate subtractive voltage of a buck connection is a string of batteries and a charger. As shown below, twenty batteries are series-connected in a string to represent the primary windings (H1 to H4), and two batteries represent the secondary windings of a 240 to 24-volt transformer. The voltage measurements are as indicated in the drawing on the next page. If a 230-volt charger is connected across the entire string of twenty-two batteries, the batteries are charged equally at 10.45 volts (230 ÷ 22 = 10.45). A load is rated to operate at 208 volts. If connected across the string of 20 batteries, it will have close to its rated voltage applied (209.1 V applied).

Subtractive-Voltage Battery String

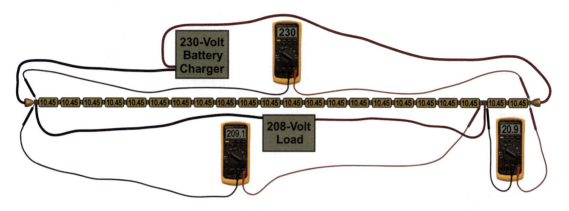

224.2 Applications

Buck-boost transformers are available in sizes as small as .05 KVA to as large as 10 KVA. Transformers in this range are single-phase insulating transformers that have two sets of primary windings and two sets of secondary windings. They are available with primary voltage ratings of 120/240 with secondary voltage ratings of 12/24 or 16/32. Transformers with 240/480 V primaries and 24/48 V secondaries are also available. Buck-boos (B-B) transformers can be field connected as autotransformers and used to slightly decrease (buck) or slightly increase (boost) a circuit voltage. An autotransformer is a transformer in which the primary and the secondary windings are electrically connected to each other. (Refer to **Yr-2 – Annex A** for the complete definition of an *autotransformer*.) Buck-boost transformers are frequently used to correct line voltage when the available line voltage is ±5-20% of that desired for the load to operate at its rated voltage. Although classified as buck-boost transformers, they can also be used as isolating (insulating) transformers to supply low voltage (12, 16, 24, 32, and 48 volts) power to control circuits or other loads such as decorative lighting. Refer to **Yr-2 – Annex D** for additional information on B-B transformers.

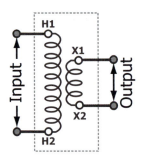

Figure 224.22

When used as an isolating transformer, a B-B transformer can carry a load that is equal to or less than the nameplate rating of the transformer. For example, when a transformer has a 240 V supply circuit and supplies power to a 24 V load, the secondary windings must carry all of the current that flows to the load. An isolation transformer, also known as an insulating transformer, has complete electrical separation between the primary (input) and the secondary (output). A simple transformer with a single primary winding and a single secondary winding is shown in Figure 224.22.

A simple B-B transformer, with a single primary winding and a single secondary winding connected as an autotransformer, is shown in Figure 224.23. This transformer is connected in a boost configuration. If the primary is rated at 240 volts and the secondary is rated at 24 volts, a 208 volt circuit connected to the input terminals would induce 20.8 volts in the secondary winding.

$$E_{sec} = E_{pri} \times (N_{sec} \div N_{pri}) = 208 \times (24 \div 240) = 20.8 \text{ V}$$

If a load is connected to the output terminals, the voltage delivered by the B-B transformer is:

$$E_{Load} = E_{pri} + E_{sec} = 208 + 20.8 = \mathbf{228.8} \text{ V} \approx 230 \text{ V}$$

Figure 224.23

B-B transformers are frequently field-connected as autotransformers. An autotransformer changes or transforms only a portion of the electrical energy it transmits. The rest of the electrical energy flows

Buck-Boost Transformers: Single-and Three- Phase

directly through the electrical connections between the primary and secondary. An isolation transformer (insulating transformer) changes (or transforms) all of the electrical energy it transmits. Although an insulating transformer could be used to transform the required voltage to a load, an autotransformer is smaller, lighter in weight, and less costly.

The key to understanding the operation and sizing of B-B transformers lies in the fact that the **secondary windings** are the only parts of the transformer that **do the work** of transforming voltage and current. When used as an autotransformer, a B-B transformer can carry much more than the nameplate rating of the transformer. In the previous illustration, only 20.8 volts are being transformed (boosted) — i.e. 208 + 20.8 = 228.8 V. This 20.8 V transformation is carried out by the secondary winding. Both secondary and primary windings are designed to operate at a maximum current as indicated on the nameplate. If this 240 – 24 V transformer has a rating of 1.0 KVA, then the primary winding has a rating of 1.0 KVA at 240 volts and the secondary winding has a rating of 1.0 KVA at 24 volts.

$$I_{rated\ pri} = 1000\ VA \div 240\ V = 4.17\ A \quad \text{and} \quad I_{rated\ sec} = 1000\ VA \div 24\ V = 41.67\ A$$

Boost Example: A 208-volt circuit is connected to a 240 – 24 volt B-B transformer. The output supplies power to a 230-volt, 9.5 kVA load. Determine the minimum rating of the B-B transformer required for this load. As determined previously, this transformer can deliver 228.8 volts to the load. This voltage is close enough to the 230-volt rating of the load for it to operate satisfactorily.

Currents in a Boost Circuit

The first step is to determine the amount of current drawn by the load at 228.8 volts. This current will also be the amount of current flowing in the secondary winding (X1-X2).

$$I_{Load} = 9500\ VA \div 228.8\ V = \mathbf{41.52\ A} = I_{sec}$$

The other currents in the boost circuit are shown above in the illustration and were determined as shown in the following calculations.

$$I_{pri} = (I_{sec} \times E_{sec}) \div E_{pri} = (41.52 \times 20.8) \div 208 = \mathbf{4.15\ A}$$
$$\text{and } I_T = I_{sec} + I_{pri} = 41.52 + 4.15 = \mathbf{45.67\ A}$$

For this simple single-winding transformer, the secondary (and the primary) must have a rating of at least:

$$P_{sec} = P_{pri} = I_{sec} \times E_{sec} = 41.52\ A \times 20.8\ V = \mathbf{863.6\ VA} \Rightarrow \mathbf{1.0\ KVA}\ \text{standard size}$$

Another method that can be used to determine the approximate rating of a B-B transformer to boost voltage to a given load is shown in the following formula.

> REQUIRED B-Boost KVA RATING = [(OVR − ICV) ÷ (OVR)] × (kVA rating of load)
> where: OVR = **O**utput **V**oltage **R**equired *and* ICV = **I**nput **C**ircuit **V**oltage

When connected as an autotransformer, the boosted kVA rating of a B-B autotransformer can be calculated using the following formula.

> B-B AUTOXFMR **Boosting** KVA RATING
> = (Total Output Voltage × Rated Secondary Current) ÷ 1000

For this boost example, the *Total Output Voltage* is 208 + 20.8 = 228.8 volts, and the *Rated Secondary Current* for this 1.0 KVA transformer is 41.667 amps (1 KVA ÷ 24 V).

B-B AUTOXFMR Boosting KVA RATING = (228.8 V × 41.667 A) ÷ 1000 = **9.533** KVA

This 1 KVA B-B transformer can supply the 9.5 kVA load with a small capacity remaining. If a 208-to-230 volt isolation transformer were used to supply the correct voltage to the load, the load would dictate that a 9.5 KVA rating (10 KVA minimum, standard rating) be used. Obviously, a 1 KVA B-B transformer would be considerable less costly than a 10 KVA isolation transformer.

Buck Example: A 208 volt circuit is connected to a 240 − 32 volt B-B transformer. The output supplies power to a 208-volt, 9.5 kVA load. Determine the minimum rating of the B-B transformer required for this load.

When using a B-B transformer to buck the voltage to a load, the bucked voltage applied to the load must first be determined. When 240 volts from the source is applied to the series-connected primary and secondary windings, as shown at the right, an equal voltage is dropped across each turn in the windings. Using the voltage ratio as a basis, the total number of turns in this buck arrangement is:

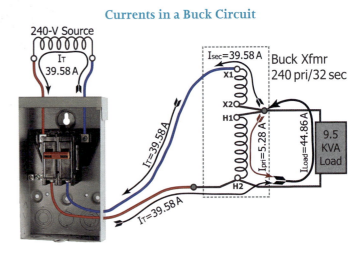

Currents in a Buck Circuit

N_{Total} = 240 turns + 32 turns = 272 turns

The number of volts dropped across each turn = 240 V ÷ 272 turns = 0.88253 V/turn

Buck-Boost Transformers: Single-and Three- Phase

The primary voltage (H1-H2) equals the voltage applied to the load. This voltage can now be determined using the number of turns in the primary winding (240 turns) and the volts/turn.

$$E_{Load} = 240 \text{ turns} \times 0.88253 \text{ V/turn} = 211.76 \text{ volts}$$

This voltage is close enough to the 208-volt rating of the load for it to operate satisfactorily.

The voltage across the secondary winding (X1-X2) can also be determined using the same method as was used to find the voltage across the primary winding.

$$E_{sec} = 32 \text{ turns} \times 0.88253 \text{ V/turn} = 28.24 \text{ volts}$$

The sum of the primary and secondary voltage drops equals the source voltage:

$$\text{Source Voltage} = E_{pri} + E_{sec} = 211.76 + 28.24 = 240 \text{ volts}$$

The next step is to determine the amount of current drawn by the load at its applied voltage of 211.76 volts. This and the other currents in the buck circuit are shown above in the illustration and were determined as shown in the following calculations.

$$I_{Load} = 9500 \text{ VA} \div 211.76 \text{ V} = \mathbf{44.86 \text{ A}}$$

The total current flowing through the source transformer can also be determined. This current will also be the amount of current flowing in the secondary winding (X1-X2).

$$I_T = 9500 \text{ VA} \div 240 \text{ V} = \mathbf{39.58 \text{ A}} = I_{sec}$$

Since the secondary current has been determined, the primary current can be calculated.

$$I_{pri} = (I_{sec} \times E_{sec}) \div E_{pri} = (39.58 \times 28.24) \div 211.76 = \mathbf{5.28 \text{ A}}$$

The total current can be verified by subtracting the bucked primary current from the load current.

$$I_T = I_{Load} + I_{pri} = 44.86 - 5.28 = \mathbf{39.58 \text{ A}}$$

Year Two (Student Manual)

For this simple single-winding transformer, the secondary (and the primary) must have a rating of at least:

$$P_{sec} = P_{pri} = I_{sec} \times E_{sec} = 39.58 \text{ A} \times 28.24 \text{ V} = \textbf{1118 VA} \Rightarrow \textbf{1.5 KVA standard size}$$

Another method that can be used to determine the approximate rating of a B-B transformer to buck voltage to a given load is shown in the following formula.

☞ REQUIRED Buck-B KVA RATING = [(ICV − OVR) ÷ (ICV)] × (kVA rating of load)
where: OVR = Output Voltage Required *and* ICV = Input Circuit Voltage

When connected as an autotransformer, the bucked kVA rating of a B-B autotransformer can be calculated using the following formula.

☞ B-B AUTOXFMR **Bucking** KVA RATING
= (Total Input Voltage × Rated Secondary Current) ÷ 1000

For this buck example, the *Total Input Voltage* is 240 volts, and the *Rated Secondary Current* for this 1.5 KVA transformer is 46.875 amps (1.5 KVA ÷ 32 V).

B-B AUTOXFMR Bucking KVA RATING = (240 V × 46.875 A) ÷ 1000 = **11.250 KVA**

This 1.5 KVA B-B transformer can supply the 9.5 kVA load with a significant capacity remaining. If a 240-to-208 volt isolation transformer were used to supply the correct voltage to the load, the load would dictate that a 9.5 KVA rating (10 KVA minimum, standard rating) be used. As with the boost example, a 1.5 KVA B-B transformer would be the better choice.

224.3 Compensation and Back-Feeding

Compensation

Transformers with nameplate ratings smaller than 3 KVA are generally designed so that the secondary voltage is slightly higher than the nameplate rating indicates. This is to correct for relatively high transformer regulation (voltage drop under load), and it is accomplished by adding a few additional turns on the secondary. This correction of the secondary voltage is called *compensation*. If these transformers did not have this compensation design, the voltage output under load would be unacceptably low. Typically, with no connected load the output voltage is 6% higher than the nameplate voltage. As the transformer is loaded the secondary voltage will drop, and should be close to nominal at full load. Industrial Control Transformers and single phase General Purpose Dry Type Transformers smaller than 3kVA generally have compensated winding ratios.

Back-Feeding

Can a 480 V pri – 120 V sec transformer be used to produce 480 volts from a 120-volt circuit? Back-feeding a small transformer results in a lower output voltage than expected. The secondary windings in these transformers are typically compensated by 6%. A small transformer with a nameplate rating of 480 – 120 might actually have a no-load voltage ratio of 480 – 126. If this transformer is connected in reverse (120 volts applied to X1 and X2), the no-load output voltage would actually be 457 volts. Due to regulation, the output voltage would drop even lower when a load is connected.

When voltage is applied to the input winding of a transformer, there can be a brief period (about a tenth of a second) of inrush current that lasts until the transformer core is stabilized. The amount of inrush current can be anywhere from zero to many times the full load current rating of the transformer. To comply with the NEC® and to avoid nuisance opening of overcurrent protective devices (OCPDs) on the primary side of the transformer when energized, careful coordination of OCPD ratings is essential. This coordination requires engineering design. Transformers that are back-fed can exhibit particularly high inrush currents. Caution should be exercised when back-feeding transformers.

Objective 224.1 Worksheet

_____ 1. When buck-boost transformers are connected to ___ the input voltage, the primary and secondary voltages are subtractive.

   a. boost   b. buck

_____ 2. When buck-boost transformers are connected to boost the input voltage, the primary and secondary voltages are ___.

   a. additive   b. subtractive

Objective 224.2 Worksheet

_____ 1. "Buck-Boost" transformers can be used ___.
   I. to raise the voltage for a load that is located at the end of a circuit where voltage drop has reduced the voltage below the rated voltage of the equipment to be operated
   II. to operate 12-volt equipment where the supply voltage is 120 volts
   III. to reduce the circuit voltage from a 240-volt supply to operate a 208-volt load

   a. II only   b. II only   c. I, II, & III   d. I & II only   e. III only   f. I & III only

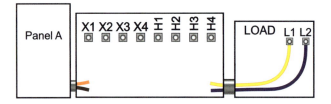

Figure 224.104

_____ 2. Locate a GROUP III Cat #31107 transformer in Yr-2 – Annex D. A 2-pole breaker in 120/240 V, 3Ø, 4W Panel A, shown above in Figure 224.104, feeds the transformer primary. Connection diagram ___, shown below, has the transformer correctly connected to supply the 48-volt LOAD.

a. 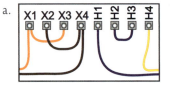   b.    c.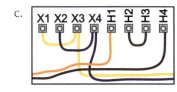

d. 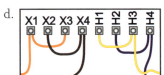   e. 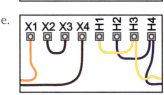   f. none of these

_____ 3. Locate a GROUP III Cat #31107 transformer in Yr-2 – Annex D. A 2-pole breaker in 120/240 V, 3Ø, 4W Panel A, shown above in Figure 224.104, feeds the transformer primary. Connection diagram ___, shown below, has the transformer correctly connected to supply the 24-volt LOAD.

a. 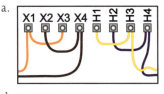   b. 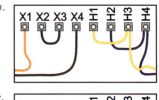   c.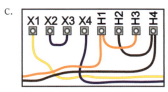

d. 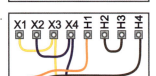   e. 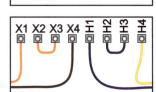   f. none of these

Buck-Boost Transformers: Single-and Three- Phase

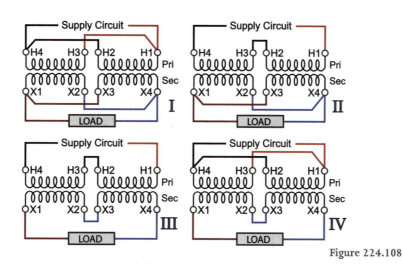

Figure 224.108

____ 4. Which connection diagram, shown above in Figure 224.108, should be used where the *Supply Circuit* is from a 1-pole breaker 277/480 volt, 3-ph, 4-W panelboard? The load is rated to operate at 24 volts. The transformer is a 240 x 480 — 24/48 volt.

    a. I   b. III   c. IV   d. II   e. none of these

____ 5. Which connection diagram, shown above in Figure 224.108, should be used where the *Supply Circuit* is from a 2-pole breaker 120/240 volt, 3-ph, 4-W panelboard? The load is rated to operate at 12 volts. The transformer is a 120 x 240 — 12/24 volt.

    a. I   b. II   c. IV   d. III   e. none of these

____ 6. Which connection diagram, shown above in Figure 224.108, should be used where the *Supply Circuit* is from a 1-pole breaker 120/240 volt, 3-ph, 4-W panelboard? The load is rated to operate at 12 volts. The transformer is a 120 x 240 — 12/24 volt.

    a. II   b. IV   c. III   d. I   e. none of these

____ 7. Which connection diagram, shown above in Figure 224.108, should be used where the *Supply Circuit* is from a 2-pole breaker 120/240 volt, 3-ph, 4-W panelboard? The load is rated to operate at 48 volts. The transformer is a 240 x 480 — 24/48 volt.

    a. I   b. III   c. IV   d. II   e. none of these

____ 8. Which connection diagram, shown above in Figure 224.108, should be used where the *Supply Circuit* is from a 2-pole breaker 120/240 volt, 3-ph, 4-W panelboard? The load is rated to operate at 32 volts. The transformer is a 120 x 240 — 16/32 volt.

    a. II   b. I   c. III   d. IV   e. none of these

_____ 9. You have a 240 volt circuit available to supply power to a .5 KVA, Group I transformer. Refer to Yr-2 – Annex D. You need to have 12 volts for a piece of equipment. Assuming that you correctly connected your 240-volt supply circuit, you will connect your 12-volt load circuit as shown in drawing ___ below.

a. 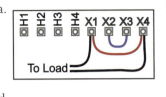 b. 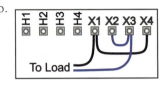 c.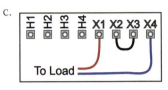

d. 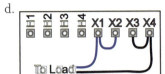 e. 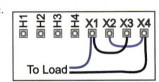 f. none of these

_____ 10. The circuit you have available to supply a 240-V, 150 A load is 208 volts. The rating for a B-B transformer that can operate the load at its rated voltage is ___ KVA.

a. 3.0   b. 37.5   c. 1.5   d. 2.0   e. 7.5   f. 5.0

_____ 11. When properly connected to deliver 12 volts from a 240 volt supply, the secondary of a 150 VA, 120 x 240–12/24 V transformer can carry a maximum of ___ amps without being overloaded.

a. 52.05   b. 41.67   c. 6.25   d. 12.5

_____ 12. You have a 240 volt circuit available to supply power to a .5 KVA, Group I transformer. Refer to Yr-2 – Annex D. You need to have 24 volts for a piece of equipment. You will connect your 240 volt supply circuit as shown in drawing ___ below.

a. 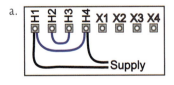 b. 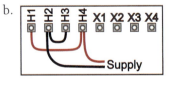 c.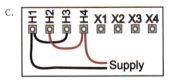

d. 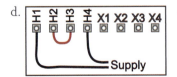 e. 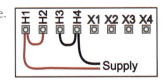 f. none of these

Buck-Boost Transformers: Single-and Three- Phase   224-625

_____ 13. When a supply voltage is to be changed from 102 V to 120 V, a "buck-boost" transformer is connected as ___ to deliver the proper voltage to the 120-volt load.

   a. an autotransformer
   b. an isolation transformer
   c. a distribution transformer
   d. none of these

_____ 14. To produce an output of 12 volts, a 120 x 240 volt primary 12/24 volt secondary transformer is connected as a(an) ___.

   a. insulating transformer   b. autotransformer

_____ 15. Buck-boost transformers are available with secondary windings that are rated to deliver voltages of ___.

   a. 24   b. 16   c. 12   d. 48   e. any of these   f. none of these

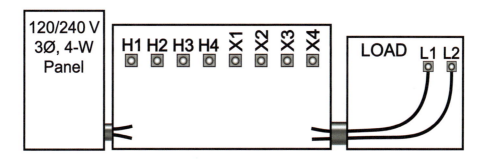

Figure 224.103

_____ 16. Refer to Figure 224.103 and the information in Yr-2 – Annex D for a 3.0 KVA GROUP I - Cat #11110 transformer. A single-pole breaker in the panel feeds the transformer primary. The LOAD is rated to operate at 24 volts. Which of the connection diagrams, shown below, has the transformer connected correctly?

a. 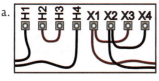   b.   c.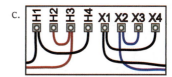

d. 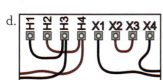   e. 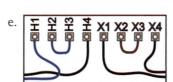   f. none of these

Year Two (Student Manual)

_____ 17. Buck-boost transformers are available with windings that are rated to accept line voltages of ____.

   a. 208
   b. 480
   c. 240
   d. 120
   e. 277
   f. any of these

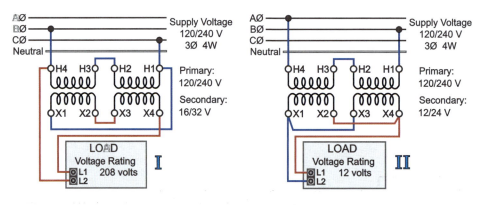

Figure 225.101

_____ 18. Which of the drawings, shown above in Figure 225.101, is connected as an isolation transformer?

   a. neither I nor II
   b. II only
   c. both I & II
   d. I only

_____ 19. Can a "buck-boost" transformer be used to operate 277 volt lighting where the supply circuit is 230 volts?

   a. Yes
   b. No

_____ 20. A ____ KVA 240 × 480 – 24/48 V transformer, that is connected to a 277 volt circuit, should be used to operate a 230 volt 1Ø motor that draws 25 amps.

   a. 1
   b. .75
   c. 1.5
   d. .5

Objective 224.3 Worksheet

_____ 1. Small transformers that are back fed can experience ___.

    a. high inrush currents
    b. nuisance tripping of primary OCPDs
    c. both a and b
    d. neither a nor b

_____ 2. The secondary voltage from a buck-boost transformer with a nameplate rating of 1.0 KVA can be expected to be about ___ volts with the 24-volt load disconnected. The supply circuit voltage is 120 volts.

    a. 25.44   b. 22.6   c. 127.2   d. 97.4   e. 112.8   f. 135.4

Lesson 224 Safety Worksheet

Cranes and Crane Safety

Cranes are a very common piece of heavy equipment that are in use on most construction sites. We could not build the large and tall buildings that are commonplace today without them. They permit us to reach out and up to incredible lengths and to lift and place the materials and tools needed to build our modern structures where we live, work, and shop.

Cranes come in many types, styles, and configurations. Rubber-tired cranes are very versatile and are often used on construction sites. They can maneuver easily around the site, lift what is necessary, and then move on to the next site. Crawler cranes are commonly used to erect large panels of precast concrete wall sections and structural steel because they can move over the rough terrain of new construction. Tower cranes are often used for high-rise construction in urban areas because they can be erected on a small footprint and can be built many stories tall.

The most common types of crane accidents involve crane booms and cables touching overhead power lines, employees caught in the rotating structure of the crane, and tipovers. If you are working in the vicinity of a crane, be extra alert for sudden movements. You could become caught in or in between the moving parts. The moving and rotating structures of cranes must be barricaded to prevent the movement of employees into these moving parts. **Never** walk or stand under any suspended load. **Never** touch or lean on any part of the crane. You could be electrocuted if the operator comes near or into contact with overhead power lines. **Cranes must be kept clear of the proximity of overhead power lines.** Only experienced crane signal persons may signal if the crane is near energized power lines. It is very important to maintain the minimum distances as shown in the OSHA Crane Standard from any energized power lines. These distances are a **minimum** of 10 feet for voltages up to 50KV and require additional distance for higher voltages.

If you are working with a crane operator as part of your work assignment, you need to know the industry standard hand signals that are required to be used. A copy of these hand signals is required to be posted on the jobsite where cranes are in use. The standard signals are included on the next page for you to learn and use. Only the person who is designated can signal the crane operator, except in an emergency where anyone can signal for an **Emergency Stop**.

Cranes, as with all other types of equipment, can only be operated by employees who are trained and authorized to operate the equipment. They must be maintained in a safe manner and records of the required inspections must be kept. Cranes must only be operated on surfaces that can safely support the load of the crane.

Proper use of Hand Signals for Cab-controlled Cranes

The operator of a cab-controlled crane should use a floor person to direct the lift, or placement, of a load and should respond to signals only from the person directing the lift. However, the operator should obey an emergency stop signal from any employee at all times. The operator of a cab-controlled crane and the floor person working with them should learn and use the industry standard signals.

There are nine industry standard hand signals that are used for communication between the operator in the crane's cab and the floor person. These signals are:

HOIST (RAISE)
▶ With forearm vertical—and forefinger pointing up—move hand in small horizontal circle.

LOWER
▶ With palm up—and fingers closed —point in direction of motion, and jerk hand horizontally.

BRIDGE TRAVEL
▶ With arm extended forward— and hand open and slightly raised—make pushing motion in direction of travel.

TROLLEY TRAVEL
▶ With palm up—and fingers closed —point in direction of motion, and jerk hand horizontally.

MOVE SLOWLY
▶ Use one hand to give any motion signal and place the other hand motionless in front of hand giving the motion signal by anyone present.

STOP
▶ With arm extended—and palm down—hold position rigidly.

EMERGENCY STOP
▶ With arm extended—and palm down—move hand rapidly to the right and left.

MULTIPLE HOIST CRANES
▶ Hold up one finger to indicate block 1, and two fingers to indicate block 2
▶ Follow with regular hand signals.

MAGNET DISCONNECTED
▶ With palms up, crane operators spreads both hands apart.

Questions

_____ 1. Cranes, as with all other types of equipment, can be operated only by ___ persons.
  I. qualified
  II. authorized
  III. trained

  a. III only
  b. II only
  c. I and III only
  d. I only
  e. II and III only
  f. I and II only

_____ 2. The most common types of crane accidents involve getting caught in moving structures and___.
  I. cave-ins
  II. tipovers
  III. electric lines

  a. I only
  b. III only
  c. I and III only
  d. II only
  e. II and III only
  f. none of these

_____ 3. Never stand or walk under any load suspended by a crane.

  a. True
  b. False

_____ 4. There are only two types of cranes.

  a. True
  b. False

_____ 5. When signaling a crane, any hand signals can be used as long as all employees agree to this method.

  a. False
  b. True

_____ 6. Cranes are used to lift and place ___.
   I. materials
   II. persons
   III. tools

   a. II and III only
   b. I only
   c. II only
   d. I and III only
   e. III only
   f. I and II only

_____ 7. You should not touch or lean against a crane at any time.

   a. False
   b. True

_____ 8. When in operation, a crane must be kept a minimum distance of ___ feet from power lines of 50 KV or less.

   a. 10
   b. 6
   c. 12
   d. 8

Lesson 225
Buck-Boost Transformers: Connection and Selection

Electrical Curriculum

Year Two
Student Manual

Purpose

The purpose of this lesson is to address the selection of the proper buck-boost transformer to supply sufficient capacity at the rated voltage of the load to be served.

Homework

(Due at the beginning of this class)

For this lesson, you should:

- Thoroughly read the material contained within Lesson 225.
- Utilize Yr-2 Annex D (Located at the back of this manual).
- Complete Objective 225.1 Worksheet.
- Complete Objective 225.2 Worksheet.
- Read and complete Lesson 225 Safety Worksheet.

- Complete additional worksheets, if available, as directed by your instructor.

Objectives

By the end of this lesson, you should:

    225.1 ....Be able to determine the appropriate buck-boost transformer and the connections required to deliver proper voltage to a load, and be able to calculate the output voltage when supply voltages are other than the table values.

    225.2 ....Be able to select and connect buck-boost transformers to deliver required voltage and power to single and three-phase loads.

225.1 Output Voltages

Output Voltages Using Single-Phase Connection Diagrams

The output voltage from a buck-boost transformer depends on both the input voltage and the connection diagram used. For single-phase transformers, Autotransformer Connection Diagrams C through J are used. (Refer to Yr-2 – Annex D.) Since these are single-phase transformers with dual windings on both the primary (H leads) and the secondary (X leads), the voltage induced in each X winding is dependent upon the voltage applied to each H winding. Note also that the H windings can be connected in either series or in parallel. Similarly, the X windings can be connected in either series or in parallel.

The selection tables for the three groups of single-phase B-B transformers are shown in Yr-2 – Annex D. Each table is separated into a "Boosting" section and a "Bucking" section. Various columns show the load (output) voltage that can be obtained given the available line (supply) voltage. At the bottom of each column, the connection diagram to use for the desired output voltage is shown.

Example 225.11:

A Group I transformer can be used to boost a voltage of 208 up to 230 volts. To attain this 230-volt output, given the 208-volt input, Connection Diagram H is to be used. Using this diagram, the supply circuit is connected to terminals (leads) H4 and H1, and the load is connected to terminals (leads) H4 and X1. Jumpers are also installed between H3 and H2, between H1 and X4, and between X3 and X2.

The Line (Input or Supply) and Load (Output) voltages shown in the table are actually calculated. Remember that these transformers consist of two winding sets. A Group I, B-B transformer has a 120 x 240 volt primary and a 12/24 volt secondary. If 120 volts is connected to H1 and H2, the output will be 12 volts between X1 and X2. However, if less than 120 volts is applied to the primary, less than 12 volts will be available on the secondary. For Connection Diagram H, if a 208-volt supply circuit is connected to H4 and H1, only half of this voltage is applied to each series-connected primary winding (Refer to Figure 225.11.)

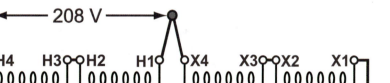

Figure 225.11   Connection Diagram H

$$208 \text{ V} \div 2 = 104 \text{ V}$$

104 volts is applied to the H4-H3 winding, and 104 volts is applied to the H2-H1 winding. From this, the voltage induced in each series-connected secondary winding can be determined using the voltage ratios:

☞ E_{X4-X3}	$= (E_{H4-H3}) \times (E_{Rated\ Sec} \div E_{Rated\ Pri})$	☞ E_{X2-X1}	$= (E_{H2-H1}) \times (E_{Rated\ Sec} \div E_{Rated\ Pri})$
E_{X4-X3}	$= (104) \times (12 \div 120)$	E_{X2-X1}	$= (104) \times (12 \div 120)$
E_{X4-X3}	$= (104) \times (12 \div 120)$	E_{X2-X1}	$= (104) \times (12 \div 120)$
E_{X4-X3}	$= (104) \times (0.1)$	E_{X2-X1}	$= (104) \times (0.1)$
E_{X4-X3}	$= 10.4$ volts	E_{X2-X1}	$= 10.4$ volts

Note that, although the table voltage shown is 230 volts, the actual calculated output in Figure 225.11 is 228.8 volts.

Example 225.12:

A Group I transformer is to be used to buck a voltage of 230 down to 208 volts. To attain this 208-volt output, given the 230-volt input, Connection Diagram I is to be used. In this case, as shown in Figure 225.12, the supply circuit is connected to terminals (leads) H4 and X1, and the load is connected to terminals (leads) H4 and H1. Diagram I requires that jumpers be installed between H3 and H2, between H1 and X4, and between X3 and X2.

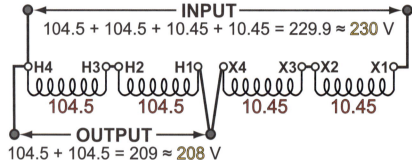

Figure 225.12 Connection Diagram - Diagram I

For simplicity, assume that the actual number of turns in each winding equals the rated voltage of that winding. When 230 volts is connected to H4 and X1, as shown in Figure 225.12, this voltage is divided equally across each turn between these two connection points. That is, between H4 and X1, there are a total of 264 turns. (120 + 120 + 12 + 12 = 264 turns). The voltage dropped across each turn equals the input voltage divided by the total number of turns:

☞ $E_{per\ turn}$	$=$ (Input Voltage $\div$ Total Turns between Input Connection Points)	
$E_{per\ turn}$	$= (230 \div 264)$	
$E_{per\ turn}$	$= .8712$ volts	

Again, Refer to Figure 225.12. Since there are 120 turns between H4 and H3, the voltage between H3 and H4 equals the sum of the voltages across these 120 turns. The voltage between H2 and H1 will be the same. These calculations are shown in the table below.

E_{H4-H3}	$=$ (Turns between H4 & H3)$\times (E_{per\ turn})$	E_{H2-H1}	$=$ (Turns between H2 & H1)$\times (E_{per\ turn})$
E_{H4-H3}	$= (120) \times (.8712)$	E_{H2-H1}	$= (120) \times (.8712)$
E_{H4-H3}	$= \mathbf{104.5}$ volts	E_{H2-H1}	$= \mathbf{104.5}$ volts

To verify that the sum of the voltages between the input points (H4 to X1) equals 230 volts, the voltage dropped across the X4-X3 and X2-X1 windings can be calculated. Since there are 12 turns between X4 and X3, the voltage between X4 and X3 equals the sum of the voltages across these 12 turns. The voltage between X2 and X1 will be the same. These calculations are shown in the table below.

$E_{X4\text{-}X3}$ = (Turns between X4 & X3) × ($E_{per\ turn}$)	$E_{X2\text{-}X1}$ = (Turns between X2 & X1) × ($E_{per\ turn}$)
$E_{H4\text{-}X3}$ = (12) × (.8712)	$E_{X2\text{-}X1}$ = (12) × (.8712)
$E_{X4\text{-}X3}$ = **10.45** volts	$E_{X2\text{-}X1}$ = **10.45** volts

The sum of the voltages between supply voltage connection points equals:

$$\text{Sum} = 104.5 + 104.5 + 10.45 + 10.45 = 229.9 \text{ or } \mathbf{230} \text{ volts} = \text{Supply Circuit Voltage}$$

Connection Diagram I requires that the load be connected between H4 and H1. Between these two points, the voltage is actually 209 volts, as shown in Figure 225.12. Note that the Group I Selection Table shows 208 volts. When an input voltage and/or an output voltage is other than that shown in the selection table, the output voltage should be calculated to determine which group and which connection diagram to use, so that rated voltage is applied to the load.

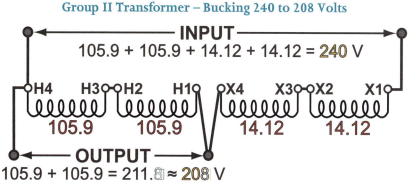

Figure 225.13   Connection Diagram I

Example 225.13:

A voltage of 240 volts is to be bucked down to 208 volts. In this case a Group II transformer is selected. The Group II, Selection Table shows to use Connection Diagram I for this installation. As shown in Figure 225.13, the supply circuit is connected to terminals (leads) H4 and X1, and the load is connected to terminals (leads) H4 and H1. Diagram I requires that jumpers be installed between H3 and H2, between H1 and X4, and between X3 and X2. Note that these are the same connections as made in Example 225.12, however, since the supply (input) voltage is now 240 volts, the voltages dropped across each turn will differ slightly. The differences are due to the higher supply voltage as well as the rated secondary voltages. The secondary of a Group II transformer is rated at 16/32 volts.

When 240 volts is connected to H4 and X1, as shown in Figure 225.13, this voltage is divided equally across each turn between these two connection points. That is, between H4 and X1, there are a total of 272 turns. (120 + 120 + 16 + 16 = 272 turns). The voltage dropped across each turn equals the input voltage divided by the total number of turns:

☞ $E_{per\ turn}$ = (Input Voltage ÷ Total Turns between Input Connection Points)

$E_{per\ turn} = (240 ÷ 272)$

$E_{per\ turn} = .8824$ volts

Again, Refer to Figure 225.13. Since there are 120 turns between H4 and H3, the voltage between H3 and H4 equals the sum of the voltages across these 120 turns. The voltage between H2 and H1 will be the same. These calculations are shown in the table below.

E_{H4-H3} = (Turns between H4 & H3)×($E_{per\ turn}$)	E_{H2-H1} = (Turns between H2 & H1)×($E_{per\ turn}$)
E_{H4-H3} = (120) × (.8824)	E_{H2-H1} = (120) × (.8824)
E_{H4-H3} = **105.9** volts	E_{H2-H1} = **105.9** volts

To verify that the sum of the voltages between the input points (H4 to X1) equals 240 volts, the voltage dropped across the X4-X3 and X2-X1 windings can be calculated. Since there are 16 turns between X4 and X3, the voltage between X4 and X3 equals the sum of the voltages across these 16 turns. The voltage between X2 and X1 will be the same. These calculations are shown in the table below.

E_{X4-X3} = (Turns between X4 & X3)×($E_{per\ turn}$)	E_{X2-X1} = (Turns between X2 & X1)×($E_{per\ turn}$)
E_{H4-X3} = (16) × (.8824)	E_{X2-X1} = (16) × (.8824)
E_{X4-X3} = **14.12** volts	E_{X2-X1} = **14.12** volts

The sum of the voltages between supply voltage connection points equals:

Sum = 105.9 + 105.9 + 14.12 + 14.12 = 240.04 or **240** volts = Supply Circuit Voltage

Connection Diagram I requires that the load be connected between H4 and H1. Between these two points, the voltage is actually 211.8 volts, as shown in Figure 225.13. Note that the Group I Selection Table shows 208 volts. Again, when an input voltage and/or an output voltage is other than that shown in the Selection Table, the output voltage should be calculated to determine which group and which connection diagram to use, so that rated voltage is applied to the load.

Example 225.14:

A voltage of 256 volts is to be bucked down to 240 volts. In this case a Group II transformer is selected. The Group II Selection Table advises to use Connection Diagram E for this installation. As shown in Figure 225.14, the supply circuit is connected to terminals (leads) H4 and X1, and the load is connected to terminals (leads) H4 and H1. Diagram E requires that jumpers be installed between H3 and H2,

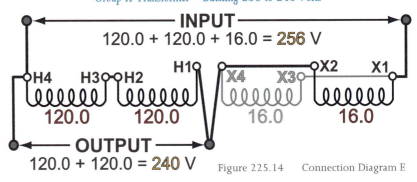

Figure 225.14   Connection Diagram E

between H1 and X4, between X4 and X2, and between X3 and X1. Note that the X windings are now connected in parallel. Again, the secondary of a Group II transformer is rated at 16/32 volts.

Equipment for General Use

When 256 volts is connected to H4 and X1, as shown in Figure 225.14, this voltage is divided equally across each turn between these two connection points. Since the X windings are in parallel, only the turns in one winding are counted. Now, between H4 and X1, there are a total of 256 turns. (120 + 120 + 16 = 256 turns). The voltage dropped across each turn equals the input voltage divided by the total number of turns:

☞ $E_{per\ turn}$ = (Input Voltage ÷ Total Turns between Input Connection Points)

$E_{per\ turn} = (256 \div 256)$

$E_{per\ turn} = \mathbf{1.0}$ volts

Again, Refer to Figure 225.14. Since there are 120 turns between H4 and H3, the voltage between H4 and H3 equals the sum of the voltages across these 120 turns. The voltage between H2 and H1 will be the same. These H winding calculations are shown in the table below.

E_{H4-H3} = (Turns between H4 & H3)×($E_{per\ turn}$)	E_{H2-H1} = (Turns between H2 & H1)×($E_{per\ turn}$)
$E_{H4-H3} = (120) \times (1.0)$	$E_{H2-H1} = (120) \times (1.0)$
$E_{H4-H3} = \mathbf{120.0}$ volts	$E_{H2-H1} = \mathbf{120.0}$ volts

To verify that the sum of the voltages between the input points (H4 to X1) equals 256 volts, the voltage dropped across the X4-X3 and X2-X1 windings can be calculated. Since there are 16 turns between X4 and X3, the voltage between X4 and X3 equals the sum of the voltages across these 16 turns. The voltage between X2 and X1 is the same, but isn't used because these windings are in parallel with X4-X3. The X winding calculations are shown in the table below.

Paralleled X windings – Use only one set	E_{X2-X1} = (Turns between X2 & X1)×($E_{per\ turn}$)
	$E_{X2-X1} = (16) \times (1.0)$
	$E_{X2-X1} = \mathbf{16.0}$ volts

The sum of the voltages between supply voltage connection points equals:

Sum = 120.0 + 120.0 + 16.0 = **256** volts = Supply Circuit Voltage

Connection Diagram E requires that the load be connected between H4 and H1. Between these two points, the voltage is actually 240 volts, as shown in Figure 225.14. Note that the Group II Selection Table does not have a column for exactly 256 to 240. The 255 to 239 column was selected because it most closely matches the input voltage and the desired output voltage.

> Had a Group I transformer been selected for this example, the column headed 252/240 volts would have been the best choice, however the output voltage would have been 244 volts. This is because the total turns for a Group I using Diagram E would have been:
>
> 120 + 120 + 12 = 252 turns
>
> Then, the volts per turn = 256 ÷ 252 = 1.0159

Year Two (Student Manual)

This would result in an output voltage, between H4 and H1, of:

[(120) × (1.0159)] + [(120) × (1.0159)] = 243.8 volts

The use of a Group II transformer results in an output voltage that more closely matches the desired 240 volts.

Example 225.15:

A load is rated to operate at 120 volts, however the available circuit is 137 volts. The supply circuit should be bucked down to 120 volts. From among the "input/output" bucked voltages shown in the selection tables for Groups I and II, the 135/119 shown for a Group II would be the better choice to use. The Group II, Selection Table shows to use Connection Diagram F. As shown in Figure 225.15, the supply circuit is connected to terminals (leads) H4 and X1, and the load is connected to terminals (leads) H4 and H1. Diagram F requires that jumpers be installed between H4 and H2, between H3 and H1, between H1 and X4, between X4 and X2, and between X3 and X1. Note that Connection Diagram F requires that both the H and the X windings be connected in parallel. Again, the secondary of a Group II transformer is rated at 16/32 volts.

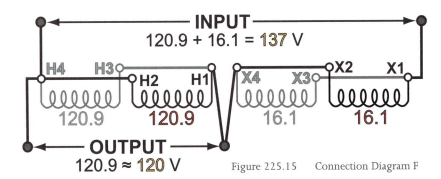

Figure 225.15 Connection Diagram F

When 137 volts is connected to H4 and X1, as shown in Figure 225.15, this voltage is divided equally across each turn between these two connection points. Since the H windings are in parallel, only the turns in one winding are counted. Likewise, since the X windings are in parallel, only the turns in one winding are counted. Now, between H4 and X1, there are a total of 136 turns. (120 + 16 = 136 turns). The voltage dropped across each turn equals the input voltage divided by the total number of turns:

☞ $E_{per\ turn}$ = (Input Voltage ÷ Total Turns between Input Connection Points)
$E_{per\ turn}$ = (137 ÷ 136)
$E_{per\ turn}$ = **1.0074** volts

Again, Refer to Figure 225.15. Since there are 120 turns between H2 and H1, the voltage between H2 and H1 equals the sum of the voltages across these 120 turns. Although the voltage between H4 and H3 is the same, it will not be used because it is in parallel with H2-H1. Note that, electrically, H4 and H2 are considered to be the same point. The H winding calculations are shown in the table below.

Paralleled H windings – Use only one set	E_{H2-H1} = (Turns between H2 & H1) × ($E_{per\ turn}$)
	E_{H2-H1} = (120) × (1.0074)
	E_{H2-H1} = **120.89** volts

To verify that the sum of the voltages between the input points (H4 to X1) equals 256 volts, the voltage dropped across the X4-X3 and X2-X1 windings can be calculated. Since there are 16 turns between X4 and X3, the voltage between X4 and X3 equals the sum of the voltages across these 16 turns. The voltage across parallel winding X2-X1 is the same, but again, it won't be used. The X winding calculations are shown in the table below.

Paralleled X windings – Use only one set	$E_{X2\text{-}X1}$ = (Turns between X2 & X1) × ($E_{\text{per turn}}$)
	$E_{X2\text{-}X1}$ = (16) × (1.0074)
	$E_{X2\text{-}X1}$ = **16.12** volts

The sum of the voltages between supply voltage connection points equals:

Sum = 120.9 + 16.1 = **137** volts = Supply Circuit Voltage

Connection Diagram F requires that the load be connected between H4 and H1. Between these two points, the voltage is actually 121 volts, as shown in Figure 225.15. Although this output isn't exactly the voltage required by the load, it is the closest possible for a Group II transformer.

Had a Group I transformer been selected for this example, the column headed 132/120 volts would have been the selected. Although Diagram F is also used for these voltages, the output voltage would have been 244 volts. This is because the total turns for a Group I would have been:

120 + 12 = 132 turns

Then, the volts per turn = 137 ÷ 132 = 1.0379

This would result in an output voltage, between H4 and H1, of:

(120) × (1.0379) = 124.5 volts

The election to use a Group II transformer was the better choice because the 121-volt output is better than the 124.5 that would have been available from a Group I transformer.

225.2 Selection and Connection of Buck-Boost Transformers

Manufacturer Supplied Data for Buck-Boost Transformer Selection

An alternative to calculating the required size of a buck-boost transformer is through the use of tables supplied by the manufacturer. These are almost universal among manufacturers, but are specific to the manufacturer. When used to buck or boost a 1Ø circuit's voltage, the correctly sized transformer is selected from the appropriate 1Ø Selection Table. When used to buck or boost a 3Ø circuit's voltage, the correctly sized transformer is selected from the appropriate 3Ø Selection Table. For 3Ø loads either 2 or 3 transformers will be required. For both 1Ø and 3Ø applications, the specified wiring diagram is referenced on the table. The selection tables and connection diagrams, along with other information, for buck-boost transformers are located in **Yr-2 – Annex D**. Be aware that larger buck-boost transformers frequently have double secondary leads (two leads for each X connection), as noted at the bottom of each B-B transformer selection table.

Selecting and Connecting a Buck-Boost Transformer for use as an Insulating Transformer

Selection

1. Read directly from the nameplate, or calculate the load in kVA.
2. Determine the available supply circuit voltage.
   a. For 120 volt circuits, use **Selection Table BB1-1** or **-2**.
   b. For 240 volt circuits, use **Selection Table BB1-1**, **-2**, or **-3**.
   c. For 480 volt circuits, use **Selection Table BB1-3**.
3. Determine the rated voltage from the equipment nameplate.
   a. For 12 or 24 V loads, go to **Selection Table BB1-1**.
   b. For 16 or 32 V loads, go to **Selection Table BB1-2**.
   c. For 24 or 48 V loads, go to **Selection Table BB1-3**.
4. In the KVA Rating column of the correct **Selection Table**, select a transformer with a KVA rating that is greater than or equal to the load determined in Step 1.
5. The Catalog # of the transformer is shown above the KVA rating found in Step 4.

Connection

6. Proceed to the BUCK-BOOST Insulating Transformer Diagram Selection Table.
   a. If you selected a transformer from Selection Table BB1-1, go to the Group I table.
   b. If you selected a transformer from Selection Table BB1-2, go to the Group II table.
   c. If you selected a transformer from Selection Table BB1-3, go to the Group III table.
7. Select the correct row that shows the input voltage (of the available circuit) and the output voltage to the load. The connection, or wiring, diagram is referenced on this row.
8. Connect the supply circuit and the load to the transformer as shown in the INSULATING TRANSFORMER – CONNECTION DIAGRAM.

Selecting and Connecting a Buck-Boost Transformer to Serve a Single-Phase Load

Selection

1. Determine the available supply circuit voltage and the rated voltage of the load. These are the *input* and *output* voltages, respectively.
2. From among the three single-phase **Selection Tables (BB1-1, -2, or -3)**, select the table, and column in this table, that has values of *input* and *output* voltages that most closely match those found in Step 1.
3. Read the kVA or current rating of the load directly from the equipment nameplate.
4. In the column selected in Step 2, go down and find the row containing a KVA or Amps value that is greater than or equal to the value found in Step 3.
5. On the row found in Step 4, go to the far left-hand column and read the Catalog # and the KVA rating of the transformer.

Connection

6. Go to the Single-Phase Autotransformer – CONNECTION DIAGRAMS.
7. Select the connection diagram that is referenced at the bottom of the column selected in Step 2.
8. Connect the supply circuit (INPUT) and the load (OUTPUT) to the transformer as shown in the CONNECTION DIAGRAM.

Selecting and Connecting a Buck-Boost Transformer to Serve a Three-Phase Load

Selection

1. Determine the available supply circuit voltage and the rated voltage of the load. These are the *input* and *output* voltages, respectively.
2. From among the three three-phase **Selection Tables (BB3-1, -2, or -3)**, select the table, and column in this table, that has values of *input* and *output* voltages that most closely match those found in Step 1.
3. Read the kVA or current rating of the load directly from the equipment nameplate.
4. In the column selected in Step 2, go down and find the row containing a KVA or Amps value that is greater than or equal to the value found in Step 3.
5. On the row found in Step 4, go to the far left-hand column and read the Catalog # and the KVA rating of the transformer.

Connection

6. Go to the **Three-Phase Autotransformer – CONNECTION DIAGRAMS**.
7. Select the connection diagram that is referenced at the bottom of the column selected in Step 2.
8. Connect the supply circuit (INPUT) and the load (OUTPUT) to the transformer as shown in the **CONNECTION DIAGRAM**.

Some manufacturers mark only LV (Low Voltage) and HV (High Voltage) connection points on their diagrams. In these instances, the INPUT circuit and the OUTPUT circuit are connected according to which circuit is lower and which is higher. For example, when boosting a 208 volt INPUT circuit to serve a 230 V load, the OUTPUT (230 V) is the higher of the two and is connected to the points marked HV, and the 208 volt INPUT, since is lower, it is connected to the LV points.

Typically, only the connections shown in the table at the right should be used to obtain 3Ø B-B transformer outputs. A delta-to-wye connection does not provide enough current capacity to accommodate unbalanced currents flowing in the neutral wire of a 4-wire circuit. OPEN-DELTA outputs are preferred. Compared to an open-delta output, a closed-delta B-B transformer output requires three transformers, results in phase shifting, is more expensive, and is electrically inferior. Generally, it is 3Ø, 3-wire loads that require voltage adjustments. On rare occasions it will be necessary to buck or boost the voltage to a 3Ø, 4-wire load. In these rare instances, it will be necessary to have a 3Ø, 4-wire supply circuit, where the supply neutral must also serve as the load neutral.

INPUT (SUPPLY) CIRCUIT from a 3Ø SYSTEM	DESIRED OUTPUT CIRCUIT	
DELTA 3-wire	WYE 3-wire	**DO NOT USE**
DELTA 3-wire	WYE 4-wire	**DO NOT USE**
OPEN DELTA 3-wire	WYE 3-wire	**DO NOT USE**
OPEN DELTA 3-wire	WYE 4-wire	**DO NOT USE**
WYE 3-wire	CLOSED DELTA 3-wire	**DO NOT USE**
WYE 4-wire	CLOSED DELTA 3-wire	**DO NOT USE**
WYE 4-wire	WYE 4-wire	OK
WYE 3-wire	OPEN DELTA 3-wire	OK
OPEN DELTA 3-wire	OPEN DELTA 3-wire	OK
CLOSED DELTA 3-wire	OPEN DELTA 3-wire	OK

Refer to the information and connection diagrams available from the various transformer manufacturers for the installation of 3Ø, 4-wire loads. Most manufacturers enclose a copy of all connection diagrams with their B-B transformers. The correct diagram, from among these manufacturer-supplied illustrations, should be used for field connections.

For all autotransformer installations refer to the NEC®. Section 450.4 requires OCPDs to be installed on each ungrounded input circuit conductor. Refer also to the permitted uses of autotransformers as directed in 450.4(B) FPN.

Buck-Boost Transformers Used for Voltage Drop

B-B transformers can be used for voltage drop correction, but only if the load is constant. Typically, these are used to boost the voltage applied to a single load at the end of a long circuit. For example, to be able to irrigate additional acreage, a farmer replaced a small irrigation pump with a larger horsepower pump. He used the existing circuit that had conductors and conduit sized for the smaller pump motor. Consequently, the motor circuit suffered voltage drop. Instead of replacing the conduit and wire, the clever electrician remedied the situation by installing a boost transformer near the pump.

Caution! Voltage drop varies with fluctuating loads. If buck-boost transformers are used to correct for voltage drop during a peak load cycle, dangerously high voltages may result under lightly loaded conditions. Buck-boost transformers should only be used for voltage drop correction when the load is constant.

Loads should be operated at their rated voltages to optimize performance and ensure that they are operational throughout their expected life. Operating a load at more or less than its rating can have a detrimental effect on the equipment, particularly motors. Although the performance of resistive loads, including heating equipment and incandescent lighting, is diminished (less heat or light output) by operation at lower than rated voltage, these loads are not damaged. At higher than rated voltage, resistive loads can be damaged by current flow in excess of the ratings. Conversely, when motors are operated at lower or higher than rated voltages, these can be damaged. When a motor is operated at a lower than rated voltage the motor torque decreases and the motor can stall and suffer overheating. At higher than rated voltage, the motor windings overheat and the life expectancy of the motor is reduced exponentially with overvoltage.

Equipment for General Use

Objective 225.1 Worksheet

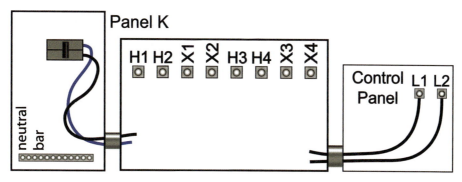

Figure 225.102

_____ 1. Panel K, shown above in Figure 225.102, is 120/240V 3Ø, 4-W. The control panel requires 208 volts. Using a 3.0 KVA Buck-Boost GROUP II transformer from Yr-2 – Annex D, the calculated voltage that would be applied to the control panel is ___ volts where the proper connections are made and the input voltage is exactly 240 volts.

    a. 229   b. 218   c. 212   d. 236   e. 208

_____ 2. Refer to Figure 225.102, shown above, and the information in Yr-2 – Annex D for a 3.0 KVA buck-boost transformer. Panel K is 120/208V 3Ø, 4-W. The control panel requires 240 volts. Using a GROUP II transformer, assuming the proper connections are made and the input voltage is exactly 208 volts, the calculated voltage that would be applied to the control panel is ___ volts.

    a. 231   b. 229   c. 236   d. 219   e. 240

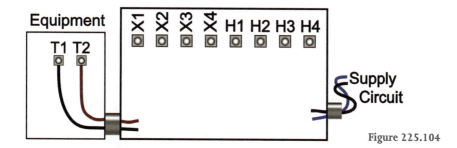

Figure 225.104

_____ 3. The supply circuit, shown above in Figure 225.104, is from a 2-P breaker in a 120/208 V, 3Ø, 4W panel. The foreign-made equipment is rated to operate at 220 volts. The supply circuit voltage is 206 volts where it enters the B-B transformer. If Connection Diagram H, from Yr-2 – Annex D, is used to connect a Group I transformer, the equipment will be supplied with ___ volts.

    a. 240   b. 227   c. 233   d. 216   e. 230   f. 220

Year Two (Student Manual)

_____ 4. The supply circuit, shown above in Figure 225.104, is from a 2-P breaker in a 120/208 V, 3Ø, 4W panel. The foreign-made equipment is rated to operate at 220 volts. The supply circuit voltage is 206 volts where it enters the B-B transformer. If Connection Diagram G, from Yr-2 – Annex D, is used to connect a Group II transformer, the equipment will be supplied with ___ volts.

   a. 233
   b. 227
   c. 240
   d. 216
   e. 220
   f. 230

Equipment for General Use

Objective 225.2 Worksheet

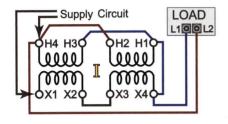 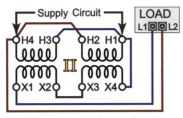 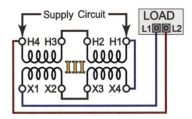

Figure 225.107

_____ 1. Which connection diagram in Figure 225.107, shown above, should be used where the Supply Circuit is from a 1-pole breaker in a 277/480 volt, 3-ph, 4-W panelboard? The load is rated to operate at 230 volts. The transformer is a 240 × 480 – 24/48 volt.

    a. III   b. II   c. I   d. none of these

_____ 2. Which connection diagram in Figure 225.107, shown above, should be used where the Supply Circuit is from 2-pole breaker in a 120/240 volt, 3-ph, 4-W panelboard? The load is rated to operate at 277 volts. The transformer is a 240 × 480 – 24/48 volt.

    a. II   b. I   c. III   d. none of these

_____ 3. The catalog number for a 120 x 240–16/32 V transformer that is rated for 5000 volt-amperes is ___.

    a. 11111   b. 21101   c. 21105   d. 11105   e. 21111

_____ 4. A 2-wire circuit from a 2-pole breaker in a 120/240 volt 3Ø, 4W panelboard has been run to supply a piece of equipment. The voltage measured at the load end of this circuit is **exactly** the rated voltage from the panel. You should connect the equipment ___ because when the equipment arrives the nameplate indicates that it is rated to operate at 208 volts.

    a. through a Group II Buck-Boost transformer
    b. directly to the circuit
    c. through a Group I Buck-Boost transformer

_____ 5. A 3Ø, 230-volt load is to be supplied with a 3Ø, **3-wire** circuit that originates from a 120/208 volt 3Ø, 4-w panelboard. The 3Ø load is rated at 30 KVA. From Yr-2 – Annex D, the smallest Group I transformer(s) you can select for this load is ___ KVA.

    a. 5   b. 3   c. 10   d. 1   e. 2

Year Two (Student Manual)

225-650　　　　　　　　　　　　　　　　　　　　　　　　　　　　　Equipment for General Use

_____ 6. A particular load has a nameplate rating of 31.2 amps at 120 volts. The circuit you have available to supply the load is 100 volts. The catalog number for a B-B transformer that can operate the load at its rated voltage is ___.

   a. 21107   b. 11106   c. 21106   d. 21105   e. 11105   f. 11112

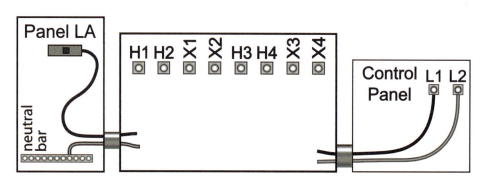

Figure 225.103

_____ 7. Figure 225.103, above, shows 120/208 volt Panel LA and a 120-volt Control Panel (CP). A circuit originated in the panel and was run a great distance out to the CP. The voltage has dropped to 95 volts at L1 & L2 in the CP. To properly boost this voltage back up to 120 volts you would select a ___ transformer from Yr-2 – Annex D.

   a. Group III   b. Group II   c. Group I   d. none of these

_____ 8. The Control Panel (CP), shown above in Figure 225.103, requires 120 volts to operate correctly, but due to the circuit length, from 120/208 volt Panel LA, the voltage has dropped to 100 volts at the CP terminals. Refer to the appropriate GROUP in Yr-2 – Annex D and determine which connection drawing, from among those shown below, is correctly connected to boost this voltage back up to 120 volts using a 500 VA transformer.

a. 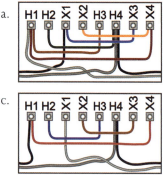   b.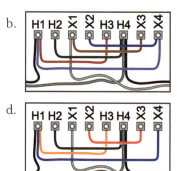

c.   d.

e. none of these

Year Two (Student Manual)

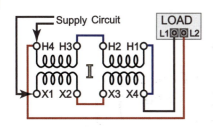 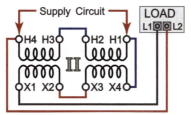 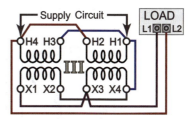

Figure 225.106

_____ 9. Using a Group II Buck-Boost transformer, which drawing, shown above in Figure 225.106, is correctly connected to supply rated voltage to the 240 volt load? The *Supply Circuit* is 208 volts.

    a. III
    b. II
    c. I
    d. none of these

_____ 10. Which connection diagram in Figure 225.106, shown above, should be used where the *Supply Circuit* is from a 2-pole breaker in a 120/240 volt, 3-ph, 4-W panelboard? The load is rated to operate at 208 volts. The transformer is a 120 × 240 – 16/32 volt.

    a. II
    b. III
    c. I
    d. none of these

_____ 11. You have a 240-volt circuit available to supply a 208-V, 65 A load. The catalog number for a B-B transformer that can operate the load at its rated voltage is ___.

    a. 21108
    b. 11109
    c. 21110
    d. 11112
    e. 21109
    f. 11108

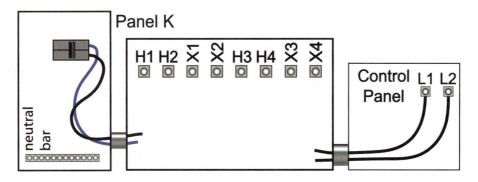

Figure 225.102

_____ 12. Refer to Figure 225.102, shown above, and the information in Yr-2 – Annex D for a 3.0 KVA buck-boost transformer. Panel K is 120/208V 3Ø, 4-W. The Control Panel requires 240 volts. Using a transformer from the appropriate GROUP, you should make your connections as shown at ____.

a. 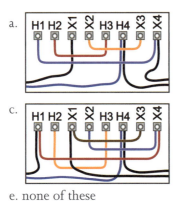 b.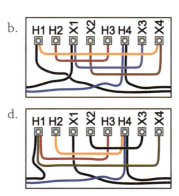

c.   d.

e. none of these

_____ 13. Refer to the information in Yr-2 – Annex D and Figure 225.102, shown above. Panel K is 120/208V 3Ø, 4-W. To supply the correct voltage to the 240-volt control panel, a ____ 3.0 KVA buck-boost transformer should be used.

a. GROUP III   b. GROUP II   c. GROUP I   d. none of these

_____ 14. Refer to Figure 225.102, shown above, and the information in Yr-2 – Annex D for a 1.5 KVA Buck-Boost transformer. The Control Panel requires **230** volts. Panel K is 120/208V 3Ø, 4-W. The Catalog number for this transformer is ____.

a. 21103   b. 11103   c. 11108   d. 21108   e. none of these

Equipment for General Use

_____ 15. The Control Panel, shown above in Figure 225.102, requires 208 volts. Panel K is 120/240V 3Ø, 4-W. Using a 3.0 KVA Buck-Boost transformer from the appropriate Group in Yr-2 – Annex D, you should make your connections as shown in Connection Diagram ____.

a. H   b. G   c. E   d. I   e. none of these

_____ 16. Refer to Figure 225.102 and the information from Yr-2 – Annex D for a 3.0 KVA buck-boost transformer. The control panel requires **230** volts. Panel K is 120/208V 3Ø, 4-W. Using a B-B transformer from the appropriate Group, you should make your connections as shown in Connection Diagram ____.

a. I   b. G   c. E   d. H   e. none of these

_____ 17. Panel K, shown in Figure 225.102, is 120/240V 3Ø, 4-W, however the Control Panel requires 208 volts. Using a 3.0 KVA Buck-Boost transformer from the appropriate GROUP in Yr-2 – Annex D, you should make your connections as shown at ____.

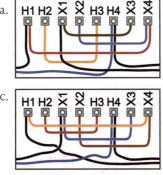
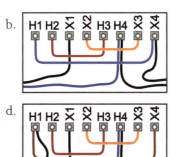

e. none of these

Scenario B-B 1

A 240-volt 3Ø load is to be supplied from a 208-volt 3Ø, 3-wire circuit. You will have to boost the circuit voltage to supply rated voltage to the load.

_____ 18. Refer to Yr-2 – Annex D and Scenario B-B 1, shown above, to answer this question. As a minimum, you will need to use ____ transformers to make your connections.

a. three
b. one
c. two

_____ 19. Refer to Scenario B-B 1, shown above, to answer this question. The better choice would be for you to should select ___ buck-boost transformers. Refer to Yr-2 – Annex D.

a. Group I   b. Group II

_____ 20. Refer to Yr-2 – Annex D and Scenario B-B 1, shown above, to answer this question. You should make your field connections using connection diagram ___.

a. A-A
b. F-F
c. B-B
d. C-C
e. G-G

_____ 21. From Yr-2 – Annex D, you have selected a catalog number 21110 transformer to boost the circuit voltage from 208 to 240 volts to serve a load. The 208 volt circuit originates in a 120/208 volt, 3Ø, 4-W panelboard. The largest breaker permitted by the NEC® for this circuit is a 2-pole ___ A.

a. 110
b. 150
c. 60
d. 125
e. 100

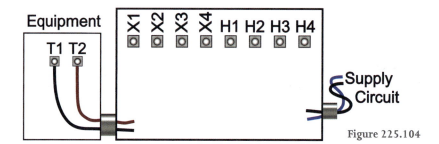

Figure 225.104

_____ 22. The equipment, shown above in Figure 225.104, is rated to operate at 240 volts and will be connected to an existing circuit. The supply circuit is from a 2-P breaker in a 120/240 V, 3Ø, 4W panel. The equipment replaces a piece of equipment that required less power. The additional current required by the new equipment has caused the circuit voltage to drop to 210 volts. You should ___.

a. select a Group I, B-B transformer and connect it to the equipment, as shown in Figure 225.104, so that rated voltage is applied to the load
b. connect the equipment *directly* to the supply circuit, and *do not* connect the transformer, shown above in Figure 225.104
c. select a Group II, B-B transformer and connect it to the equipment, as shown in Figure 225.104, so that rated voltage is applied to the load

Equipment for General Use

_____ 23. The supply circuit, shown on the previous page in Figure 225.104, is from a 2-Pole breaker in a 120/240 V, 3Ø, 4W panel. The foreign-made equipment is rated to operate at 220 volts. You should select a transformer from the appropriate Group and connect it as shown at ___, below.

a.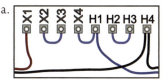
b.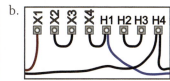
c.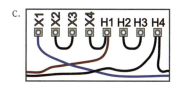

d.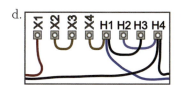
e.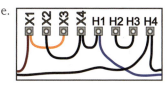
f. none of these

_____ 24. When used to buck or boost a 120-volt supply circuit, A 1.5 KVA, 120 x 240 – 12/24 V transformer can be connected to supply ___ different voltage outputs. The primary windings are supplied with rated, or close to rated, voltage.

a. 4
b. 2
c. 8

_____ 25. A ___ KVA 120 x 240–12/24 V transformer has two X1 leads.
I. 10
II. 7.5

a. I only
b. II only
c. neither I nor II
d. both I & II

Match the 1Ø connection diagram, shown in Yr-2 – Annex D, to the statement. The connection diagrams may apply to more than one statement.

    a. Connection Diagram C   e. Connection Diagram D
    b. Connection Diagram E   f. Connection Diagram F
    c. Connection Diagram G   g. Connection Diagram H
    d. Connection Diagram I   h. Connection Diagram J

_____ 26. Supply circuit voltage to the Group II transformer is 208, but load needs 240 volts.

_____ 27. Group III transformer has 503 volt supply circuit, but load needs 480 volts.

_____ 28. Group II transformer has 240 volt supply circuit, but load needs 208 volts.

_____ 29. Supply circuit voltage to the Group I transformer is 110, but load needs 120 volts.

_____ 30. Supply circuit voltage to the Group III transformer is 438, but load needs 480 volts.

_____ 31. Group I transformer has 133 volt supply circuit, but load needs 120 volts.

Select the smallest 120 × 240 – 12/24 volt transformer you could install to meet the requirements in the statement.

    a. 0.05 KVA   h. 1.5 KVA
    b. 0.10 KVA   i. 2.0 KVA
    c. 0.15 KVA   j. 3.0 KVA
    d. 0.25 KVA   k. 5.0 KVA
    e. .050 KVA   l. 7.5 KVA
    f. 0.75 KVA   m. 10.0KVA
    g. 1.0 KVA

_____ 32. A 208-V supply circuit connected to a ___ transformer would supply enough power to a 230-volt, 13-amp motor

_____ 33. A 99-V supply circuit connected to a ___ transformer would supply enough power to a 120-volt, 5.5-amp load

_____ 34. A 96-V supply circuit connected to a ___ transformer would supply enough power to a 120-volt, 300-amp load

Lesson 225 Safety Worksheet

Excavations and Trenching

Accidents involving excavations or trenching are usually very serious and often deadly. You only have to watch your local news broadcast or read a local paper to learn of these tragedies. Far too many workers either have not been trained in the hazards involved or knowingly take the risks that too often turn deadly. On average as many as 200 deaths per year are due to these accidents.

Working in an unprotected excavation is just plain risky. The walls of the excavation can suddenly cave-in, often trapping workers by engulfing them in earth. The weight of the earth on your chest is too heavy to allow you to breathe. If you are totally engulfed, your fellow workers may not know exactly where your body and extremities are. If they elect to try to remove you by excavating more earth, they run the risk of amputating your extremities or worse. The time that rescue can be successful is very short, before unconsciousness sets in, and then death due to asphyxiation.

A broad knowledge of soils and soil types is necessary to be able to identify the dangers involved in excavating. Various types of soils react differently to excavating. Solid rock can be very stable, whereas granular soils are very unstable, and other types in between. Other elements affect the stability of excavations as well. Water from the surrounding surfaces and from the ground in the excavations can create a very dangerous condition, possibly indicating potential cave-in. Any spoils removed from the excavation must be placed well back from the excavation to prevent it from falling into the excavation, possibly onto workers in the excavation.

The exact location of any underground utilities must be determined prior to opening any excavation. This OSHA regulation reinforces state laws requiring the same. Most states have a one-call center to coordinate this procedure with the utility companies. These centers will need information provided by you concerning the exact location you will be excavating. You will provide additional information as to where on the property you need the locations marked. They will tell you which utilities will be notified and which will not be notified. The center will also provide you with a ticket or case number for reference. You must provide 24 to 48 hours for this to be completed. Verify which time frame is correct for your area.

If you have to work in an excavation in other than stable rock, there must be protection from cave-in by either some form of benching, sloping, shoring, or protective shields. Any shoring or shields shall be installed in accordance with the manufacturers instructions and have strength to withstand the forces that could be placed on it.

Egress and access to the excavation must be provided and maintained. In trenches that are 4 feet or more in depth must have a ramp, stairway or ladder within 25 feet of workers in the trench. For trenches that are more than 6 feet deep, a walkway with proper guardrails must be provided for crossing the trench. **You are not permitted to jump over the trench**. Reflective vests must be worn if you are in a vicinity of vehicular traffic.

Equipment for General Use

Questions

_____ 1. If you are working in a trench ___ feet or more in depth, you must have a means of egress within 25 feet of where you are working.

    a. 3   b. 5   c. 4   d. 6

_____ 2. The exact ___ of all underground utilities must be determined prior to the opening of any excavation.

    a. routing   b. depth   c. dimensions   d. all of these

_____ 3. Most states have a ___ center to notify the utilities of your planned excavation.

    a. one-call   b. drop off   c. one-stop   d. data

_____ 4. A broad knowledge of soils and soil types is necessary to be able to ___ the hazards involved in excavating.

    a. count   b. quantify   c. identify   d. eliminate

_____ 5. Many workers ___ involved with excavation work.

    a. take deadly risks while
    b. are not trained in the hazards
    c. either a or b
    d. neither a nor b

_____ 6. Working in excavations requires the use of ___ if the excavation is not made in stable rock.

    a. shoring
    b. sloping
    c. Benching
    d. protective shields
    e. any of these
    f. none of these

_____ 7. Accidents involving excavating and trenching contribute to an average of ___ deaths per year.

    a. 30   b. 200   c. 30,000   d. 2,000

Lesson 226
Generators and Transfer Switches

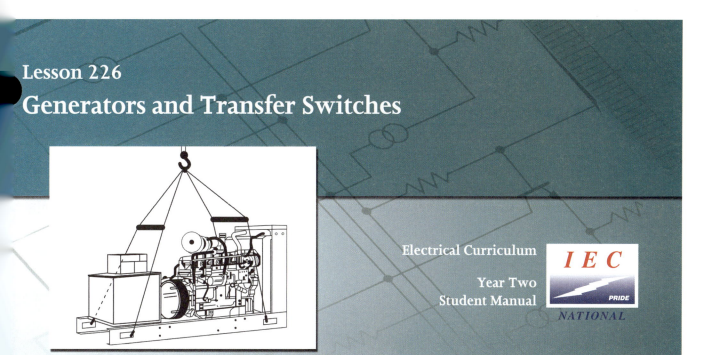

Electrical Curriculum

Year Two
Student Manual

Purpose

This lesson will introduce the student to applications, installations, NEC® requirements for generators and transfer switches.

Homework

(Due at the beginning of this class)

For this lesson, you should:

- Thoroughly read the material contained within Lesson 226.
- Re-read "Generators – NEC® 250.34" in Electrical Systems Chapter 6.
- Read "Generators – NEC® Article 445" in Electrical Systems Chapter 11.
- Read the following in the NEC®:
  - Section 700.1 with its FPN No. 3
  - Section 701.1 and 701.2 with its FPN
  - Section 702.1 and 702.2 with its FPN
- Complete Objective 226.1 Worksheet.
- Complete Objective 226.2 Worksheet.
- Complete Lesson 226 NEC® Worksheet.
- Read and complete Lesson 226 Safety Worksheet.

- Complete additional assignments or worksheets if available and directed by your instructor.

Objectives

By the end of this lesson, you should:

    226.1 ....Understand the basic purpose of standby generator sets and the installation criteria for these systems.
    226.2 ....Understand the electrical and transfer switch installation requirements for generator systems.

Objective 226.1   Generator Sets

Overview

Having reliable electric power is important to almost all businesses but absolutely required for many others. Ordinary businesses have electrical needs that must be met to maintain day-to-day business functions. For example, lighting, heating and air conditioning, and computers all require electric power. If this power is not reliable, then the business is interrupted, thereby affecting safety, customer service, production, and even employee morale.

Businesses such as hospitals, surgery centers, computer data centers, refrigerated warehouses, and others where operations need to be maintained, must have consistently reliable electric power.

Many businesses have "standby generators" installed to provide electric power if and when the utility power is not reliable, or to provide electric power when an emergency situation causes disruption of available utility power. These emergency situations may be the result of extreme weather conditions such as ice and snow-storms, heat waves, tornadoes, hurricanes, floods, wind, lightning, motor vehicle accidents, fires, catastrophes, and more. When these situations occur, standby generators are designed to transfer the electric power that they generate to the residence or business to which they are connected. Some systems start automatically, while others are manually operated. These units are available in many sizes (output power ratings) and many output voltages, from small units that are used to provide power to a residence or small business, to very large units that provide power for utility company distribution. Generators with output power ratings given in KW are generally used to power loads with power factors that are close to 1.0. Generators with power output ratings in KVA are used to power inductive loads, such as motors.

These generator sets come from many manufacturers and several countries. They may be fueled by natural gas, propane (liquefied petroleum gas – LPG), or diesel fuel. Typically these units have a fuel-propelled prime mover for propulsion along with an alternator to produce the electricity. Although the prime mover is typically an internal combustion engine, it could be an electric motor that is supplied power from an alternate power source, such as solar cells. As applied to generators, a *prime mover* is an engine or motor and its function is to turn the alternator so that a voltage is generated. Additional peripheral equipment and controls are required to make the entire system work properly. More sophisticated generator sets can be used to co-generate electricity with the utility companies. These units can be used to reduce the peak demand seen by the customer or to reduce the possible damages from brown-outs and other conditions. They can even be used to sell power to the utility company during times of peak load or other situations.

Generator sets require adherence to many standards and codes for safe and efficient operation. Standards that apply to installation, maintenance and safe operation (specifically for United States installations) are as follows:

- NFPA 54 National Fuel Gas Code
- NFPA 70 National Electrical Code
- NFPA 99 Standard for Health Care Facilities
- NFPA 100 Life Safety Code
- NFPA 110 Emergency and Standby Power Systems.

- UL 486A-486B Wire Connectors
- UL 486E Equipment Wiring Terminals for Use with Aluminum and/or Copper Conductors
- UL 2200 Stationary Engine Generator Assemblies

Safety considerations are very important and must be followed. The manufacturer's installation instructions provide a list of safety precautions. Additional information can be found on the manufacturer's website, which is listed in the instruction manual that accompanies the Gen-Set. Extreme caution must be exercised around all rotating and moving parts. Refer to the manufacturer's glossary or list of abbreviations for an explanation of specific equipment terms.

Information regarding dimensions, weight, exhaust outlet size, battery rating, fuel supply lines, and air requirements are all key considerations for efficient and safe generator installation.

Selection Criteria

Generator Sets need to be selected by a number of criteria. The full load requirements to be provided by the generator set must be determined and calculated. The correct voltage necessary must be selected. Is the unit going to be installed indoors or will it be installed outdoors in an enclosure? If an enclosure is required, a heater may be necessary to keep the enclosure heated to allow the generator and related equipment to be at optimum operating condition. What will be the fuel source, diesel, propane or natural gas? Will the fuel need a storage tank? Will this tank be installed underground? If diesel is the fuel source, then filters and auxiliary items are required. Will the unit require a special concrete (inertia) isolation pad? What kind of a skid or base is supplied with the unit? Is this unit being provided for standby power, as addressed in NEC® Article 702; or is it a legally required standby (emergency) system, as addressed in NEC® Article 701?

Cooling

Large internal combustion engines and connected alternator units (generator sets), such as the one shown in Figure 226.101, require efficient cooling systems. Just as your car may overheat if its radiator is starved for fresh air or has insufficient coolant in the radiator, these generator sets must be provided with engineered cooling systems. Some radiators are attached to the skid the unit is mounted on while others may be several feet from the working machinery. All conduits and other piping must be carefully planned and installed in such a manner as to not inhibit the installation or operation of the cooling, fuel, or exhaust systems.

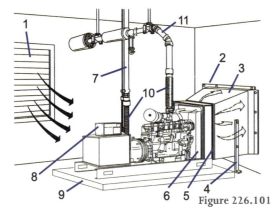

Radiator-Cooled Generator Set Installation

Figure 226.101

1. Air inlet opening
2. Ductwork mounting flange
3. Air outlet duct
4. Support legs
5. Flexible section with radiator duct flange
6. Pusher fan
7. Electrical conduit
8. Electrical Controller
9. Mounting base
10. Flexible sections
11. Exhaust piping

Generators and Transfer Switches

Enclosed rooms or compartments may have auxiliary exhaust fans that move excess heat from the equipment and surrounding areas. Duct work and louvers may be required to channel the air flow both into and out of the generator location. Remote radiators require electrically powered cooling fans and may be some distance from the generator set.

When increasing the length of radiator piping, larger diameter pipe may be required. Surge tanks for radiator coolant may be required for horizontally mounted radiators to provide venting, expansion and filling makeup functions.

Exhaust systems require a series of components that include a silencer, water trap, drain petcock and flexible section to provide safe passage of exhaust gasses. These system components are positioned for effective operation and should be avoided when installing electrical conduits. Exhaust piping can have temperatures in excess of 500° F and require proper ventilation and clearances from combustible materials, as well as electrical system conductors and components.

Excess horizontal and vertical exhaust piping distances produce increased back pressure. This condition may require up-sizing of the exhaust piping, reducing the number of elbows in the line or re-routing the entire system.

Fuel Systems Requirements

Diesel fueled units, such as the one shown in Figure 226.102, often have fuel storage tanks underground and may be located some distance from the generator skid. Some smaller gen-sets may have a sub-base fuel tank mounted on or under the mounting skid. Since diesel fuel is less volatile than liquefied petroleum gas (LPG), natural gas, or gasoline, day tanks or transfer tanks may be located near the generator unit. Since the fuel that is pumped from the main storage tank may have a long distance to travel, a day tank is placed near the generator unit. This tank allows for easier transfer of fuel to the engine and provides for easier access for servicing filters and draining sediment from the fuel lines. The day tank may supply up to four hours of fuel and may serve to increase or decrease the fuel temperature for optimum performance. Number 2 diesel fuel is also used in many oil-fired heating units or boilers. The same fuel supply and storage may be used for both systems.

1. Injector return line
2. Day tank vent
3. Day tank
4. Auxiliary fuel pump
5. Tank drain
6. Electric fuel level control switch
7. Fuel supply line from day tank to engine connection
8. Fuel supply line from main fuel tank to day tank
9. Overflow line
10. Foot valve
11. Main fuel storage tank
12. Fuel tank vent
13. Tank filling inlet

Diesel Fuel System

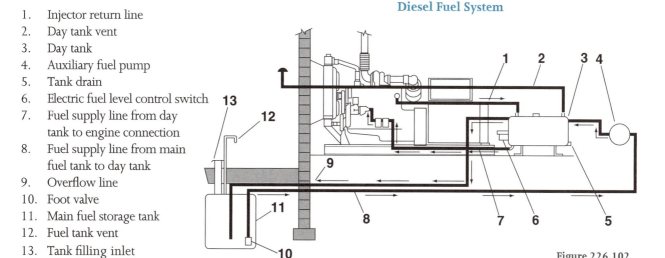

Figure 226.102

Black iron piping of the smallest size will deliver adequate fuel without injecting air into the fuel line. Diesel fuel reacts with galvanized pipes and tanks and causes clogs in filters and fuel pumps.

Compared to diesel fuel, the use of gasoline as a fuel has more restrictive requirements for storage locations and volumes. Gasoline deteriorates after six months and should be stored in smaller quantities. For these reasons, gasoline is typically not used for large generating equipment.

While providing conveniences, liquid petroleum (LP) and natural gas fuels create other hazards when used on equipment that operates inside of or near buildings. LP gas is heavier than air and may accumulate in low areas that are poorly ventilated, which could create explosive conditions.

Some portions of generator systems may be hazardous (classified) locations. In these locations, electrical installations are made according to the NEC®.

Installation Lifting and Placement Requirements

The generator set is a complex and heavy machine. Some can be the size of a small locomotive and yet be placed into tight quarters by large lifting cranes. Lifting the generator set requires attention to detail so as not to damage components such as shrouds, radiators, and fuel tanks during the installation process.

Lifting eyes and lifting bars are placed for the specific purpose of proper lifting and not allowing slippage or damage to the unit. Spreader bars, shown at point 1 in Figure 226.103, may be necessary to protect the generator set. Some equipment may be mounted over a sub-base fuel tank which may be installed separately or together as one unit.

Vibration isolation and inertia absorption requires substantial mounting bases equipped with vibration isolators. Some large units may have concrete inertia bases up to two feet thick, resting on a gravel bed beneath the base.

Generator Set with Lifting Hooks in Skid

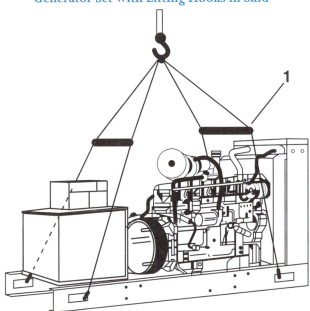

Figure 226.103

Generators and Transfer Switches

Objective 226.2 Generator Systems – Electrical Equipment and Connections

NEC® Overview

NEC® Article 445 addresses the installation of generators and Article 250 addresses the grounding and bonding of generators.

NEC® Article 700 addresses emergency systems. The requirements set forth in this article are for those systems **essential for safety to life**. Wiring for emergency systems must be kept entirely independent of other wiring.

NEC® Article 701 addresses legally required standby systems. The requirements set forth in this article are intended to provide electric power to aid in fire fighting, rescue operations, control of health hazards, and similar operations. The requirements for legally required standby systems are similar to the requirements for emergency systems, with some differences. When normal power is lost, legally required systems must be able to supply standby power in 60 seconds or less, instead of the 10 seconds or less required of emergency systems. Wiring for legally required standby systems is permitted to occupy the same raceways, cables, boxes, and cabinets as other general wiring.

NEC® Article 702 addresses optional standby systems. The requirements set forth in this article are those in which failure can cause physical discomfort to personnel or customers, serious interruption of an industrial process, damage to process equipment, or disruption of business operations.

Batteries

The energy source required to start a generator set is a properly sized, tested and maintained battery or batteries. An enclosed generator, with the battery system compartment cover removed, is shown in Figure 226.201. The placement of batteries should be:

- In a clean, dry, and stable temperature location
- Provided with easy access to battery caps for checking electrolyte levels
- Close to the generator set to keep cables short

Batteries are usually of the lead-acid type and should be able to crank single 45-second cycles for sets below 15KW and 15-second cycles separated by 15 second rests for larger models. Nickel-cadmium batteries are sometimes used because of their long life (20 years).

Battery chargers may be slow trickle chargers powered by an AC source and may also include engine driven charging units to restore the charge used in a normal cranking cycle.

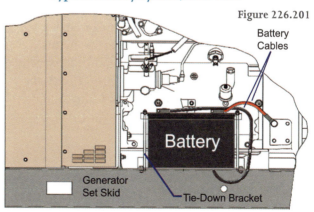

Typical Battery System, Side View

Figure 226.201

Electrical Connections

Electrical connections to generators may be provided through a transfer switch and/or other accessories. All AC circuits from a generator must include a circuit breaker or fuse protection. The circuit breaker or fuse must open all ungrounded conductors.

> **CAUTION:** The generator set must **always** be disabled before any servicing or repair operations are performed. Disconnect the battery cables (negative lead first) to prevent accidental starting. Disable or turn off all master controls to the generator set before servicing. This is one of the requirements of Lock-out/Tag-out so that a *zero energy state* is achieved.

Accessory packages for generator sets include audio-visual alarms, bus bar kits, common failure relay kits, controller (customer) connection kits, float/equalize battery charger kit, line circuit breaker, low fuel switch, remote annunciator kit, remote emergency stop kit, run relay kit, Safe-guard breaker, single-relay dry contact kit, and ten-relay dry contact kit. These accessories serve to enhance safety and notification of operation and/or failure of the generator set.

Typical Electrical Conduit Feeds

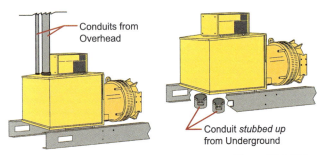

Figure 226.202

The connections to the control panel and load terminals of the generator set may feed from above or from beneath the generator set, as shown in Figure 226.202. Rough-in details should be obtained from manufacturer drawings, spec sheets, and drawings provided for the installation. Be aware of the movement or vibration that may occur during operation of the generator set. This may require portions of the electrical raceways to be flexible.

Bus bar kits are offered for most generator sets and may aid in the electrical installation. Three load busses and one neutral bus are available in such kits.

Grounding and Grounded Conductor Connections

A permanently installed AC generator is considered to be a separately derived systems when the grounded conductor is bonded to the generator frame. In this instance, a transfer switch would be required to switch the neutral so that there would be no direct electrical connection between the normal system neutral and the generator neutral.

Grounding requirements for separately derived systems are addressed in NEC® Section 250.30. Review Electrical Systems Chapter 6, Separately Derived Systems – NEC® 250.30.

Generator Set Equipment Grounding Connection

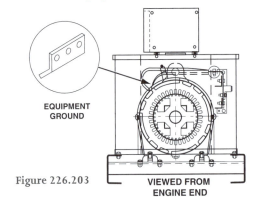

Figure 226.203

Generators and Transfer Switches

Most installations require the grounded conductor (neutral) to be grounded. With the installation of a system bonding jumper, the system grounding conductor is connected to the equipment grounding connector on the alternator. An external connection point for connection to the grounding electrode is typically available, as shown in Figure 226.203.

Applications, system requirements and the type of transfer switch used, determines the grounding of the neutral at the generator set.

NEC® 250.20(D) requires that separately derived systems be grounded as specified in 250.30. FPN No. 1 to 250.20(D) states that: *An alternate ac power source such as an on-site generator is not a separately derived system if the neutral is solidly interconnected to a service-supplied system neutral.* A generator that is **NOT** a separately derived system is shown in Figure 226.204(A). In this drawing note that the neutral conductor from the generator to the load is not disconnected by the transfer switch. There is a direct electrical connection between the normal system (service) grounded conductor (neutral) and the generator neutral through the neutral bar in the transfer switch, thereby grounding the generator neutral through the main bonding jumper in the service panel. Because the generator is grounded by connection to the normal system ground, it is NOT a

3-Pole Transfer Switch NOT a Separately Derived System

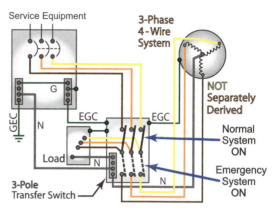

Figure 226.204(A)

separately derived system, and there are no requirements for grounding the neutral at the generator. Under these conditions, it is necessary to run an equipment grounding conductor from the service equipment to the 3-pole transfer switch and from the 3-pole transfer switch to the generator.

A generator that IS a separately derived system is shown in Figure 226.204(B). Here, the grounded conductor (neutral) is connected to the switching contacts of a 4-pole transfer switch. Therefore, the generator system **does not** have a direct electrical connection to the normal service supply system grounded conductor (neutral), and the system supplied by the generator is considered

Separately Derived System and a 4-Pole Transfer Switch

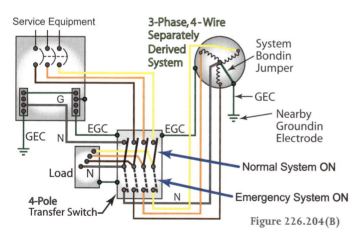

Figure 226.204(B)

separately derived. This separately derived system (3-phase, 4-wire, wye-connected system that supplies line-to-neutral loads) is required to be grounded in accordance with NEC® 250.20(B) and 250.20(D). The methods for grounding the system are specified in 250.30(A).

NEC® Section 250.30(A)(1) requires separately derived systems to have a system bonding jumper connected between the generator frame and the grounded circuit conductor (neutral), as shown in Figure 226.204(B). The grounding electrode conductor from the generator is required to be connected to a grounding electrode. This grounding electrode should be located as close to the generator as practicable. If the generator is in a building, the preferred grounding electrode is required to be either an effectively grounded structural metal member, or the first 5 ft of water pipe into a building where the piping is effectively grounded. (Note that the exception to 250.52(A)(1) permits the grounding connection to the water piping beyond the first 5 ft.) For buildings or structures in which the preferred electrodes are not available, the choice can be made from any of the grounding electrodes specified in 250.52(A)(3) through 250.52(A)(7).

Grounding requirements for portable and vehicle-mounted generators are addressed in NEC® Section 250.34. Review Electrical Systems Chapter 6, Generators – NEC® 250.34.

Pin-and-Sleeve Male Inlet

4P-5W
3ØY277/480
100 AMP   Side View

Figure 226.205

Bay Colony Elementary School Plans

Note on the BCES plans that provisions are made for connection of a portable generator to supply emergency power to Panel HG-Transformer XG-Panel LG. Panel HG serves emergency and exit lights in the cafetorium and the kitchen. Panel LG supplies the refrigeration rack, the lights, and the door heaters for the walk-in cooler/freezer. Refer to the panel schedules to determine the circuits to the lighting and equipment. Connection of the portable generator will maintain power to the cooler/freezer and the emergency ballasts in the lights. The generator is connected through a "receptacle" that is shown on the exterior wall on the east side of Area D (Drawing E3.4). The "receptacle" is actually called a pin-and-sleeve male inlet. A Leviton 5100B7W inlet is shown in Figure 226.205. FYI: *The approximate cost of this one item is* $500. On the One-Line Diagram (E5.1), the disconnect switch/inlet is shown to be 100A, 4Ø, 5W, N3R. The inlet rating is actually: 100A, 3Ø, 4-Pole, 5-wire, with the fifth wire being the equipment grounding conductor.

Combination Inlet/ Safety Switch

Figure 226.206

BCES Drawing E5.1 should be corrected to show: *100A, 3Ø, 4-Pole, 5-wire.*

The inlet for BCES can be a part of the disconnecting means (safety switch), as shown in Figure 226.206, or it can be mounted in a box adjacent to the disconnect switch. In either case, the safety switch then feeds the manual transfer switch (located adjacent to Panel HG).

Refer to the Fixture Schedule on BCES Dwg E4.1. The CE luminaires (fixtures) are shown to be Lithonia type 2SPG-**3**32-A12125-277-GEB10**EL**. The part number indicates that these are **3**-lamp fixtures with an emergency (battery backup) ballast. Remark 5 requires that the emergency ballast be connected only to the center lamp in this CE fixture. As shown on the lighting plan (E2.4) these CE fixtures are to be installed in the cafetorium where normal circuits D-15 or D-17 are connected to the 2 outer (outboard) lamps and emergency circuit HG-1 is connected to the inner (inboard) lamp. Note that, although the outboard lamps are switched with the room lighting switches, the inboard lamp is on 24/7, because circuit HG-1 is connected directly from the panel. Consequently, during periods when normal power is available, the inboard lamps serve as "nightlights". The battery in the emergency ballasts powers the inboard lamps when normal power is lost. Now the inboard lamps serve as emergency lights so that people can safely egress the cafetorium via the exit doors on the north wall.

Again, refer to the Fixture Schedule on BCES Dwg E4.1. The VE luminaires (fixtures) are shown to be Lithonia type 2SRT-G-**4**32-A12125-277-GEB10**EL**. These are special sealed troffer (lay-in) luminaires. These are suitable for use in contamination-controlled environments (such as food preparation areas). Care should be taken to install the correct fixtures as shown on the plans. The part number indicates that these are **4**-lamp fixtures with an emergency (battery backup) ballast. Remark 4 is shown on the Fixture Schedule for these fixtures. THIS IS INCORRECT. For this installation, the center lamps will be required to operate as emergency lights.

> **Make the following corrections BCES Drawing E4.1:**
> - In the Remarks column, change Remark 4 to Remark 8 on the VE row
> - Write a new Remark 8 below the table:
>   "ONE 2-LAMP ELECTRONIC BALLAST FOR OUTBOARD LAMPS AND ONE BODINE #850 (OR EQUAL) 2-LAMP EMERGENCY BALLAST SUITABLE FOR 90 MIN OPERATION FOR INBOARD LAMPS

As shown on the lighting plan (E2.4) these VE fixtures are to be installed in the kitchen. Keyed Note 8 indicates that these fixtures are to be connected to the unswitched night light circuit. Emergency circuit HG-1 is to be connected to the two outer (outboard) lamps and to the two inner (inboard) lamps. During periods when normal power is available, both the inboard and outboard lamps serve as "nightlights". The battery in the emergency ballasts powers only the inboard lamps when normal power is lost. Again, the inboard lamps serve as emergency lights so that people can safely egress the kitchen via the exit door on the east wall. When the generator is plugged into the inlet, the walk-in cooler/freezer will be powered, and all four lamps in the VE fixtures will be ON to permit personnel to have a safely illuminated access to the walk-ins.

Transfer Switches

Transfer switches are used to protect critical electrical loads against loss of power. The load's normal power source is backed up by a secondary (emergency) power source. A transfer switch is connected to both the normal and emergency sources and supplies the load with power from one of these two sources. In the event that power is lost from the normal source, the transfer switch transfers the load to the secondary source. Transfer can be automatic or manual, depending upon the type of transfer switch equipment being used. Once normal power is restored, the load is transferred back to the normal power source.

Transfer switches typically have three states, or positions, ON-OFF-ON. When in one ON position, the load is supplied from the normal source of power, generally the utility. When in the other ON position, the load is supplied from the standby power source, typically a generator. Both power sources can be disconnected from the load if the switch is in the OFF position. Transfer switches are available in various ampere ratings and voltage ratings. These are also specified to be shipped with the appropriate number of poles (2-pole for single-phase, and 3-pole for 3-phase circuits). Note that, as previously discussed, an additional pole will be required if the standby generator is a separately derived system.

Manual Transfer Switches

Manual transfer switches are referred to as Double-Throw, center-OFF switches, if equipped with an OFF position.

Transfer switches can be of the 2-pole type, such as that shown in Figures 226.207(A) and (B). This type would be used on a 240-volt, single-phase circuit. This switch would be a DPDT (double-pole, double-throw) switch. If the load circuit is 3-phase, a 3PDT switch would be required. Both of these switches have a neutral kit available for either factory or field installation. The common point within these switches is where the load circuit is connected. The "Line 1" connections are where the normal power is connected, and the "Line 2" connections are where the standby by power is connected.

Double-Pole (Manual Transfer) Safety Switch

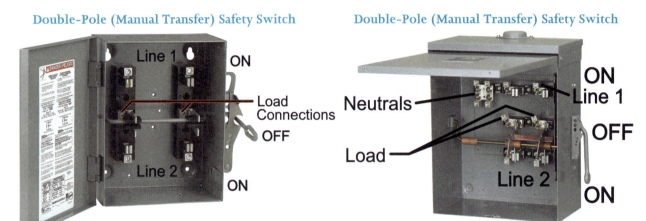

Figure 226.207(A)   Figure 226.207(B)

Automatic Transfer Switches

The first automatic transfer switch (ATS) was introduced in 1920. ATS's come in many sizes and several types that are designated by their application. The ATS is used to transfer loads between alternate power sources in a safe and efficient manner. These double-throw switches provide reliable operation while coordinating the starting, running, and stopping of emergency engine generator sets. A typical ATS is shown in Figure 226.208.

Automatic Transfer Switch (ATS)

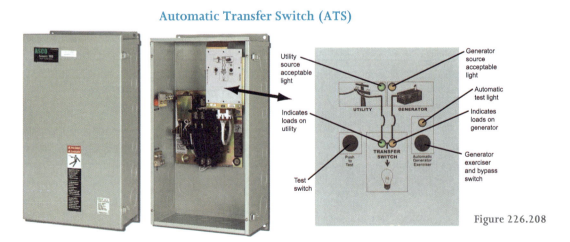

Figure 226.208

Generators and Transfer Switches

As with manual switches, an automatic transfer switch is connected to both the normal and emergency sources and supplies the load with power from one of these two sources. In the event that power is lost from the normal source, the automatic transfer switch's intelligence system initiates the transfer when normal power falls below a preset voltage or frequency. If the emergency source is a standby generator, the transfer switch initiates generator starting and transfers to the emergency source when sufficient generator voltage is available. When normal power is restored, the transfer switch automatically transfers back and initiates engine shutdown.

An automatic transfer switch consists of three basic elements:

- Main contacts to connect and disconnect the load to and from the source of power.
- A transfer mechanism to affect the transfer of the main contacts from source to source.
- Intelligence/supervisory circuits to constantly monitor the condition of the power sources and thus provide the intelligence necessary for the switch and related circuit operation.

A transfer switch is a critical component of any emergency or standby power system. When the normal (preferred) source of power is lost, a transfer switch quickly and safely shifts the load circuit from the normal source of power to the emergency (alternate) source of power. This permits critical loads to continue running with minimal or no outage. After the normal source of power has been restored, the re-transfer process returns the load circuit to the normal power source. Transfer switches are available with different operational modes including:

- **Manual -** Manually operated transfer switches are designed for a variety of standby power applications for critical loads. In the event of a primary power source interruption, the user can manually transfer the load circuits to the standby power source. Once primary power has been restored, the user can manually transfer the load circuits back to the primary power source.

- **Non-automatic -** Non-Automatic Transfer Switches are designed for a variety of standby power applications for critical loads. In the event of a primary power source interruption, the user can manually transfer the load circuits to the standby power source through the use of an external pushbutton. Once primary power has been restored, the user can manually transfer the load circuits back to the primary power source through the use of an external pushbutton.

- **Automatic -** Automatic Transfer Switches are designed for standby power applications for both critical and non-critical loads. They monitor both Source 1 (Normal) and Source 2 (Emergency) power sources. In the event of a Source 1 power interruption, these switches will automatically transfer the load circuits to the Source 2 power source. Once the Source 1 power source has been restored, the process is automatically reversed.

- **Bypass isolation -** The bypass isolation switch is designed for applications where maintenance, inspection, and testing must be performed while maintaining continuous power to the load. This is typically required in critical life support systems and standby power situations calling for safe system maintenance with no power disruptions. Such a design allows for the quick removal of the different switching devices for inspection, maintenance, or replacement.

- **Soft load** - Soft Load Automatic Transfer Switches are closed transition transfer switches. Unlike traditional open transition switches that provide a break-before-make operation, the closed transition soft load switch allows two power sources, usually the utility and a generator set, to be paralleled indefinitely. This permits the load, inductive or resistive, to be gradually and seamlessly transferred from one source to another. All of this is accomplished through the make-before-break operation of the switch with no power interruption to the load.

- **Maintenance Bypass** – A Maintenance Bypass Switch provides a simple and effective means for bypassing uninterruptible power supplies while maintaining continuity of power to critical computer loads. A maintenance bypass switch is a requirement on every uninterruptible power supply (UPS) installation in order to accommodate the maintenance and testing of the UPS system.

The power switching operation of transfer switches may be separated into the three key categories.

- **Open Transition — Break-before-Make operation-** The Open Transition ATS makes the transfer of power when the normal source has failed. There is a time delay while the engine generator is starting. This switch disconnects from one source before connecting to the emergency source.

- **Closed Transition — Make-before-Break operation-** This is a test scenario when the test switch is operated. During the transfer to the emergency source the engine generator set starts. Both sources come within acceptable parameters and then the emergency contacts close. Following a time delay of more than 100 ms, the normal source contacts open.

The advantage of this transition is that there is passive synchronization without in-phase monitoring and the transfer occurs without interruption of the power source and has minimal switching transients.

- **Closed Transition Soft Load —** Both sources are paralleled and can remain so indefinitely- These ATS units match and lock frequency, phase angle, and voltage. They add load by KW and power factor and ramp up the load to a set point. They may remain connected or they can disconnect both sources. The generator and utility sources operate in parallel. This transfer switch type is used for smooth non-disruptive planned transfer.

Utility — Generator

Transfer switches are traditionally applied between a utility and a generator set for emergency and standby power systems. *See Figure 226.209.*

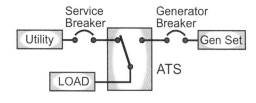

Figure 226.209

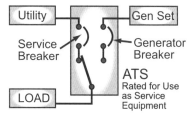

Figure 226.210

Generators and Transfer Switches

Some automatic transfer switches are suitable for use as service equipment and contain OCPDs. *See Figure 226.210.*

Generator — Generator

Transfer switches are sometimes applied between two generator sets for prime power use, often in remote installations. In such applications, source power is periodically alternated between the generator sets to equally share run-time. *See Figure 226.211.*

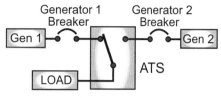

Figure 226.211

Generators and Transfer Switches

Objective 226.1 Worksheet

_____ 1. Exhaust systems for generator sets require a ___ to provide safe passage of exhaust gasses.

    a. drain petcock
    b. water trap
    c. flexible section
    d. silencer
    e. all of these

_____ 2. Typically generator sets are fueled by ___ or diesel fuel.

    a. wind
    b. natural gas
    c. water
    d. any of these
    e. none of these

_____ 3. Standards for the installation, maintenance and safe operation of generator sets are for ___.

    a. new installations only
    b. foreign installations
    c. US and foreign installations worldwide
    d. new and retro-fit installations
    e. US installations

_____ 4. A prime mover is the ___ component of a generator set.

    a. fuel
    b. engine or motor
    c. starting battery
    d. alternatore. all of these
    f. none of these

_____ 5. More sophisticated gen-sets can be used to ___ electricity with the utility companies.

    a. cooperate
    b. convert
    c. co-generate
    d. use
    e. all of these
    f. none of these

_____ 6. Horizontally mounted radiators on generators sets may require ___.

   a. overflow tubes
   b. drain petcocks
   c. antifreeze
   d. surge tanks
   e. pressure caps

_____ 7. Lifting generator sets during installation requires special care to avoid damage to ___.

   a. fuel tanks   b. shrouds   c. radiators   d. all of these   e. none of these

_____ 8. Selection criteria considerations include ___ requirements.

   a. full load
   b. isolation pad
   c. voltage
   d. fuel source
   e. all of these
   f. none of these

_____ 9. Day fuel tanks can supply up to ___ of fuel to the generator unit.

   a. 2 days   b. 10 minutes   c. four hours   d. 60 seconds   e. none of these

_____ 10. The output rating of generators is given on the nameplate in ___.
   I. KVA   II. KW

   a. II only   b. neither I nor II   c. I only   d. either I or II

_____ 11. Installation of electrical conduit must maintain adequate clearance from generator ___.

   a. cooling systems
   b. fuel systems
   c. exhaust systems
   d. all of these
   e. none of these

Generators and Transfer Switches

_____ 12. Compared to gasoline, diesel fuel is ___ volatile.

a. equally   b. less   c. more

_____ 13. Some of the standards required to install gen-sets are ___.

a. NFPA54
b. UL2200
c. NFPA100
d. NFPA70
e. NFPA110
f. all of these

_____ 14. Exhaust systems for generator sets require proper ventilation and clearances from combustible materials because ___.

a. of carbon-monoxide gases
b. the piping temperatures exceed 500° F
c. the piping temperatures are subnormal
d. vibration is excessive
e. noise exceeds 100 decibels

_____ 15. In generator set exhaust systems, excess horizontal and vertical exhaust piping distances produce increased back pressure and may require ___.

a. the number of elbows to be reduced
b. an increase of exhaust piping size
c. re-routing the entire system
d. any of these
e. none of these

_____ 16. A large concrete base, up to 2 feet thick, used to support and mount the generator unit, is called a(an) ___.

a. inertia base
b. ufer base
c. grounding electrode base
d. foundation
e. footing base

_____ 17. Businesses such as ___, where operations need to be maintained, must have consistent reliable electric power.

   a. hospitals
   b. surgery centers
   c. refrigerated warehouses
   d. computer data centers
   e. all of these
   f. none of these

_____ 18. Compared to diesel, gasoline is a less desirable fuel for large generators because ___.

   a. storage locations have more restrictive requirements
   b. it should be stored in smaller quantities
   c. it deteriorates after six months
   d. all of these

_____ 19. When operated within or near buildings, the use of liquefied petroleum (LP) and natural gas to fuel generator sets ___.

   a. is more desirable than using diesel
   b. is safer than diesel because these are lighter than air
   c. may create an explosive hazard

_____ 20. Diesel-fueled generator sets may have fuel storage tanks underground or distant from the generator set and may require ___.

   a. a day tank
   b. a high pressure fuel pump
   c. an oversized fuel line
   d. a secondary fuel pump
   e. all of these
   f. none of these

Objective 226.2 Worksheet

_____ 1. When servicing a generator set, always ___.
   I. disconnect the battery cables
   II. disable or turn off all master controls
   III. disconnect the fuel line

   a. I, II, & III   b. II & III only   c. I & II only   d. III only   e. I only

_____ 2. A(An) automatic transfer switch in the ___ mode allows two power sources to be paralleled indefinitely, with no power interruption to the load in the event of power loss from either source.

   a. soft load
   b. manual
   c. automatic
   d. bypass isolation
   e. any of these
   f. none of these

_____ 3. A ___ switch is a requirement on every UPS installation in order to accommodate the maintenance and testing of the UPS.

   a. bypass isolation   b. isolation   c. manual isolation   d. maintenance bypass

_____ 4. Transfer switches can be used to transfer load power between ___.

   a. a utility and a generator
   b. a generator and a second generator
   c. either a or b
   d. neither a nor b

_____ 5. The first Automatic Transfer Switch (ATS) was introduced in ___.

   a. 1924   b. 1898   c. 1938   d. 1920   e. 1910

_____ 6. Generator set circuit breakers and fuses must open all ___ output circuit conductors.

   a. ungrounded   b. grounding   c. grounded

Generators and Transfer Switches

_____ 7. When servicing the generator set, always disconnect the battery cable ___ first.

   a. negative   b. positive

_____ 8. Automatic transfer switches protect electrical loads against ___.

   a. low frequency   b. low voltage   c. loss of power   d. all of these   e. none of these

_____ 9. Automatic transfer switches consist of ___.

   a. main contacts
   b. transfer mechanism
   c. intelligence/supervisory circuits
   d. all of these

_____ 10. UPS is an abbreviation for ___.

   a. utility parallel system
   b. user process station
   c. utility passive synchronization
   d. uninterruptible power supply
   e. unity process server f. none of these

_____ 11. Optional accessories for generator sets may include ___.

   a. audio visual alarms
   b. common failure relay kits
   c. line circuit breakers
   d. low fuel switches
   e. all of these

_____ 12. A(An) ___ switch provides a simple and effective means for bypassing UPS systems while maintaining continuity of power to critical computer loads.

   a. bypass isolation
   b. manual isolation
   c. maintenance bypass
   d. isolation

_____ 13. Batteries, which provide the energy source required to start the generator set, should be located ___.

   a. in a clean, dry and temperature stable location
   b. not more than 100 feet from the generator set
   c. both a and b
   d. neither a nor b

_____ 14. A(An) ___ ATS is designed for applications where maintenance, inspection and testing must be performed while maintaining continuous power to the load.

   a. maintenance bypass
   b. manual bypass
   c. bypass isolation
   d. isolation

_____ 15. Transfer switches are categorized as ___.

   a. open transition
   b. closed transition
   c. closed transition soft load
   d. any of these

_____ 16. Make-before-Break operation is utilized by ___ type automatic transfer switches.

   a. open transition   b. closed transition

_____ 17. An automatic transfer switch (ATS) transfers power to electrical loads when ___ are detected.

   a. low current
   b. low voltage
   c. either a or b
   d. neither a nor b

_____ 18. A transfer switch is a ___ switch.

   a. three-way
   b. single-throw
   c. four-way
   d. double-throw

Generators and Transfer Switches

_____ 19. A typical standby system ATS ___.

    a. disconnects the normal power source connections when the normal power source fails
    b. transfers the load back to the normal power source when the normal source is restored
    c. transfers the electrical load to the generator when the normal source fails
    d. signals the generator to stop when the normal power source is restored
    e. all of these

The remaining questions on this worksheet require the use of the BCES Plans.

_____ 20. When the room lighting switch in the BCES kitchen is turned OFF, there are ___ lamps that remain ON in the Type V luminaires.

    a. 0   b. 2   c. 4   d. 3   e. 1

_____ 21. The portable generator required to supply power to BCES must have an output rating of ___ volts.

    a. 120/208 3Ø, 4-wire
    b. 120/240 3Ø, 4-wire
    c. 480 3Ø, 3-wire
    d. 277/480 3Ø, 4-wire

_____ 22. The BCES walk-in cooler/freezer is automatically powered when normal power is lost.

    a. True   b. False

_____ 23. The Type CE luminaires in the BCES cafetorium contain a total of ___ lamps.

    a. 3   b. 2   c. 1   d. 4

_____ 24. Refer to the BCES plans. Between the transfer switch and the disconnect switch for the generator input, you are required to install ___ conductor(s).
    I. three ungrounded   II. four ungrounded   III. one grounded   IV. one grounding

    a. I & IV only
    b. II, III, & IV only
    c. I & III only
    d. II & III only
    e. II & IV only
    f. I, III, & IV only

_____ 25. The receptacle shown on BCES, for the connection of a portable generator, is actually called a pin-and-sleeve ___.

   a. male inlet   b. connector   c. receptacle   d. female inlet   e. none of these

_____ 26. When the room lighting switch in the BCES kitchen is turned OFF, there are ___ lamps that remain ON in the Type VE luminaires.

   a. 4   b. 1   c. 2   d. 0   e. 3

_____ 27. The Type VE luminaires in the BCES kitchen contain a total of ___ lamps.

   a. 1   b. 4   c. 2   d. 3

_____ 28. When the room lighting switches (a and b) in the BCES cafetorium are turned OFF, there are ___ lamps that remain ON in the Type CE luminaires.

   a. 3   b. 1   c. 2   d. 4   e. 0

Generators and Transfer Switches

Lesson 226 NEC® Worksheet

_____ 1. When grounding the frame of a portable generator, the frame shall be grounded in accordance with NEC® Section ___

   a. 250.52(A)(2)   b. 250.96   c. 250.32(B)   d. 250.34

_____ 2. Transfer equipment for generators, including automatic transfer switches, are covered by NEC® Section ___.

   a. 701.7   b. 700.6   c. 702.6   d. all of these   e. none of these

_____ 3. Transfer equipment for emergency systems shall be designed and installed to prevent the inadvertent interconnection of ___ sources of supply in any operation of the transfer equipment.

   a. utility and permanent
   b. temporary and generator set
   c. normal and emergency
   d. grounded and ungrounded

_____ 4. Audible and visual signal devices on emergency system transfer equipment shall be provided where practicable for ___.

   a. indicating that the battery charger is not functioning
   b. indicating derangement of the emergency source
   c. indicating a ground fault in a solidly grounded wye emergency systems of more than 150 volts to ground and circuit protective devices rated 1000 amperes or more
   d. indicating that the battery is carrying the load
   e. all of these

_____ 5. For emergency systems, where internal combustion engines are used as the prime mover, an on-site fuel supply shall have sufficient fuel supply for not less than ___ hours' full-demand operation of the system.

   a. 12   b. 4   c. 6   d. 2

_____ 6. The ampacity of conductors from the generator to the first distribution device(s) containing overcurrent protection shall not be less than ___ percent of the nameplate current rating of the generator.

   a. 115   b. 200   c. 150   d. 125

Year Two (Student Manual)

_____ 7. Legally required standby systems, as classified by the AHJ, would generally be required for which of the following?

   a. sewage disposal in hospitals
   b. HVAC equipment in hospitals
   c. ventilation and smoke removal equipment in office buildings
   d. all of these
   e. none of these

_____ 8. Emergency system equipment shall be designed and located so as to minimize the hazards that might cause complete failure due to flooding, fires, and ____.

   a. icing   b. vandalism   c. both a and b   d. neither a nor b

_____ 9. On legally required standby systems, the automatic transfer switch shall be ____.

   a. manually operated and manually held
   b. electrically operated and mechanically held
   c. electrically operated and electrically held
   d. manually operated and electrically held

_____ 10. The neutral is bonded to the frame and connected to a ground rod near a 3Ø, 4-wire 120/208 volt optional standby generator. The transfer switch supplying power to a building system from this generator is required to have ____ poles.

   a. 4   b. 3   c. 2

_____ 11. When the generator is connected to BCES, this system meets the NEC® definition of a(n) ____ system

   a. emergency   b. optional standby   c. legally required standby

_____ 12. The optional standby system wiring ____ be permitted to occupy the same raceways, cables, boxes, and cabinets with other general wiring.

   a. shall not   b. shall

_____ 13. Emergency systems shall have adequate capacity and rating for all loads to operate ____.

   a. at 125% of load
   b. separately
   c. successively
   d. simultaneously
   e. independently

Generators and Transfer Switches

_____ 14. Optional standby systems, as classified by the AHJ, would generally be required for which of the following?

   a. refrigeration equipment in food stores
   b. HVAC equipment in office buildings
   c. computer equipment in office buildings
   d. all of these
   e. none of these

_____ 15. Emergency systems, as classified by the AHJ, would generally be required for which of the following?

   a. exit and egress lighting in hotels
   b. elevators in hospitals
   c. 9-1-1 communication systems
   d. all of these
   e. none of these

_____ 16. Generator overcurrent protection is covered in NEC® Section ___.

   a. 445.14   b. 445.17   c. 445.10   d. 445.12

Lesson 226 Safety Worksheet

Accident Factors

The principal causes of accidents on construction sites are **Unsafe Acts** and **Unsafe Conditions.** These cause many painful injuries and too many deaths.

You have control of unsafe acts. You may be using tools or equipment that are inadequate for the job, defective or damaged. **You** may choose to ignore warning signs or not to wear the necessary Personal Protective Equipment that could save **you** from injury. **You** may be working too close to energized circuits or parts that should have been de-energized. Perhaps **you** are working next to unprotected sides and edges. Maybe **you** are handling materials incorrectly or not heeding the warning labels on chemicals. **You** may even be placing your body or body parts into dangerous moving parts of equipment.

Notice all of these things that **you** have control over. Don't take chances and hope that an accident won't happen to **YOU.**

The next cause is **Unsafe Conditions.** Unsafe conditions can be things such as no guardrails where they should be installed, not enough illumination to be able to see clearly, poor ventilation, improper grounding or lack of GFCI protection. Perhaps there are no fire extinguishers or maybe the labels are missing on containers of hazardous chemicals. Poor housekeeping also leads to slips, trips and falls. Any or all of the conditions may have been caused by some other contractor, or by some other employee or even created by the general contractor.

Who ever has caused these conditions has put **you** in a position that puts **you** at risk for an accident. Correct these conditions yourself or bring them to the attention of whomever is responsible and advise your supervisor to follow-up to make sure they are corrected. Learn from any accidents and make sure that they do not happen again.

Generators and Transfer Switches

Questions

_____ 1. Regardless of who created an unsafe condition, you should ___.
   I. correct it   II. report it   III. ignore it

   a. II and III only
   b. II only
   c. I and III only
   d. I, II, and III
   e. I only
   f. I and II only

_____ 2. ___ has/have control of unsafe acts on construction sites.

   a. OSHA   b. The States   c. General Contractors   d. You

_____ 3. After witnessing any unsafe acts or seeing any unsafe conditions you should ___.
   I. discuss issues   II. learn from them   III. prevent future issues

   a. II only
   b. I, II, and III
   c. I and III only
   d. III only
   e. I and II only
   f. I only

_____ 4. The principal cause(s) of accidents on construction sites is/are ___.
   I. unsafe acts   II. unsafe rules   III. unsafe conditions

   a. I and II only
   b. I only
   c. II only
   d. III only
   e. II and III only
   f. I and III only

_____ 5. No guardrails, not enough illumination, poor ventilation and no GFCI are ___.

   a. unsafe conditions   b. unsafe acts

_____ 6. Unsafe acts can include ___.
   I. using defective tools
   II. ignoring warning signs
   III. not using PPE
   IV. working near energized circuits

   a. III only
   b. II and III only
   c. I, II, and III only
   d. I, II, III, and IV
   e. I only
   f. I and IV only

_____ 7. Placing your body or body parts into dangerous moving parts of machines is an ___.

   a. unsafe act   b. unsafe condition

Lesson 227
Mid-Term Exam

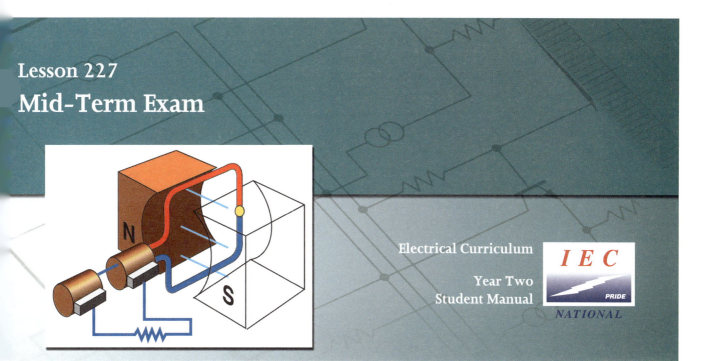

Electrical Curriculum

Year Two
Student Manual

Purpose
The second semester mid-term exam is scheduled for this lesson.

Homework
(Due at the beginning of this class)

For this lesson, you should:

- Review all your worksheets, homework, and quizzes.
- Complete your "Practice Exam."
- Be prepared to clarify any questions or problem areas you encountered during your review.

Objectives
This lesson will determine your proficiency in the subject matter from the previous lessons.
The instructor should review next week's material beforehand to determine if the students will require additional instruction before assigning the worksheets for that lesson.

Lesson 228
Electric Motors – DC and 1Ø AC

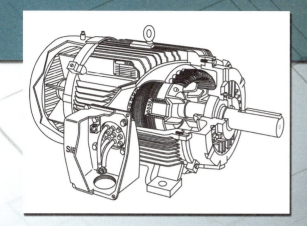

Electrical Curriculum

Year Two
Student Manual

Purpose

This lesson will discuss the basic operation, construction, and connection of motors, both DC and single-phase AC. This lesson will also discuss the various types of these motors that are available as well as how to interpret the information provided by manufacturers.

Homework

(Due at the beginning of this class)

For this lesson, you should:

- Thoroughly read the material contained within Lesson 228.
- Complete Objective 228.1 Worksheet.
- Complete Objective 228.2 Worksheet.
- Read and complete Lesson 228 Safety Worksheet.

- Complete additional assignments or worksheets if available and directed by your instructor.

Objectives

By the end of this lesson, you should:

228.1 .... Understand the manufacturer provided information, operation, construction, and required connections for DC motors.

228.2 .... Understand the manufacturer provided information, operation, construction, and required connections for single-phase AC motors.

228.1 Direct Current (DC) Motors

An electric motor is a machine that converts electrical energy to mechanical energy. There are two main groups of electrical motors: DC and AC motors. This lesson will discuss DC and 1Ø AC motors, and how to control them.

Note: There are two theories regarding the flow of current. The *Electron Flow Theory* states that current flows from negative to positive, while *Conventional Flow Theory* states that current flows from positive to negative. This lesson reflects the use of Electron Flow Theory.

Magnetic Fields

Between the poles of a magnet, there exists a magnetic field. The direction of the magnetic field is called *Magnetic Flux*. Magnetic flux moves from the north pole to the south pole, as shown in Figure 228.11.

Lines of Magnetic Flux Flow from the North to the South Pole

Figure 228.11

Current Flow

Now, let's consider a loop of wire (conductor) with an electric current flowing through it. A magnetic field surrounds the wire, as shown in Figure 228.12A and B.

Understanding the direction of the magnetic flux around the conductor is critical to understanding motor motion. The direction of the magnetic flux can be determined using the *Left Hand Flux Rule*. To apply this rule, imagine grasping the wire with your left hand, making sure your thumb points in the direction of the current flow. Your fingers will curl around the wire in the direction of the magnetic flux.

In Figure 228.12A, the current is flowing into the page, so the lines of flux rotate counterclockwise around the wire. In Figure 228.12B, the current is flowing out of the page, so the lines of flux rotate clockwise around the wire.

Left Hand Flux Rule — Lines of Magnetic Flux Surround a Conductor

Figure 228.12A

Figure 228.12B

Induced Motion

When this current-carrying conductor loop is placed between the poles of a magnet, the magnetic fields around the conductors and between the magnets both distort. In Figure 228.13, the conductor will tend to rotate clockwise. A simple method for determining the direction of motion is the *Right Hand Motor Rule*. To apply this rule, refer to Figure 228.13. The index finger points in the direction of the magnetic flux (N to S), the middle finger points in the direction of current flow through the conductor, and the thumb points in the direction of the conductor loop movement. The rotational force exerted on each half of the conductor loop depends on the strength of the magnetic field between the poles of the magnet, and the strength of the current through the loop. The strengths of these magnetic fields determine the amount of force applied to the loop. This rotational (turning) force is referred to as *torque*.

Right-Hand Rule for Motors – Conductor Loop Rotates Clockwise

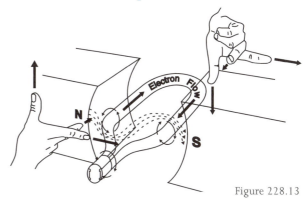

Figure 228.13

Magnetic Flux Strength

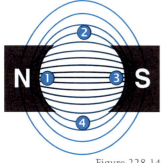

Figure 228.14

As shown in Figure 228.14, the strength of the magnetic flux (field) between the two magnetic poles is at a maximum at Points 1 and 3, and at a minimum at Points 2 and 4.

In Figure 228.15A the conductor loop is in Position 1 and is shown connected to the *commutator*. A commutator is a device used in a DC motor to reverse the current through the loop every one-half rotation. The loop/commutator assembly is called the *armature* and rotates inside the motor housing. In Position 1 the loop is located in the strongest part of the magnetic field between the magnet's poles. The maximum torque is developed at this point. If voltage is applied to the loop when in Position 1, the loop is forced to rotate away from this position.

In Figure 228.15B the conductor loop has rotated into Position 2. At this point torque is at its minimum because the loop has now rotated into the weakest part of the magnetic flux field. Were it not for the inertia (momentum), the loop would stop in Position 2. However, since the loop is moving, it passes through Position 2 and is attracted toward Position 3.

Position 1

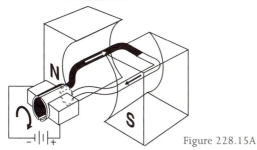

Figure 228.15A

Position 2

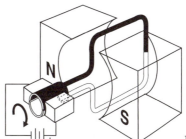

Figure 228.15B

Year Two (Student Manual)

Note that Position 2 is called the neutral position because if voltage is applied to the loop when it starts out in Position 2, the torque is insufficient to cause the loop to rotate away from this position. Zero torque is produced when the loop is in Position 2.

In Figure 228.15C the conductor loop is moving past Position 3. Recall that in Position 1 current was flowing from the "black" section of the conductor loop and toward the "white" section of the loop. The commutator has caused this current flow to reverse direction. Were it not for this change in current flow, when the loop got close to Position 3, it would reverse rotation and travel back toward Position 2. Since the direction of current flow has changed direction, so have the magnetic fields surrounding the two halves of the conductor loop. Once again, as when in Position 1, the loops are forced to continue in a clockwise rotation. The loop passes through the weak magnetic fields (Points 2 and 4 in Figure 228.14) and are attracted to the poles. And again, the commutator reverses the current direction when the loop approaches these poles so that the loop is forced to continue its rotation clockwise.

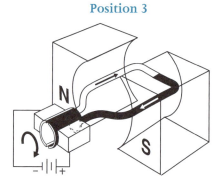

Figure 228.15C

Voltage Applied to the Armature

The conductors from the DC power supply are not directly connected to the commutator. If so, the conductors would turn with the armature and would be pulled apart. Instead, conductive, carbon brushes are held against the commutator with springs, and the DC supply conductors are attached to the brushes.

DC Motor Speed and Torque

As stated previously, zero torque is produced when a conductor loop is in the neutral positions (¼ turn and ¾ turn). With no torque applied, the rotational speed drops. Torque and Speed are shown in Figure 228.16. Since most driven loads require that the motor turn at a uniform speed, this is not acceptable. Additionally, should the simple motor try to start when the single loop is in the neutral position, no torque is applied to start the rotation.

Practical DC Motors

A practical DC motor utilizes many coils of wire for each armature loop. This increases the strength of the magnetic field surrounding the loop. This stronger magnetic field increases the motor's torque. Additional off-setting loops are also utilized as shown in Figure 228.17. And finally, the commutator consists of many

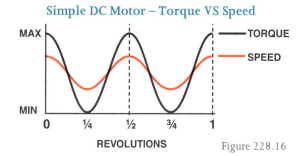

Figure 228.16

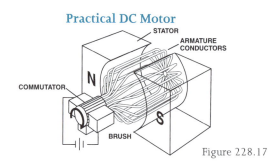

Figure 228.17

segments. Each loop is connected to a different set of commutator segments. In a practical DC motor, the armature is never in a neutral position, and the torque is always at its maximum. When current flows through the brushes, all of the loops act together, producing full torque at all times. Also, notice that the brushes are larger than the gaps between the commutator segments. This means that contact with the commutator is maintained at every instant of rotation of the armature.

DC Motor – Cutaway View

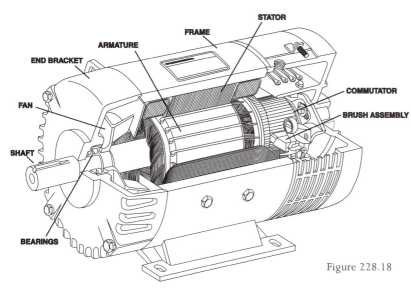

Figure 228.18

A cutaway view of a large DC motor is shown in Figure 228.18. Field windings and pole pieces are bolted to the frame. The armature is inserted between the field windings. The armature is supported by bearings and end brackets. Carbon brushes are spring-held against the commutator. A fan cools the windings. The shaft is connected to the driven load.

In Figure 228.19 the armature has been removed from the motor housing.

Each pole of the stator is an electromagnet and consists of a coil of wire (the Field Winding) wrapped around a pole piece. The stator's field windings are connected to a DC power supply. The power supplied to the brushes could be from the same source, or a different DC power supply.

DC Motor – Armature and Stator

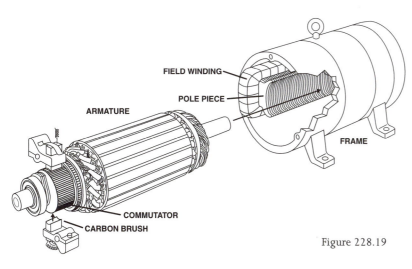

Figure 228.19

Stators Using Electromagnets

Small DC motors, such as those used in cordless hand tools, typically utilize permanent magnets for the stator poles. Power from the battery pack is applied to the armature in these small motors.

The stator poles in larger motors are typically electromagnets. Field windings are mounted on pole pieces to form electromagnets. Electromagnets work very similarly to permanent magnets. To make one, simply wrap an iron rod with insulated wire and run DC current through the wire. The iron rod develops a magnetic field, and North and South magnetic poles.

The electromagnet has two advantages over the permanent magnet.
- By adjusting the amount of current flowing through the wire, the strength of the electromagnet can be controlled.
- By changing the direction of current flow, the poles of the electromagnetic can be reversed. Switching the leads on the DC power source would change the direction of current flow.

We have already discussed three of the four major components that make up a DC motor: the armature, the brushes, and the commutator. The fourth component is the stator, or stationary part. It consists of an even number of field poles. Each field pole is made by wrapping many coils of wire around two pole pieces. The two pole pieces are directly opposite each other and are attached inside the motor frame. The coils of wire are the field winding. Together these form electromagnetic field poles (one North and one South). The stator contains many sets of these field poles. When current passes through the field winding, the pole pieces are magnetized.

Types of DC Motors

Different types of DC motors are utilized as required by the load. The different types are listed below.

Permanent Magnet Motors

The permanent magnet motor uses a magnet to supply field flux. Permanent magnet DC motors have excellent starting torque capability with good speed regulation. A disadvantage of permanent magnet DC motors is they are limited to the amount of load they can drive. These motors can be found on low horsepower applications. Another disadvantage is that torque is usually limited to 150% of rated torque to prevent demagnetization of the permanent magnets.

- **Series Motors**

In a series DC motor the field is connected in series with the armature. The field is wound with a few turns of large wire because it must carry the full armature current. This motor is also called a universal motor because it can be used in DC or AC applications. A characteristic of series motors is the motor develops a large amount of starting torque. However, speed varies widely between no load and full load. Series motors cannot be used where a constant speed is required under varying loads. Additionally, the speed of a series motor with no load increases to the point where the motor can become damaged. Some load must always be connected to a series-connected motor. Series-connected motors generally are not suitable for use on most variable speed drive applications. The schematic for a Series Motor is shown in Figure 228.20. Note that this motor has two leads for the armature – marked A1 and A2. It also has two leads for the series field winding – marked S1 and S2.

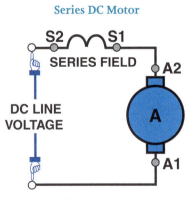

Series DC Motor

Figure 228.20

- **Shunt Motors**

   In a shunt motor the field is connected in parallel (shunt) with the armature windings. The shunt-connected motor offers good speed regulation. The field winding can be separately excited or connected to the same source as the armature. An advantage to a separately excited shunt field is the ability of a variable speed drive to provide independent control of the armature and field. The schematic for a Shunt Motor is shown in Figure 228.21. Note that this motor has two leads for the armature – marked A1 and A2, and it has two leads to the shunt field winding – marked F1 and F2. The power for the shunt winding can be from the same power supply used for the armature, or it can be supplied from a separate power source. When supplied from a separate power source it is said to be separately excited.

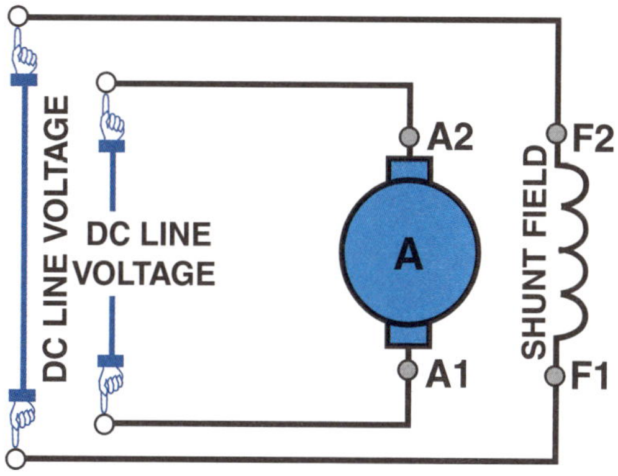

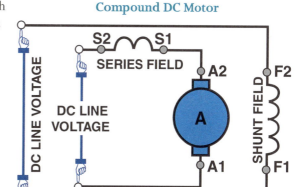

Figure 228.21

- **Compound Motors**

   Compound motors have a field connected in series with the armature and a separately excited shunt field. The series field provides better starting torque and the shunt field provides better speed regulation. However, the series field can cause control problems in variable speed drive applications. The schematic for a Shunt Motor is shown in Figure 228.22. This motor also has two leads for the armature – marked A1 and A2. Since it is a compromise between a series and a shunt motor it has leads for these two field windings: S1, S2, F1 and F2. The power for the shunt winding can be from the same power supply used for the armature, or it can be supplied from a separate power source. When supplied from a separate power source it is said to be separately excited.

Figure 228.22

Reversing a DC Motor

For a permanent magnet motor, the motor's direction of rotation can be reversed by changing polarity of the DC power leads to the armature leads. For example, when the reverse switch on a cordless drill is changed, the + and – leads from the battery to the armature leads are swapped.

For larger motors that use field windings instead of permanent magnets, the direction of rotation of a DC motor may be reversed using one of these methods:

- Reversing the direction of the current through the field
- Reversing the direction of the current through the armature

The standard used in the electrical industry is to reverse the current through the armature. This is accomplished by reversing the armature connections only. The polarity connections to the field windings remain unchanged.

DC Motor Ratings

The nameplate of a DC motor provides important information necessary for correctly applying a DC motor with a DC drive. The following specifications are generally indicated on the nameplate:

- Manufacturer's Type and Frame Designation
- Horsepower at Base Speed (HP)
- Maximum Ambient Temperature (Max Ambient)
- Insulation Class (Insul Class)
- Base Speed at Rated Load (RPM)
- Rated Armature Voltage (Volts)
- Rated Field Voltage (FLD Volts)
- Armature Rated Load Current (Arm Amps)
- Winding Type (Shunt, Series, Compound, Permanent Magnet)
- Enclosure (Encl)

Typical DC Motor Nameplate

SIEMENS

HP	10	RPM	1180	VOLTS	500
ARM AMPS	17.0	WOUND	SHUNT		
FLD AMPS	1.4/2.8	FLD OHMS 25C	156		
INSUL CLASS F	DUTY	CONT	MAX AMBIENT	40° C	
PWR SUP CODE	C		FLD VOLTS	300/150	

| TYPE | E | ENCL | DP | INSTR | |
| MOD | | | SER | | |

NP36A424835AP

DIRECT CURRENT MOTOR
Made in U.S.A.

Volts

The typical armature voltage, in the United States, for industrial DC motors is either 250 VDC or 500 VDC.

Horsepower

$$\text{Power} = (\text{Force} \times \text{Distance}) \div (\text{Time})$$

For a motor, the torque developed (measured in lb-ft) causes the shaft to rotate an angular distance.

The formula for power, delivered by a motor, becomes:

$$\text{Horsepower} = (\text{Torque} \times \text{Speed}) \div (5250)$$

Where Torque is in lb-ft and Speed is in Revolutions Per Minute (RPMs)

It can be seen from the formula that a change of torque, or speed (RPMs), or both will cause a corresponding change in horsepower (HP).

Power in an electrical circuit is measured in watts (W) or kilowatts (kW). Variable speed drives and motors manufactured in the United States are generally rated in horsepower (HP); however, it is becoming common practice to rate equipment using the International System of Units (SI units) of watts and kilowatts. The HP or KW nameplate rating of a motor is the OUTPUT power it can deliver to a driven load at its nameplate (NP) rated speed (in RPMs). Due to losses within the motor, including heat and friction, the input power (electrical power: E × I) required is always greater than the output power (mechanical power: HP) delivered. Stated mathematically this is:

INPUT POWER (in VA) > OUTPUT POWER (in HP or KW)

Motor design determines the efficiency *rating*. This rating may appear on the motor nameplate but applies only when the motor is operated at its optimum design criteria. The nameplate efficiency may appear as a percentage (for example: 87%) or it may appear as a decimal number (for example: .87). The actual *operating* efficiency varies with the torque required to drive the load. Proper selection of a motor for the application is a complex, engineering process and will not be addressed here. Typical DC motor efficiencies range from less than 50% for small motors to well over 90% for very large motors. DC Motor efficiency can be approximated using the following formula.

☞ Eff = (OUTPUT POWER) ÷ (INPUT POWER)

= (HP × 746) ÷ ($I_{line} \times E_{line}$)

Of course, efficiency is found in decimal form and can be converted to percent by multiplying it by 100. A derivation of the above formula can be utilized to calculate line current (I_{line}). In the formula that follows, efficiency should be entered as a decimal number (%EFF ÷ 100).

☞ I_{line} = (HP × 746) ÷ (E_{line} × Eff)

Insulation Class

The National Electrical Manufacturers Association (NEMA) has established insulation classes to meet motor temperature requirements found in different operating environments. The insulation classes are A, B, F, and H.

Before a motor is started the windings are at the temperature of the surrounding air. This is known as ambient temperature. NEMA has standardized an ambient temperature of 40°C (104°F) for all classes. The maximum *temperature rise* for a Class A motor is 60°C, a Class B motor is 80°C, and a Class H motor is 125°C.

Temperature will rise in the motor as soon as it is started. The combination of ambient temperature and allowed temperature rise equals the maximum winding temperature in a motor. A motor with Class F (commonly used) insulation, for example, has a maximum temperature rise of 105°C. The maximum winding temperature is 145°C (40°C ambient + 105°C rise).

The operating temperature of a motor is important to efficient operation and long life. Operating a motor above the limits of the insulation class reduces the motor's life expectancy. A 10°C increase in the operating temperature can decrease the life expectancy of a motor by as much as 50%. In addition, excess heat increases brush wear.

DC Motor Applications

DC motors offer several advantages in applications that operate at low speed, such as cranes and hoists. Advantages include low speed accuracy, short-time overload capability, size, and torque control. In some applications, DC motors are utilized in operations that require tension to be maintained at standstill. DC motors are often preferred in the high horsepower applications required in the mining and drilling industry.

Additionally, DC motors are used in industrial applications that require either variable speed control, high torque, or both. Because the speed of most DC motors can be controlled smoothly and easily from zero to full speed, DC motors are used in many acceleration and deceleration applications.

DC motors are ideal in applications where momentarily higher torque output is needed. The DC motor can deliver three to five times its rated torque for short periods of time. Compared to AC motors of equal HP ratings, a DC motor has a very high starting torque. For these reasons, DC motors are used to run large machine tools, cranes and hoists, printing presses, elevators, shuttle cars and automobile "starters".

228.2 Single-Phase Alternating Current (AC) Motors

AC Motor Construction and Types

AC motors, as shown in Figure 228.31, are used worldwide in many applications to transform electrical energy into mechanical energy. The main parts of the motor are the:

- Terminal Box
- Rotor
- Shaft
- Stator
- Enclosure

AC Motor

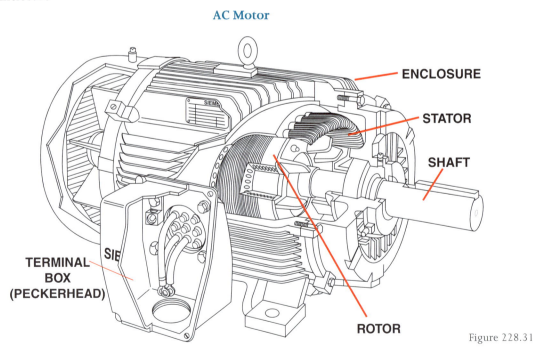

Figure 228.31

Electric Motors – DC and 1Ø AC

The *terminal box* is where circuit conductors are connected. This box is frequently referred to as the *motor peckerhead*. The term *motor peckerhead* most likely originated due to the fact that, after the wiring was complete, the terminal box looked like a woodpecker's nest. Care must be exercised when terminating conductors within the box because the terminations must be well insulated from each other and the box walls and cover. If not, vibration can cause short circuits or ground faults.

The *rotor* is mounted on the *shaft* as one assembly. This assembly is installed so that it rides on bearings. This enables the assembly to rotate within the motor.

The *shaft* is connected to a driven load. This connection is made directly using couplings, or pulleys can be installed on the shaft and belts drive the load, or gears can be installed on the shaft and the load.

Wire leads (shown inside the terminal box in Figure 228.31) are internally connected to the *stator* windings. The stator assembly is mounted to the motor enclosure. On some motors the wire leads are mounted to posts inside the terminal box. These posts are threaded "studs" that require nuts and washers to secure the connections. In these cases, field connections are made by installing terminal rings or forks on the circuit conductors. These are then installed on the terminal posts within the terminal box.

The *enclosure* provides a mounting structure and protects the internal motor components. The enclosure consists of a frame (or yoke) and two end brackets (or bearing housings). The stator is mounted inside the frame. The rotor fits inside the stator with a slight air gap separating it from the stator. There is no direct physical connection between the rotor and the stator. The enclosure protects the internal parts of the motor from water and other environmental elements. The degree of protection depends upon the type of enclosure. Enclosure types will be addressed in the next lesson. Bearings, mounted on the shaft, support the rotor and allow it to turn. Some motors use a fan, also mounted on the rotor shaft, to cool the motor when the shaft is rotating.

AC motors may be split into two main groups: single-phase and polyphase. Single-phase motors are addressed in this lesson and three-phase motors are addressed in the next lesson.

While there are only three general types of DC motors, there are many different AC motor types. This is because each type is confined to a narrow band of operating characteristics. These characteristics include torque, speed, and electrical service (single-phase or polyphase). These operating characteristics are used to determine a given motor's suitability for a given application.

Types of single-phase motors include:

- Shaded-pole
    - Referred to as "disposable" – replacement is less expensive than repairs
    - Very low starting torque
    - Very inefficient (less than 20%)
    - Typically used in multi-speed household fans

- Capacitor
  - Capacitor start/induction run
    - Most widely used among single-phase industrial types
    - Moderate starting torque
    - Low starting current (4 to 5 times FLA)
    - Utilized for belt-drive, direct-drive, and geared applications
  - Permanent split capacitor
    - Low starting torques
    - Very low starting current (less than twice FLA)
    - Utilized for fans, blowers, gate openers, and garage door openers
  - Capacitor start/capacitor run
    - High efficiency
    - Most costly among cap motors
    - Typically utilized in high-torque applications requiring 1 to 10 HP
- Split-phase
  - Also referred to as induction-start/induction-run
  - Relatively inexpensive among single-phase industrial types
  - Low starting torque
  - Very high starting current (7 to 10 times FLA)
  - Generally used for small fans, blowers, grinders, power tools
  - Typically available in sizes from 1/20 to 1/3 HP
- Universal (AC-DC)
  - Modified DC series motor
  - Very noisy due to the use of brushes
  - Highest horsepower-per-pound ratio of any ac motor
  - Operates at very high speeds
  - Capable of variable speed control
  - Typically used in household appliances, such as vacuum cleaners, circular saws, routers, etc.
- Synchronous
  - Precise, constant-speed motors – operate in absolute synchronism with line frequency
  - Typically used in clocks and as timing motors
  - Typically available in sizes less than 1/10 HP

AC Motor Operation

What makes an AC motor different from a DC motor? In a DC motor, electrical power is conducted directly to the armature through brushes and a commutator. An AC motor does not need a commutator to reverse the polarity of the current, because AC changes polarity each half cycle. Another difference is that, where the DC motor works by changing the polarity of the current running through the armature (the rotating part of the motor), the AC motor works by changing the polarity of the current running through the stator (the stationary part of the motor). The rotating part of an AC motor is referred to as the rotor. The rotor turns inside the stator.

Polarity Changes in a Magnetic Field

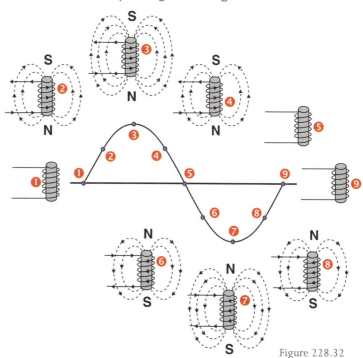

Figure 228.32

A single-phase power system has one coil in the generator (power source). Therefore, one alternating voltage is generated and this is applied to the stator winding of an AC motor. The current flowing through the stator winding creates an electromagnet. The magnetic field of an electromagnet has the same characteristics as a natural magnet, including a north and south pole. However, when the direction of current flow through the electromagnet reverses, the polarity of the electromagnet reverses. The polarity of an electromagnet connected to an AC source changes at the **frequency** of the AC source. This is demonstrated in Figure 228.32. At Time ❶, there is no current flow, and no magnetic field is produced. At Time ❷, current is flowing in a positive direction, and a magnetic field builds up around the electromagnet. Note that the south pole is on the top and the north pole is on the bottom. At Time ❸, current flow is at its peak positive value, and the strength of the electromagnetic field has also peaked. At Time ❹, current flow decreases, and the magnetic field begins to collapse. At Time ❺, no current is flowing and no magnetic field is produced. At Time ❻, current is increasing in the negative direction. Note that the polarity of the electromagnetic field has changed. The north pole is now on the top, and the south pole is on the bottom. The negative half of the cycle continues through Times ❼ and ❽, returning to zero at Time ❾. For a 60 Hz, AC power supply, this process repeats 60 times a second.

AC Motor Speed

The speed of the rotating magnetic field in the stator is referred to as the synchronous speed (SS) of the motor. Synchronous speed is equal to 120 times the frequency (f), divided by the number of motor poles (p). For example, the synchronous speed for a two-pole motor operated at 60 Hz is 3600 RPM.

☞ SS = (120 × f) ÷ p

= (120 × 60) ÷ 2 = 3600 RPM

Synchronous speed decreases as the number of poles increases. The table shows the synchronous speed at 60 Hz for several different pole numbers.

No. of Poles	Synchronous Speed
2	3600
4	1800
6	1200
8	900
10	720

A voltage can be induced across a conductor by merely moving it through a magnetic field. This same effect is caused when a stationary conductor encounters a changing magnetic field. This electrical principle is critical to the operation of AC induction motors because this causes the shaft (rotor) to turn inside the stator. When the magnetic field around the stator windings rotates, as illustrated in Figure 228.32, it causes voltage to be induced and currents to flow in the rotor. The rotor is constructed of bars and resembles a squirrel cage. Consequently, motors with rotors of this construction are referred to as squirrel-cage induction motors. The rotor's construction enables currents to flow. These rotor currents, in turn, set up electromagnetic fields around the squirrel cage rotor. At any given point in time, the magnetic fields for the stator windings are exerting forces of attraction and repulsion against the various rotor bars. This causes the rotor to rotate, but not exactly at the motor's synchronous speed. The term *synchronous* means that the rotor's rotation is synchronized with the magnetic field, and the rotor's speed is the same as the motor's synchronous speed. To operate correctly, an induction motor relies on the difference between rotor speed and synchronous speed. This difference is call *slip*. The rotor speed is the shaft speed (in RPMs) and is shown on the motor's nameplate.

Although it is theoretically possible to control the speed of an AC motor by varying the voltage, is not a good way to change the speed of the motor. In fact, if the voltage is changed by more than ±10%, the motor may be damaged. If less than nameplate voltage is applied, torque and full load speed decrease while the full load current and winding temperature increase. Regardless of applied voltage, a motor still wants to deliver rated HP. Keep this in mind before connecting a 208-volt circuit to a motor rated to operate at 230 volts. If more than nameplate voltage is applied, torque and full load speed increase while the full load current and winding temperature decrease. The two practical methods of AC motor speed control are to change the number of poles or to change the frequency.

Multi-speed AC motors are designed with windings that may be reconnected to utilize different numbers of poles. They are operated at a constant frequency. Two-speed motors usually have one winding that may be connected to provide two speeds, one of which is half the other. Motors with more than two speeds usually include many windings. These can be connected in many ways to provide different speeds. For example, a simple, oscillating, table fan may have a switch to vary the rotor speed. The switch connections within the fan assembly vary the pole connections to change speeds.

Another method to vary the speed of an AC motor is to change the frequency. Changing the frequency requires a device, called a variable (or adjustable) frequency drive (VFD), to be inserted upstream from the motor. This device converts the incoming 60 Hz into any desired frequency, allowing the motor to run at virtually any speed. For example, by adjusting the frequency to 30 Hz, the motor can be made to run only half as fast. VFDs are typically used only on large motors due to the cost of the required equipment.

AC Motor Power Ratings

Although manufactured in smaller sizes, single-phase motors are readily available in sizes ranging from 1/6 to 10 hp. Motors with ratings less than one horsepower are called fractional horsepower motors. These are generally used to operate mechanical devices and machines requiring a relatively small amount of power. AC motors manufactured in the United States are generally rated in horsepower, but motors manufactured in many other countries are generally rated in kilowatts (kW).

The KW rating of a motor can be determined when the HP rating is known by using the following formula:

☞ **Input Power (in kW) = 0.746 × Power (in HP)**

For example, a motor rated at 25 HP is equivalent to a motor rated at 18.65 kW.

$$0.746 \times 25 \text{ HP} = 18.65 \text{ kW}$$

The HP rating of a motor can be determined when the KW rating is known by using the following formula:

☞ **Output Power (in HP) = 1.34 × Input Power (in kW)**

As with DC motors the efficiency of an AC motor can be approximated. However, the power factor (PF) must also be considered. The following formula is used to find the efficiency of an AC motor.

☞ **Eff = (OUTPUT POWER) ÷ (INPUT POWER)**

$$= (HP \times 746) \div (I_{line} \times E_{line} \times PF)$$

Of course, efficiency is found in decimal form and can be converted to percent by multiplying it by 100. A derivation of the above formula can be utilized to calculate line current (I_{line}). In the formula that follows, efficiency should be entered as a decimal number (%EFF ÷ 100).

☞ $I_{line} = (HP \times 746) \div (E_{line} \times Eff \times PF)$

Because each motor type has its own characteristics of horsepower, torque and speed, different motor types are more suited for different applications.

AC Motor Design Characteristics

Some motors are manufactured to International Electrotechnical Commission (IEC) standards. IEC is another organization responsible for electrical standards. IEC standards perform the same function as NEMA standards, but differ in many respects. In many countries, electrical equipment is commonly designed to comply with IEC standards. In the United States, although IEC motors are sometimes used, NEMA motors are more common. Keep in mind, however, that many U.S.-based companies build products for export to countries that follow IEC standards.

NEMA develops standards for a wide range of electrical products, including AC motors. Some motors are built to meet specific application requirements or are larger than the largest NEMA frame size and are commonly referred to as *above NEMA* motors.

The basic characteristics of each AC motor type are determined by the design of the motor and the supply voltage used. These design types are classified and given a letter designation, which can be found on the nameplate of motor types listed as "NEMA Design."

The most commonly used AC NEMA Design motor is the NEMA B.

Recall that torque is the force required to cause the motor shaft to turn. Once a load is connected to the motor, torque terms are further defined.

NEMA Design	Starting Torque	Starting Current	Breakdown Torque	Full Load Slip	Typical Applications
A	Normal	Normal	High	Low	Machine Tool, Fan, and Centrifugal Pump
B	Normal	Low	High	Low	Machine Tool, Fan, and Centrifugal Pump
C	High	Low	Normal	Low	Loaded Compressor Loaded Conveyor
D	Very High	Low	-	High	Punch Press

Starting torque is sometimes referred to as *locked rotor torque* because the rotor has not yet started turning or has stalled or stopped (locked). Locked rotor torque is produced when full power (nameplate rated current and voltage) is applied to the stator. *Breakdown torque* is the maximum torque a motor can produce. If a motor is loaded such that breakdown torque is exceeded, the rotor speed decreases.

Two additional terms frequently associated with motors are *pull-up torque* and *full-load torque*. Full-load torque is required to deliver NP (nameplate) rated horsepower to a driven load at NP rated speed. Pull-up torque is required to bring the rotor up to NP rated speed.

Electricians generally don't select motors for an installation. However, it IS the responsibility of the electrician to replace a motor with another motor that has the exact same operating characteristics, mounting requirements, physical dimension requirements, and enclosure type suitable for the operating environment.

Single-Phase AC Motor Connections

Compared to 3-phase motors, 1-phase motors are less efficient and more expensive to operate. The output power of a 5 HP 1Ø motor equals the output power of a 5 HP 3Ø motor. However, the input power required for a single-phase motor is slightly higher because the efficiency of a single-phase motor is less than that of a three-phase motor. Additionally, for single and three-phase motors of equal horsepower, circuit conductors and the related circuit equipment can be smaller when a three-phase motor is installed.

Compared to DC and 3-phase motors, single-phase motors are difficult to start and must employ special starting methods. Because of this, single-phase motors contain both run and start windings. Typically, T1 and T4 are connected to the run winding and are "brought out" by the manufacturer into the terminal box. T5 and T8

Single-Voltage Capacitor Motor

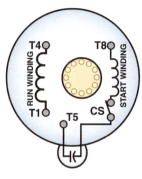

Figure 228.33

are the leads brought out into the box from the start winding. Some motors have a centrifugal switch (CS) mounted on the shaft inside the motor enclosure. When the motor is started, the CS switch disconnects the start winding after the motor approaches NP rated speed, and when almost stopped, the switch closes again. The windings and motor leads for a single-voltage capacitor motor, as viewed from the back end with the shaft into the page, are shown in Figure 228.33.

Not all 1Ø motors are manufactured to be field-reversible. If the manufacturer intends these motors to be reversible, the connection for reverse operation will be shown on the nameplate. Typically, the connections at T5 and T8 are reversed to cause the motor shaft (rotor) to reverse direction. This causes the current flow through the start winding to be opposite that flowing in the run winding. Note that the original current flow direction through the run winding has not been changed. Although motors should be connected according to the manufacturer's diagram, the connection table for a typical 1Ø motor is shown below. Similar markings appear on motor nameplates. Swapping the L1 and L2 connections changes the direction of current flow through the start *and* the run winding(s). Consequently, the direction of rotation remains unchanged.

Refer to Figure 228.33. Notice that, when L1 is connected to T1 and T5 for clockwise (CW) rotation, the instantaneous current flow through the start and run windings is "up" (from the bottom of the drawing). When the motor is rotating counter-clockwise (CCW), L1 is connected to T1 and T8 and the instantaneous current flow is "up" through the run winding and "down" through the start winding.

1Ø, Single-Voltage AC Motor Connections	
Connect L1 to (T1 & T5)	Clockwise Rotation
Connect L2 to (T4 & T8)	
Connect L1 to (T1 & T8)	Counterclockwise Rotation
Connect L2 to (T4 & T5)	

Although many single-phase motors are rated to operate at a single voltage (115, 208, 230, 277, 460, etc), some are rated for operation at either of two different circuit voltages. These are referred to as dual-voltage motors. They are commonly available with nameplate voltage ratings of 120/240 and 230/460 volts. Dual-voltage motors contain two separate run windings. Run Winding 1 leads are marked T1 & T2, and Run Winding 2 leads are marked T3 & T4. The motor windings and leads for a split-phase, dual-voltage motor are shown in Figure 228.34.

Dual-Voltage Split-Phase Motor

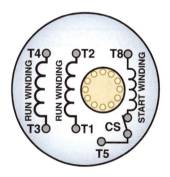

Figure 228.34

When a circuit with the higher voltage is to be connected, the two run windings are connected in series. When the lower voltage is applied, the run windings are connected in parallel. Just as with dual-voltage transformers, each winding (Run 1, Run 2, and Start) are rated to operate at the lower nameplate voltage. The connection table for a typical 1Ø dual-voltage motor is shown below. Again, similar markings appear on motor nameplates.

Remember, L1 and L2 can be swapped without affecting the direction of rotation.

A typical single-phase, dual-voltage motor nameplate appears in Figure 228.43. Note that some fractional HP, single-phase motors contain internal thermal protection. These are switches that open when the motor overheats (typically, due to overload) and remove power from the windings. In these cases, additional leads will be present in the terminal box and are marked as "P" leads. The nameplate connection information for a typical thermally-protected, single-phase motor appears in Figure 228.44.

1Ø, Dual-Voltage AC Motor Connections		
Low Voltage	Connect L1 to (T1, T3 & T5)	Clockwise Rotation
	Connect L2 to (T2, T4 & T8)	
Low Voltage	Connect L1 to (T1, T3 & T8)	Counterclockwise Rotation
	Connect L2 to (T2, T4 & T5)	
High Voltage	Connect L1 to (T1)	Clockwise Rotation
	Connect L2 to (T4 & T8)	
	Connect together (T2, T3 & T5)	
High Voltage	Connect L1 to (T1)	Counterclockwise Rotation
	Connect L2 to (T4 & T5)	
	Connect together (T2, T3 & T8)	

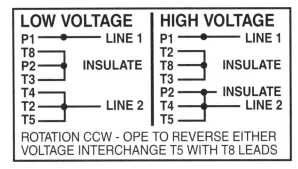

Figure 228.44

Figure 228.43

The AMPS indicated on a motor nameplate is how much current a fully-loaded motor will draw at full speed is achieved after NP rated voltage is applied. This can be measured with a clamp-on ammeter. For example, if a load is belt-driven, the belt tension is adjusted so that the motor is drawing less than or equal to NP amps.

Overloading a motor is the leading cause for motor overheating and ultimate failure. The failure occurs when the stator winding insulation deteriorates to the point where a fault occurs, or when a winding burns in two (opens). Misapplication is one cause of overload. If an undersized motor is used in a particular application, the motor will be overloaded. Another cause of overload is if a motor seizes up is forced to slow down. This could be due to obstruction, bearing wear or failure, or driven load linkage misalignment. Environmental conditions can also cause motor failure due to overheating. These include high ambient temperature, obstructed ventilation openings, and high altitude. At elevations above 3300 feet the air is thinner and has less cooling capacity. In either of these situations, a larger motor will be required. Dirt and fibers can clog ventilation openings on a motor. The correct motor enclosure type should be selected based on the installation environment. Yet another factor that can lead to overheating is the frequent starting and stopping a motor. A typical motor draws five or six times the NP FLA (full-load amps) while starting. This high current causes heating and, if not allowed to cool down before restarting, the windings will overheat and fail. Most motors are rated for continuous duty. Meaning that they are designed to be started and run for several hours before stopping and restarting.

Electric Motors – DC and 1Ø AC

Objective 228.1 Worksheet

_____ 1. A 10°C increase in the operating temperature can decrease the life expectancy of a DC motor by as much as ___.

   a. 1/3   b. ½   c. 10%

_____ 2. The brushes in a DC motor are made of ___.

   a. carbon   b. rubber   c. silicon   d. steel

_____ 3. A DC motor consists of a field circuit and an armature circuit.

   a. True   b. False

_____ 4. A commutator is a device that changes the direction of current flow through the armature loop in a(an) ___ motor(s).

   a. both AC and DC   b. AC   c. DC

_____ 5. ___ is the turning force exerted on the armature in a motor.

   a. Induction   b. Distortion   c. Torque   d. Static

_____ 6. If a motor has field leads that are marked F1 and F2 this is a ___ motor.

   a. DC motor   b. 1-ph AC

_____ 7. Each pole in a large DC motor is an electromagnet.

   a. True   b. False

_____ 8. Power to a DC motor is directly connected to ___.

   a. the brushes   b. the armature   c. both a and b   d. neither a nor b

Year Two (Student Manual)

_____ 9. A ___ motor can operate on either AC or DC power.

    a. 3-phase   b. split-phase   c. universal   d. capacitor-start

_____ 10. DC motors manufactured in the United States are generally rated in ___.

    a. KVA   b. KW   c. HP

_____ 11. The starter on your motor vehicle is actually a ___.

    a. DC motor   b. 3-Ph generator   c. 1Ø generator   d. none of these

_____ 12. If a motor has field leads that are marked S1 and S2 this is a ___ motor.

    a. DC motor   b. 1-ph AC

_____ 13. A(n) ___ is a rotating output device that converts electrical power into a rotating, mechanical force.

    a. solenoid   b. transformer   c. capacitor   d. motor

_____ 14. DC motors have very high starting torques compared to AC motors of the same HP.

    a. True   b. False

_____ 15. ___ is a type of DC motor.

    a. Shunt
    b. Compound
    c. Series
    d. Permanent magnet
    e. all of these

_____ 16. The circuit connected to the brushes in a DC motor is always the same circuit that supplies power to the stator windings.

    a. False   b. True

_____ 17. The speed of a DC motor can be easily controlled.

     a. False   b. True

_____ 18. What is the efficiency of a 50 hp DC motor that draws 80 amps at 500 volts?

     a. 80%   b. 42%   c. 107%   d. 125%   e. 93%

_____ 19. The industry standard used to reverse a DC motor is to reverse the ___ leads.

     a. armature   b. field   c. supply circuit

_____ 20. If a motor has field leads that are marked A1 and A2 this is a ___ motor.

     a. DC motor   b. 1-ph AC

_____ 21. The circuit conductors of a DC motor are connected directly to the ___.

     a. commutator   b. brushes   c. armature

Objective 228.2 Worksheet

_____ 1. Many single-phase AC motors can be operated at ___ volts.

   a. 240   b. 277   c. 480   d. 120   e. any of these

_____ 2. ___ provides standards for the construction of motors and other electrical equipment.

   a. NEMA   b. IEC   c. both a and b   d. neither a nor b

_____ 3. Which of the following rotate together as one assembly within an AC motor?

   a. stator/rotor   b. rotor/shaft   c. stator/shaft   d. none of these

_____ 4. The function of a centrifugal switch in a 1Ø motor is to to drop out the start winding when a critical speed is attained.

   a. False   b. True

_____ 5. Loads connected to a 1Ø AC motor can be ___-driven.

   a. directly   b. gear   c. belt   d. any of these

_____ 6. For a 1Ø AC motor, the field connections to the circuit conductors are made inside the ___.

   a. stator   b. terminal box   c. enclosure   d. commutator

_____ 7. A 3Ø motor is a type of polyphase motor.

   a. False   b. True

_____ 8. NEMA Design C motors are characterized by ___ starting current.

   a. low
   b. normal
   c. high

Electric Motors – DC and 1Ø AC 228-719

_____ 9. ___ torque is the torque a motor produces when the rotor is stationary and full power is applied to the motor.

   a. Breakdown   b. Locked rotor   c. Pull-up   d. Full-load

_____ 10. NEMA Design B motors are characterized by ___ starting torque.

   a. high   b. very high   c. normal

_____ 11. A single-phase AC motor has a nameplate rating of 230 volts. Connecting a 208-volt circuit to the motor will have no adverse effect on the motor.

   a. False   b. True

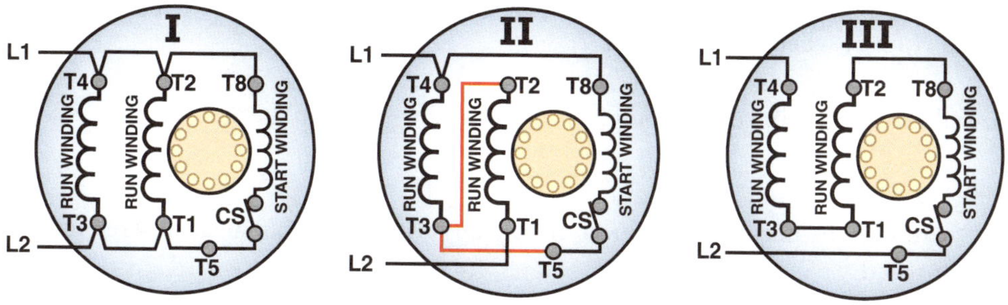

Figure 228.202

_____ 12. Refer to Figure 228.202, shown above. Which 1Ø motor connection diagram is correct for a 120/240 volt dual-voltage, split-phase motor where the branch circuit voltage is 240 volts and CW rotation is required?

   a. I   b. II   c. III   d. none of these

_____ 13. Refer to Figure 228.202, shown above. Which 1Ø motor connection diagram is correct for a 120/240 volt dual-voltage, split-phase motor where the branch circuit voltage is 120 volts and CCW rotation is required?

   a. II   b. III   c. I   d. none of these

_____ 14. ___ speed is the theoretical speed of a motor based on the motor's number of poles and the line frequency.

   a. Pull-up   b. Breakdown   c. Locked rotor   d. Synchronous

_____ 15. Which of the following are types of 1-ph AC motors?

a. capacitor
b. split-phase
c. synchronous
d. shaded-pole
e. all of these

_____ 16. The shaft speed (in RPM) shown on an AC squirrel-cage motor nameplate is the synchronous speed of the motor.

a. False   b. True

_____ 17. ___ torque is the torque required to produce the nameplate rated output power (horsepower) and speed (rpm's) when driving a load.

a. Full-load   b. Locked rotor   c. Pull-up   d. Breakdown

_____ 18. ___ torque is the maximum torque a motor can produce without an abrupt reduction in motor speed.

a. Breakdown   b. Full-load   c. Pull-up   d. Locked rotor

_____ 19. Large horsepower single-phase AC motors are generally not used where the conditions permit the use of a three-phase motors because 1-ph motors ___.
I. are less efficient   II. require larger circuit conductors

a. II only   b. both I & II   c. neither I nor II   d. I only

_____ 20. The speed of a single-phase AC motor can be easily controlled by changing the ___.
I. frequency   II. number of poles   III. voltage

a. II only
b. III only
c. I or II only
d. I or II or III
e. I only
f. II or III only

Electric Motors – DC and 1Ø AC

_____ 21. The part of an AC motor that rotates is the ___.

   a. commutator   b. armature   c. stator   d. rotor

_____ 22. Single-phase AC motors are readily available in hp ratings of 1/6 to ___.

   a. 2   b. 5   c. 1/2   d. 1   e. 10

_____ 23. The difference between the shaft speed (in RPM) and the synchronous speed of an AC motor is the called ___.

   a. frequency   b. slip   c. variance   d. differential

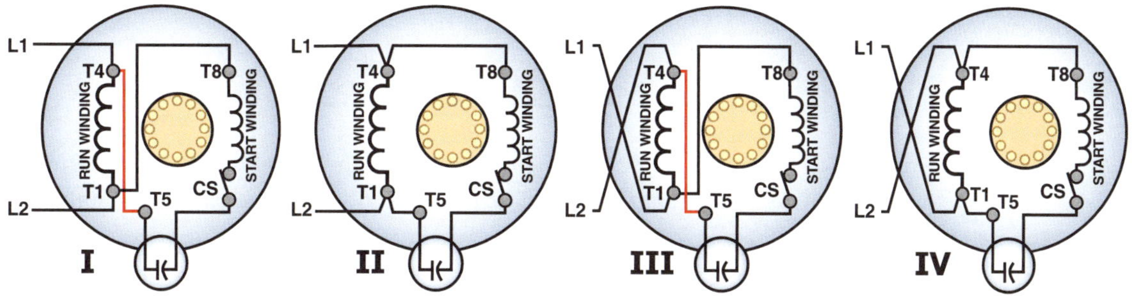

Figure 228.201

_____ 24. Which 1Ø capacitor motor connection diagram, shown above in Figure 228.201, is correct where clockwise rotation is desired?

   a. II only
   b. IV only
   c. II & IV only
   d. I & III only
   e. III & IV only
   f. I & II only

_____ 25. An induction motor is a motor that has no physical electrical connection to the rotor.

   a. False   b. True

_____ 26. For an AC motor, the motor leads in the peckerhead originate at the ___.

   a. rotor   b. stator   c. commutator   d. none of these

Year Two (Student Manual)

_____ 27. All single-phase AC motors can be reversed in the field by changing connections in the peckerhead.

   a. True   b. False

_____ 28. In high altitudes the size of a motor, to drive a given load, may have to be increased because ___.

   a. the circuit conductors can't carry as much current
   b. the motor windings can't cool properly and the windings are overheated

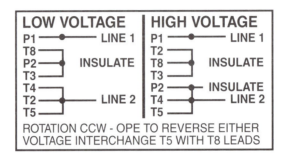

Figure 228.203

_____ 29. Refer to the nameplate from a 115/230 volt motor shown above in Figure 228.203. The motor leads have been terminated on a terminal strip. T-strip connections from a 240-volt supply circuit are correctly shown at ___ (below).

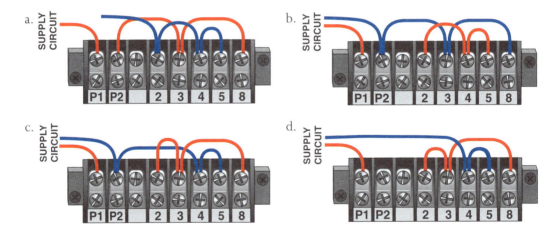

_____ 30. Refer to Figure 228.203, shown above. If you properly connected this single phase motor, but the shaft was rotating in the wrong direction, you would ___.

   a. swap L1 and L2 connections
   b. swap the T5 and T8 leads
   c. swap P1 and P2 leads
   d. none of these

Electric Motors – DC and 1Ø AC

_____ 31. Refer to Figure 228.203, shown on the previous page. The motor nameplate is for a 1-phase dual-voltage (115-208/230) motor. Select the correct connection of the leads where L1 and L2 are the branch circuit conductors that are fed from a 1-pole breaker in a 120/208 volt 3Ø, 4-wire panelboard.

    a. L1 to P1 to T8 to P2 to T3; and L2 to T4 to T2 to T5
    b. L1 to P1 to T2 to T8 to T3; and L2 to P2 to T4 to T5
    c. L1 to P1; T8 to P2 to T3; and L2 to T4 to T2 to T5
    d. L1 to P1; T2 to T8 to T3; and L2 to T4 to T5
    e. L1 to P1; T2 to T8 to T3; and L2 to P2 to T4 to T5

_____ 32. An AC motor operates by changing the direction of current through the ___.

    a. commutator   b. armature   c. rotor   d. stator

_____ 33. Which of the following are fractional horsepower motors?

    a. ½   b. 1   c. 5   d. 2   e. all of these

Lesson 228 Safety Worksheet

Concrete and Masonry Construction

Concrete and masonry construction is very common in today's world. Concrete is used in many ways. Among the most common uses are for slabs, footings, foundations, walls and building piers. This makes many types of construction quicker, more stable and cost effective to build.

Cement is the principal ingredient in concrete along with water, sand and aggregates. Cement is an alkaline material which is hazardous and causes skin reactions and problems, such as dermatitis. Do not allow wet cement to touch your skin. Wear suitable PPE, such as gloves, boots and eye protection. Sometimes more protection is required than just safety glasses, perhaps goggles or a face shield is necessary to keep the cement out of your eyes.

Be aware of backing concrete trucks and the chutes moving from side to side. Also watch for any pinch points on the chutes and sections of the chutes. More than one finger has been amputated in these pinch points.

After concrete has been placed, no loads are permitted on the structure until it can be determined that it is capable of safely supporting the loads to be placed upon it. Any protruding reinforcing steel (rebar) shall be guarded to protect employees from impalement. This can be accomplished by installation of caps that are designed for this type of protection. These are not the small orange caps often seen on construction sites, but rather larger ones that have a larger size and usually a steel insert inside of them to provide extra resistance to penetration in case someone falls on them.

Do not work under concrete buckets that are being hoisted in the air. The concrete could fall from this elevated bucket and create a dangerous or fatal condition

Masonry or poured walls and the related form work required, should use limited access zones during their construction. This zone should be 4 feet wider than the height of the wall and extend the entire length of the wall. Only workers engaged in the construction of the wall are permitted to work in the zone. The limited access zone shall remain active until the wall is supported.

Remember, when working on scaffolding of other trades, if you are going to have to work on their scaffold, it must be constructed correctly and safely before you climb it. A competent person in scaffolding should verify its integrity.

Electric Motors – DC and 1Ø AC

Questions

_____ 1. Concrete is primarily made of ___.
   I. water   II. sand   III. composite   IV. cement   V. aggregates

   a. I and IV only
   b. I, II, IV, and V only
   c. II, III, IV, and V only
   d. I, III, and IV only
   e. I, II, and V only
   f. II, III, and V only

_____ 2. Concrete mixer trucks are dangerous because of ___.
   I. dust   II. backing   III. moving chutes   IV. pinch points

   a. I, II, and III only
   b. II and IV only
   c. I, III, and IV only
   d. II and IV only
   e. II, III, and IV only
   f. I and IV only

_____ 3. Using any scaffolding erected by other trades still requires an inspection by a ___.

   a. competent person   b. superintendent   c. supervisor

_____ 4. If you are working with wet cement products, you must wear ___.
   I. boots   II. gloves   III. eye protection

   a. II and III only
   b. I and II only
   c. I, II and III
   d. none of these

_____ 5. Wet cement is a hazardous material containing ___ materials.

   a. atomic
   b. alkaline
   c. bromide
   d. nuclear

_____ 6. Typical building construction utilizes ___ for footings, slabs, walls and foundations.

   a. concrete   b. Styrofoam   c. gypsum   d. styrene

_____ 7. A limited access zone, for concrete or masonry construction, is intended to keep other employees ___.

   a. out of the zone
   b. working longer
   c. in the zone

At last, it's easier to save electrical energy.
And yours.

lighting CONTROL

Introducing Siemens i-3 Control Technology™: simple, cost-effective lighting control to reduce energy costs. Easy set up via network, USB interface, and integrated touch panel means you can have a Siemens i-3 system up and running in minutes. Easy installation saves on labor. And modular design based on off-the-shelf components saves on maintenance.

For more information or to contact a sales representative, call, e-mail or visit our web site.

SIEMENS

Siemens Energy & Automation Inc. • 1-800-964-4114, Ref. Code: I3IEC • info@sea.siemens.com • www.sea.siemens.com/i-3

Lesson 229
Electric Motors – Polyphase

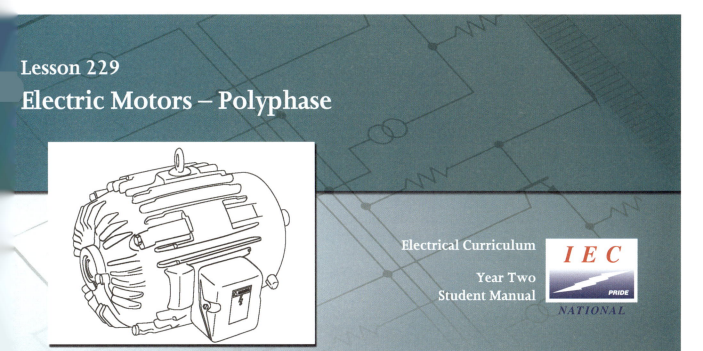

Electrical Curriculum

Year Two
Student Manual

IEC PRIDE
NATIONAL

Purpose

This lesson will discuss the basic operation, construction, and connection of three-phase motors. This lesson will also discuss the various types of these motors that are available as well as how to interpret the information provided by manufacturers.

Homework

(Due at the beginning of this class)

For this lesson, you should:

- Thoroughly read the material contained within Lesson 229.
- Locate and read Yr-2 – Annex M
- Complete Objective 229.1 Worksheet.
- Complete Objective 229.2 Worksheet.
- Complete Objective 229.3 Worksheet
- Read and complete Lesson 229 Safety Worksheet.

- Complete additional assignments or worksheets if available and directed by your instructor.

Objectives

By the end of this lesson, you should:

    229.1 ....Understand the basic 3-phase operation of wound-rotor, synchronous, and squirrel-cage induction motors.

    229.2 ....Understand the differences and be able to identify and properly connect wye and delta-wound squirrel cage induction motors. Additionally, after completing the lab in Yr-2 – Annex M, be able to identify the leads of unmarked 9-lead motors.

    229.3 ....Understand the manufacturer provided information, be able to perform calculations given this information, and be aware of the enclosure types and various starting methods employed for 3-phase motors.

Electric Motors – Polyphase

229.1 Three-Phase Motor Types and Operation

Three-phase, or polyphase, motors run on three-phase power. This power is delivered by a three-phase power system. A three-phase power system has three coils in the generator (power source). Therefore, three separate and distinct voltages will be generated.

Three-phase motors vary in size from fractional horsepower to several thousand horsepower. These motors have a fairly constant speed characteristic but a wide variety of torque characteristics. They are made for practically every standard voltage and frequency and are very often dual-voltage motors. Types of three-phase motors include: wound rotor, synchronous, and squirrel-cage induction.

Wound Rotor Motor

A major difference between the wound-rotor motor and the squirrel-cage rotor is that the conductors of the wound rotor consist of wound coils instead of bars. These coils are connected through slip rings and brushes to external variable resistors, as shown in Figure 229.11. The rotating magnetic field induces a voltage in the rotor windings. Increasing the resistance of the rotor windings causes less current to flow in the rotor windings, decreasing rotor speed. Decreasing the resistance causes more current to flow, increasing rotor speed. These motors are occasionally used when some speed control is desired.

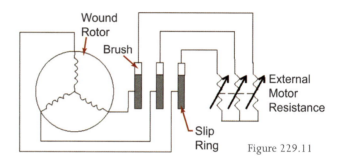

Figure 229.11

Synchronous Motor

Another type of three-phase AC motor is the **synchronous motor**. The synchronous motor is not an induction motor. One type of synchronous motor is constructed somewhat like a squirrel cage rotor. In addition to rotor bars, coil windings are also used. The coil windings are connected to an external DC power supply by slip rings and brushes. When the motor is started, AC power is applied to the stator, and the synchronous motor starts like a squirrel cage rotor. DC power is applied to the rotor coils after the motor has accelerated. This produces a strong constant magnetic field in the rotor which locks the rotor in step with the rotating magnetic field. The rotor therefore turns at synchronous speed, which is why this is called a synchronous motor. As previously mentioned, some synchronous motors use a permanent magnet rotor. This type of motor does not need a DC power source to magnetize the rotor. Synchronous motors are sometimes used instead of induction motors because synchronous motors can be made to act as capacitive loads. The addition of these capacitive loads serves to correct the poor power factor that is caused by inductive loads on the same circuit (branch circuit, feeder, or service). Synchronous motors are also utilized when constant speed is required because, regardless of the load imposed on the motor, the rotor speed always equals the synchronous speed.

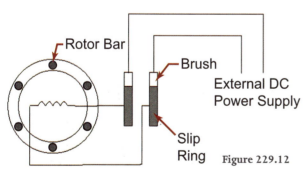

Figure 229.12

Squirrel-Cage Motor

The squirrel cage induction motor is the most widely used among AC motors. Squirrel cage motors are usually chosen over other types of motors because of their simplicity, ruggedness and reliability. There are only two points of mechanical wear on the squirrel cage motor: the two bearings located in each end of the motor enclosure. Because of these features, squirrel-cage motors have practically become the accepted standard for AC, all-purpose, constant speed motor applications. Without a doubt, the squirrel-cage motor is the workhorse of the industry. Ninety percent of the AC motors in use for industrial applications are squirrel cage induction types.

Three-Phase Motor Operation

Just as with a single-phase induction motor, a three-phase induction motor depends upon the electrically rotating magnetic field around the stator windings.

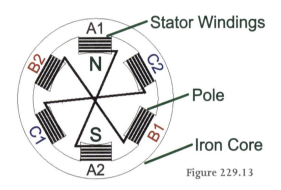

Figure 229.13

Three-phase power can be thought of as three different single-phase power supplies. They are called AØ, BØ, and CØ. In a three-phase motor, each phase of the power supply is connected to its own set of field windings on the stator poles. Each pole of a set is located directly across from the other pole of the set. Each pole set is offset equally from the pole set for each of the other two phases. Figure 229.13 shows the electrical configuration of stator windings. In this example, six windings are used, two for each of the three phases. Because each phase winding has two poles, this is a 2-pole stator. The coils are wound around the soft iron core material of the stator. When current is applied, each winding becomes an electromagnet, with the two windings for each phase operating as the opposite ends of one magnet. In other words, the coils for each phase are wound in such a way that, when current is flowing, one winding is a north pole and the other is a south pole. For example, when **A1** is a north pole, **A2** is a south pole and, when current reverses direction, the polarities of the windings also reverse. Pole **B1** is offset 120° from Pole **A1**, Pole **C1** is offset 120° from Pole **B1**, and Pole **C1** is offset 120° from Pole **A1**.

When the 3Ø supply circuit is connected to the stator, the three currents start at different times. BØ starts 120° later than AØ and CØ starts 120° later than BØ. This is shown on the sine wave graph in Figure 229.14, which indicates the way the magnetic field will point at various times in the cycle.

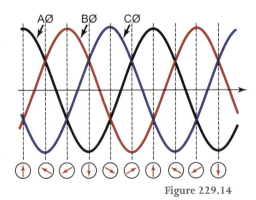

Figure 229.14

Introducing these different phase currents into three field coils 120° apart on the stator produces a rotating magnetic field, and the magnetic poles are in constant rotation. The magnetic poles chase each other, simultaneously inducing electric currents in the rotor. The induced rotor currents set up their own magnetic fields, in opposition to the magnetic field that caused the currents. The resulting attractions and repulsions provide the torque to turn the motor, and keep it turning. Due to the nature of the 3Ø supply circuit, the rotating magnetic field is created naturally in the stator of a three-phase AC induction motor.

Electric Motors – Polyphase

If each magnetic pole were to "light up" whenever it was energized, the effect would appear as though the lights were "running" around the stator, much as the lights on some electric signs simulate a running border.

The operation of one motor revolution is described below.

First, the A poles of the stator are magnetized by AØ. Then, the B poles are magnetized by BØ. The rotor turns, due to the induced current. Then, the C poles are magnetized by CØ. The rotor turns, due to the induced current. The rotor has completed one-half turn at this point. Now, the A poles of the stator are magnetized again, *but the current flow is in the opposite direction*. This causes the magnetic field to continue to rotate, and the rotor follows. Then, the B poles are magnetized by BØ. The rotor turns, due to the induced current. Then, the C poles are magnetized by CØ. The rotor turns, due to the induced current. The rotor has completed one full revolution at this point, and the process repeats itself.

All three-phase motors are wound with a number of coils. These coils are connected to produce three separate windings, sometimes called phases. Each winding must have the same number of coils. The number of coils in each winding must be one-third of the total number of coils in the stator. Therefore, if a three-phase motor has 36 coils, each winding will consist of 12 coils.

Sometimes motor leads are labeled with a "T" preceding the lead number, although some manufacturers omit the "T". Also, some motors have built-in thermal protection to protect the motor against overload. Thermally-protected, dual-voltage motors may have "P" leads that require field connection. These leads are labeled with a "P" preceding the lead number.

The 3-phase circuit conductors, typically referred to as L1, L2, and L3, are connected to the appropriate "T" leads, within the terminal box (peckerhead), to supply power to the stator windings.

3-Phase Motor Rotation

The direction of rotation of a three-phase motor can be reversed by changing the direction of current flow through the stator windings. This can be accomplished by swapping the connections of any two of the three line (circuit) conductors. For example, swapping the L1 and L2 connections will accomplish this change in rotation. However, swapping L1 and L3 or L2 and L3 will also work. These changes can be made at any convenient location within the circuit. The "T" or "L" leads could be swapped inside the peckerhead, the load-side connections could be swapped in the motor disconnect switch, or the circuit conductors could be swapped at the branch circuit breaker terminals. The changes can be made anywhere in the circuit, up to and including the service connections.

For existing installations, when electrical distribution equipment is replaced, the electrician must ensure that existing rotation is checked prior to disconnection of any equipment. After replacing any equipment, the electrician must verify that the provided rotation is correct before re-energizing the distribution equipment. A hand-held phase rotation tester (phase sequence indicator) is

Phase Sequence Indicator

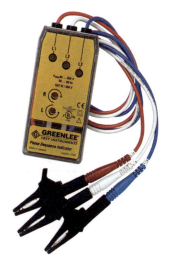

Figure 229.15(A)

shown in Figure 229.15(A). For example, if a subpanel is to be replaced, the electrician must verify that all feeder conductors in the subpanel are properly identified prior to disconnection. If the tester indicates counterclockwise rotation, the electrician must verify that CCW rotation is present at these lugs prior to energizing the new panel. Then the electrician must check the phase rotation provided to each 3-phase motor branch circuit in the subpanel. These branch circuit conductors must be identified so that, when reconnected, will provide the same rotation to the motor. These procedures will ensure that any motor circuits connected in the subpanel will have the proper rotation when reconnected to the power. Another tester available is a motor rotation indicator, as shown in Figure 229.15(B). This tester is used on a de-energized motor and can determine if the motor requires clockwise or counterclockwise rotation to correctly drive the connected load. The electrician can then use a phase rotation tester, as shown in Figure 229.15(A), to check rotation of the provided motor branch circuit. Testers with both capabilities are available.

Motor Rotation Indicator

Figure 229.15(B)

229.2 Single-Speed, 3Ø Induction Motor Configurations and Connections

Although 3Ø multi-speed and other special application 3Ø motors exist, only single-speed 3Ø induction motors are addressed in this objective.

6-Lead, Single-Voltage, 3Ø Motor Configurations

For a 6-lead motor, each of the three windings has two leads (end connections). One winding has leads labeled T1 and T4. The second winding has leads labeled T2 and T5. The third winding has leads labeled T3 and T6. These 3-winding motors have six leads that are brought out into the peckerhead (terminal box). The leads are connected so that the windings are connected in either a Wye (Y) or Delta (Δ) configuration. In either case, for the motor to operate properly, the nominal voltage of the three-phase circuit supplying power to the motor must equal the rated voltage and frequency of the motor. Occasionally, a wye-connected motor is referred to as star-connected. All 3Ø motors are either Y-wound or Δ-wound.

Just as with 3Ø transformer connections, a Wye (Y) configuration is when one end of each of the three windings is connected together. For a wye connection, T4, T5, and T6 are connected together and the remaining leads (T1, T2, and T3) are connected to the 3Ø branch circuit conductors. The branch circuit *line* conductors are typically referred to as L1, L2 and L3 to represent Line 1, Line 2, and Line 3, respectively. A wye connection is shown in Figure 229.16(A). For example, if this *wye-connected* motor's nameplate indicates that a 480-volt, 3-phase circuit is required to operate the motor, 277 volts (480 ÷ 1.732) is applied to each of the individual *windings*.

Electric Motors – Polyphase

Again, as with 3Ø transformer connections, a delta (Δ) configuration is when each winding is wired end-to-end to form a completely closed loop circuit. For this 6-lead, 3-winding motor, T1 is connected to T6, T2 is connected to T4, and T3 is connected to T5. The branch circuit conductors (L1, L2, and L3) are then connected to T1, T2, and T3. A delta connection is shown in Figure 229.16(B). For example, if this *delta-connected* motor's nameplate indicates that a 480-volt, 3-phase circuit is required to operate the motor, 480 volts is applied to each of the individual *windings*.

3-Lead, Single-Voltage, 3Ø Motor

Some motors have only three leads available in the terminal box. With these, the manufacturer has made the wye or delta connections internally and only T1, T2, and T3 are field-connected to the circuit conductors. Since the configuration can not be changed in the field, these 3-lead, factory-connected motors are referred to as wye-wound, as shown in Figure 229.17(A), or delta-wound, as shown in Figure 229.17(B).

Lead Numbering for 3Ø Motors

Three-phase motor leads are numbered spirally, toward the center. This is illustrated in Figure 229.18 for both wye and delta-wound 6-lead motors.

6-Lead Motors

Wye-Wound

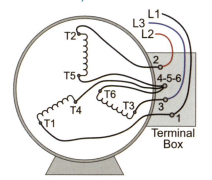

Figure 229.16(A)

Delta-Wound

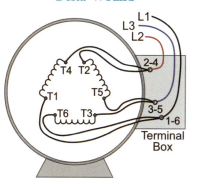

Figure 229.16(B)

Factory-Connected, 3-Lead Motors

Wye-Wound

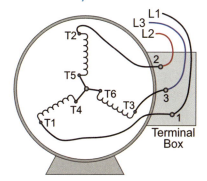

Figure 229.17(A)

Delta-Wound

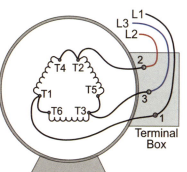

Figure 229.17(B)

Numbering for 6-Lead Motors

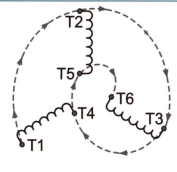
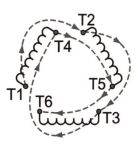

Figure 229.18

Numbering for 12-Lead Motors

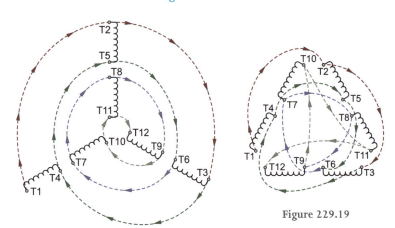

Figure 229.19

Figure 229.19 shows the lead numbering for 12-lead wye and delta-wound motors.

12-Lead, Dual-Voltage, 3Ø Motor Configurations

Many three-phase motors are made so that they can be connected to either of two voltages. These dual-voltage motors are rated to be connected to either of the rated voltages shown on the nameplate.

Usually, the dual-voltage rating of industrial motors is 230/460V. With these motors, each winding is rated at the lower of the two voltages. However, the nameplate **for any motor** must **always** be checked for proper voltage ratings.

Dual-voltage, three-phase motors consist of six individual windings, two for each phase. Again, each individual winding has two leads that are brought out into the terminal box. Since there are two windings for each phase, these motors have twelve leads. Just as with dual-voltage transformers, when supplied with the lower of the two nameplate-rated voltages, the windings for each phase are connected in parallel. When the higher voltage is applied, the windings are connected in series. At either voltage, the **same** amount of current flows through each winding. For a 230/460V motor, if the circuit voltage is 240 (nominal), the windings for each phase are parallel-connected so that 240 volts is applied to each of the two windings. If the circuit voltage is 480 volts (nominal), the windings for each phase are series-connected so that 240 volts is applied to each of the two windings.

At either voltage, the motor will use the same amount of power and deliver the same HP output. However, as with dual-voltage, single-phase motors, when the higher voltage is used, the current in the circuit conductors is halved. With lower current, smaller wire, conduit, and OCPDs can be used. However, at the higher voltage, equipment rated for the circuit voltage must be used. Keep in mind that the equipment cost may be double the cost of the lower voltage equipment.

230/460 V, 12-Lead, Wye-Wound Motor Connections

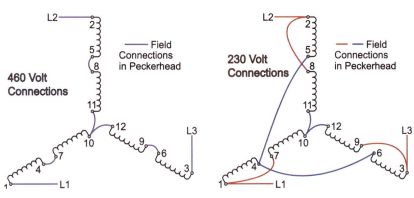

Figure 229.20

The low and high voltage connection diagrams, as made inside the peckerhead (terminal box), for a 12-lead, dual-voltage, wye-wound motor are shown in Figure 229.20.

Electric Motors – Polyphase

Connections, as they typically appear, on a 12-lead, dual-voltage, wye-wound motor nameplate are shown in Figure 229.21.

The low and high voltage connection diagrams, as made inside the peckerhead (terminal box), for a 12-lead, dual-voltage, delta-wound motor are shown in Figure 229.22.

Connections, as they typically appear, on a 12-lead, dual-voltage, delta-wound motor nameplate are shown in Figure 229.23.

9-Lead, Dual-Voltage, 3Ø Motor

The most commonly used dual-voltage, 3Ø motors are those with only nine leads available in the terminal box. With these, the manufacturer has made the required T10, T11, and T12 connections internally. Nine-lead, dual-voltage motors can be either wye or delta wound. Wye-wound motors are more popular in the United States because the line-to-ground insulation requirements are less demanding than the requirements for delta-wound motors. This results in wye-wound motors being less costly to manufacture. Delta-wound motors are typically found in equipment manufactured in Europe, and are especially popular in Germany and Italy.

Nameplate Connections for a 230/460 V, 12-Lead, Wye-Wound Motor

Nameplate Connections for a 230/460 V, 12-Lead, Delta-Wound Motor

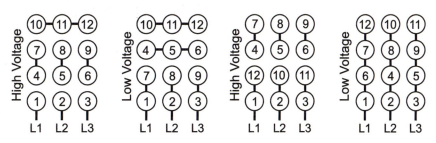

Figure 229.21

Figure 229.23

230/460 V, 12-Lead, Delta-Wound Motor Connections

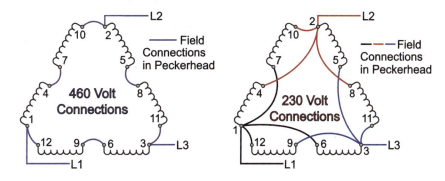

Figure 229.22

230/460 V, 9-Lead, Wye-Wound Motor Connections

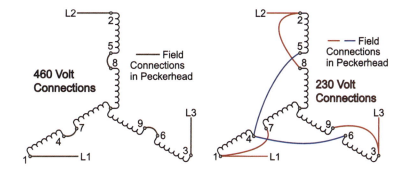

Figure 229.24

The typical nameplate diagram and the required field connections for a 9-lead, wye-wound, dual-voltage motor are shown in Figures 229.24 and 229.25.

The typical nameplate diagram and the required field connections for a 9-lead, delta-wound, dual-voltage motor are shown in Figures 229.26 and 229.27.

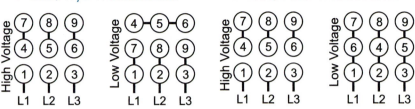

Nameplate Connections for a 230/460 V, 9-Lead, Wye-Wound Motor

Figure 229.25

Nameplate Connections for a 230/460 V, 9-Lead, Delta-Wound Motor

Figure 229.27

230/460 V, 9-Lead, Delta-Wound Motor Connections

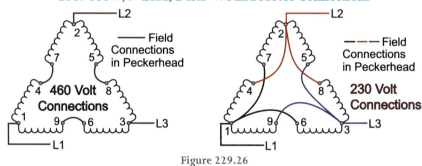

Figure 229.26

Motor Lead Continuity

Individual motor windings and their leads are insulated from each other and from the motor housing. A simple ohmmeter can be used to make cursory checks of insulation and continuity.

Between each end of a winding (T1 and T4, for example) an ohmmeter should indicate between 0.2 and 20.0 Ω. For larger motor sizes, larger wire is used for the windings. The use of larger wire results in smaller ohmmeter readings (closer to the 0.2 ohm reading). Of course, resistance measurements on smaller motors will yield readings closer to the upper end of the range (closer to the 20 Ω reading). For a 9-lead, delta-wound motor, an ohmmeter reading taken between T4 and T9 should result in twice the reading between T1 and T4. If the reading taken between T1 and T4 was 9 Ω, the reading between T4 and T9 would be 18 Ω.

Insulation resistance checks between any lead, of any motor, and the motor housing should be infinity (∞). Of course, on a digital ohmmeter, ∞ appears as OL when the meter is set to read Ω.

After a period of time in service, due to dust, moisture, vibration, etc, motor leads and windings suffer some deterioration of their insulation. As a part of scheduled maintenance, motors are tested with megohmmeters (meggers). During these tests, a test voltage (usually high-voltage DC) is imposed on the motor windings. The test results are interpreted to determine if the motor can be pulled from service and simply cleaned or if the motor must be sent to a motor repair facility for rewinding.

Identifying the Leads of an Unmarked 3-Phase Motor

On occasion the lead markings of a 3-phase motor are lost. Generally, this occurs when a motor is removed from service or relocated, and proper procedures are not followed. Since 9-lead motors are most popularly used, procedures have been developed to remark the leads. Although one procedure involves applying 3-phase line voltage to some of the windings, working with line voltage can be dangerous. Instead, a small 6 or 12-volt battery can be used. Refer to Yr-2 – Annex M, for the complete lab procedure.

NEMA versus IEC 3-Phase Motor Lead Markings

As the economy becomes more global, the electrician must be able to recognize the differences between those motors manufactured to NEMA (National Electrical Manufacturers Association) standards versus those manufactured to IEC (International Electrotechnical Commission) standards. In addition to the differences in voltage ratings, the IEC motor leads are numbered differently. The comparisons, as adopted by the IEC in 2003, appear in the table shown below.

3-Phase Motor Type	NEMA Lead Identification	IEC Lead Identification
3-Lead, Y and Δ-Wound	T1 T2 T3	U1 V1 W1
6-Lead, Y and Δ-Wound	T1-T4 T2-T5 T3-T6	U1-U2 V1-V2 W1-W2
9-Lead, Y and Δ-Wound	T1-T4 T2-T5 T3-T6 T7 T8 T9	U1-U2 V1-V2 W1-W2 U5 V5 W5
12-Lead, Y and Δ-Wound	T1-T4 T2-T5 T3-T6 T7-T10 T8-T11 T9-T12	U1-U2 V1-V2 W1-W2 U5-U6 V5-V6 W5-W6

229.3 Motor Nameplates, Starting Methods, Enclosure Types, and Calculations

3-Phase Motor Nameplate

The nameplate of a motor provides important information necessary for proper application. Shown at the right is the nameplate of a typical three-phase motor.

Other information, included on the nameplate for this motor, is detailed below and on the following pages.

SIEMENS
HIGH EFFICIENT

ORD.NO.	1LA02864SE41	E NO.			
TYPE	RGZESD	FRAME	286T		
H.P.	30.00	SERVICE FACTOR	1.15		3 PH
AMPS	35.0	VOLTS	460		
R.P.M.	1765	HERTZ	60		
DUTY	CONT 40°C AMB			DATE CODE	
CLASS INSUL	F	NEMA DESIGN	B	K.V.A. CODE	G	NEMA NOM. EFF.	93.0
SH. END BRG	50BC03JPP3	OPP. END BRG.	50VC03JPP3		

Made in Mexico by SIEMENS

H.P.	Horsepower is the amount of output mechanical power delivered by the motor. This motor is rated to deliver 30 horsepower when fully loaded and operating at rated voltage and frequency.
VOLTS	This is a single-voltage motor and is designed to operate at 460 volts (nominal).
AMPS	When fully loaded and operating at rated voltage and frequency, this motor will pull 35.0 amps through the branch circuit conductors. Sometimes noted as FLA or FLC.
R.P.M.	This is the shaft speed, given in RPM (revolutions per minute), at which the motor develops rated horsepower at rated voltage and frequency. This base speed is an indication of how fast the output shaft will turn the connected equipment when fully loaded. This motor has a base speed of 1765 RPM at a rated frequency of 60 Hz. If the motor is operated at less than full load, the output speed will be slightly greater than the base speed.
DUTY	This motor is rated to be operated continuously when installed in a location where the ambient temperature (AMB) does not exceed 40°C. If a motor is to be operated for 1 hour or more within a 24-hour period, a motor with a continuous-duty rating should be used. Motors rated for continuous duty (24 hr) are the most commonly used type.
CLASS INSULATION	NEMA defines motor insulation classes to describe the ability of motor insulation to handle heat. The four insulation classes are A, B, F, and H. All four classes identify the allowable temperature rise from an ambient temperature of 40° C (104° F). Classes B and F are the most commonly used.

Electric Motors – Polyphase

AMB	Ambient temperature is the temperature of the surrounding air. This is also the temperature of the motor windings before starting the motor, assuming the motor has been stopped long enough. Temperature rises in the motor windings as soon as the motor is started. The combination of ambient temperature and allowed temperature rise equals the maximum rated winding temperature. If the motor is operated at a higher winding temperature, service life will be reduced. A 10° C increase in the operating temperature above the allowed maximum can cut the motor's insulation life expectancy in half. The standard ambient temperature is 40°C, or 104°F. This value was selected as one that would seldom be exceeded for any appreciable length of time in the majority of cases. Motors are usually designed for this temperature unless there is a definite requirement for a machine with some other value. Motors for use in abnormally hot places are usually designed to accommodate the higher ambient by having a lower winding temperature rise, and are sometimes designed in a larger frame size. The opposite is also true. Operation of motors in very cold ambients can result in severe duty on the motor component parts. Arctic duty motors are available that are rated for operation down to minus 60°C.
NEMA DESIGN	NEMA uses letters (A, B, C, and D) to identify motor designs based on torque characteristics. The motor in this example is a Design B motor. Some motor nameplates note this as simply DESIGN. • Design A is the least common type. These have normal starting torque (typically 150-170% of rated) and relatively high starting current. The breakdown torque is the highest of all the NEMA types. It can handle heavy overloads for a short duration. A typical application is the powering of injection molding machines. • Design B is the most common type of AC induction motor sold. It has a normal starting torque, similar to Design A, but offers low starting current. The locked rotor torque is good enough to start many loads encountered in the industrial applications. The motor efficiency and full-load PF are comparatively high, contributing to the popularity of the design. The typical applications include pumps, fans and machine tools. • Design C has high starting torque and low starting current. These are used for driving heavy loads such as conveyors, crushers, stirring machines, agitators, reciprocating pumps, compressors, etc. These motors are intended for operation near full speed without great overloads. • Design D has the highest starting torque. The starting current and full-load speed are low. This motor is suitable for applications with changing loads and subsequent sharp changes in the motor speed. The speed regulation is poor, making the design suitable only for punch presses, cranes, elevators and oil well pumps.
KVA CODE	Often noted on nameplates and specifications as Locked Rotor Code (or simply CODE), this code is an indicator of how much starting current (locked rotor current) is required to start the motor. NEC© Article 430 includes a table to approximate the locked rotor current (LRA or LRC), given HP.

NEMA NOMINAL EFF.	Motor efficiency is the percentage of the energy supplied to the motor that is converted into mechanical energy at the motor's shaft when the motor is continuously operating at full load with the rated voltage applied. Generally, larger motors are more efficient than smaller motors. Premium efficiency 3-phase motors have efficiencies ranging from 86.5% at 1 hp to 95.8% at 300 hp. Although not always shown on the nameplate, efficiency ratings are available on manufacturer's specification data sheets.
HERTZ	This is the line frequency at which this motor is rated to operate.
SERVICE FACTOR	Service factor (SF) is a number that is multiplied by the rated HP of the motor to determine the HP at which the motor can be operated. Therefore, a motor designed to operate at or below its nameplate HP rating has a SF of 1.0. Some motors are designed for a SF higher than 1.0, so that they can, occasionally and for short periods of time, exceed their rated horsepower. For example, this 30 HP motor has a SF of 1.15 and can be operated at 34.5 HP (15% higher than its nameplate HP). Keep in mind that any motor operating continuously above its rated HP will have a reduced service life. The service factor was established for operation at rated voltage, frequency, ambient and sea level conditions.
PH	Single or three-phase is indicated in this location.
FRAME	NEMA has standardized motor dimensions for a range of frame sizes. Standardized dimensions include bolt-hole size, mounting base dimensions, shaft height, shaft diameter, and shaft length. Use of standardized dimensions allows existing motors to be replaced without reworking the mounting arrangement. When replacing an existing motor, another with the same frame size should be used.

Another motor nameplate is shown in Figure 229.28. Notice that all the information shown on the Siemens nameplate is not shown on the nameplate for this General Purpose motor. General Purpose type motors are typically used because they are much less expensive than premium, energy efficient motors that have high power factors. The manufacturer's specification data sheet, shown in the following table, provides additional information. Note the efficiency and power factor for this 1 hp motor. Low efficiency and low power factor are typical of small (hp rating) motors.

General Purpose Motor Nameplate

SER FAC	NEMA DES	HZ	INDUCTION MOTOR
1.25	B	60	
Horsepower	RPM	PH	MODEL
1	3450	3	TEFC
TIME HRS	INS CL	FRAME	MAX AMB
24	B	56	40° C
VOLTS		AMPS	
230/460			3.6/1.8
Electric Motor Company			

Figure 229.28

Electric Motors – Polyphase

Horsepower:	1	Service Factor:	1.25
Voltage:	230/460	Rating:	40C AMB-CONT
Hertz:	60	Locked Rotor Code:	H
Phase:	3	NEMA Design Code:	B
Full Load Amps:	3.6/1.8	Insulation Class:	B
Usable at 208 Volts:	N/A	Full Load Efficiency:	75.5
RPM:	3450	Power Factor:	71
Frame Size:	56	Enclosure:	TEFC

Motor Circuit Voltage

Each motor winding is rated to operate at its rated voltage. Although motors with other voltage ratings are available, within this lesson only those motors rated to operate at 208-230, 230, 230/460, or 460 volts will be considered. Whether the supply circuit is from a wye or delta source (transformer or generator) is IRRELEVENT. A motor only requires that the nominal line-to-line voltage of the circuit be the same as the nameplate rated voltage. For example, a 3Ø, 480-volt, delta-wound motor can be supplied by a 3-phase, 3-wire, 480-volt circuit that originates in a 277/480-volt, 3ph, 4W panelboard. Although the motor circuit is from a wye system, the line-to-line (line) voltage applied to the motor is 480 volts. Conversely, a 3Ø, 240-volt, wye-wound motor can be supplied by a 3-phase, 3-wire, 240-volt circuit that originates in a 120/240-volt, 3ph, 4W panelboard. Although the motor circuit is from a delta system, the line-to-line (line) voltage applied to the motor is 240 volts.

208-230/460 V Motor Nameplate

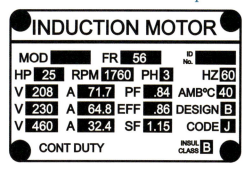

Figure 229.29

3-Phase Motors Operated at 208 Volts, Line-to-Line

Some three-phase motors can be operated at 208 volts without damaging the motor. Only those motors with this information included on the nameplate should be operated at 208 volts.

For example, if a motor nameplate reads: "VOLTS 230" or "VOLTS 230/460" then it is **not** rated to be operated at 208 volts. If operated at 208 volts, the windings will be damaged and the life of the motor could be severely shortened.

Conversely, if the motor nameplate reads: "VOLTS 208-230" or "VOLTS 208-230/460", as shown in Figure 229.29, then it **is** rated to be operated at 208 volts. In those cases where a motor with a straight voltage rating of "230" or "230/460" is to be operated at 208 volts, a boost transformer may be required.

Refer to Figure 229.29 which shows the nameplate for a 25 hp, dual-voltage motor. At any of the rated voltages shown on the nameplate (208 or 230 or 460) this motor will deliver about 25 horsepower. To accomplish this, the motor will require that sufficient line current flow through the windings, according to the voltage applied to each winding. The formula for calculating line current for a three-phase motor, given hp, is:

☞ $I_{line} = (HP \times 746) \div (E_{line} \times Eff \times PF \times 1.732)$

For this 25 hp motor, when the circuit voltage is 460 volts, the formula yields a line current of 32.4 amps.

$I_{line} = (25 \times 746) \div (460 \times .86 \times .84 \times 1.732) = 32.4$ amps at 460 volts

When the circuit voltage is 460, the windings are connected in *series* so that 32.4 amps of current flows through each of the two windings in a set. This is the rated current for each winding in this motor.

When the circuit voltage is 230 volts, the formula yields a line current of 64.8 amps.

$I_{line} = (25 \times 746) \div (230 \times .86 \times .84 \times 1.732) = 64.8$ amps at 230 volts

When the circuit voltage is 230, the windings are connected in *parallel* so that 32.4 amps of current flows through each of the two windings in a set. Half of the total line current (64.8 ÷ 2) flows through each of the two parallel-connected windings. Again, 32.4 amps is the rated current for each winding in this motor.

This particular motor is rated to operate at 208 volts without damage to the motor windings. When the circuit voltage is 208 volts, the formula yields a line current of 71.7 amps.

$I_{line} = (25 \times 746) \div (208 \times .86 \times .84 \times 1.732) = 71.66$ amps at 208 volts

When the circuit voltage is 208, the windings are also connected in *parallel* so that 35.8 amps of current flows through each of the two windings in a set. Half of the total line current (71.66 ÷ 2) flows through each of the two parallel-connected windings. Although the rated current for each of this motor's windings is 32.4 amps, the manufacturer permits operation at 208 volts. Note that most manufacturers stipulate in the specifications that motor life may be somewhat shortened by the slightly increased current when operated at 208 volts.

Common Low Voltage Starting Methods for Induction Motors

Note that reduced voltage starting methods will be studied in depth during the 4th year of the curriculum.

- Across-the-Line – Can be used on either a delta or wye-wound motor. Its use is limited to those systems or services of sufficient capacity to stand the high starting current without excessive voltage drop. Large HP motors generally utilize one of the "reduced voltage" starting methods listed below. When motors are started "across-the-line", line voltage is applied directly to the motor leads.
- Part Winding – Requires a special part-winding motor.
- Primary Resistor – Can be used on either a delta- or wye-wound motor.
- Primary Reactor – Can be used on either a delta- or wye-wound motor
- Neutral Reactor – Requires a wye-wound motor.

- Wye-Delta (Star-Delta) – Requires a 6-lead or 12-lead, delta-wound motor.
- Soft Start (SCR Phase Modulation) – Can be used on either a delta- or wye-wound motor.
- Autotransformer – Can be used on either a delta- or wye-wound motor.
- VFD (Variable Frequency Drive) – Can be used on either a delta- or wye-wound motor.

Motor Enclosure Types

A motor's enclosure not only holds the motors components together, it also protects the internal components from moisture and contaminants. The degree of protection depends on the enclosure type. In addition, the type of enclosure affects the motor's cooling.

Heat is one of the most destructive stresses causing premature motor failure. Overheating occurs because of motor overloading, low or unbalanced voltage at the motor terminals, excessive ambient temperatures, or poor cooling caused by dirt or lack of ventilation. If the heat is not dissipated, insulation failure and possibly lubrication and bearing failure can damage a motor.

Moisture should be kept from entering a motor. Water from splashing or condensation seriously degrades an insulation system. Non-conducting contaminants are readily converted into good leakage current conductors. The proper type of motor should be chosen for use in a damp environment. Most motors can be equipped with drains or breathers to allow moisture to drain from the motor. Space heaters are also available to prevent moisture condensation in the motor during times the motor is not running. Non-conducting contaminants such as factory dust and sand gradually promote over-temperature by restricting cooling air circulation, and may erode the winding insulation.

There are two general categories of enclosures: open and totally enclosed. Four types of commonly used motor enclosures are described below.

Open enclosures permit cooling air to flow through the motor. One type of open enclosure is the open drip proof (ODP) enclosure. This enclosure has vents that allow for air flow. Fan blades attached to the rotor move air through the motor when the rotor is turning. The vents are positioned so that liquids and solids falling from above at angles up to 15° from vertical cannot enter the interior of the motor when the motor is mounted on a horizontal surface. When the motor is mounted on a vertical surface, such as a wall or panel, a special cover may be needed. ODP enclosures should be used in environments free from contaminates, including splashing liquid.

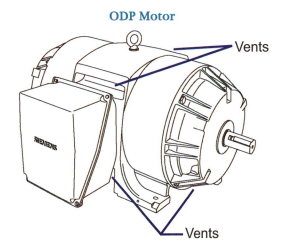

In some applications, the air surrounding the motor contains corrosive or harmful elements which can damage the internal parts of a motor. A totally enclosed non-ventilated (TENV) motor enclosure limits the flow of air into the motor, but is not airtight.

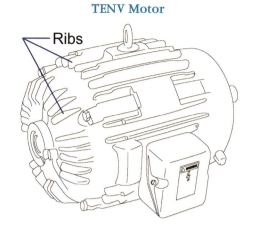
TENV Motor

However, a seal at the point where the shaft passes through the housing prevents water, dust, and other foreign matter from entering the motor along the shaft. Most TENV motors are fractional horsepower. However, integral horsepower TENV motors are used for special applications. The absence of ventilating openings means that all the heat from inside the motor must dissipate through the enclosure by conduction. These larger horsepower TENV motors have an enclosure that is heavily ribbed to help dissipate heat more quickly. TENV motors can be used indoors or outdoors.

A totally enclosed fan-cooled (TEFC) motor is similar to a TENV motor, but has an external fan mounted opposite the drive end of the motor. The fan blows air over the motor's exterior for additional cooling. The fan is covered by a shroud to prevent anyone from touching it. TEFC motors can be used in dirty, moist, or mildly corrosive environments.

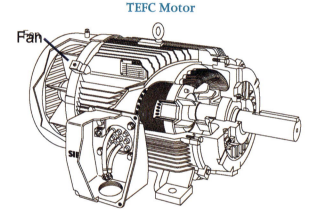
TEFC Motor

Hazardous duty applications are commonly found in chemical processing, mining, foundry, pulp and paper, waste management, and petrochemical industries. In these applications, motors have to comply with the strictest safety standards for the protection of life, machines and the environment. This often requires use of explosion proof (XP) motors. An XP motor is similar in appearance to a TEFC motor, however, most XP enclosures are cast iron. In the United States, the application of motors in hazardous locations is subject to the NEC© and standards set by UL and various regulatory agencies.

You should never specify or suggest the type of hazardous location classification. It is the user's responsibility to comply with all applicable codes and to contact local regulatory agencies to define hazardous locations.

Motor Calculations

A table of motor calculation formulas is shown below. Along with the formulas used for three-phase motors, the formulas used for DC and single-phase AC motors are also shown.

Electric Motors – Polyphase

To find line current: Given HP Rating	☞ $I_{line} = (HP \times 746) \div (E_{line} \times Eff)$	DC
	☞ $I_{line} = (HP \times 746) \div (E_{line} \times Eff \times PF)$	1Ø
	☞ $I_{line} = (HP \times 746) \div (E_{line} \times Eff \times PF \times 1.732)$	3Ø
To find line current: Given kW Rating	☞ $I_{line} = (kW \times 1000) \div (E_{line})$	DC
	☞ $I_{line} = (kW \times 1000) \div (E_{line} \times PF)$	1Ø
	☞ $I_{line} = (kW \times 1000) \div (E_{line} \times PF \times 1.732)$	3Ø
To find line current: Given kVA Rating	--	DC
	☞ $I_{line} = (kVA \times 1000) \div (E_{line})$	1Ø
	☞ $I_{line} = (kVA \times 1000) \div (E_{line} \times 1.732)$	3Ø
To find kW:	☞ $kW = (I_{line} \times E_{line}) \div (1000)$	DC
	☞ $kW = (I_{line} \times E_{line} \times PF) \div (1000)$	1Ø
	☞ $kW = (I_{line} \times E_{line} \times PF \times 1.732) \div (1000)$	3Ø
To find kVA:	--	DC
	☞ $kVA = (I_{line} \times E_{line}) \div (1000)$	1Ø
	☞ $kVA = (I_{line} \times E_{line} \times 1.732) \div (1000)$	3Ø
To find HP output:	☞ $HP = (I_{line} \times E_{line} \times Eff) \div (746)$	DC
	☞ $HP = (I_{line} \times E_{line} \times Eff \times PF) \div (746)$	1Ø
	☞ $HP = (I_{line} \times E_{line} \times Eff \times PF \times 1.732) \div (746)$	3Ø
To find Efficiency:	☞ $Eff = (HP \times 746) \div (I_{line} \times E_{line})$	DC
	☞ $Eff = (HP \times 746) \div (I_{line} \times E_{line} \times PF)$	1Ø
	☞ $Eff = (HP \times 746) \div (I_{line} \times E_{line} \times PF \times 1.732)$	3Ø

Electric Motors – Polyphase

Objective 229.1 Worksheet

_____ 1. Motors with built-in thermal protection have field leads labeled with a(an) ___ preceding the lead number.

    a. P   b. A   c. M   d. none of these

_____ 2. If a 3-phase motor is connected to power but is not rotating in the desired direction the rotation can be changed by swapping any two "hot" circuit conductors ___.

    a. in the motor terminal box
    b. at the ckt breaker feeding the disconnect
    c. in the safety switch
    d. any of these
    e. none of these

_____ 3. Three-phase motors are available with ratings of ___ hp.

    a. less than one   b. 1 to 100   c. several thousand   d. all of these

_____ 4. ___ is a basic type of 3-phase motor.
    I. Squirrel cage induction   II. Wound rotor   III. Synchronous

    a. I & II only
    b. I, II, & III
    c. I only
    d. II & III only
    e. I & III only

_____ 5. When motor-driven equipment is first energized, ___.

    a. start-up protocols must be followed
    b. the electrical inspector must be present
    c. voltage must be ramped up (slowly increased) to the rated equipment voltage
    d. all of these

_____ 6. Three separate winding sets are contained in the stator of a 3-phase motor. These are set at ___ degrees apart.

    a. 90   b. 180   c. 180   d. 120   e. 45

_____ 7. A ___ motor utilizes a bank of variable resistors that are external to the motor.

   a. synchronous b. squirrel-cage induction c. wound rotor

_____ 8. A set of wires (black, red, and blue) are installed between a 3-pole safety switch and a 3Ø motor. The direction of rotation of the motor can be changed by swapping the ___ wire connections on the load side of the switch.

   a. black and red b. black and blue c. red and blue d. either a or b or c

_____ 9. Generally, only ___ is authorized to initially energize air-handling and other HVAC equipment.

   a. a journeyman electrician
   b. the MEP engineer
   c. the mechanical contractor
   d. none of these

_____ 10. Prior to disconnecting a 3-phase service drop or lateral, the phase rotation at the service equipment terminals should be checked with a meter and noted. When the service equipment is reconnected, the connection should be made so that the rotation present at the terminals is ___.

   a. opposite to the noted phase rotation
   b. the same as the noted phase rotation

_____ 11. The most commonly used 3-phase motor is the ___ motor.

   a. synchronous b. wound rotor c. squirrel-cage induction

_____ 12. A rotating magnetic field rotates in the ___ of a 3-phase AC motor.

   a. rotor b. stator

_____ 13. A ___ motor rotates at a constant speed, regardless of how heavily loaded.

   a. squirrel-cage induction b. wound rotor c. synchronous

Objective 229.2 Worksheet

Scenario 229-201

Select the resistance measurement that should be indicated between the points on the 9-lead wye-wound 230/460 volt dual-voltage 3Ø motor. The motor has not been connected to any circuit leads. Nor have any leads been connected to each other.

_____ 1. Refer to Scenario 229-201, shown above. Between T7 and T8 your meter should indicate ____ Ω.

    a. 0.2 to 10 Ω   b. ∞ Ω (or OL on a digital meter)

_____ 2. Refer to Scenario 229-201, shown above. Between T1 and T2 your meter should indicate ____ Ω.

    a. 0.2 to 10 Ω   b. ∞ Ω (or OL on a digital meter)

_____ 3. Refer to Scenario 229-201, shown above. Between T3 and T6 your meter should indicate ____ Ω.

    a. ∞ Ω (or OL on a digital meter)   b. 0.2 to 10 Ω

_____ 4. Refer to Scenario 229-201, shown above. Between T8 and the motor housing your meter should indicate ____ Ω.

    a. ∞ Ω (or OL on a digital meter)   b. 0.2 to 10 Ω

_____ 5. Refer to Scenario 229-201, shown above. Between T1 and T4 your meter should indicate ____ Ω.

    a. ∞ Ω (or OL on a digital meter)   b. 0.2 to 10 Ω

Scenario 229-202

Select the resistance measurement that should be indicated between the points on a 9-lead, dual-voltage, delta-wound, 3Ø motor. The motor has not been connected to any circuit leads. Nor have any leads been connected to each other.

_____ 6. Refer to Scenario 229-201, shown above. Between T4 and T9 your meter should indicate ____ Ω.

    a. ∞ Ω (or OL on a digital meter)
    b. 0.2 to 10 Ω

Electric Motors – Polyphase

229-751

_____ 7. Refer to Scenario 229-201, shown above. Between T3 and T6 your meter should indicate ___ Ω.

    a. ∞ Ω (or OL on a digital meter)   b. 0.2 to 10 Ω

_____ 8. To re-identify the leads of a 9-lead motor, the procedure can involve the use of ___.

    a. 3Ø line voltage
    b. a 6 or 12-volt battery
    c. either of these
    d. neither of these

_____ 9. A 3-ph motor with six windings will have ___ leads in the peckerhead.

    a. 6   b. 9   c. either 6 or 12   d. 12   e. either 6 or 9 or 12

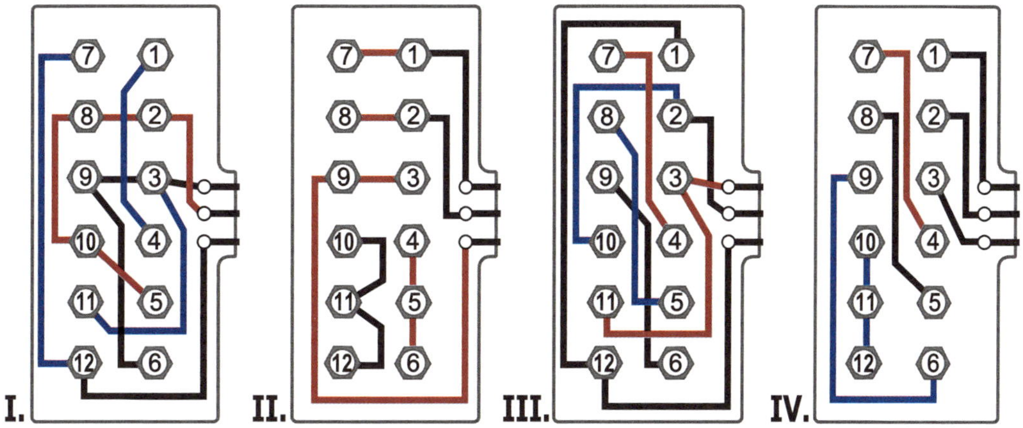

Figure 229.212

_____ 10. Among the terminal box (peckerhead) connection drawings shown above in Figure 229.212, Diagram ___ is correctly drawn for a 12-lead, delta-wound, dual-voltage motor connected to high (480V) voltage.

    a. IV   b. I   c. II   d. III   e. none of these

_____ 11. Among the terminal box (peckerhead) connection drawings shown above in Figure 229.212, Diagram ___ is correctly drawn for a 12-lead, wye-wound, dual-voltage motor connected to high (480V) voltage.

    a. IV   b. III   c. II   d. I   e. none of these

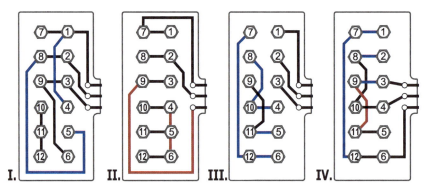

Figure 229.211

_____ 12. Among the terminal box (peckerhead) connection drawings shown above in Figure 229.211, Diagram ___ is correctly drawn for a 12-lead, wye-wound, dual-voltage motor connected to low (240V) voltage.

a. IV b. II c. I d. III e. none of these

_____ 13. Among the terminal box (peckerhead) connection drawings shown above in Figure 229.211, Diagram ___ is correctly drawn for a 12-lead, delta-wound, dual-voltage motor connected to low (240V) voltage.

a. II b. III c. IV d. I e. none of these

_____ 14. When the windings of a 3Ø dual-voltage motor are connected for the lower voltage, the winding sets are connected in ___.

a. series b. parallel

_____ 15. The motor winding leads in Drawing ___ are correctly numbered.

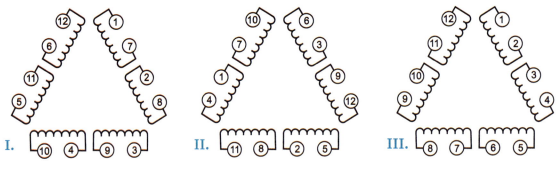

a. I only b. II only c. III only d. none of these

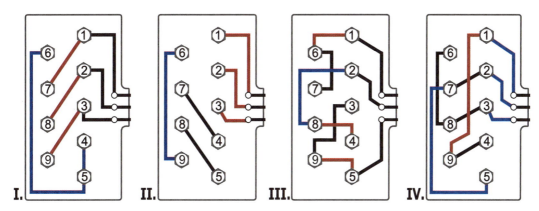

Figure 229.214

_____ 16. Among the terminal box (peckerhead) connection drawings shown above in Figure 229.214, Diagram ___ is correctly drawn for a 9-lead, wye-wound, dual-voltage motor connected to low (240V) voltage.

   a. III   b. IV   c. I   d. II   e. none of these

_____ 17. Among the terminal box (peckerhead) connection drawings shown above in Figure 229.214, Diagram ___ is correctly drawn for a 9-lead, delta-wound, dual-voltage motor connected to high (480V) voltage.

   a. I   b. II   c. III   d. IV   e. none of these

_____ 18. Routine maintenance of motors includes tests performed with a ___.

   a. megger   b. 6 or 12-volt battery   c. 3Ø AC generator   d. none of these

_____ 19. The terminal box for a 230/460-volt motor is shown below. If the leads are connected for the lower voltage, as shown, this is a ___-wound motor.

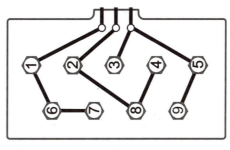

   a. neither Δ nor Y   b. Δ   c. either Δ or Y   d. Y

20. A motor has a nameplate wiring diagram as shown below. Where connected to the available 208-volt, 3-phase branch circuit, select the correct connection of the leads.
L1=AØ   L2=BØ   L3=CØ

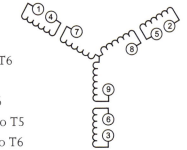

a. L1 to T1 to T7; L2 to T2 to T8; L3 to T3 to T9; and T4 to T5 to T6
b. L1 to T1; L2 to T2; L3 to T3; T7 to T4; T8 to T5, and T9 to T6
c. L1 to T1; L2 to T2; L3 to T3; T7 to T8 to T9; and T4 to T5 to T6
d. L1 to T1 to T7 to T6; L2 to T2 to T8 to T4; and L3 to T3 to T9 to T5
e. L1 to T1 to T7 to T4; L2 to T2 to T8 to T5; and L3 to T3 to T9 to T6

21. Among the terminal box (peckerhead) connection drawings shown below, Diagram ____ is correctly drawn for a 9-lead, wye-wound, dual-voltage motor connected to high (480V) voltage.

a.
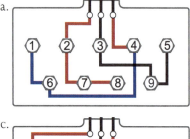

b.

c.
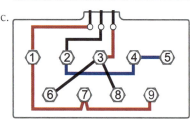

d.
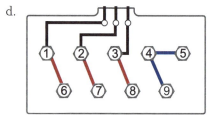

e. none of these

22. The motor winding leads in Drawing ____ are correctly numbered.

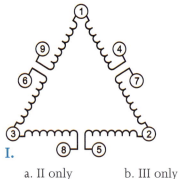
I.

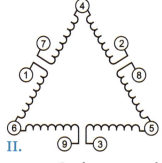
II.

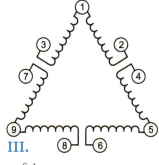
III.

a. II only   b. III only   c. I only   d. none of these

Electric Motors – Polyphase

23. A 460-volt 3Ø motor could operate correctly where circuit conductors are run from the 120/240 V, 3Ø, 4-wire panelboard shown above in Figure 229.201. The proper circuit should have ___.

a. a neutral plus 3"hots" connected to the breaker in spaces 4-6-8
b. 3 "hots" connected to the breakers in spaces 15-17-19
c. 3 "hots" connected to the breaker in spaces 16-18-20
d. none of these

Figure 229.201

24. The motor winding leads in Drawing ___ are correctly numbered.

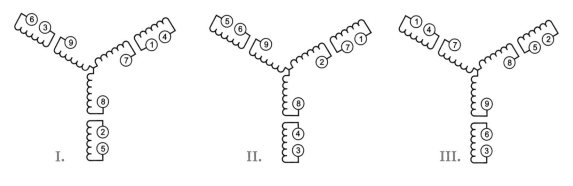

a. III only   b. II only   c. I only   d. none of these

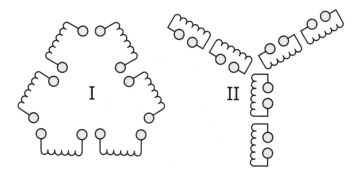

Figure 229.202

25. The winding connections shown at I in Figure 229.202 above, are for a ___-wound motor.

a. 12-lead wye
b. 9-lead wye
c. 9-lead delta
d. 6-lead delta
e. 6-lead wye
f. 12-lead delta

_____ 26. The motor winding leads in Drawing ___ are correctly numbered.

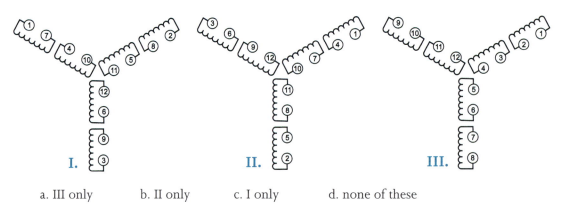

a. III only   b. II only   c. I only   d. none of these

Match the NEMA motor lead markings to the equivalent IEC markings.

a. U1-U2   b. V1-V2   c. W1-W2   d. U5-U6   e. V5-V6   f. W5-W6

_____ 27. NEMA motor leads T3-T6 are equal to IEC motor leads ___.

_____ 28. NEMA motor leads T1-T4 are equal to IEC motor leads ___.

_____ 29. NEMA motor leads T8-T11 are equal to IEC motor leads ___.

_____ 30. Correctly number the leads on the motor shown below.

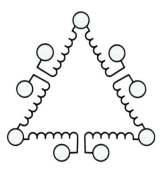

_____ 31. Correctly number the leads on the motor shown below.

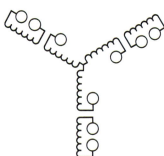

Objective 229.3 Worksheet

_____ 1. A 25 hp motor has a nameplate SF = 1.10 and FLA = 60. Without damage, this motor can be overloaded so that it pulls ___ amps, for a short period of time, at rated voltage.

    a. 69   b. 27.5   c. 66   d. 60   e. 54

MOTORS UNLIMITED
AC INDUCTION MOTOR

Thermally Protected	NO		Type ☐
STYLE	K	SERIAL	347456
FRAME	56	TYPE	
HP	25 PH 3	HOUSING	T.O.
RPM	1760	SERVICE FACTOR	1.15
CYCLES	60	S. F. AMPS	
VOLTS	208-230/460	AMPS	68.4-62.2/31.1
DEGREES C RISE	40	CODE	D
HOURS	24		

Figure 229.302

_____ 2. The motor nameplate, shown above in Figure 229.302, indicates that this motor is rated to operate at ___ volts.

    I. 120   II. 208   III. 230   IV. 480

    a. I, II, or III only
    b. II or III only
    c. II, III, or IV only
    d. I or III only
    e. III or IV only
    f. any of these

_____ 3. Refer to Figure 229.302, shown above. The actual rotor speed of this motor is ___ rpm.

    a. 900   b. 2400   c. 1200   d. 1800   e. none of these

_____ 4. Refer to Figure 229.302, shown above. This motor will use ___ power when operated at 230 volts rather than if operated at 460 volts.

    a. the same   b. more   c. less

_____ 5. Refer to Figure 229.302, shown above. When operated at 480 volts, rather than if operated at 240 volts, this motor will require (or permit) ___ size circuit conductors.

    a. the same   b. larger   c. smaller

Electric Motors – Polyphase

_____ 6. The DESIGN letter noted on a motor nameplate indicates ___.

   a. torque characteristics
   b. duty rating
   c. insulation class
   d. starting current

E FAC	NEMA DES	HZ	INDUCTION MOTOR
1.15	C	60	
Horsepower	RPM	PH	MODEL
1	3525	1	TEFC
TIME HRS	INS CL	FRAME	MAX AMB
24	B	56	40° C
VOLTS		AMPS	
115/230		13.9/6.9	
Electric Motor Company			

SER FAC	NEMA DES	HZ	INDUCTION MOTOR
1.15	C	60	
Horsepower	RPM	PH	MODEL
1	3525	3	TEFC
TIME HRS	INS CL	FRAME	MAX AMB
24	B	56	40° C
VOLTS		AMPS	
230/460		4.0/2.0	
Electric Motor Company			

Figure 229.304

_____ 7. Refer to Figure 229.304, shown above. Which of these motors is more efficient (± 2%) when operated at 230 volts?

   a. They are equally efficient.
   b. The 1Ø motor is more efficient.
   c. The 3Ø motor is more efficient.

_____ 8. Refer to Figure 229.304, shown above. Which of these motors is dual-voltage?

   a. Both of the motors.
   b. Neither of the motors.
   c. The 1Ø motor only.
   d. The 3Ø motor only.

SER FAC	NEMA DES	HZ	INDUCTION MOTOR
1.15	C	60	
Horsepower	RPM	PH	MODEL
3	3525	3	TEFC
TIME HRS	INS CL	FRAME	MAX AMB
24	B	56	40° C
VOLTS		AMPS	
230/460		9.2/4.6	
Electric Motor Company			

Figure 229.303

_____ 9. Refer to Figure 229.303, shown above. Each winding will carry ___ amps of current if operated at 460 volts.

   a. 4.6   b. 18.4   c. 9.2   d. 2.3

____ 10. Refer to Figure 229.303, shown on the previous page. Each winding will carry ___ amps of current if operated at 230 volts.

   a. 18.4   b. 4.6   c. 2.3   d. 9.2

____ 11. Refer to Figure 229.303, shown on the previous page. Each winding is rated to carry ___ amps of current.

   a. 4.6
   b. 18.4
   c. 2.3
   d. 9.2

____ 12. A continuous-duty motor should be used if the motor is to be operated for 1 hour or more within a 24-hour period.

   a. True   b. False

____ 13. Motor overheating can be caused by ___.

   a. unbalanced voltage at the motor terminals
   b. overloading the motor
   c. poor ventilation
   d. any of these

____ 14. When a 230/460 volt dual-voltage 3Ø motor is properly connected to a 480-volt circuit, the current through each winding is ___.

   a. more than it would be for a 240-volt circuit
   b. less than it would be for a 240-volt circuit
   c. the same as it would be for a 240-volt circuit

____ 15. Horsepower is the amount of ___ by a motor.

   a. output power produced   b. input power required

____ 16. Efficiency is always shown on a motor's nameplate.

   a. False   b. True

Electric Motors – Polyphase

17. On a motor nameplate, the amount of line current required at rated voltage and frequency appears adjacent to the marking ___.

    a. FLC   b. AMPS   c. FLA   d. any of these

18. A 208-230/460V, delta-wound motor can be supplied with a 3-phase circuit from a ___ service panel.

    a. 277/480V, 3Ø, 4W wye
    b. 120/208V, 3Ø, 4W wye
    c. 120/240V, 3Ø, 4W delta
    d. any of these

19. Large hp motors are generally not started across-the-line.

    a. True   b. False

20. A 230/460V, delta-wound motor can be supplied with a 3-phase circuit from a ___ service panel.

    a. 240V, 3Ø, 3W delta
    b. 277/480V, 3Ø, 4W wye
    c. 120/240V, 3Ø, 4W delta
    d. any of these

21. FRAME, on a motor nameplate, indicates physical characteristics of the motor. These characteristics include ___.

    a. mounting bolt-hole sizes
    b. mounting base dimensions
    c. shaft diameter
    d. all of these

22. The CODE letter noted on a motor nameplate indicates ___.

    a. insulation class
    b. duty rating
    c. torque characteristics
    d. starting current

_____ 23. A 3Ø motor has is pulling 20.6 amps when connected to a 480-volt circuit. You know that this type of motor has a pf of about .82 and an efficiency of about 65%. The motor is delivering about ___ hp.

   a. 9.3   b. 7.1   c. 13.9   d. 14.9   e. 18.8   f. 12.2

_____ 24. Generally, the efficiency of larger motors is ___ that of smaller motors.

   a. more than   b. less than   c. equal to

_____ 25. A 3 hp, 240-volt, 1Ø motor has FLA = 14 on its nameplate. The power factor is about 91%. The efficiency of this motor is about ___%.

   a. 30.6   b. 42.3   c. 60.6   d. 73.2   e. 52.9   f. 66.6

_____ 26. NEMA DESIGN ___ motors are the most common type in use.

   a. A   b. B   c. C   d. D

_____ 27. For operation at temperatures near –60°C, a(an) ___ motor should be used.

   a. continuous duty   b. arctic duty   c. totally enclosed

_____ 28. A motor in a ___ enclosure can be used outdoors.

   a. TENV   b. ODP   c. TEFC   d. any of these

_____ 29. KVA CODE and LRA are equivalent terms, as used on a motor nameplate.

   a. True
   b. False

_____ 30. Low efficiency and low power factor are typical of motors with ___ hp ratings.

   a. large
   b. small

Lesson 229 Safety Worksheet

Permit-Required Confined Spaces

There are large numbers of confined spaces where you work that can be dangerous to anyone who enters into them. In this safety lesson you will learn what dangers exist and how you can be safe when you have to work in these spaces.

Definitions:

Acceptable entry conditions: conditions that must exist in a permit space to allow entry and to ensure that employees involved with a permit-required confined space entry can safely enter into and work within the space.

Attendant: an individual stationed outside one or more permit spaces who monitors the authorized entrants and who performs all attendant's duties assigned in the employer's permit space program.

Authorized entrant: an employee who is authorized by the employer to enter a permit space.

Confined space is a space that:

(1) Is large enough and so configured that an employee can bodily enter and perform assigned work; and
(2) Has limited or restricted means for entry or exit (for example, tanks, vessels, silos, storage bins, hoppers, vaults, and pits are spaces that may have limited means of entry.) and
(3) Is not designed for continuous employee occupancy.

Emergency: any occurrence (including any failure of hazard control or monitoring equipment) or event internal or external to the permit space that could endanger entrants.

Engulfment: the surrounding and effective capture of a person by a liquid or finely divided (flowable) solid substance that can be aspirated (breathed) to cause death by filling or plugging the respiratory system or that can exert enough force on the body to cause death by strangulation, constriction, or crushing.

Entry: the action by which a person passes through an opening into a permit-required space. Entry includes ensuing work activities in that space and is considered to have occurred as soon as any part of the entrant's body breaks the plane of an opening into the space.

Entry permit (permit): the written or printed document that is provided by the employer to allow and control entry into a permit space and that contains the information required.

Entry supervisor: the person (such as the employer or foreman) responsible for determining if acceptable entry conditions are present at a permit space where entry is planned, for authorizing entry and overseeing entry operations, and for terminating entry as required. NOTE: An entry supervisor also may serve as an attendant or as an authorized entrant; as long as that person is trained and equipped for each role they may fill. Also the duties of entry supervisor may be passed from one individual to another during the course of an entry operation.

Hazardous atmosphere: an atmosphere that may expose employees to the risk of death, incapacitation, impairment of ability to self-rescue (that is, escape unaided from a permit space), injury, or acute illness from one or more of the following causes:

(1) Flammable gas, vapor, or mist in excess of 10 percent of its lower flammable limit (LFL);
(2) Airborne combustible dust at a concentration that meets or exceeds its LFL;
(3) Atmospheric concentration of any substance for which a dose or a permissible exposure limit is published in Subparts G and Z of the OSHA standards;
(4) Atmospheric oxygen concentration below 19.5% or above 23.5% levels.

Hot work permit: the employer's written authorization to perform operations (for example, welding, riveting, cutting, burning or heating) capable of providing a source of ignition.

Immediately dangerous to life or health (IDLH): any condition that poses immediate or delayed threat to life or that would cause irreversible, adverse, health effects or that would interfere with an individual's ability to escape unaided from a permit space.

Inerting: the displacement of the atmosphere in a permit space by a noncombustible gas (such as nitrogen) to such an extent that the resulting atmosphere is noncombustible. NOTE: This procedure produces an IDLH oxygen-deficient atmosphere.

Isolation: the process by which a permit space is removed from service and completely protected against the release of energy and material into the space by such means as: blanking or blinding; misaligning or removing sections of lines, pipes, or ducts; a double block and bleed system, lockout of all sources of energy; or blocking or disconnecting all mechanical linkages.

Line breaking: the intentional opening of a pipe, line, or duct that is or has been carrying flammable, corrosive, or toxic material, an inert gas, or any fluid at a volume, pressure, or temperature capable of causing injury.

Non-permit confined space: a confined space that does not contain, or with respect to atmospheric hazards, has potential to contain any hazard capable of causing death or serious physical harm.

Oxygen deficient atmosphere: an atmosphere containing less than 19.5% oxygen by volume.

Oxygen enriched atmosphere: an atmosphere that contains more than 23.5% oxygen by volume.

Permit-required confined space: a confined space that has one or more of the following characteristics:

(1) Contains or has potential to contain a hazardous atmosphere;
(2) Contains a material that has potential for engulfing an entrant;
(3) Has an internal configuration such that an entrant could be trapped or asphyxiated by inwardly converging walls or by a floor which slopes downward and tapers to a smaller cross-section; or
(4) Contains any other recognized serious safety or health hazard.

Permit-required confined space program: the employer's overall program for controlling, and where appropriate, for protecting employees from, permit space hazards and for regulating employee entry into permit spaces.

Permit system: the employer's written procedure for preparing and issuing permits for entry and for returning the permit space to service following termination of entry.

Prohibited condition: any condition in a permit space that is not allowed by permit during the period when entry is authorized.

Rescue service: the personnel designated to rescue employees from permit spaces.

Retrieval system: the equipment (including a retrieval line, chest or full-body harness, wristlets, if appropriate, a lifting device or anchor) used for non-entry rescue of persons from permit spaces.

Testing: the process by which the hazards that may comfort entrants of a permit space are identified and evaluated. Testing includes specifying the tests that are to be performed in the space. NOTE: Testing enables employers both to advise and implement adequate control measures for the protection of authorized entrants and to determine if acceptable entry conditions are present immediately prior to, and during, entry.

Some of the dangers of confined spaces are hazardous atmosphere, oxygen deficient atmosphere, oxygen enriched atmosphere, immediately dangerous to life and health conditions (IDLH), limited means of egress in case emergency exit is necessary, possible engulfment from the contents of the confined space, the risks of possible energized electrical circuits and more. There may be hazardous vapors that can put life at risk. Liquids from pipes in the confined space may be present. Testing of the atmosphere is required to be sure that hazardous vapors are not present prior to entering into a confined space. This is accomplished by using the correct meter to test for these vapors and also testing the percentage of oxygen present. These could be hazardous chemicals or even water or sewer liquids.

These types of confined spaces require the owner to identify and to post danger signs to warn of these dangers. These signs must have clear instructions regarding these dangers, such as **"Danger Confined Space - Entry With Permit Only."** If these types of hazards are present, a permit is required before you are permitted to enter.

To enter these **permit required confined spaces**, a written permit must be completed by the supervisor, the person who will perform the entry, any rescue personnel, and support people involved. The permit (see sample permit form following) will have a lot of information, such as the information from the atmospheric testing, what PPE is required, what type of rescue equipment will be required, how will communications be assured, are all requirements of lock-out and NFPA70E being followed. Do you have the required fire extinguisher(s) on hand? Have all the necessary signatures been obtained and authorizations completed? Have acceptable entry conditions been met? Does the attendant know and understand what is required of them?

A Confined Space Decision Flow Chart (example follows) should be used to make sure all steps are followed to be safe when entry is made. All of this information and more is found in OSHA CFR1910.146 General Industry Standards.

By using this information, you can protect yourself or others from the dangers involved in Confined Space entry.

CONFINED SPACE ENTRY PERMIT – SAMPLE
Permit valid for one shift only

Job Site:	Location of Confined Space:	Date:

Description of work to be performed:

Expected duration of work:

Nature of Hazards Encountered: (check)
- ☐ Oxygen deficiency
- ☐ Oxygen enrichment
- ☐ Flammable gases
- ☐ Toxic gases
- ☐ Other chemical hazards
- ☐ Mechanical hazards
- ☐ Electrical hazards
- ☐ Other materials within space

Preparation of space: (check)
- ☐ Lockout/Tagout
- ☐ Ventilation
- ☐ Multiple contractor's entry coordinated
- ☐ Emergency #'s
  - ○ _____
- ☐ Emergency equipment available
  - ○ Lifelines
  - ○ Respirators
  - ○ PPE
  - ○ Retrieval equipment
  - ○ Ventilation
  - ○ Communication equipment

Names of employees entering space:
_____

Names of attendants:
_____

Atmospheric testing:

Testing equipment: (list type)
_____

Equipment calibration: (date)
_____

Test	Result	Time	Limit
Oxygen			19.5-23.5%
Flammable			0%
Other (H_2S)			OSHA PEL
Other (CO)			OSHA PEL
Other (Toxic)			OSHA PEL

Authorization:

Entry Supervisor signature: _____

Signature of person performing atmospheric test: _____

Employee supervisor signature: _____

Area/Department supervisor: _____

Other: List special requirements here
_____

Confined Space Entry

Sample Test

Date: _____

Employee taking test: _____

1. Who may enter a confined space?

2. Is it OK for the attendant to leave while workers are in the confined space? Why or why not?

3. What types of hazards could be present within a confined space?

4. What is the primary responsibility of the attendant?

5. How do I know if a work area is a confined space?

6. What equipment is required to be worn by employees entering confined spaces?

7. What types of things need to be tested before entering the space?

8. What should the attendant do if the persons within the confined space collapse?

APPENDIX A TO §1910.146 - PERMIT-REQUIRED CONFINED SPACE DECISION FLOW CHART

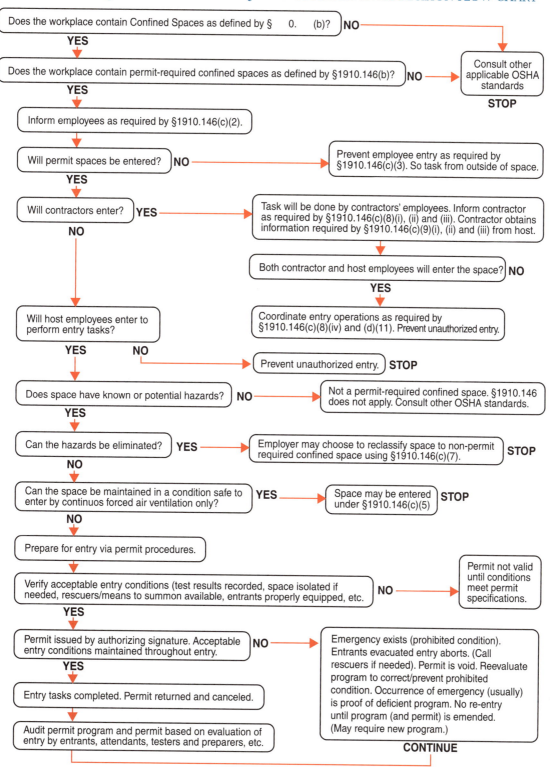

Questions

_____ 1. Acceptable entry conditions means the conditions that must exist in a ___ space to allow entry.

    a. small   b. restricted   c. permit   d. enclosed   e. encapsulated

_____ 2. Non-permit confined space is a confined space that ___ has potential to contain any hazard capable of causing death or serious physical harm.

    a. may   b. will   c. does   d. does not   e. should

_____ 3. Authorized entrant is a(n) ___ who is authorized by the employer to enter a permit space.

    a. worker
    b. employee
    c. assistant
    d. partner
    e. journeyman
    f. apprentice

_____ 4. ___ means any condition that poses immediate or delayed threat to life or that would cause Irreversible, adverse, health affects or that would interfere with an individual's ability to escape unaided from a permit space.

    a. ICMC   b. IDLH   c. IBM   d. IEC

_____ 5. There are large numbers of ___ spaces where we work.

    a. constricted
    b. confined
    c. concentrated
    d. comfortable
    e. common

_____ 6. ___ means the written or printed document that is provided by the employer to allow and control entry into a permit space and that contains the information required.

    a. entry form   b. entry permit   c. entry supervisor   d. hot work permit

Electric Motors – Polyphase

_____ 7. ___ is a(n) individual stationed outside permit space(s) who monitors the authorized entrants.

    a. employer
    b. supervisor
    c. employee
    d. foreman
    e. attendant
    f. observer

_____ 8. Confined space requirements are found in OSHA CFR___.

    a. 1910.146
    b. 1926.400
    c. 1910.1200
    d. 1926.20
    e. 1919.147
    f. 1926.500

_____ 9. Owners are required to identify and post danger ___ to warn workers of confined spaces.

    a. signs
    b. warnings

____ 10. A confined space is a space that is large enough and so configured that an employee can ___ enter and perform assigned work.

    a. bodily
    b. partially
    c. possibly
    d. not

____ 11. To enter a ___, a written permit must be completed by the supervisor, the entrant and any rescue personnel involved.

    a. permit required confined space
    b. permit required constricted space
    c. constricted space
    d. confined space

_____ 12. Confined Spaces are spaces that by definition are large enough and so configured that an employee can bodily enter and perform assigned work, and has ___ means for entry or exit.

    a. limited
    b. restricted
    c. limitless
    d. redundant
    e. both a and b
    f. both a and d
    g. both b and c
    h. both a and c

_____ 13. Permit-required confined spaces are confined spaces that have one or more hazards such as___.

    a. hazardous atmosphere
    b. engulfment materials
    c. inwardly converging walls
    d. sloping floors
    e. a and d only
    f. b and d only
    g. a, b and c only
    h. a,b,c and d

_____ 14. Testing is ___ to be sure that hazardous vapors are not present prior to entering into a confined space.

    a. required   b. requested   c. suggested   d. recommended

Why Buy Accubid?

estimating, billing, and project management software

- Accubid has over 20,000 users putting Accubid's estimating, project management, and service management software to the test in over 4,400 companies throughout North America
- Accubid estimating software is used by 9 out of the top 10 electrical contractors, and 36 out of the top 50*
- Accubid has been recognized as the estimating software of choice among specialty contractors in the last three CFMA surveys**
- Accubid has won the Technology's Hottest Companies award from Constructech magazine for the last four years running
- Accubid was the only software company to receive IEC's Innovative Product of the year award at the 2007 IEC National Conference
- Accubid is the only estimating software company to offer an unconditonal money-back guarantee
- Accubid is the only estimating software company to offer a 45-minute guaranteed support response time
- Accubid is the only software company where you can trade one product for another Accubid product

Discover for yourself why Accubid continues to be the contractor's choice.

Call for an online demonstration

total solutions for contractors
1-800-222-8243
www.accubid.com

* October 15, 2007 issue of Engineering News-Record (ENR) magazine.
** Statistics excerpted from the CFMA's 2002, 2004 and 2006 Information Technology Surveys for the Construction Industry with the permission of the Construction Financial Management Association, Princeton, NJ, 609-452-8000. CFMA neither evaluated nor ranked software in terms of performance. The survey should not be construed as the advice of CFMA.

Lesson 230
Motors: General Knowledge and Sizing Branch Circuit Conductors

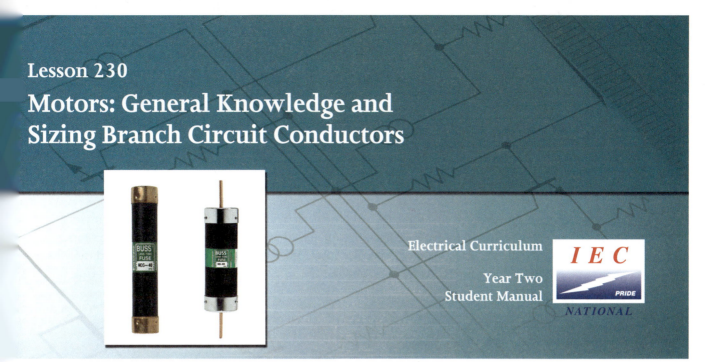

Electrical Curriculum

Year Two
Student Manual

IEC

Purpose

This lesson will discuss how to interpret motor nameplate information to determine the required size of motor branch circuit conductors as required by the NEC©. This lesson will also discuss general motor information, some of which is a review of Lessons 228 and 229, but also includes information presented in Electrical Systems

Homework

(Due at the beginning of this class)

For this lesson, you should:

- Thoroughly read the material contained within Lesson 230.
- Read only the following in Electrical Systems
- **MOTORS – ARTICLE 430:**
  - Introductory paragraphs
  - Ampacity and Motor Ratings – 430.6
  - Markings on Motors and Multimotor Equipment – 430.7
  - Branch Circuit – Single Motor – 430.22
  - Sizing Conductors
  - Other Motors
  - Multispeed Motors – 430.22(A)
  - Duty-Cycle Motors – 430.22(E)
  - Wound-Rotor Motors – 430.23
  - Synchronous Motors
- Complete Objective 230.1 Worksheet.
- Complete Objective 230.2 Worksheet.
- Complete Objective 230.3 Worksheet.
- Complete Objective 230.4 Worksheet.
- Read and complete Lesson 230 Safety Worksheet.

- Complete additional assignments or worksheets if available and directed by your instructor.

Objectives

By the end of this lesson, you should:

230.1 ....Be able to integrate the knowledge gained from previous lessons with material presented in this lesson to be able to answer application-type general motor questions.

230.2 ....Be able to determine the proper size branch circuit conductors for single-phase motors.

230.3 ....Be able to determine the proper size branch circuit conductors for three-phase motors.

230.4 ....Be able to determine the proper size motor branch circuit conductors as adjusted for voltage drop and ampacity corrections, and be able to determine the proper size flexible cord for motor connections.

Motors: General Knowledge and Sizing Branch Circuit Conductors

230.1 Application of General Motor Knowledge

Although the proper abbreviation for horsepower is *hp*, horsepower most commonly appears on motor nameplates as *HP*.

SFA appears on some, but not all motor nameplates. SFA, as it appears on the nameplate in Figure 230.11, represents *Service Factor Amps*. SFA and FLA should not be confused. FLA represents how much current the motor is designed to draw at the nameplate rated horsepower. SFA is the absolute, maximum current that this motor can draw. SFA is sometimes referred to as the "speed limit" or the "redline" of a motor. The motor is not designed to handle any more current than SFA, and chronically exceeding SFA can shorten the motor's life.

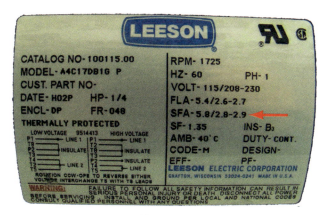

Figure 230.11

Recall that service factor (SF) is provided by the manufacturer. The motor can deliver:

(NP HP) × (SF)

for a short period of time, without damaging the motor. For this motor, the motor could deliver:

(NP HP) × (SF) = (0.25) × (1.35) = 0.3375 hp

When this motor is **occasionally** required to deliver 0.3375 hp, the current (at nameplate rated voltage) equals the values shown at SFA.

Advisory Notes for IEC 2nd Year Apprenticeship Motor Studies:

- *Keep these notes in a convenient location for frequent reference during the remaining 2nd Year lessons.*

- All conductors are THHN/THWN copper, unless noted otherwise (U.N.O.). Note that 8 AWG and larger conductors are manufactured with THWN-2 ratings. As shown in NEC® Table 310.13, THWN-2 conductors are rated for 90°C in wet locations while THWN conductors are rated for 75°C.

- Overload (OL) protection for motors is provided by heaters that are installed in the protection section part of a motor starter. Heaters will be selected from a heater selection table for a particular starter. Follow any additional instructions given on the table. This skill will be required in the field.

- Although the NEC® presents methods used to calculate the size required for overloads, this method is typically utilized only on typical state and local jurisdiction licensing examinations. If this method is applied to the actual heater selection tables in the field, an INCORRECT heater will be selected. DO NOT use the NEC® METHOD for 2nd year apprenticeship questions and problems. The NEC® method will be covered in 4th year in preparation for licensing examinations.

Year Two (Student Manual)

- All motor BC OCPDs should be sized using Exception 1 to NEC® 430.52(C)(1), unless SPECIFICALLY INSTRUCTED to do otherwise within the question. This, of course, is because, by utilizing the exception, designers of motor circuits avoid nuisance tripping of CBs and opening of fuses.
- NEC® 430.52(C)(1) Exception 2 – will **NOT** be applied in 2nd year! This exception can NOT be applied unless the motor has actually been started (or tried to be started) OR if, prior to installation, an <u>engineer</u> has determined that the situation requires application of the exception.
- Motors and motor circuits operating at over 600 volts will not be covered.
- Only full-voltage starting is covered in 2nd year. Starting methods such as wye-start, delta-run will be covered, more appropriately, in 4th year.
- Sizing OCPDs and conductors for control circuits will be covered in 4th year.
- "Instantaneous Trip Circuit Breakers" are available only in listed equipment for specific applications. These CBs provide ONLY short-circuit protection and can't protect circuit conductors against overload. Therefore, the typical application of these is within a combination starter where the OL protection within the starter also protects the circuit conductors. Circuit breaker discussions will be confined to Inverse Time Circuit Breakers.
- "torque," "ac adjustable," "wound rotor," & "synchronous" motors require that special provisions be made if ever these are encountered. Unless noted otherwise, motor questions and problems refer to squirrel-cage, induction motors.
- OCPDs
    - One-time fuse = NEC® term: nontime delay fuse (Refer to Figures 230.12A and 230.12B)
    - **ITCB** = Inverse Time Circuit Breaker
    - **DETD** = Dual-Element Time Delay fuse

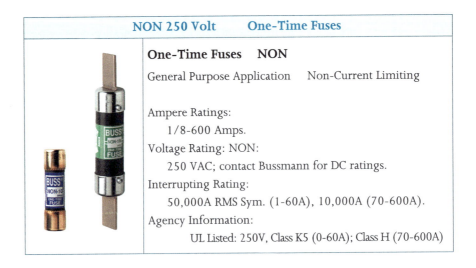

Figure 230.12A

Motors: General Knowledge and Sizing Branch Circuit Conductors

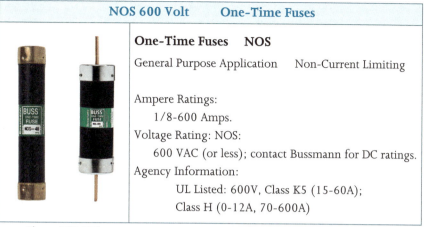

NOS 600 Volt One-Time Fuses

One-Time Fuses   NOS

General Purpose Application   Non-Current Limiting

Ampere Ratings:
  1/8-600 Amps.
Voltage Rating: NOS:
  600 VAC (or less); contact Bussmann for DC ratings.
Agency Information:
  UL Listed: 600V, Class K5 (15-60A);
  Class H (0-12A, 70-600A)

Figure 230.12B

230.2   Single-Phase Motor Branch Circuit Conductors

230.3   Three-Phase Motor Branch Circuit Conductors

230.4   Motor Branch Circuit Conductors – Special Conditions

Objective 230.1 Worksheet

_____ 1. Motors have about the same efficiency as transformers.

   a. True   b. False

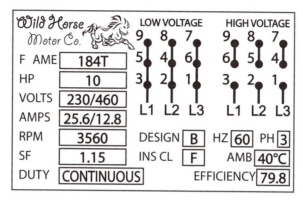

Figure 230.123

_____ 2. Refer to the nameplate shown above in Figure 230.123. The shaft speed on this motor is ___ rpm.

   a. 3600   b. 3096   c. 4094   d. 3560   e. 1840

_____ 3. Refer to the nameplate shown above in Figure 230.123. For this motor you would expect to have perfect insulation ($\infty\ \Omega$) between leads ___.

   a. 7 & 9   b. 3 & 8   c. 3 & 6   d. 2 & 5

_____ 4. Refer to the nameplate shown above in Figure 230.123. According to the manufacturer, this motor can be operated for 20 hours without damage.

   a. False   b. True

_____ 5. Refer to the nameplate shown above in Figure 230.123. For this motor you would expect to measure between 0.2 and 20 Ω between leads ___.

   a. 1 & 5   b. 2 & 8   c. 7 & 9   d. 1 & 9

_____ 6. Refer to the nameplate shown above in Figure 230.123. This motor is ___ wound.

   a. wye   b. delta

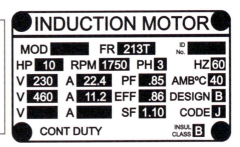

Branch Circuit Supply:
3-Pole breaker in a 277/480 V, 3Ø, 4W panel
Branch Circuit Terminations:
75°C rated
Conductors:
Copper THHN/THWN

Figure 230.224

_____ 7. Refer to Figure 230.224, shown above. Each winding is rated to carry ___ amps of current.

    a. 28   b. 22.4   c. 35   d. 14   e. 11.2

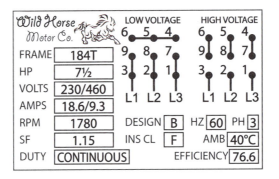

Figure 230.122

_____ 8. Refer to the nameplate shown above in Figure 230.122. For this motor you would expect to measure between 0.2 and 20 Ω between leads ___.

    a. 2 & 7   b. 1 & 5   c. 7 & 8   d. 3 & 8

_____ 9. Refer to the nameplate shown above in Figure 230.122. Drawing ___ below shows the correct lead numbering and winding connections for the motor.

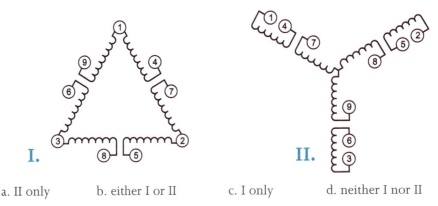

    a. II only   b. either I or II   c. I only   d. neither I nor II

Motors: General Knowledge and Sizing Branch Circuit Conductors

____ 10. Refer to the nameplate shown above in Figure 230.122. This motor can be supplied power from a ____-connected power supply.
I. wye
II. delta

a. II only   b. neither I nor II   c. either I or II   d. I only

____ 11. Refer to the nameplate shown above in Figure 230.122. This motor is ____ wound.

a. delta   b. wye

____ 12. Refer to the nameplate shown above in Figure 230.122. This motor should draw ____ amps when connected to Circuit 4-6-8 that is fed from a 3-pole breaker in a 120/240 volt, 3Ø, 4-wire panelboard.

a. 18.6   b. 22   c. 11   d. 9.3

MOTORS UNLIMITED
AC INDUCTION MOTOR

Thermally Protected NO		Type ☐
STYLE K	SERIAL 7456085	
FRAME 56	TYPE	
HP 3	PH 1	HOUSING T.O.
RPM 1760	SERVICE FACTOR 1.15	
CYCLES 60	S.F. AMPS	
VOLTS 115/230	AMPS 30.4/15.2	
DEGREES C RISE 40	CODE D	HOURS 24

Figure 230.115

____ 13. Refer to the motor nameplate shown above in Figure 230.115. This motor is a ____ motor.

a. DC   b. 1Ø   c. 3Ø

____ 14. Refer to the motor nameplate shown above in Figure 230.115. This motor is rated to operate ____ without damage when fully loaded.

a. for 5 minutes   b. for 15 minutes   c. for 60 minutes   d. continuously

____ 15. Refer to the motor nameplate shown above in Figure 230.115. This motor can be safely operated at 208 volts, when fully loaded, without damage to the motor.

a. True   b. False

16. Refer to the motor nameplate shown on the previous page in Figure 230.115. According to the service factor, this motor can be safely operated at 230 volts and pull ___ amps, for a short period of time, without damage to the motor.

    a. 30.4   b. 34.96   c. 15.2   d. 19.0   e. 17.48

17. A 3-phase, 15 HP, 230/460 volt motor has a nameplate FLA = 38/19. If operated at 230 volts, ___ amps of current would flow in each motor winding.

    a. 42
    b. 19
    c. 32.9
    d. 21
    e. 38

18. FLA is an abbreviation for ___.

    a. Frequently Loaded Armature
    b. Forward Lagging Angle
    c. Flowing Liquid Actuator
    d. Full Load Amperage

19. Nine-lead 3Ø motors are usually internally (factory) connected as either a ___ configuration.

    a. torque or wound
    b. delta or wye
    c. series or parallel
    d. start or stop

20. FLC (full load current) and FLA (full load amps) are the same thing.

    a. False   b. True

21. Torque motors and AC adjustable motors make up about 10% of all large motor applications. Of the remaining 90%, most are ___.

    a. synchronous
    b. wound-rotor
    c. squirrel-cage induction

Motors: General Knowledge and Sizing Branch Circuit Conductors

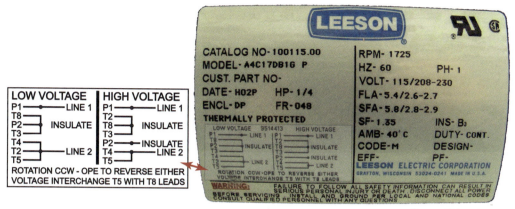

Figure 230.110

_____ 22. Refer to the motor nameplate shown above in Figure 230.110. This motor _____ have built-in overload protection.

a. does not   b. does

_____ 23. Refer to the motor nameplate shown above in Figure 230.110. If the supply circuit is 120 volts, you would connect as shown at _____ where the motor leads are connected to the T-strips.

a.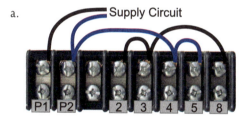
b.
c.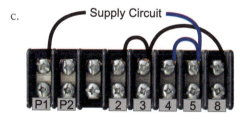
d.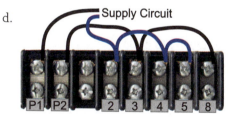

_____ 24. Refer to the motor nameplate shown above in Figure 230.110. When the motor is connected to a 240-volt circuit, the motor branch circuit current is _____ it would be for a 120-volt circuit.

a. more than
b. less than
c. the same as

_____ 25. Refer to the motor nameplate shown on the previous page in Figure 230.110. For connection on the jobsite, the manufacturer has provided ___ motor leads.

a. 9
b. 8
c. 7
d. 3
e. 6

_____ 26. Refer to the motor nameplate shown on the previous page in Figure 230.110. When the windings of the motor are connected for "high voltage", the winding sets are connected in ___.

a. parallel
b. series

_____ 27. Refer to the motor nameplate shown on the previous page in Figure 230.110. If you properly connected the single phase motor but the shaft was rotating in the wrong direction, you would swap the ___.

a. L1 & L2 connections
b. P2 & T2 leads
c. T2 & T4 leads
d. P1 & P2 leads
e. none of these

_____ 28. When sizing conductors for a 1-phase AC motor, ___ FLA is used.

a. nameplate
b. Table 430.249
c. Table 430.250
d. Table 430.247
e. Table 430.248

_____ 29. When sizing conductors for a DC motor, ___ FLA is used.

a. Table 430.247
b. Table 430.249
c. nameplate
d. Table 430.248
e. Table 430.250

Motors: General Knowledge and Sizing Branch Circuit Conductors

_____ 30. When replacing a defective motor in a piece of equipment, the replacement motor must have ___.
   I. the same frame size
   II. the same speed

   a. neither I nor II
   b. II only
   c. both I and II
   d. I only

_____ 31. A 3-phase, 15 HP, 230/460 volt motor has a nameplate FLA = 38/19. This motor will actually draw ___ amps of line current if operated at 230 volts.

   a. 19
   b. 32.9
   c. 38
   d. 42
   e. 21

_____ 32. Dual-voltage 3-phase motors have ___ windings per phase.

   a. 2
   b. 1
   c. 3
   d. 4
   e. 6

_____ 33. Motor installation requirements are provided in Article ___ in the NEC®.

   a. 445
   b. 440
   c. 430
   d. 460
   e. 470

_____ 34. A 3-phase motor can be reversed by swapping any two circuit conductor connections ___.

   a. at the line side of the starter
   b. in the terminal box on the motor
   c. at the load side of the safety switch
   d. at the load side of the starter
   e. any of these

Motors: General Knowledge and Sizing Branch Circuit Conductors

Objective 230.2 Worksheet

_____ 1. Motor Nameplate Information: 208-230/115 V 1Ø 7½ hp
Supply: 2-pole breaker in a 120/208 V, 3-ph, 4-W panel
Minimum size circuit conductors are ___ AWG.
All conductors are THHN/THWN copper and all terminations are rated at 75° C.

   a. 8
   b. 4
   c. 12
   d. 10
   e. 6

MOTORS UNLIMITED
AC INDUCTION MOTOR
Thermally Protected NO   Type ☐
STYLE K   SERIAL 7456085
FRAME 56   TYPE
HP 3   PH 1 HOUSING T.O.
RPM 1760   SERVICE FACTOR 1.15
CYCLES 60   S.F. AMPS
VOLTS 115/230   AMPS 30.4/15.2
DEGREES C RISE 40   CODE D   HOURS 24

Figure 230.115

_____ 2. Refer to the motor nameplate shown above in Figure 230.115. The smallest THHN/THWN branch circuit conductors that can be installed between the 1-pole circuit breaker in the 120/240 volt, 3Ø, 4-wire panelboard and the motor are ___ AWG. All terminations are rated at 75° C.

   a. 10
   b. 6
   c. 12
   d. 4
   e. 8

SER FAC 1.15	NEMA DES C	HZ 60	INDUCTION MOTOR
Horsepower 10	RPM 3525	PH 1	MODEL TEFC
TIME HRS 24	INS CL B	FRAME 256U	MAX AMB 40° C
VOLTS 230/460		AMPS 46/23	
Electric Motor Company			

Branch Circuit Supply:
2-Pole breaker in a
277/480 V, 3Ø, 4W panel
Branch Circuit Terminations:
75°C rated
Conductors:
THHN/THWN Copper

Figure 230.223

_____ 3. The smallest circuit conductors permitted for the motor circuit for Figure 230.223 (above) are ___ AWG.

   a. 12
   b. 8
   c. 10
   d. 4
   e. 6

Year Two (Student Manual)

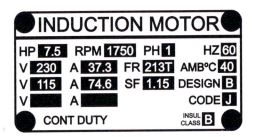

Figure 230.108

_____ 4. Refer to the motor nameplate shown above in Figure 230.108. The smallest THHN/THWN branch circuit conductors that can be installed between the 2-pole circuit breaker in the 120/240 volt, 3Ø, 4-wire panelboard and the motor are ___ AWG. All terminations are rated at 75° C.

a. 6   b. 4   c. 10   d. 8   e. 3

_____ 5. Motor Nameplate Information: 115/230 V   1Ø   1½ hp
Supply: 1-pole breaker in a 120/240 V, 1-ph, 3-W panel
Minimum size circuit conductors are ___ AWG.
All conductors are THHN/THWN copper and all terminations are rated at 75° C.

a. 6   b. 12   c. 8   d. 10   e. 14

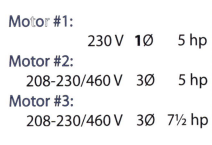

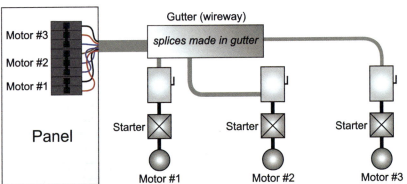

Figure 230.261

All conductors are THHN/THWN copper and all terminations are rated at 75° C. This means that the wire connected to the motor must have the correct ampacity according to the 75° column. However, inside the conduit between the panel and the wireway ampacity corrections are made, as usual, in the 90° column. Use Exception 1 to 430.52(C)(1) to size OCPD's. The specifications require that separate equipment grounding conductors be installed in each raceway. The engineer who wrote the specs means that a metal raceway can't be used as the only EGC. But remember that only one EGC must be installed in each conduit.
Refer to the drawing shown above to answer questions that refer to this Figure.

_____ 6. Refer to Figure 230.261. The smallest circuit conductors that can be installed between the wireway and the safety switch for Motor #1 are ___ AWG.

a. 6   b. 8   c. 10   d. 12   e. none of these

Motors: General Knowledge and Sizing Branch Circuit Conductors 230-789

_____ 7. Refer to Figure 230.261. The panel is ___.

   a. 120/208 V, 3Ø, 4W
   b. 120/240 V, 3Ø, 4W
   c. 277/480 V, 3Ø, 4W
   d. 120/240 V, 1Ø, 3W

```
MOTORS UNLIMITED
AC INDUCTION MOTOR
Thermally Protected  NO              Type ☐
STYLE     K            SERIAL  623565
FRAME    56            TYPE
HP  1          PH  1   HOUSING  T.O.
RPM  1760              SERVICE FACTOR  1.15
CYCLES  60             S.F. AMPS
VOLTS        115/230   AMPS      14.2/7.1
DEGREES C RISE  40     CODE     D
HOURS  24
```
Figure 230.111

_____ 8. Refer to the motor nameplate shown above in Figure 230.111. The smallest THHN/THWN branch circuit conductors that can be installed between the 1-pole circuit breaker in the 120/240 volt, 3Ø, 4-wire panelboard and the motor are ___ AWG. All terminations are rated at 75° C.

   a. 18   b. 16   c. 10   d. 14   e. 12

_____ 9. A THHN/THWN copper circuit supplies power to a 2 hp, 115/230 V, 1Ø motor from a 1-pole breaker in a 120/208 V, 3-ph, 4-W panel. The minimum size circuit conductors permitted are ___ AWG. All terminations are rated at 75° C.

   a. 6   b. 12   c. 14   d. 10   e. 8

```
SER FAC   NEMA DES   HZ     INDUCTION
1.15         B       60      MOTOR
Horsepower  RPM      PH     MODEL
   5        3525      1      TEFC
TIME HRS  INS CL  FRAME    MAX AMB
   24        B      256U     40° C
VOLTS              AMPS
     115/230              50.8/25.4
         Electric Motor Company
```
Figure 230.109

_____ 10. Refer to the motor nameplate shown above in Figure 230.109. The smallest THHN/THWN branch circuit conductors that can be installed between the 2-pole circuit breaker in the 120/240 volt, 3Ø, 4-wire panelboard and the motor are ___ AWG. All terminations are rated at 75° C.

   a. 12   b. 8   c. 4   d. 6   e. 10

Motors: General Knowledge and Sizing Branch Circuit Conductors

Objective 230.3 Worksheet

____ 1. Motor Nameplate Information: 208-230/460 V   3Ø 3 hp
   Supply: 3-pole breaker in a 277/480 V, 3-ph, 4-W panel
   Minimum size circuit conductors are ____ AWG.
   All conductors are THHN/THWN copper and all terminations are rated at 75° C.

   a. 6   b. 14   c. 10   d. 8   e. 12

_____ 2. What size THHN copper transformer secondary conductors will be required to feed a 3-phase, 30 hp, 480-volt motor load? The secondary conductors terminate at a fusible disconnect with 70 amp fuses. The Δ-Δ transformer primary is supplied by a 12,470-volt circuit protected with 10 amp fuses.

   a. 4   b. 3   c. 10   d. 6   e. 8

Motor #1:
   230 V   1Ø   5 hp
Motor #2:
   208-230/460 V   3Ø   5 hp
Motor #3:
   208-230/460 V   3Ø   7½ hp

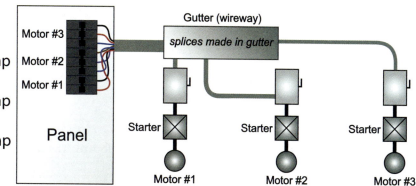

Figure 230.271

All conductors are THHN/THWN copper and all terminations are rated at 75° C. This means that the wire connected to the motor must have the correct ampacity according to the 75° column. However, inside the conduit between the panel and the wireway ampacity corrections are made, as usual, in the 90° column. Use Exception 1 to 430.52(C)(1) to size OCPDs. The specifications require that separate equipment grounding conductors be installed in each raceway. The engineer who wrote the specs means that a metal raceway can't be used as the only EGC. But remember that only one EGC must be installed in each conduit. The small wireway is located near the motor location. Refer to the drawing shown below to answer questions that refer to this Figure.

_____ 3. Refer to Figure 230.271. The smallest circuit conductors that can be installed between the wireway and the safety switch for Motor #1 are ____ AWG.

   a. 10   b. 14   c. 8   d. 12   e. none of these

_____ 4. Refer to Figure 230.271. The smallest circuit conductors that can be installed between the wireway and the safety switch for Motor #4 are ____ AWG.

   a. 10   b. 12   c. 14   d. 8   e. none of these

Year Two (Student Manual)

_____ 5. Motor Nameplate Information: 208-230/460 V   3Ø 5 hp
       Supply: 3-pole breaker in a 120/208 V, 3-ph, 4-W panel
       Minimum size circuit conductors are ___ AWG.
       All conductors are THHN/THWN copper and all terminations are rated at 75° C.

   a. 14   b. 8   c. 6   d. 10   e. 12

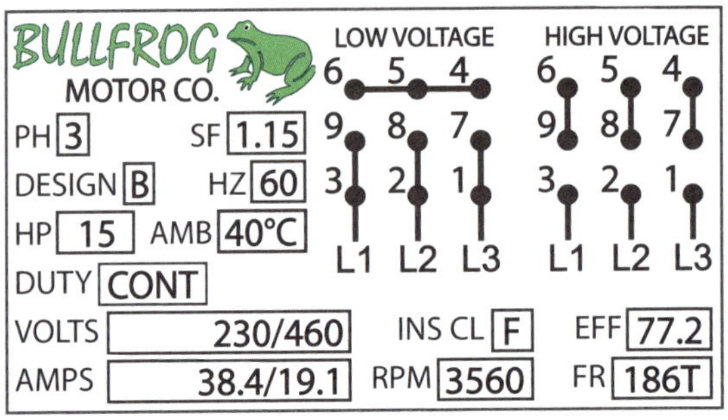

Figure 230.128

BULLFROG MOTOR CO.
PH 3   SF 1.15
DESIGN B   HZ 60
HP 15   AMB 40°C
DUTY CONT
VOLTS 230/460   INS CL F   EFF 77.2
AMPS 38.4/19.1   RPM 3560   FR 186T

_____ 6. Refer to the nameplate shown above in Figure 230.128. The branch circuit for the motor is supplied from a 277/480 volt, 3Ø, 4-wire panelboard. The minimum size branch circuit conductors for the motor are ___ AWG.

   a. 4   b. 8   c. 3   d. 10   e. 6

_____ 7. Refer to the nameplate shown above in Figure 230.128. The branch circuit for the motor is supplied from a 120/240 volt, 3Ø, 4-wire panelboard. The minimum size branch circuit conductors for the motor are ___ AWG.

   a. 10   b. 3   c. 4   d. 8   e. 6

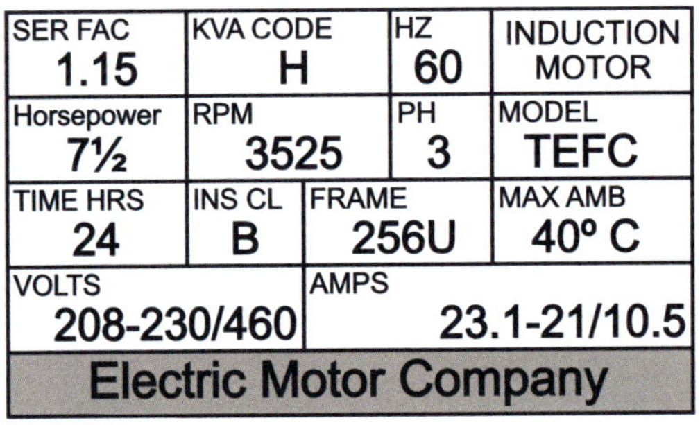

Figure 230.127

SER FAC	KVA CODE	HZ	INDUCTION MOTOR
1.15	H	60	
Horsepower	RPM	PH	MODEL
7½	3525	3	TEFC
TIME HRS	INS CL	FRAME	MAX AMB
24	B	256U	40° C
VOLTS		AMPS	
208-230/460		23.1-21/10.5	
Electric Motor Company			

_____ 8. Refer to the nameplate shown above in Figure 230.127. The branch circuit for the motor is supplied from a 120/208 volt, 3Ø, 4-wire panelboard. The minimum size branch circuit conductors for the motor are ___ AWG.

   a. 12   b. 8   c. 14   d. 6   e. 10

Motors: General Knowledge and Sizing Branch Circuit Conductors

_____ 9. Motor Nameplate Information: 208-230/460 V   3Ø 100 hp
Supply: 3-pole breaker in a 120/240 V, 3-ph, 4-W panel
Minimum size circuit conductors are ___ AWG.
All conductors are THHN/THWN copper and all terminations are rated at 75° C.

    a. 350   b. 300   c. 250   d. 4/0   e. none of these

_____ 10. Motor Nameplate Information: 208-230/460 V   3Ø 200 hp
Supply: 3-pole breaker in a 120/208 V, 3-ph, 4-W panel
Minimum size circuit conductors are ___ AWG or kcmil where 3 paralleled sets of conductors are used.
All conductors are THHN/THWN copper and all terminations are rated at 75° C.

    a. 250   b. 3/0   c. 2/0   d. 1/0   e. none of these

_____ 11. Motor Nameplate Information: 208-230/460 V   3Ø 5 hp
Supply: 3-pole breaker in a 120/240 V, 3-ph, 4-W panel
Minimum size circuit conductors are ___ AWG.
All conductors are THHN/THWN copper and all terminations are rated at 75° C.

    a. 6   b. 8   c. 14   d. 12   e. 10

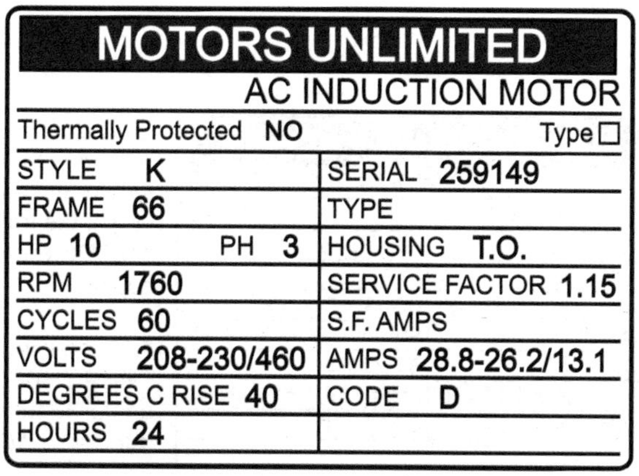

MOTORS UNLIMITED	
AC INDUCTION MOTOR	
Thermally Protected NO	Type ☐
STYLE K	SERIAL 259149
FRAME 66	TYPE
HP 10 PH 3	HOUSING T.O.
RPM 1760	SERVICE FACTOR 1.15
CYCLES 60	S.F. AMPS
VOLTS 208-230/460	AMPS 28.8-26.2/13.1
DEGREES C RISE 40	CODE D
HOURS 24	

Figure 230.113

_____ 12. Refer to the motor nameplate shown above in Figure 230.113. The smallest THHN/THWN branch circuit conductors that can be installed between the 3-pole circuit breaker in the 120/240 volt, 3Ø, 4-wire panelboard and the motor are ___ AWG. All terminations are rated at 75° C.

    a. 4   b. 12   c. 10   d. 8   e. 6

____ 13. The nameplate for a 15 hp, three-phase motor shows an FLA of 20 amps at 460 volts. What minimum size copper THWN/THHN wire is required for this motor? All conductors are THHN/THWN copper and all terminations are rated at 75° C.

a. 6 b. 12 c. 10 d. 8 e. 14

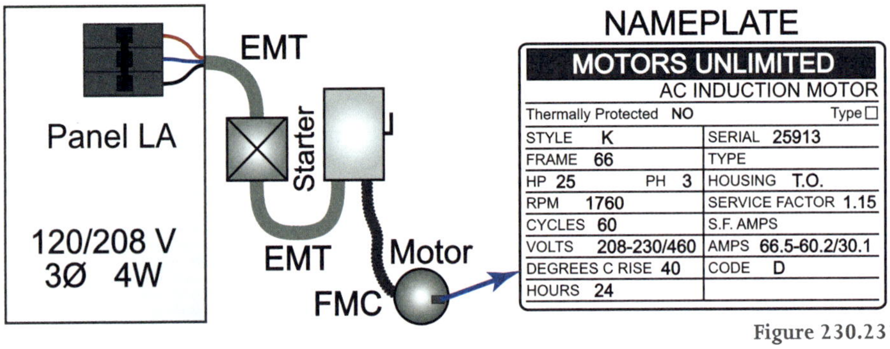

Panel LA
120/208 V
3Ø 4W
EMT
Starter
EMT
Motor
FMC
NAMEPLATE
MOTORS UNLIMITED
AC INDUCTION MOTOR
Thermally Protected NO
Type
STYLE K
SERIAL 25913
FRAME 66
TYPE
HP 25
PH 3
HOUSING T.O.
RPM 1760
SERVICE FACTOR 1.15
CYCLES 60
S.F. AMPS
VOLTS 208-230/460
AMPS 66.5-60.2/30.1
DEGREES C RISE 40
CODE D
HOURS 24

Figure 230.231

____ 14. Refer to the information shown above in Figure 230.231. The minimum permitted size branch circuit conductors are ___ AWG.

a. 8 b. 3 c. 4 d. 10 e. 6

____ 15. *Motor Nameplate Information:* 208-230/460 V 3Ø200 hp
Supply: 3-pole breaker in a 277/480 V, 3-ph, 4-W panel
Minimum size circuit conductors are ___ AWG or kcmil where 3 paralleled sets of conductors are used. All conductors are THHN/THWN copper and all terminations are rated at 75° C.

a. 3/0 b. 1/0 c. 250 d. 2/0 e. none of these

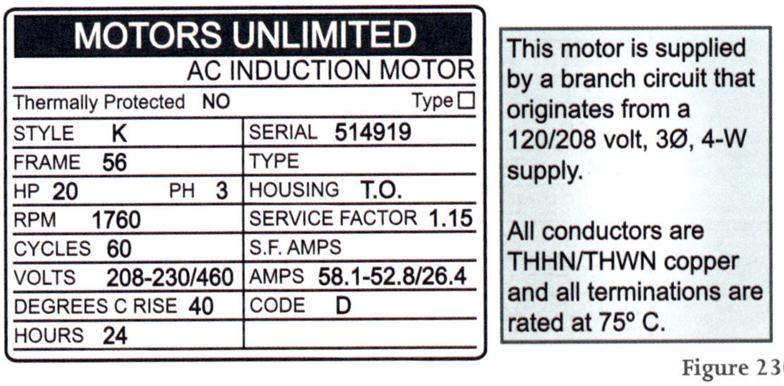

MOTORS UNLIMITED
AC INDUCTION MOTOR
Thermally Protected NO
Type
STYLE K
SERIAL 514919
FRAME 56
TYPE
HP 20
PH 3
HOUSING T.O.
RPM 1760
SERVICE FACTOR 1.15
CYCLES 60
S.F. AMPS
VOLTS 208-230/460
AMPS 58.1-52.8/26.4
DEGREES C RISE 40
CODE D
HOURS 24
This motor is supplied by a branch circuit that originates from a 120/208 volt, 3Ø, 4-W supply.
All conductors are THHN/THWN copper and all terminations are rated at 75° C.

Figure 230.131

____ 16. Refer to the nameplate and information shown above in Figure 230.131. The minimum size BC conductors are ___ AWG.

a. 6 b. 3 c. 4 d. 8 e. none of these

Objective 230.4 Worksheet

_____ 1. The nameplate for a 15 HP, three-phase motor shows an FLA of 35 amps at 230 volts. What minimum size SOO cord is required to connect the motor to the branch circuit? All terminations are rated at 75° C.

 a. 8 b. 4 c. 10 d. 6 e. 12

_____ 2. Motor Nameplate Information: 115/230 V   1Ø   1½ hp
Supply: 1-pole breaker in a 120/240 V, 1-ph, 3-W panel
SJO cord is used to connect the motor. The minimum size conductors permitted in the cord are ___ AWG.

 a. 8 b. 6 c. 14 d. 10 e. 12

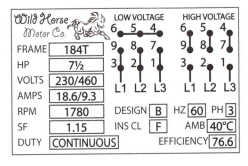

Figure 230.122

_____ 3. Refer to the nameplate shown above in Figure 230.122. The branch circuit for the motor is supplied from a 120/240 volt, 3Ø, 4-wire panelboard. What is the minimum size (3 wire + ground) SO cord that could be used as a flexible connection for the motor?

 a. 6 b. 10 c. 14 d. 12 e. 8

Scenario 230-28

Motor Nameplate Information: 5 hp   208-230/460 V   3Ø
Branch Circuit Supply: 3-pole breaker in a 120/208 V, 3-ph, 4-W panel
Branch Circuit Terminations: 75° C rated
Conductors: THHN/THWN copper

_____ 4. The minimum size SO cord is permitted for the motor, shown above in Scenario 230-28, are ___ AWG.

 a. 14
 b. 12
 c. 8
 d. 6
 e. 10

Figure 230.271

Motors: General Knowledge and Sizing Branch Circuit Conductors

All conductors are THHN/THWN copper and all terminations are rated at 75° C. This means that the wire connected to the motor must have the correct ampacity according to the 75° column. However, inside the conduit between the panel and the wireway ampacity corrections are made, as usual, in the 90° column. Use Exception 1 to 430.52(C)(1) to size OCPDs. The specifications require that separate equipment grounding conductors be installed in each raceway. The engineer who wrote the specs means that a metal raceway can't be used as the only EGC. But remember that only one EGC must be installed in each conduit. The small wireway is located near the motor location. Refer to the drawing shown below to answer questions that refer to this Figure.

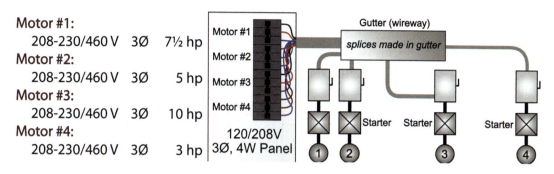

Motor #1:
  208-230/460 V 3Ø   7½ hp
Motor #2:
  208-230/460 V 3Ø   5 hp
Motor #3:
  208-230/460 V 3Ø   10 hp
Motor #4:
  208-230/460 V 3Ø   3 hp

Figure 230.271

_____ 5. Refer to Figure 230.271. Between the panel and the wireway, the minimum size circuit conductors for Motor #3 are ___ AWG. Count the number of circuit conductors in the conduit and determine ampacity inside the conduit from the 90° column.

  a. 10   b. 6   c. 8   d. 12   e. none of these

_____ 6. Refer to Figure 230.271. Between the panel and the wireway, the minimum size circuit conductors for Motor #1 are ___ AWG. Count the number of circuit conductors in the conduit and determine ampacity inside the conduit from the 90° column.

  a. 6   b. 8   c. 12   d. 10   e. none of these

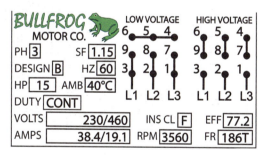

Figure 230.128

_____ 7. Refer to the nameplate shown above in Figure 230.128. The branch circuit for the motor is supplied from a 120/240 volt, 3Ø, 4-wire panelboard. What is the minimum size (3 wire + ground) STO cord that could be used as a flexible connection for the motor?

  a. 3   b. 6   c. 10   d. 8   e. 4

All conductors are THHN/THWN copper and all terminations are rated at 75° C. This means that the wire connected to the motor must have the correct ampacity according to the 75° column. However, inside the conduit between the panel and the wireway ampacity corrections are made, as usual, in the 90° column. Use Exception 1 to 430.52(C)(1) to size OCPD's. The specifications require that separate equipment grounding conductors be installed in each raceway. The engineer who wrote the specs means that a metal raceway can't be used as the only EGC. But remember that only one EGC must be installed in each conduit. Refer to the drawing shown below to answer questions that refer to this Figure.

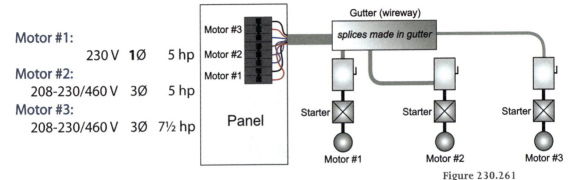

Motor #1:
    230 V 1Ø 5 hp

Motor #2:
    208-230/460 V 3Ø 5 hp

Motor #3:
    208-230/460 V 3Ø 7½ hp

Figure 230.261

_____ 8. Refer to Figure 230.261. Between the panel and the wireway, the minimum size circuit conductors for Motor #2 are ___ AWG. Count the number of circuit conductors in the conduit and determine ampacity inside the conduit from 90° column.

    a. 10   b. 4   c. 8   d. 6   e. 12

_____ 9. Refer to Figure 230.261. Between the panel and the wireway, the minimum size circuit conductors for Motor #1 are ___ AWG. Count the number of circuit conductors in the conduit and determine ampacity inside the conduit from 90° column.

    a. 4   b. 10   c. 8   d. 12   e. 6

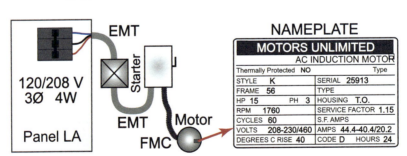

Figure 230.230

_____ 10. Refer to the information shown above in Figure 230.230. What are the minimum permitted size branch circuit conductors? The motor is located 160 feet from the panelboard and the specifications limit the voltage drop to 3% at this motor. Use K=10.86.

    a. 3   b. 4   c. 2   d. 6   e. 1

Lesson 230 Safety Worksheet

Compressed Gas Cylinders

Compressed Gas cylinders are frequently used on construction sites for a number of uses. These purposes can be using CO_2 (carbon dioxide) for our conduit fishing systems, Oxygen and Acetylene for welding or brazing, Propane for heating equipment and sometimes as fuel for aerial lift equipment or material handling equipment. The plumbers and HVAC contractors use several different compressed gases in their work. They may use map gas or if mig or tig welding is performed, argon or argon/CO_2 may be used.

The OSHA regulations require that employers determine that gas cylinders under their control are in a safe condition and are visually inspected. Various other regulations can be found in information from the Compressed Gas Association. Additional information can be found in OSHA subparts related to welding or fire prevention and protection.

The most frequent compressed gas hazards to which we are exposed are propane, Oxy/Acetylene and CO_2.

Propane is flammable and must be treated accordingly. Keep propane tanks at least 6 feet away from any sources of heat or ignition. Heaters should be at least 20 feet away from other heaters. When these tanks are refilled on site, they must be at least 50 feet away from nearest building. These tanks must be stored outdoors; they cannot be stored indoors. See OSHA CFR1926.153 (k)(1) and table F-3 for exact distances that must be maintained based on the amount of propane stored. Storage areas shall have at least one fire extinguisher rated at least 20-B: C rating. If propane is stored with other flammable gasses, they must be identified by contents. Protection from motor vehicle damage must be provided.

Welding tanks must have their protective caps in place before they can be moved. Regulators and gauges must be removed prior to moving the cylinders. If cylinders are hoisted they must be secured on skids, cradles or a pallet. They shall not be hoisted by slings or chokers. Cylinders shall be rolled on their bottom edges. If they are transported by vehicle they must be vertical and secured. Oxygen cylinders must be stored at least 20 feet from fuel gas cylinders, or a fire resistant barrier five feet tall and at least one hour fire rated must separate these cylinders. These cylinders must not be stored near any highly flammable materials. Store them in a well ventilated area, do not store them in a closed location, such as a locker or gang box. Keep the welding tanks far enough away from sparks or hot slag from any welding operations. Do not take any fuel gas or oxygen cylinders into any confined space. For additional information on welding and cutting operations, see CFR1926.350 in the OSHA standards.

CO_2 as are other gases is compressed at high pressure inside the tanks. The valve caps tanks must be in place to protect them from damage, if you are going to move them. If the valve is broken off, the tank can be propelled similar to a missile and can cause much damage or cause serious personal injury. CO_2 displaces oxygen and is very dangerous if used in areas where ventilation is poor. Never use CO_2 in confined spaces if good ventilation cannot be maintained. This would include when we are working in manholes and vaults, using our conduit fishing systems. CO_2 is tasteless and odorless which makes it hard to detect and thus the dangers can overcome or even cause death in these conditions.

Motors: General Knowledge and Sizing Branch Circuit Conductors

Questions

_____ 1. Employers are required to ___ that gas cylinders under their control are in safe condition and are visually inspected.

    a. state
    b. determine
    c. decide

_____ 2. CO_2 displaces ___ and is very dangerous if used in areas where ventilation is poor.

    a. nitrogen
    b. oxygen
    c. map gas
    d. carbon monoxide
    e. propane
    f. hydrogen

_____ 3. Keep propane tanks ___ feet away from any sources of ignition or heat sources.

    a. 20
    b. 5
    c. 6
    d. 4
    e. 12
    f. 15

_____ 4. Plumbers and HVAC contractors may use additional types of compressed gases. They could be ___.

    a. CO_2
    b. naptha
    c. map gas
    d. argon
    e. a and d only
    f. b and d only
    g. a, c and d only
    h. all of these

_____ 5. Oxygen cylinders must be stored at least ___ feet from fuel gas cylinders, or a fire resistant barrier at least 5 feet high and have a one hour fire rating.

   a. 12   b. 15   c. 18   d. 25   e. 20   f. 10

_____ 6. Oxygen cylinders cannot be stored near any highly ___ materials.

   a. non-corrosive
   b. corrosive
   c. concentrated
   d. flammable
   e. non-flammable
   f. explosive

_____ 7. Compressed gas cylinders can be used for ___ purposes.

   a. oxygen
   b. acetylene
   c. CO_2
   d. propane
   e. b and d only
   f. a and c only
   g. a, b and d
   h. all of these

_____ 8. Propane is ___ and must be treated accordingly.

   a. corrosive
   b. heavy
   c. flammable
   d. poisonous
   e. caustic
   f. all of these

_____ 9. Heaters must be kept at least ___ feet from other heaters.

   a. 6
   b. 10
   c. 20
   d. 15
   e. 12
   f. 4

Motors: General Knowledge and Sizing Branch Circuit Conductors   230-803

___ 10. The most frequent compressed gas hazards that we are exposed to are Oxy/Acetylene, propane, and ___.

  a. chlorine
  b. ammonia
  c. CO_2
  d. all of these

___ 11. Further information regarding regulations for compressed gas storage and use are available from ___.

  a. NAFTA
  b. OSHA
  c. American Gas Association
  d. Compressed Gas Association
  e. b and d only
  f. a, b and c only
  g. a and c only
  h. none of these

___ 12. The requirements of distances from structures for propane storage can be found ___.

  a. NFPA70E
  b. OBBC
  c. NFPA72
  d. OSHA
  e. NFPA70
  f. IEC

___ 13. Welding tanks can be stored in ___ location(s).

  a. well-ventilated
  b. any
  c. barricaded
  d. locked
  e. confined space
  f. closed

Lesson 231
Motor Branch Circuit Overcurrent Protective Devices: Short Circuit and Ground Fault Protection

Electrical Curriculum

Year Two
Student Manual

Purpose

This lesson will discuss how to interpret motor nameplate information to determine the required size or rating of motor branch circuit overcurrent protective devices (OCPDs), as required by the NEC©, to provide short circuit and ground fault protection for conductors. Lesson 232 will address overload protection.

Homework

(Due at the beginning of this class)

For this lesson, you should:

- Thoroughly read the material contained within Lesson 231.
- Read only the following in Electrical Systems
- **MOTORS – ARTICLE 430:**
    Branch Circuit – Single Motor – 430.22
    Sizing Raceways
    Sizing Overload Protection
    Sizing Fuses and Circuit Breakers
    Sizing Disconnects
- Refer to "Advisory Notes for IEC 2nd Year Apprenticeship Motor Studies" from Lesson 230.
- Complete Objective 231.1 Worksheet.
- Complete Objective 231.2 Worksheet.
- Read and complete Lesson 231 Safety Worksheet.

- Complete additional assignments or worksheets if available and directed by your instructor.

Objectives

By the end of this lesson, you should:

231.1 ....Be able to determine the proper size, or rating, of a short-circuit and ground-fault protective device for a single-phase motor branch circuit, and understand that calculated values for overload protection, as provided in the NEC©, are not utilized for actual installations.

231.2 ....Be able to determine the proper size, or rating, of a short-circuit and ground-fault protective device for a three-phase motor branch circuit.

231.1 OCPDs for Single-Phase Motors

The NEC® overcurrent protective device requirements for motor branch circuits differ from those for general-purpose branch circuits. For general-purpose branch circuits, the OCPD provides conductor protection against short circuits, ground faults, and overloads. However, the OCPD for a motor branch circuit provides conductor protection against only short circuit and ground fault (SC & GF) currents. The conductors of a motor branch circuit are protected against overload by a separate device that, typically, responds only to overload current and opens to remove the overloaded motor from the circuit. The opening of the circuit protects both the motor and the circuit conductors against overload. Overload protection is addressed in the next lesson.

Refer to Figure 430.1 in the NEC®. Note that motor branch-circuit short-circuit and ground-fault protection requirements are located in Part IV of Article 430. Note also that motor overload protection requirements are addressed *separately* and are located in Part III of Article 430.

The SC & GF protective device, fuse(s) or circuit breaker, that is connected to the motor branch circuit must be selected so that it opens when subjected to ground faults or short circuits, but must remain closed while the motor is starting. The starting current of a typical motor is about six times the FLA value shown on the motor's nameplate. The SC & GF protective device opens very quickly in response to high current levels caused by fault currents (SC or GF). The higher the current level, the faster the device opens. The time/current curve is exponential.

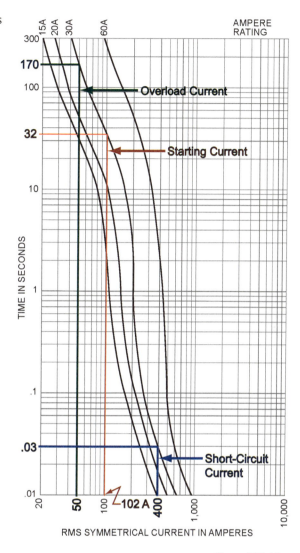

Figure 231.11

Typical time/current curves for 15, 20, 30, and 60 amp fuses are plotted logarithmically and are shown in Figure 231.11. Although the rating of the protective device is exceeded, when starting current is flowing, the device will not open because the higher starting current flows only for a short period of time.

Example:

Given a 2 horsepower motor operating at 120 volts, the table FLA is found to be 17 amps. Using 17 amps, 12 AWG wire is selected for the branch circuit conductors. To determine the size DETD fuses to be used for the motor circuit, refer to NEC® Table 430.52.

$$17 \text{ A} \times 1.75 = 29.75 \Rightarrow 30 \text{ A fuse}$$

Using the rule of thumb, LRA (starting current) = 17 × 6 = 102 A. Refer to the red line on Figure 231.11 and note that the 30 A fuse can carry 102 amps of starting current for about 32 seconds. Since the circuit will probably have to carry starting current for less than a second, this fuse will be acceptable. The duration of starting current is a function of loading, among other factors, and a protective device must be selected to remain closed during the starting period. Referring again to Figure 232.11, the blue line shows a short circuit current of 400 amps. At 400 amps the 30 amp fuse will open in about 0.03 seconds. Once again refer to Figure 232.11. The green line shows the motor suffering from overload and 50 amps is flowing through the circuit conductors. At this current level, the 30 amp fuse won't open for almost 3 minutes (170 seconds). If the conductors are required to carry 50 amps for this period of time, the conductor insulation, the conductor itself, and the motor may be damaged. Because the protective device won't respond quickly to overload conditions, separate overload protection is required for this circuit so that the conductors and motor can be protected.

231.2 OCPDs for Three-Phase Motors

Student Notes

Motor Branch Circuit Overcurrent Protective Devices: Short Circuit and Ground Fault Protection

Objective 231.1 Worksheet

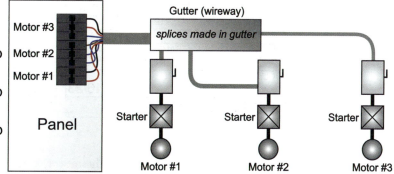

Branch Circuit Supply:
2-Pole breaker in a
120/240 V, 3Ø, 4W panel
Branch Circuit Terminations:
75°C rated
Conductors:
Copper THHN/THWN

SER FAC 1.15	NEMA DES C	HZ 60	INDUCTION MOTOR
Horsepower 10	RPM 3525	PH 1	MODEL TEFC
TIME HRS 24	INS CL B	FRAME 256U	MAX AMB 40° C
VOLTS 230/460	AMPS		46/23
Electric Motor Company			

Figure 230.121

_____ 1. When a safety switch is installed adjacent to the motor shown above in Figure 230.121, the maximum size DETD fuses permitted are ___ A.

   a. 100   b. 150   c. 90   d. 125   e. 80

Motor #1:
    230 V   1Ø   5 hp
Motor #2:
  208-230/460 V   3Ø   5 hp
Motor #3:
  208-230/460 V   3Ø   7½ hp

Figure 230.261

All conductors are THHN/THWN copper and all terminations are rated at 75° C. This means that the wire connected to the motor must have the correct ampacity according to the 75° column. However, inside the conduit between the panel and the wireway ampacity corrections are made, as usual, in the 90° column. Use Exception 1 to 430.52(C)(1) to size OCPD's. The specifications require that separate equipment grounding conductors be installed in each raceway. The engineer who wrote the specs means that a metal raceway can't be used as the only EGC. But remember that only one EGC must be installed in each conduit.

Refer to the drawing shown below to answer questions that refer to this Figure.

Three motors are to be connected by the electrician. Each motor will have a separate circuit breaker in the panel. The branch circuits are to be installed in one conduit between the panel and a small wireway near the motor location. The nameplate information for each of the motors is as follows:

    Motor #1:   230 V   1Ø   5 hp
    Motor #2:   208-230/460 V   3Ø   5 hp
    Motor #3:   208-230/460 V   3Ø   7½ hp

_____ 2. Refer to Figure 230.261. The largest permitted breaker that can be installed in the panel for Motor 1 is ___ A.

    a. 40   b. 50   c. 70   d. 60   e. 35

_____ 3. Motor Nameplate Information: 115/230 V   1Ø   3 hp
Supply: 1-pole breaker in a 120/240 V, 1-ph, 3-W panel
The maximum size ITCB permitted is ___ A.

    a. 90   b. 60   c. 100   d. 110   e. none of these

_____ 4. A 1 hp, 115/230 V, 1Ø motor circuit is fed from a single-pole breaker in a 120/240 V, 1-ph, 3-W panel. The maximum permitted rating of an ITCB is ___ A.

    a. 25   b. 50   c. 45   d. 40   e. 30

SER FAC	NEMA DES	HZ	INDUCTION MOTOR
1.15	C	60	
Horsepower	RPM	PH	MODEL
10	3525	1	TEFC
TIME HRS	INS CL	FRAME	MAX AMB
24	B	256U	40° C
VOLTS		AMPS	
230/460		46/23	
Electric Motor Company			

Branch Circuit Supply:
  2-Pole breaker in a
  277/480 V, 3Ø, 4W panel
Branch Circuit Terminations:
  75°C rated
Conductors:
  THHN/THWN Copper

Figure 230.223

_____ 5. When a safety switch is installed adjacent to the motor, as shown above in Figure 230.223, the maximum size DETD fuses permitted are ___ A.

    a. 60   b. 40   c. 70   d. 45   e. 80

_____ 6. Refer to Figure 230.223, shown above. An ITCB with a maximum rating of ___ A is the largest permitted to be used for the branch circuit OCPD.

    a. 80   b. 45   c. 70   d. 40   e. 60

_____ 7. Motor Nameplate Information: 115/230 V   1Ø   2 hp
Supply: 1-pole breaker in a 120/208 V, 3-ph, 4-W panel
The maximum size DETD fuses permitted are ___ A.

    a. 50   b. 70   c. 40   d. 45   e. none of these

Motor Branch Circuit Overcurrent Protective Devices: Short Circuit and Ground Fault Protection   231-813

Scenario 230-27

Motor Nameplate Information: 3 hp   208 V   1Ø
Branch Circuit Supply: 2-pole breaker in a 120/208 V, 3-ph, 4-W panel

_____ 8. Refer to Scenario 230-27, shown above. A safety switch is used as the disconnecting means for the motor. The maximum size nontime-delay fuses permitted to be installed are ___ A.

   a. 35   b. 60   c. 45   d. 50   e. 30

_____ 9. Motor Nameplate Information: 115/230 V   1Ø   3 hp
   Supply: 1-pole breaker in a 120/240 V, 1-ph, 3-W panel
   The maximum size DETD fuses permitted are ___ A.

   a. 70   b. 40   c. 100   d. 60   e. none of these

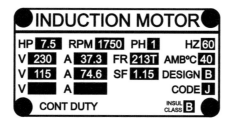

Figure 230.108

_____ 10. Refer to the motor nameplate shown above in Figure 230.108. The largest 1-pole circuit breaker that can be installed in a 120/208 volt, 3Ø, 4-wire panelboard for this motor is a ___ A.

   a. 70   b. 125   c. 80   d. 100   e. 200

_____ 11. A single-pole breaker in a 120/240 V, 1-ph, 3-W panel supplies power to a motor circuit consisting of THHN/THWN copper conductors. When DETD fuses are installed in a safety switch for a 115/230 V, 1Ø, 1 hp motor, these fuses can have a rating of not greater than ___ A.

   a. 30   b. 45   c. 40   d. 25   e. 50

Scenario 230-26

Motor Nameplate Information: 208 V   1Ø   2 hp
Branch Circuit Supply: 2-pole breaker in a 120/208 V, 3-ph, 4-W panel

_____ 12. Refer to Scenario 230-26, shown above. A safety switch will be installed as the disconnecting means for the motor. One-time fuses with a maximum permitted rating of ___ A would be installed in this switch.

   a. 40   b. 20   c. 30   d. 35   e. 25

Motor Branch Circuit Overcurrent Protective Devices: Short Circuit and Ground Fault Protection   231-815

Objective 231.2 Worksheet

Branch Circuit Supply:
3-Pole breaker in a
120/208 V, 3Ø, 4W panel
Branch Circuit Terminations:
75°C rated
Conductors:
Copper THHN/THWN

SER FAC	PH	KVA CODE	INDUCTION
1.15	3	H	MOTOR
Horsepower	RPM	NEMA DESIGN	MODEL
10	3525	C	TEFC
HZ	HRS	INS CL	FRAME	MAX AMB
60	24	B	256U	40° C
VOLTS		AMPS	
208-230/460		28.8-26.2/13.1	
Electric Motor Company			

Figure 230.225

_____ 1. DETD fuses are to be installed in a safety switch adjacent to a motor. See Figure 230.225 (above). Fuses with a rating not exceeding ___ A are permitted.

a. 50   b. 60   c. 70   d. 100   e. none of these

_____ 2. Using the information shown above in Figure 230.225, DETD fuses with the maximum permitted rating were installed in the disconnecting means for the motor. Given this information, the smallest permitted circuit conductors are ___ AWG.

a. 14   b. 12   c. 6   d. 8   e. 10

_____ 3. A 200 hp, three-phase motor is connected to a 480-volt circuit. What are the maximum size DETD fuses permitted?

a. 600   b. 450   c. 400   d. 500   e. 300   f. 350

Wild Horse Motor Co.
FRAME 184T
HP 7½
VOLTS 230/460
AMPS 18.6/9.3
RPM 1780
SF 1.15
DUTY CONTINUOUS
DESIGN B   HZ 60   PH 3
INS CL F   AMB 40°C
EFFICIENCY 76.6

Figure 230.122

_____ 4. Refer to the nameplate shown above in Figure 230.122. The branch circuit for the motor is supplied from a 120/240 volt, 3Ø, 4-wire panelboard. The maximum size ITCB that can be used on the motor branch circuit is ___ A.

a. 50   b. 45   c. 70   d. 40   e. 60

Year Two (Student Manual)

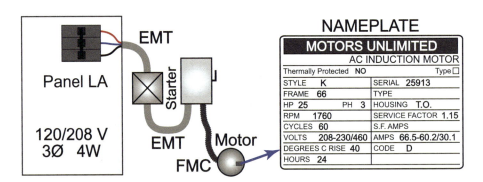

Figure 230.231

_____ 5. Refer to the nameplate shown above in Figure 230.231. As the BC OCPD, the largest DETD fuses you can use in an existing 400 amp safety switch are rated at ___ amps.

a. 100   b. 110   c. 175   d. 125   e. 150

Scenario 230-28

Motor Nameplate Information: 5 hp   208-230/460 V   3Ø
Branch Circuit Supply: 3-pole breaker in a 120/208 V, 3-ph, 4-W panel

_____ 6. Nontime-delay fuses with a rating of ___ A are the largest permitted to be installed in a disconnect switch for the motor shown above in Scenario 230-28.

a. 25   b. 45   c. 40   d. 50   e. 30

_____ 7. DETD fuses with a rating of ___ A are the largest permitted to be installed in a disconnect switch for the motor shown above in Scenario 230-28.

a. 40   b. 30   c. 25   d. 50   e. 45

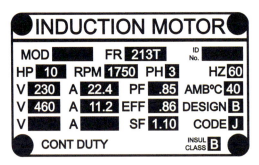

Figure 231.114

Motor Branch Circuit Overcurrent Protective Devices: Short Circuit and Ground Fault Protection

_____ 8. Refer to the motor nameplate shown on the previous page in Figure 231.114. The largest DETD fuses permitted to be installed in a safety switch for this motor are ____ A, when the circuit line voltage is 240.

   a. 35   b. 70   c. 50   d. 45   e. 40

_____ 9. Refer to the motor nameplate shown on the previous page in Figure 231.114. The largest DETD fuses that can be installed in a safety switch for this motor are ____ A, when the circuit line voltage is 480.

   a. 35   b. 15   c. 10   d. 30   e. 25

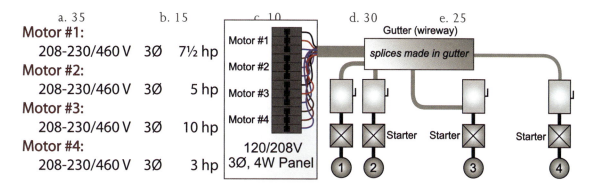

Motor #1:
   208-230/460 V 3Ø   7½ hp
Motor #2:
   208-230/460 V 3Ø   5 hp
Motor #3:
   208-230/460 V 3Ø   10 hp
Motor #4:
   208-230/460 V 3Ø   3 hp

Figure 230.271

All conductors are THHN/THWN copper and all terminations are rated at 75° C. This means that the wire connected to the motor must have the correct ampacity according to the 75° column. However, inside the conduit between the panel and the wireway ampacity corrections are made, as usual, in the 90° column. Use Exception 1 to 430.52(C)(1) to size OCPDs. The specifications require that separate equipment grounding conductors be installed in each raceway. The engineer who wrote the specs means that a metal raceway can't be used as the only EGC. But remember that only one EGC must be installed in each conduit. The small wireway is located near the motor location. Refer to the drawing shown below to answer questions that refer to this Figure.

_____ 10. Refer to Figure 230.271. The largest permitted breaker that can be installed in the panel for Motor 3 is ____ A.

   a. 60   b. 45   c. 70   d. 50   e. none of these

_____ 11. Refer to Figure 230.271. The largest permitted breaker that can be installed in the panel for Motor 1 is ____ A.

   a. 50
   b. 60
   c. 45
   d. 40
   e. none of these

SER FAC	KVA CODE	HZ	INDUCTION MOTOR
1.15	H	60	
Horsepower	RPM	PH	MODEL
7½	3525	3	TEFC
TIME HRS	INS CL	FRAME	MAX AMB
24	B	256U	40° C
VOLTS	AMPS		
208-230/460	23.1-21/10.5		
Electric Motor Company			

Figure 230.127

_____ 12. Refer to the nameplate shown above in Figure 230.127. The branch circuit for the motor is supplied from a 277/480 volt, 3Ø, 4-wire panelboard. The maximum size motor branch circuit time-delay fuses that can be installed in a safety switch are ___ A.

a. 40 b. 25 c. 30 d. 20 e. 35

_____ 13. Refer to the nameplate shown above in Figure 230.127. The branch circuit for the motor is supplied from a 120/208 volt, 3Ø, 4-wire panelboard. The maximum size ITCB that can be used on the motor branch circuit is ___ A.

a. 50
b. 60
c. 40
d. 45
e. 70

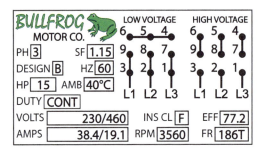

Figure 230.128

_____ 14. Refer to the nameplate shown above in Figure 230.128. The branch circuit for the motor is supplied from a 480/277 volt, 3Ø, 4-wire panelboard. The maximum size ITCB that can be used on the motor branch circuit is ___ A.

a. 40
b. 60
c. 45
d. 70
e. 50

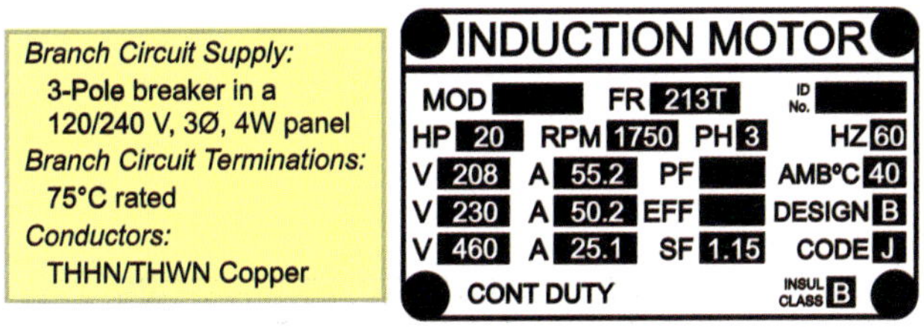

Motor Branch Circuit Overcurrent Protective Devices: Short Circuit and Ground Fault Protection

Figure 230.222

_____ 15. When a safety switch is installed adjacent to the motor shown above in Figure 230.222, the maximum size one-time fuses permitted are ___ A.

a. 175   b. 125   c. 90   d. 100   e. 150

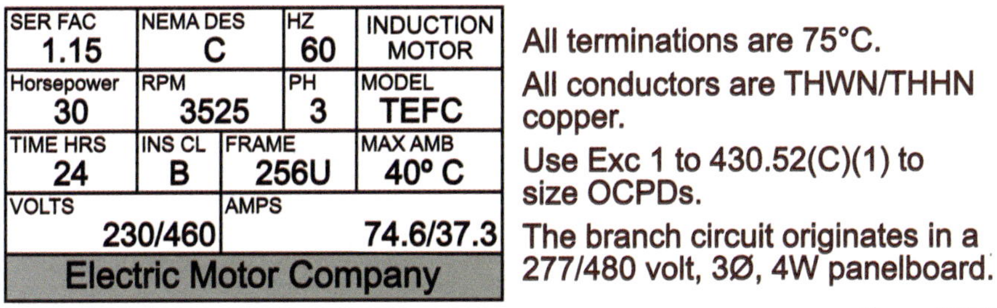

Figure 230.251

_____ 16. For the installation in Figure 230.251, the maximum size DETD fuses permitted to be installed in a safety switch used as the disconnecting means are ___ A.

a. 40   b. 100   c. 50   d. 125   e. 70

_____ 17. For the installation in Figure 230.251, the maximum size one-time fuses permitted to be installed in a safety switch used as the disconnecting means are ___ A.

a. 50   b. 125   c. 100   d. 40   e. 70

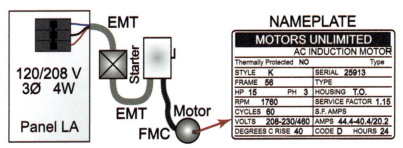

Figure 230.230

_____ 18. Refer to the information shown above in Figure 230.230. The largest 3-pole breaker you can install in Panel LA is a ___ amp.

a. 125   b. 110   c. 100   d. 80   e. 90

19. Refer to the information shown above in Figure 230.230. The largest time-delay fuses you can install in the safety switch are ___ amp.

   a. 125   b. 60   c. 80   d. 90   e. 110

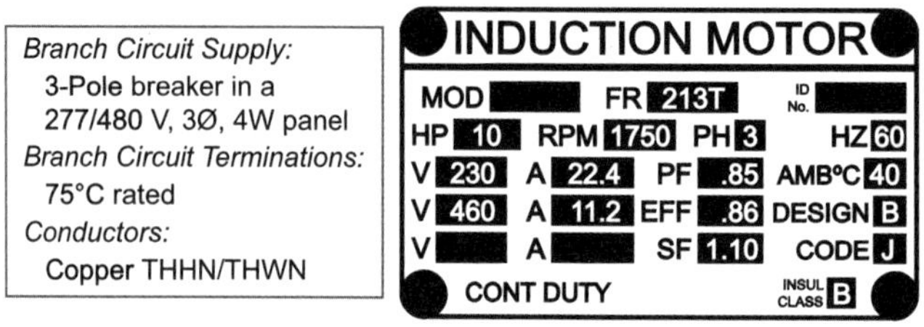

Branch Circuit Supply:
3-Pole breaker in a 277/480 V, 3Ø, 4W panel
Branch Circuit Terminations:
75°C rated
Conductors:
Copper THHN/THWN

INDUCTION MOTOR
MOD ___   FR 213T   ID No. ___
HP 10   RPM 1750   PH 3   HZ 60
V 230   A 22.4   PF .85   AMB°C 40
V 460   A 11.2   EFF .86   DESIGN B
V ___   A ___   SF 1.10   CODE J
CONT DUTY   INSUL CLASS B

Figure 230.224

20. Using the information shown above in Figure 230.224, a circuit breaker with a rating of not more than ___ A is the largest ITCB that can be installed in the panel.

   a. 40   b. 25   c. 45   d. 35   e. 30

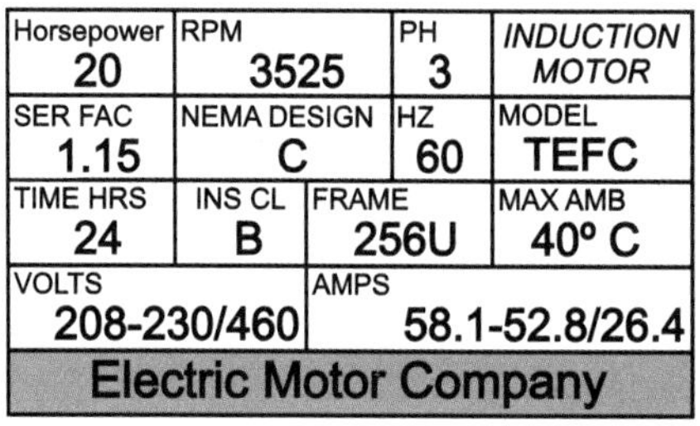

Horsepower	RPM	PH	INDUCTION MOTOR
20	3525	3	
SER FAC	NEMA DESIGN	HZ	MODEL
1.15	C	60	TEFC
TIME HRS	INS CL	FRAME	MAX AMB
24	B	256U	40° C
VOLTS		AMPS	
208-230/460		58.1-52.8/26.4	
Electric Motor Company			

Figure 230.112

21. Refer to the motor nameplate shown above in Figure 230.112. The largest 3-pole circuit breaker that can be installed in a 120/208 volt, 3Ø, 4-wire panelboard for this motor is a ___ A.

   a. 150   b. 60   c. 70   d. 125   e. 100

Lesson 231 Safety Worksheet

Lead Awareness

Lead has been poisoning workers for thousands of years. In the construction industry, traditionally most over-exposures to lead have been found in the trades, such as plumbing, welding, painting and demolition of structures.

In building construction, lead is frequently used for roofs, cornices, tank linings, and electrical products. In plumbing, soft solder is used primarily for soldering copper pipe and fittings. It is an alloy of tin and lead. Soft solder has been banned for many uses in the United States. The uses of lead-based paints in residential applications have also been banned by the Consumer Product Safety Commission. However, since lead-based paints inhibit rust and corrosion of iron and steel, it is still used on bridges, railway equipment, ships, lighthouses and other structures.

Significant lead exposures can come from removing paint from surfaces previously coated with lead-based paint, such as in bridge repair, residential renovation and demolition. There is an increase in highway and bridge repair and replacement, residential lead abatement and residential remodeling. The potential for exposure to lead-based paint has become more common. The trades most likely to be exposed to this increase are iron workers, demolition workers, abatement workers, plumbers, HVAC workers, electricians and carpenters performing renovation type work.

OSHA's requirements for exposure to lead must contain a program that at minimum has these following elements:

- Hazard determination, including exposure assessment;
- Engineering and work practice controls;
- Respiratory protection;
- Protective clothing and equipment;
- Housekeeping;
- Medical surveillance and provisions for medical removal;
- Training;
- Signs;
- Recording.

To implement the worker protection program properly, the employer needs to designate a competent person, i.e., one who is capable of identifying existing and predictable hazards or working conditions which are hazardous or dangerous to employees, in accordance with the general safety and health provisions of OSHA's construction standards. The competent person must have the authorization to take prompt corrective measures to eliminate such problems. Qualified medical personnel must be available to advise the employer and employees on the health effects of employee lead exposure and supervise the medical surveillance program.

See the following OSHA Fact Sheet for additional information regarding lead exposure.

OSHA FactSheet

Protecting Workers from Lead Hazards

Cleaning up after a flood requires hundreds of workers to renovate and repair, or tear down and dispose of, damaged or destroyed structures and materials. Repair, renovation and demolition operations often generate dangerous airborne concentrations of lead, a metal that can cause damage to the nervous system, kidneys, blood forming organs, and reproductive system if inhaled or ingested in dangerous quantities. The Occupational Safety and Health Administration (OSHA) has developed regulations designed to protect workers involved in construction activities from the hazards of lead exposure.

How You Can Become Exposed to Lead

Lead is an ingredient in thousands of products widely used throughout industry, including lead-based paints, lead solder, electrical fittings and conduits, tank linings, plumbing fixtures, and many metal alloys. Although many uses of lead have been banned, lead-based paints continue to be used on bridges, railways, ships, and other steel structures because of its rust- and corrosion-inhibiting properties. Also, many homes were painted with lead-containing paints. Significant lead exposures can also occur when paint is removed from surfaces previously covered with lead-based paint.

Operations that can generate lead dust and fumes include:

- Demolition of structures;
- Flame-torch cutting;
- Welding;
- Use of heat guns, sanders, scrapers, or grinders to remove lead paint; and
- Abrasive blasting of steel structures

OSHA has regulations governing construction worker exposure to lead. Employers of construction workers engaged in the repair, renovation, removal, demolition, and salvage of flood-damaged structures and materials are responsible for the development and implementation of a worker protection program in accordance with Title 29 Code of Federal Regulations (CFR), Part 1926.62. This program is essential to minimize worker risk of lead exposure. Construction projects vary in their scope and potential for exposing workers to lead and other hazards. Many projects involve only limited exposure, such as the removal of paint from a few interior residential surfaces, while others may involve substantial exposures. Employers must be in compliance with OSHA's lead standard at all times. A copy of the standard and a brochure — Lead in Construction (OSHA 3142) — describing how to comply with it, are available from OSHA Publications, P.O. Box 37535, Washington, D.C. 20013-7535, (202) 693-1888(phone), or (202) 693-2498(fax); or visit OSHA's website at www.osha.gov.

Major Elements of OSHA's Lead Standard

- A permissible exposure limit (PEL) of 50 micrograms of lead per cubic meter of air, as averaged over an 8-hour period.
- Requirements that employers use engineering controls and work practices, where feasible, to reduce worker exposure.
- Requirements that employees observe good personal hygiene practices, such as washing hands before eating and taking a shower before leaving the worksite.
- Requirements that employees be provided with protective clothing and, where necessary, with respiratory protection accordance with 29 CFR 1910.134.

- A requirement that employees exposed to high levels of lead be enrolled in a medical surveillance program.

Additional Information

For more information on this, and other health-related issues impacting workers, visit OSHA's Web site at www.osha.gov.

This is one in a series of informational fact sheets highlighting OSHA programs, policies or standards. It does not impose any new compliance requirements. For a comprehensive list of compliance requirements of OSHA standards or regulations, refer to Title 29 of the Code of Federal Regulations. This information will be made available to sensory impaired individuals upon request. The voice phone is (202) 693-1999; teletypewriter (TTY) number: (877) 889-5627.

For more complete information:

Occupational Safety and Health Administration

U.S. Department of Labor
www.osha.gov
(800) 321-OSHA

DSTM 9/2005

231-826 Motor Branch Circuit Overcurrent Protective Devices: Short Circuit and Ground Fault Protection

Questions

_____ 1. Over-exposure to lead can cause damage to the ___ if ingested or inhaled.

   a. nervous system
   b. kidneys
   c. blood organs
   d. reproductive organs
   e. all of these
   f. none of these

_____ 2. A worker in the construction industry can be exposed to lead in sufficient quantities to suffer from overexposure.

   a. True   b. False

_____ 3. In building construction, lead could be encountered when working with ___.

   a. special sheetrock used for an X-Ray lab
   b. plumbing fixtures
   c. tank linings
   d. electrical fittings
   e. all of these

_____ 4. In plumbing ___ is used primarily for sweating copper pipe and fittings.

   a. brass
   b. soft solder
   c. bronze
   d. tin
   e. aluminum
   f. steel

_____ 5. ___ has been banned for many uses in the United States.

   a. Bronze
   b. Brass
   c. Soft solder
   d. Aluminum
   e. Steel
   f. Tin

Year Two (Student Manual)

6. The use of ___-based paints in residential applications have been banned by the Consumer Product Safety Commission.

   a. lead   b. copper   c. latex   d. oil   e. epoxy   f. acrylic

7. Because ___-based paints inhibit rust and corrosion of iron and steel, these paints are still used on bridges, railway equipment, ships, lighthouses and other structures.

   a. lead   b. copper   c. latex   d. oil   e. epoxy   f. acrylic

8. Significant lead exposures can result from sandblasting surfaces previously coated with lead-based paint.

   a. True   b. False

9. Hazard determination, engineering and work practice controls, and ___ are part of OSHA's requirements for lead exposure control.
   I. respiratory protection
   II. protective clothing

   a. both I and II
   b. II only
   c. I only
   d neither I nor II

10. On a jobsite when lead exposure is an issue, the employer needs to designate a competent person to implement worker protection.

    a. True   b. False

11. Operations that can generate lead dust and fumes from___.

    a. flame cutting
    b. abrasive blasting
    c. sanding
    d. demolition
    e. welding
    f. all of these

_____ 12. The OSHA regulation covering lead in construction is in CFR ___.

    a. 1926.500
    b. 1926.350
    c. 1926.62
    d. 1926.20
    e. 1926.450f. 1926.100

Lesson 232
Motor: Branch Circuit Grounding Conductors, Disconnects, Starters, and Overload Protection

Electrical Curriculum

Year Two
Student Manual

Purpose
This lesson will discuss NEC© requirements and how to determine the required sizes or ratings of motor branch circuit equipment grounding conductors (EGCs), disconnecting means, starters, and overload protection.

Homework
(Due at the beginning of this class)
For this lesson, you should:

- Thoroughly read the material contained within Lesson 232.
- Read only the following in Electrical Systems
- **MOTORS – ARTICLE 430:**
    Motor Controllers – Article 430, Part VII
        All
    Motor Control Centers – Article 430, Part VIII
        All
    Disconnecting Means – Article 430, Part IX
        All
- Refer to "Advisory Notes for IEC 2nd Year Apprenticeship Motor Studies" from Lesson 230.
- Complete Objective 232.1 Worksheet.
- Complete Objective 232.2 Worksheet.
- Complete Objective 232.3 Worksheet.
- Complete Objective 232.4 Worksheet.
- Complete Objective 232.5 Worksheet.
- Complete Objective 232.6 Worksheet.
- Complete Objective 232.7 Worksheet.
- Read and complete Lesson 232 Safety Worksheet.
- Locate and become familiar with Yr-2 – Annex N
- Complete additional assignments or worksheets if available and directed by your instructor.

Motor: Branch Circuit Grounding Conductors, Disconnects, Starters, and Overload Protection

Objectives

By the end of this lesson, you should:

232.1 ....Be able to determine the minimum size equipment grounding conductor (EGC) required for a single-phase motor branch circuit.

232.2 ....Be able to determine the minimum size equipment grounding conductor (EGC) required for a three-phase motor branch circuit and determine the minimum size raceway required to enclose the circuit conductors and the EGC.

232.3 ....Understand the function of overload heaters contained within motor starters, and the purpose of a motor controller, including those controllers contained within MCCs and combination starters.

232.4 ....Understand the various devices that can be used as a disconnecting means for a motor and know the various standard sizes and ratings of safety switches, as shown in Yr-2 – Annex N.

232.5 ....Be able to select a safety switch, used as the disconnecting means for a motor, based on the circuit and motor requirements.

232.6 ....Be able to determine the minimum NEMA size motor starter required, as shown in Yr-2 – Annex N, based on circuit voltage and motor ratings.

232.7 ....Be able to select overload "heater" for installation in a motor starter, using heater selection tables from various manufacturers. These selection tables are included in Yr-2 – Annex N.

Errata to Electrical Systems: In the box, located in the lower left-hand corner on page 306 of the 2005 Edition, note that a combination starter does not *always* include a branch circuit OCPD as the author states. A combination starter is an enclosure that contains a switch to open the circuit conductors, and a motor starter. The enclosure CAN also include an OCPD.

More Advisory Notes for IEC 2nd Year Apprenticeship Motor Studies:

- **Keep these notes in a convenient location for frequent reference during the remaining 2nd Year lessons.**

- U.N.O., the smallest permitted NEMA size starters, for the motor's hp rating and the circuit voltage, are used.

- All Cutler-Hammer (Eaton) starters are of the non-compensating, enclosed type.

- As used on construction documents and drawings, as well as within the curriculum and its assessment bank questions, NF is an industry abbreviation for no-fuse, non-fused, or unfused when used to describe a safety switch (disconnecting means).

232.1 EGCs for Single-Phase Motor Circuits

The minimum size equipment grounding conductor required for a motor circuit is sized based on NEC® Section 250.122. Section 250.122(A) states that Table 250.122 is to be utilized and that the EGC is sized based on the rating of the OCPD protecting the circuit. However, the rating of the OCPD protecting motor circuits typically exceeds the ampacity of the motor circuit conductors. Section 250.122(A) also states that the size of the EGC never has to be larger than the circuit conductors. Consequently, the EGC for a motor circuit is frequently smaller than that shown in Table 250.122.

232.2 EGCs for Three-Phase Motor Circuits

232.3 Motor Controllers and MCCs

Do not confuse a controller with a disconnecting means, although a disconnecting means can sometimes also be the controller. NEC® Section 430.84 only requires that enough circuit conductors be opened to stop the motor. For a 3Ø motor, only two ungrounded circuit conductors need to be opened. For a 1Ø motor, only one ungrounded circuit conductor needs to be opened.

Don't forget that NEC® Section 430.102(A) requires that a controller have a disconnecting means that is within sight of the controller. This disconnecting means is not to be confused with the disconnecting means required for the motor by NEC® Section 430.102(B).

232.4 Disconnecting Means for Motors

Calculating the minimum ampere rating of a disconnecting means (safety switch) for a motor differs from the actual process of selecting a safety switch. NEC® Section 430.108 requires that any disconnecting means in the circuit comply with the requirements of 430.109 and 430.110. Section 430.109(A)(1) requires that a safety switch have rating in horsepower. Since safety switches are also used in other than motor circuits, they have both a horsepower and an ampere rating. Section 430.110(A) requires that a disconnecting means (safety switch) have an ampere rating that is at least 115% of the Table FLA of the motor. In many cases the calculated ampere rating of a safety switch is not sufficient for the motor's horsepower. Each manufacturer provides hp ratings for each safety switch. These differ between fused and no-fuse switches. Standard ampere and corresponding hp ratings for Square D safety switches are shown on the Square D Digest pages (included in Year 2 Annex N). Note that these switches have voltage ratings and are available as either fused or non-fused. Another consideration is the ampere rating of the fuses to be installed in a switch. Certainly, if 110 Amp fuses are to be installed in a switch, a 200 Amp switch (or larger) will have to be selected.

Single-Pole – Single-Throw Toggle Switch

Grounding: Self Grounding
Amperage: 20 Amp
Voltage: 120/277 Volt
HP Rating: 1HP-120V
              2HP-277V

Double-Pole – Single-Throw, Toggle Switch and Lockout Covers

Grounding: Self Grounding
Amperage: 20 Amp
Voltage: 120/277 Volt
HP Rating: 1HP-120V
              2HP-240V-277V

Grounding: Non-Grounding
Amperage: 60 Amp
Voltage: 600 Volt
HP Rating: 10HP-240V
              15HP-480V
              20HP-600V

 N3R Lockable Enclosure

Three-Pole – Single-Throw, Motor-Rated Switch and Enclosures

Grounding: Non-Grounding
Amperage: 30 Amp
Voltage: 600 Volt
HP Rating: 3HP-120V
              7½HP-240V
              15HP-480V
              20HP-600V

 N3R Lockable Enclosure

 N1 Lockable Enclosure

Properly rated switches can frequently be used as the disconnecting means for motors. Several different types of switches with lockable covers or enclosures are shown below. The ability to lock a disconnecting means can meet the requirements of the exceptions in NEC® Section 430.102.

Motor: Branch Circuit Grounding Conductors, Disconnects, Starters, and Overload Protection 232-833

Standard sizes and ratings of safety switches are shown in Yr-2 – Annex N. The pages showing these safety switches are excerpted from Square D's Digest.

Digest page 3-2 shows General (Light) Duty safety switches, both fused and non-fused. General Application information for these switches is shown at the top of the Digest page. The far-left column shows the *System*, as indicated by the number of poles (blades) available. The next column shows the ampere rating (*Amps*) of the switch. The ampere ratings of these switches are standardized throughout the industry. The type of fuse holder is shown in the next column, labeled *Fuse*. These 240-volt switches are rated in both amps and horsepower. In the *Horsepower Ratings* columns for the fusible switches, the option is available to install either one-time or time-delay fuses. Utilize the appropriate column based on the fuse type (one-time or DETD) and phase (single or three-phase). A switch is selected based on its capacity to interrupt circuit current. For a motor this requires that the horsepower rating of the switch be equal to or greater than that of the motor. For example, a 10 HP, 240-volt, single-phase motor would require a fused switch that has a rating of 10 hp, when DETD fuses are used. This 2-Pole switch also has a rating of 60 amps. If a no-fuse switch is desired, it also has 10 hp and 60 amp ratings.

Digest page 3-4 shows Heavy Duty, 240-volt, fused safety switches. Application information for these switches is shown at the top of the Digest page. For all switches, if a neutral conductor is not included in the motor circuit, switches with blades only (no neutral connection) should be selected.

Digest page 3-5 shows Heavy Duty, 600-volt, fused safety switches. Digest page 3-6 shows Heavy Duty, 600-volt, unfused safety switches.

Year Two (Student Manual)

232.5 Selection of Safety Switches Used as Motor Disconnects

Remember that safety switches are horsepower rated for use with motor loads but the ampere ratings listed permit their use with other than motor loads or a combination of various loads (including motors).

Be aware that, although Section 430.110(A) in the NEC® requires that a disconnecting means (safety switch) have an ampere rating that is at least 115% of the table FLA, the exception permits an unfused (no-fuse) switch with a lesser ampere rating to be used provided the horsepower rating of the switch is at least that of the motor.

232.6 NEMA Size Motor Starters

Before a heater can be selected to protect a motor, the NEMA size starter must be determined based on the nameplate HP rating and operating voltage. NEMA starters are generally available in sizes ranging from Size 00 (pronounced two-aught) to Size 7. Size 00 is the smallest. Larger sizes, typically Size 5 and larger, require the use of current transformers so that full line current doesn't pass through the heaters. Some manufacturers also make odd-size starters, such as 1¾ and 2½.

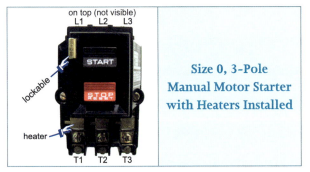

Size 0, 3-Pole Manual Motor Starter with Heaters Installed

Starters can be magnetically or manually operated. Magnetically-operated starters require a control circuit to cause them to open and close. Magnetic starters and control circuits will be studied in 3rd year. Manually-operated starters are opened and closed by pushing the START and STOP buttons on the starter. A manually-operated starter is shown at the right. This particular starter is rated to operate a 3-phase motor at up to 480 volts. The three circuit conductors are connected to L1, L2, and L3 on the top of the starter. The conductors to the motor are connected to T1, T2, and T3. This starter is another type of 3-pole, single-throw switch (3PST). When closed (START button depressed), the switch contacts within the starter CLOSE so that L1 is connected to T1, L2 is connected to T2, and L3 is connected to T3. When stopped (STOP button depressed), the switch contacts OPEN so that L1 is *disconnected* from T1, L2 is *disconnected* from T2, and L3 is *disconnected* from T3. Each of the three heaters senses overload current in that conductor. When an overload occurs, in any conductor, that heater causes all three to trip. When tripped, all three contact sets within the starter open, disconnecting the "L" contacts from the "T" contacts. Also note that this starter is lockable. When a lockout ring and padlock, or a small padlock, is inserted through the hole in the lock tab, the START button cannot be pushed in.

To determine the starter size required, first locate the *NEMA Size Motor Starters* table in Yr-2 – Annex N. Look in the appropriate column (Single Phase or Three Phase) for the circuit voltage utilized, and find a row that has a horsepower rating listed that is at least as large as the motor. For example, a 5 hp, 3Ø motor operating at 240 volts would require a starter that has a rating of 7½ hp. Look to the left in this row and see that a Size 1 starter is required. In many cases a larger size starter can be utilized if the heaters (thermal protection) will fit into the larger starter. Unless noted otherwise, the smallest permitted starter should be selected.

232.7 Selection of Overload Protection for Motors

Heater selection tables are typically included when a starter is shipped. The ability to select heaters (thermal protectors) for installation in starters is a valuable skill that is frequently required in the field. Several heaters are shown at the right.

For thermal unit (heater) selection for Square D starters, refer to Tables 14 (Size 00 starter), Table 15 (Size 0 and 1 starters), Table 58 (Size 2 starter), Table 16 (Size 3 starter), and Table 61 (Size 4 starter). These tables have column headings that indicate when 1, 2, or 3 thermal units (T.U.) are to be installed. Unless otherwise noted in the worksheet or exam question, the minimum quantity of thermal units, as required by the NEC®, are to be installed in the starter. Note that the heaters selected for a Square D Size 0 starter will also fit into a Size 1 starter. For example, a three-phase, 7 ½ hp motor with a nameplate FLA of 9.5 amps will be operated at 480 volts. In this case a NEMA size 1 starter is required. For a Square D starter, refer to Table 15 in Yr-2 – Annex N. Of course, this 3Ø motor requires three heaters. Note that the Size 1 portion of the table does not list a range of FLAs that includes 9.5 amps. Since a Square D Size 1 starter can use Size 0 heaters, the heaters can be selected from the Size 0 portion of the table. The range shown for 9.33-10.2 amps includes 9.5 amps. Three heaters with a part number of B 15.5 should be selected for the Size 1 starter. Note that for the Square D heaters in Yr-2 – Annex N, when a service factor of 1.0 to 1.10 is encountered, the nameplate FLA should be multiplied by 90% before entering the table.

For any manufacturer's selection tables, always follow any special instructions included with the tables. Heater selection tables from Siemens and Cutler-Hammer are also included in Yr-2 – Annex N. Note on the SIEMENS Table 233 selection table, that for motors with a SF of 1.0 a heater that is one size smaller should be selected. That is, if the motor nameplate FLA is 18.1 amps and the heaters are to be installed in a Size 3 starter, E60 heaters would be selected. However, if the SF is 1.0 then the next smaller (E57) should be selected.

On occasion (It could be that an older inefficient motor was replaced with a new energy efficient motor that requires less current.) existing starters are used for new motors and the starter may be a larger NEMA size than is actually required. However, the heaters must be sized for the new motor and the heaters must fit into that starter. The existing starter may be far too large to allow the correct heaters to fit and the starter must be replaced. For example, if an Eaton Cutler-Hammer (C-H) Size 3 starter is to be used for a 3Ø motor, the smallest heater that can be installed is an FH68 that is rated for 11.9 to 13.0 amps. Note that a Size 3, C-H starter can use heaters ranging from 11.9 amps (FH68) to 91 amps (FH89), and a Size 4 starter can use heaters ranging from 11.9 amps (FH68) to 133 amps (FH94).

Objective 232.1 Worksheet

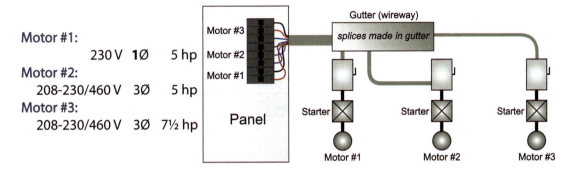

Motor #1:
  230 V 1Ø 5 hp
Motor #2:
  208-230/460 V 3Ø 5 hp
Motor #3:
  208-230/460 V 3Ø 7½ hp

Figure 230.261

All conductors are THHN/THWN copper and all terminations are rated at 75° C. This means that the wire connected to the motor must have the correct ampacity according to the 75° column. However, inside the conduit between the panel and the wireway ampacity corrections are made, as usual, in the 90° column. Use Exception 1 to 430.52(C)(1) to size OCPD's. The specifications require that separate equipment grounding conductors be installed in each raceway. The engineer who wrote the specs means that a metal raceway can't be used as the only EGC. But remember that only one EGC must be installed in each conduit. Refer to the drawing shown below to answer questions that refer to this Figure.

_____ 1. Refer to Figure 230.261. The minimum size equipment grounding conductor that can be installed in the conduit between the panel and the wireway is ___ AWG. The breakers are the largest permitted. Determine the maximum required for each circuit, then select the largest from among these. For this problem, assume that the nipple length between the panel and the wireway is 22 inches.

   a. 12   b. 8   c. 6   d. 4   e. 10

Scenario 230-26

Motor Nameplate Information: 208 V 1Ø   2 hp
Branch Circuit Supply: 2-pole breaker in a 120/208 V, 3-ph, 4-W panel
Branch Circuit Terminations: 75° C rated
Conductors: THHN/THWN copper

_____ 2. Refer to Scenario 230-26, shown above. The minimum permitted size circuit conductors are used for this circuit. Utilizing 430.52(C)(1) Exc. 1, one-time fuses with a maximum rating were installed in the disconnect for the motor. For this installation the minimum size EGC permitted is ___ AWG.

   a. 10   b. 8   c. 6   d. 12   e. 14

_____ 3. Refer to Scenario 230-26, shown above. Utilizing 430.52(C)(1) Exc. 1, the maximum size ITCB permitted was installed. For this installation the minimum size EGC permitted is ___ AWG. The minimum permitted size circuit conductors are used for this circuit.

   a. 12   b. 8   c. 6   d. 14   e. 10

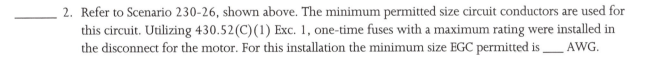

4. Refer to Figure 230.121, shown above. A safety switch is installed near the motor with the maximum permitted size DETD fuses. [430.52(C)(1) Exc 1 was used.] When the smallest permitted circuit conductors are used between the switch and the motor, the smallest size EGC permitted is ___ AWG.

   a. 14
   b. 12
   c. 10
   d. 6
   e. 8

5. Refer to Figure 230.121, shown above. When the maximum permitted size ITCB has been installed for the branch circuit [using 430.52(C)(1) Exc 1], and the smallest permitted circuit conductors are installed, the smallest size EGC permitted is ___ AWG.

   a. 12
   b. 10
   c. 6
   d. 8
   e. 14

6. A 115/230 V, 1Ø, 3 hp motor is fed from a circuit supplied by a 1-pole breaker in a 120/240 V, 1-ph, 3-W panel. The safety switch adjacent to the motor has the smallest permitted DETD fuses installed. Utilize 430.52(C)(1) Exc 1. The minimum size EGC permitted is ___ AWG copper where minimum permitted size circuit conductors are used for this circuit.

   a. 12
   b. 6
   c. 8
   d. 10
   e. 14

Objective 232.2 Worksheet

Scenario 230-28

Motor Nameplate Information: 5 hp   208-230/460 V   3Ø
Branch Circuit Supply: 3-pole breaker in a 120/208 V, 3-ph, 4-W panel

_____ 1. The minimum size EGC permitted for the installation shown above in Scenario 230-28 is ___ AWG where the smallest permitted circuit conductors were used. Also the ITCB used in the panel has the maximum rating permitted by 430.52(C)(1) Exc. 1.

     a. 8   b. 10   c. 6   d. 12   e. 14

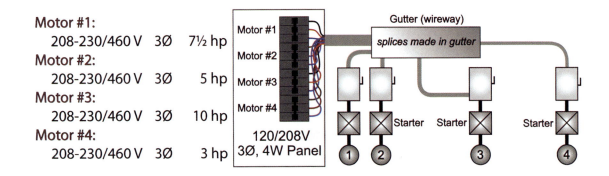

Figure 230.271

All conductors are THHN/THWN copper and all terminations are rated at 75° C. This means that the wire connected to the motor must have the correct ampacity according to the 75° column. However, inside the conduit between the panel and the wireway ampacity corrections are made, as usual, in the 90° column. Use Exception 1 to 430.52(C)(1) to size OCPDs. The specifications require that separate equipment grounding conductors be installed in each raceway. The engineer who wrote the specs means that a metal raceway can't be used as the only EGC. But remember that only one EGC must be installed in each conduit. The small wireway is located near the motor location. Refer to the drawing shown below to answer questions that refer to this Figure.

_____ 2. Refer to Figure 230.271. The smallest EGC that can be installed between the safety switch And Motor #4 is ___ AWG.

     a. 10   b. 8   c. 14   d. 12   e. 6

_____ 3. Refer to Figure 230.271. The minimum size equipment grounding conductor that can be installed in the conduit between the panel and the wireway is ___ AWG. The breakers are the largest permitted.

     a. 12   b. 4   c. 10   d. 8   e. 6

Motor: Branch Circuit Grounding Conductors, Disconnects, Starters, and Overload Protection

_____ 4. Refer to Figure 230.271. The smallest EGC that can be installed between the safety switch and Motor #1 is ___ AWG.

a. 10   b. 14   c. 6   d. 8   e. 12

_____ 5. The minimum size EMT that can be used between the panel and the wireway, shown above in Figure 230.271, is Trade Size ___.

a. 1 ¼   b. ¾   c. 1   d. 1 ½   e. ½

_____ 6. You installed a 3-pole 300 A breaker in a panel to feed a motor. Although the plans required only 3 AWG circuit conductors to the motor, you ran 1 AWG because of the long length of the circuit. The smallest EGC you can run with the circuit is ___ AWG.

a. 6   b. 1   c. 4   d. 2   e. 3

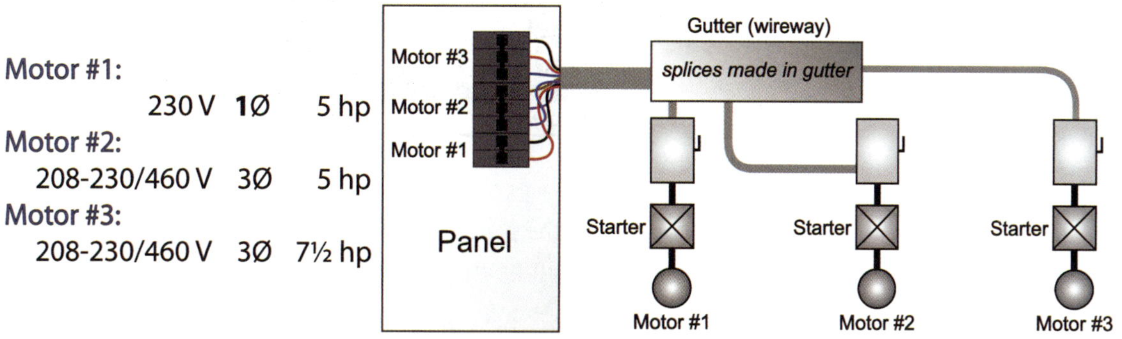

Motor #1:
  230 V 1Ø 5 hp
Motor #2:
  208-230/460 V 3Ø 5 hp
Motor #3:
  208-230/460 V 3Ø 7½ hp

Figure 230.261

All conductors are THHN/THWN copper and all terminations are rated at 75° C. This means that the wire connected to the motor must have the correct ampacity according to the 75° column. However, inside the conduit between the panel and the wireway ampacity corrections are made, as usual, in the 90° column. Use Exception 1 to 430.52(C)(1) to size OCPD's. The specifications require that separate equipment grounding conductors be installed in each raceway. The engineer who wrote the specs means that a metal raceway can't be used as the only EGC. But remember that only one EGC must be installed in each conduit.
Refer to the drawing shown below to answer questions that refer to this Figure.

_____ 7. Refer to Figure 230.261. The minimum size EMT that can be used between the panel and the wireway is Trade Size ___.

a. ¾   b. 1   c. ½   d. 1 ½   e. 1 ¼

_____ 8. Refer to Figure 230.261. The minimum size LFMC that can be used between the safety switch and Motor #3 is Trade Size ___.

    a. 1   b. ½   c. 1 ½   d. ¾   e. 1 ¼

MOTORS UNLIMITED
AC INDUCTION MOTOR

Thermally Protected NO	Type ☐
STYLE K	SERIAL 514919
FRAME 56	TYPE
HP 20 PH 3	HOUSING T.O.
RPM 1760	SERVICE FACTOR 1.15
CYCLES 60	S.F. AMPS
VOLTS 208-230/460	AMPS 58.1-52.8/26.4
DEGREES C RISE 40	CODE D
HOURS 24	

This motor is supplied by a branch circuit that originates from a 120/208 volt, 3Ø, 4-W supply.

All conductors are THHN/THWN copper and all terminations are rated at 75° C.

Figure 230.131

_____ 9. Refer to the nameplate and information shown above in Figure 230.131. Smallest EGC permitted with an ITCB is ___ AWG. 430.52(C)(1) Exc 1 was used to determine the largest permitted breaker.

    a. 8   b. 6   c. 4   d. 10   e. none of these

Branch Circuit Supply:
3-Pole breaker in a 277/480 V, 3Ø, 4W panel
Branch Circuit Terminations:
75°C rated
Conductors:
Copper THHN/THWN

INDUCTION MOTOR

MOD	FR 213T	ID No.
HP 10	RPM 1750 PH 3	HZ 60
V 230 A 22.4	PF .85	AMB°C 40
V 460 A 11.2	EFF .86	DESIGN B
V A	SF 1.10	CODE J
CONT DUTY		INSUL CLASS B

Figure 230.224

_____ 10. Using the information shown above in Figure 230.224 and 430.52(C)(1) Exc 1, an ITCB with a maximum permitted rating was installed in the panel. Given this information and that the smallest permitted circuit conductors are used, the minimum size EGC permitted is ___ AWG.

    a. 6   b. 14   c. 8   d. 12   e. 10

Branch Circuit Supply:
3-Pole breaker in a 120/240 V, 3Ø, 4W panel
Branch Circuit Terminations:
75°C rated
Conductors:
Copper THHN/THWN

MOTORS UNLIMITED
AC INDUCTION MOTOR

Thermally Protected NO	Type ☐
STYLE K	SERIAL 591379
FRAME 56	TYPE
HP 30 PH 3	HOUSING T.O.
RPM 1760	SERVICE FACTOR 1.15
CYCLES 60	S.F. AMPS
VOLTS 208-230/460	AMPS 82.1-74.6/37.3
DEGREES C RISE 40	CODE D HOURS 24

Figure 230.226

_____ 11. Using the information shown above in Figure 230.226 and 430.52(C)(1) Exc 1, DETD fuses with the maximum permitted rating were installed in the disconnecting means for the motor. Given this information and that the smallest permitted circuit conductors are used, the minimum size EGC permitted is ___ AWG.

    a. 6   b. 3   c. 2   d. 8   e. 4

Motor: Branch Circuit Grounding Conductors, Disconnects, Starters, and Overload Protection 232-843

_____ 12. Using the information shown above in Figure 230.226 and 430.52(C)(1) Exc 1, nontime-delay fuses with the maximum permitted rating were installed in the disconnecting means for the motor. Given this information and that the smallest permitted circuit conductors are used, the minimum size EGC permitted is ___ AWG.

   a. 8   b. 3   c. 2   d. 4   e. 6

_____ 13. Given the information shown above in Figure 230.226 and that the smallest permitted circuit conductors are used, the minimum size EGC permitted is ___ AWG. An ITCB was installed in the panel using 430.52(C)(1) Exc 1.

   a. 6   b. 3   c. 2   d. 4   e. 8

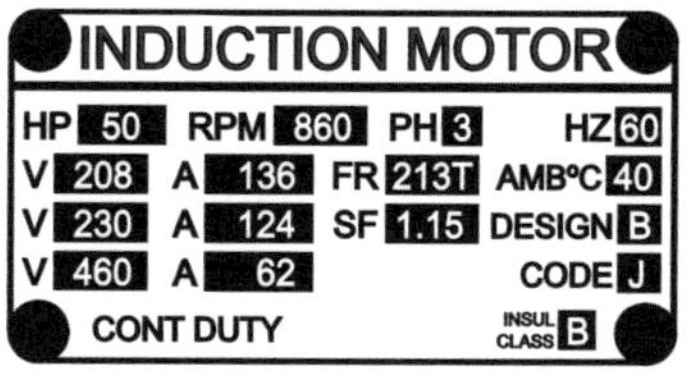

INDUCTION MOTOR
HP 50	RPM 860	PH 3	HZ 60
V 208	A 136	FR 213T	AMB°C 40
V 230	A 124	SF 1.15	DESIGN B
V 460	A 62		CODE J
CONT DUTY		INSUL CLASS B	

The branch circuit originates in a 120/240 volt, 3Ø, 4W panelboard.
All terminations are 75°C.
All conductors are THWN/THHN copper.
Use Exc 1 to 430.52(C)(1) to size OCPDs.

Figure 230.252

_____ 14. For the installation in Figure 230.252, the smallest EGC permitted (using DETD fuses) is ___ AWG. 430.52(C)(1) Exc 1 was used to determine the largest permitted fuses.

   a. 1   b. 3/0   c. 3   d. 2/0   e. 4

Year Two (Student Manual)

Student Notes

Objective 232.3 Worksheet

_____ 1. THHN/THWN 14 AWG conductors are installed between a 3-pole, 40 A circuit breaker in a panel and a motor starter. These conductors are also installed between the starter and the motor. When "heaters" (motor overload protectors) are installed in the motor starter these protect the ___ against overload.

   a. conductors   b. motor   c. both a and b   d. neither a nor b

_____ 2. Motor controllers must be marked with ___.
   I. maximum rated voltage
   II. maximum hp or current rating

   a. both I & II   b. I only   c. II only   d. neither I nor II

_____ 3. All motor circuits require ___ protection.

   a. ground-fault   b. short-circuit   c. overload   d. all of these

_____ 4. A combination starter is a ___.
   I. safety switch
   II. motor starter

   a. both I and II   b. neither I nor II   c. II only   d. I only

_____ 5. A(n) ___ is an assembly of one or more enclosed sections having a common power bus and principally containing motor control units.

   a. MCA   b. CCM   c. ACM   d. MCC

_____ 6. The overcurrent protection for an MCC may be located ahead of or within the MCC.

   a. False   b. True

_____ 7. Heaters are installed in motor starters to protect ___.
   I.   the motor against overload
   II. the branch circuit conductors against overload
   III. the motor against short-circuits and ground-faults

   a. I & II only   b. III only   c. II & III only   d. I & III only

Motor: Branch Circuit Grounding Conductors, Disconnects, Starters, and Overload Protection

Objective 232.4 Worksheet

_____ 1. A toggle switch (standard AC only "light" switch) can be used as the disconnecting means for a stationary motor rated at 2 hp or less and 300 volts or less as long as the switch has an ampere rating that is at least ___ % of the motor ampere rating. **Note:** the NEC® considers a toggle switch to be a general-use snap switch. See the definition in Article 100.

   a. 200 b. 125 c. 80 d. 50

_____ 2. Refer to the Plan View of the electrical and equipment rooms below. A circuit breaker in Panel HA feeds the starter and the motor in the equipment room. There is a full-height wall between the two rooms. Which of the safety switches must be installed to satisfy the minimum requirements of the NEC®?

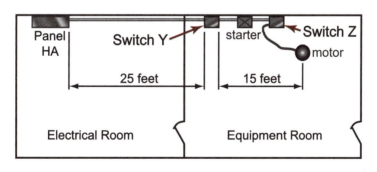

   a. Neither Y nor Z b. Z only c. Both Y and Z d. Y only

_____ 3. Can a 15 amp, 277 volt AC only toggle switch (standard "light" switch) be used as the disconnect for a stationary 1.5 hp, 230 volt, single-phase motor? **Note:** the NEC® considers a toggle switch to be a general-use snap switch. See the definition in Article 100.

   a. Yes b. No

_____ 4. Can a double-pole single-throw (light) switch rated for 250 volts and 15 amps be used as a disconnect for a 1.5 HP, 230-volt, single-phase, stationary motor? **Note:** the NEC® considers a toggle switch to be a general-use snap switch. See the definition in Article 100.

   a. No   b. Yes

_____ 5. The smallest safety switch (standard size) that can be installed for a circuit is a ___ Amp switch, when 35 amp fuses are to be installed.

   a. 60   b. 40   c. 35   d. 100   e. 30

_____ 6. When a switch is used for the disconnecting means for a 1-phase motor operating at 240 volts, the switch can be of the ___ type.

   I. SPST   II. DPST

   a. either I or II   b. II only   c. neither I nor II   d. I only

_____ 7. A branch circuit for a motor originates from a breaker in a panelboard. The circuit runs from the panel, to a motor starter, and finally to the motor. The panel is located in one room and the starter and the motor are located in another room. The controller (starter) and the motor are "within sight" of each other. A safety switch (or other type of disconnect) is ___.

   a. required between the panel and the starter in the room with the motor
   b. required between the starter and the motor in the room with the motor
   c. both a and b
   d. neither a nor b

I L6-30
   30A
   Locking
   Plug and
   Connector
   250V - 2 hp

II 30A Non-fused
    3P Safety Switch
    5 hp - 120/208-240V
    15 hp - 480-600V

Figure 232.104

Motor: Branch Circuit Grounding Conductors, Disconnects, Starters, and Overload Protection   232-849

_____ 8. Refer to Figure 232.104, shown on the previous page. ___ could be used as a disconnecting means for a 3 horsepower, 240-volt, single-phase motor.

a. either I or II
b. neither I nor II
c. II only
d. I only

_____ 9. Refer to Figure 232.104, shown on the previous page. ___ could be used as a disconnecting means for a 2 horsepower, 240-volt, single-phase motor.

a. I only   b. neither I nor II   c. either I or II   d. II only

I — 30 A 240V Non-fused Safety Switch 3Ø-7½ hp 1Ø-3 hp

II — L6-30 30A Locking Plug and Connector 250V - 2 hp

Figure 232.103

_____ 10. Refer to Figure 232.103, shown above. ___ could be used as a disconnecting means for a 5 horsepower, 240-volt, single-phase motor.

a. I only   b. neither I nor II   c. II only   d. either I or II

_____ 11. A 3Ø safety switch that is rated at 600 volts can be used as the disconnecting means for a ___ volt, 3Ø motor.
I. 208
II. 240
III. 480

a. I or II or III   b. II or III only   c. II only   d. III only   e. I or II only

_____ 12. Which of the following is (are) standard size safety switches?
I. 40 A   II. 50 A   III. 60 A

a. III only   b. I & II only   c. I only   d. II & III only   e. I & II & III

Year Two (Student Manual)

Figure 232.101

_____ 13. Refer to Figure 232.101, shown above. ___ could be used as a disconnecting means for a 2 horsepower, 240-volt, single-phase motor.

    a. II only   b. I or II or III   c. II or III only   d. I or II only   e. III only

_____ 14. A branch circuit for a motor originates from a breaker in a panelboard. The circuit runs from the panel, to a motor starter, and finally to the motor. The panel and the starter are located in a room, but are not "within sight" of each other. The motor is located outside. A safety switch (or other type of disconnect) is ___.

    a. required outside and "within sight" of the motor
    b. required between the panel and the starter within sight of the starter
    c. both a and b
    d. neither a nor b

_____ 15. A 3Ø safety switch that is rated at 250 volts can be used as the disconnecting means for a ___ motor.
    I. 240 volt, 1Ø
    II. 277 volt, 1Ø
    III. 480 volt, 3Ø

    a. I only
    b. III only
    c. I or II or III
    d. I or II only
    e. II or III only

_____ 16. A NEMA Type ___ enclosure provides a degree of protection against windblown dust. Refer to NEC® Table 430.91.

    a. 3R   b. 1   c. 3

Motor: Branch Circuit Grounding Conductors, Disconnects, Starters, and Overload Protection

232-851

Student Notes

Student Notes

Year Two (Student Manual)

Objective 232.5 Worksheet

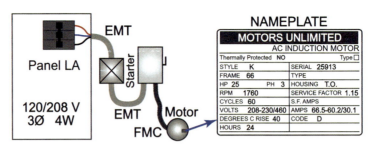

Figure 230.231

_____ 1. Refer to the information shown above in Figure 230.231. If you were to use a non-fused, Square D safety switch, it must have a rating of at least ___ A.

    a. 200   b. 125   c. 150   d. 60   e. 100

_____ 2. Refer to the information shown above in Figure 230.231. To be able to contain time-delay fuses, the smallest, Square D, heavy duty safety switch has a ___ amp rating.

    a. 60   b. 125   c. 150   d. 200   e. 100

Wild Horse Motor Co.

FRAME	184T
HP	10
VOLTS	230/460
AMPS	25.6/12.8
RPM	3560
SF	1.15
DUTY	CONTINUOUS

DESIGN B   HZ 60   PH 3
INS CL F   AMB 40°C
EFFICIENCY 79.8

Figure 230.123

_____ 3. Refer to the motor nameplate shown above in Figure 230.123. The branch circuit for the motor is supplied from a 120/240 volt, 3Ø, 4-wire panelboard. The smallest, non-fused, Square D safety switch (standard size Heavy Duty) that can be installed for this motor is a ___ A.

    a. 200   b. 30   c. 100   d. 60   e. 80

_____ 4. Refer to the motor nameplate shown above in Figure 230.123. The branch circuit for the motor is supplied from a 120/240 volt, 3Ø, 4-wire panelboard. The smallest, non-fused, Square D safety switch (standard size General Duty) that can be installed for this motor is a ___ A.

    a. 100   b. 200   c. 30   d. 60   e. 80

Motor: Branch Circuit Grounding Conductors, Disconnects, Starters, and Overload Protection

```
┌─────────────────────────────────────┬──────────────────────┐
│         MOTORS UNLIMITED            │ This motor is supplied│
│         AC INDUCTION MOTOR          │ by a branch circuit that│
│ Thermally Protected  NO    Type ☐   │ originates from a    │
│ STYLE    K      │ SERIAL  514919    │ 120/208 volt, 3Ø, 4-W│
│ FRAME    56     │ TYPE              │ supply.              │
│ HP  20     PH 3 │ HOUSING  T.O.     │                      │
│ RPM      1760   │ SERVICE FACTOR 1.15│ All conductors are  │
│ CYCLES   60     │ S.F. AMPS         │ THHN/THWN copper     │
│ VOLTS 208-230/460│ AMPS 58.1-52.8/26.4│ and all terminations are│
│ DEGREES C RISE 40│ CODE   D         │ rated at 75° C.      │
│ HOURS    24     │                   │                      │
└─────────────────────────────────────┴──────────────────────┘
```

Figure 230.131

_____ 5. Refer to the nameplate and information shown above in Figure 230.131. According to the NEC®, a safety switch (when used as the disconnecting means for this motor) must have a minimum standard amp rating of ___.

    a. 100   b. 30   c. 60   d. 200   e. none of these

_____ 6. Refer to the nameplate and information shown above in Figure 230.131. The smallest, Square D, NF safety switch that can be used has an ampere rating of ___ amps. A switch with the lowest permitted voltage rating is used.

    a. 30   b. 60   c. 100   d. 20   e. 200

_____ 7. Refer to the nameplate and information shown above in Figure 230.131. A Square D safety switch with a minimum rating of ___ Amps is the smallest that can be used to hold DETD fuses. A switch with the lowest permitted voltage rating is used.

    a. 30   b. 100   c. 60   d. 200   e. none of these

_____ 8. Refer to the nameplate and information shown above in Figure 230.131. The minimum hp rating of any manufacturer's safety switch must be ___ hp at the applied voltage.

    a. 30   b. 20   c. 15   d. 25

_____ 9. Refer to the nameplate and information shown above in Figure 230.131. The NEC® minimum calculated ampere rating of the disconnecting means is ___ amps.

    a. 58.1   b. 59.4   c. 68.3   d. 62.1   e. 74.3

Year Two (Student Manual)

```
                  MOTORS UNLIMITED
                         AC INDUCTION MOTOR        This motor is supplied
         Thermally Protected  NO         Type □    by a branch circuit that
         STYLE    K        SERIAL  514919          originates from a
         FRAME   56        TYPE                    120/208 volt, 3Ø, 4-W
         HP  20      PH  3  HOUSING  T.O.          supply.
         RPM    1760       SERVICE FACTOR  1.15
         CYCLES  60        S.F. AMPS               All conductors are
         VOLTS   208-230/460  AMPS  58.1-52.8/26.4 THHN/THWN copper
         DEGREES C RISE  40  CODE  D               and all terminations are
         HOURS  24                                 rated at 75° C.
```

Figure 230.251

_____ 10. For the installation in Figure 230.251, the smallest, Square D, NF safety switch that can be used has a hp rating of ___. A switch with the lowest permitted voltage rating is used.

    a. 40   b. 50   c. 60   d. 100   e. 30

_____ 11. For the installation in Figure 230.251, the minimum size (std amp rating), Square D safety switch required to hold DETD fuses is rated at ___ amps. A switch with the lowest permitted voltage rating is used.

    a. 100   b. 60   c. 200   d. 30   e. none of these

```
         SER FAC | PH  | KVA CODE    | INDUCTION
          1.15   |  3  |             |   MOTOR
         Horsepower | RPM | NEMA DESIGN | MODEL
             15     | 3525|      C      | TEFC
         HZ | HRS | INS CL | FRAME      | MAX AMB
         60 | 24  |   B    | 256U       |  40° C
         VOLTS              | AMPS
         208-230/460        | 44.4-40.4/20.2
                Electric Motor Company
```

Figure 232.119

_____ 12. Refer to the motor nameplate shown above in Figure 232.119. The smallest, Square D, general duty, fused safety switch (standard size) that can be installed for this motor is a ___ Amp. The circuit line voltage is 240. Fuses are DETD.

    a. 100   b. 80   c. 50   d. 30   e. 60

_____ 13. Refer to the motor nameplate shown above in Figure 232.119. The smallest, Square D, general duty, non-fused safety switch (standard size) that can be installed for this motor is a ___ Amp. The circuit line voltage is 240.

    a. 30   b. 60   c. 50   d. 80   e. 100

Motor: Branch Circuit Grounding Conductors, Disconnects, Starters, and Overload Protection

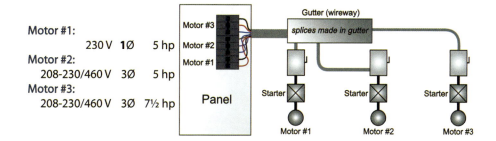

Figure 230.261

All conductors are THHN/THWN copper and all terminations are rated at 75° C. This means that the wire connected to the motor must have the correct ampacity according to the 75° column. However, inside the conduit between the panel and the wireway ampacity corrections are made, as usual, in the 90° column. Use Exception 1 to 430.52(C)(1) to size OCPD's. The specifications require that separate equipment grounding conductors be installed in each raceway. The engineer who wrote the specs means that a metal raceway can't be used as the only EGC. But remember that only one EGC must be installed in each conduit. Refer to the drawing shown below to answer questions that refer to this Figure.

_____ 14. Refer to Figure 230.261. The minimum standard size, Square D, non-fused safety switch that can be used between the wireway and Motor #2 has a hp rating of ___. A switch with the lowest permitted voltage rating is used.

  a. 3   b. 7 ½   c. 10   d. 15   e. none of these

_____ 15. Refer to Figure 230.261. The minimum standard size, Square D, non-fused safety switch that can be used between the wireway and Motor #1 has a hp rating of ___. A switch with the lowest permitted voltage rating is used.

  a. 7 ½   b. 10   c. 3   d. 15   e. none of these

_____ 16. Refer to Figure 230.261. The minimum standard size, Square D, non-fused safety switch that can be used between the wireway and Motor #1 has an ampere rating of ___ amps. A switch with the lowest permitted voltage rating is used.

  a. 30   b. 60   c. 100   d. 200   e. none of these

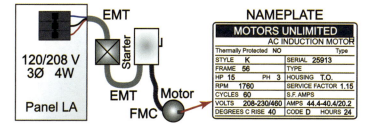

Figure 230.230

_____ 17. Refer to the information shown above in Figure 230.230. The smallest, fused, Square D safety switch (standard size General Duty) that can be installed for this motor is a ___ A. The largest permitted DETD fuses, per Exc 1 to 430.52(C)(1), are to be installed in the switch.

  a. 60   b. 200   c. 80   d. 30   e. 100

Objective 232.6 Worksheet

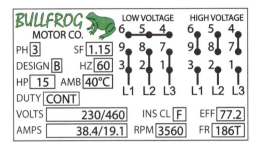

Figure 230.128

_____ 1. Refer to the nameplate shown above in Figure 230.128. The branch circuit for the motor is supplied from a 480/277 volt, 3Ø, 4-wire panelboard. A NEMA size ___ starter is the smallest that could be used.

    a. 00   b. 1   c. 0   d. 2   e. 3

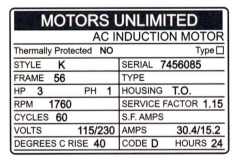

Figure 230.112

_____ 2. Refer to the motor nameplate shown above in Figure 230.112. The smallest size NEMA starter that can be used for this motor is ___, when the motor circuit is supplied from a 480/277 volt, 3Ø, 4-W panel.

    a. 2   b. 4   c. 1   d. 0   e. 3

MOTORS UNLIMITED	
AC INDUCTION MOTOR	
Thermally Protected NO	Type ☐
STYLE K	SERIAL 7456085
FRAME 56	TYPE
HP 3 PH 1	HOUSING T.O.
RPM 1760	SERVICE FACTOR 1.15
CYCLES 60	S.F. AMPS
VOLTS 115/230	AMPS 30.4/15.2
DEGREES C RISE 40	CODE D HOURS 24

Figure 230.115

_____ 3. Refer to the motor nameplate shown above in Figure 230.115. The smallest size NEMA starter that can be used for this motor is ___, when the motor circuit is supplied from a 2-pole breaker in a 120/240 volt, 3Ø, 4-W panel.

    a. 3   b. 2   c. 0   d. 00   e. 1

Motor: Branch Circuit Grounding Conductors, Disconnects, Starters, and Overload Protection

_____ 4. Refer to the motor nameplate shown on the previous page in Figure 230.115. The smallest size NEMA starter that can be used for this motor is ___, when the motor circuit is supplied from a 1- pole breaker in a 120/240 volt, 3Ø, 4-W panel.

a. 1 b. 00 c. 3 d. 2 e. 0

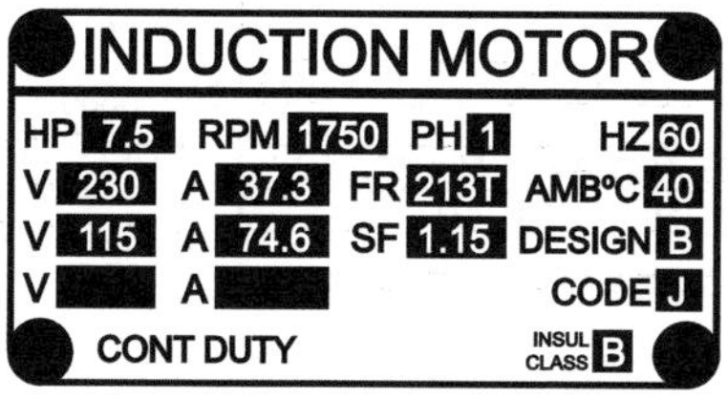

INDUCTION MOTOR
HP 7.5 RPM 1750 PH 1 HZ 60
V 230 A 37.3 FR 213T AMB°C 40
V 115 A 74.6 SF 1.15 DESIGN B
V ___ A ___ CODE J
CONT DUTY INSUL CLASS B

Figure 230.108

_____ 5. Refer to the motor nameplate shown above in Figure 230.108. The smallest size NEMA starter that can be used for this motor is ___, when the motor circuit is supplied from a 2-pole breaker in a 120/240 volt, 3Ø, 4-W panel.

a. 2 b. 1 c. 0 d. 00 e. 3

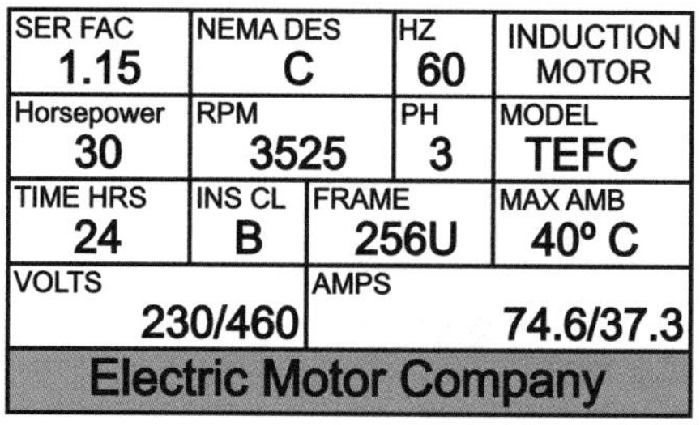

SER FAC	NEMA DES	HZ	INDUCTION
1.15	C	60	MOTOR
Horsepower	RPM	PH	MODEL
30	3525	3	TEFC
TIME HRS	INS CL	FRAME	MAX AMB
24	B	256U	40° C
VOLTS		AMPS	
230/460		74.6/37.3	
Electric Motor Company			

All terminations are 75°C.
All conductors are THWN/THHN copper.
Use Exc 1 to 430.52(C)(1) to size OCPDs.
The branch circuit originates in a 277/480 volt, 3Ø, 4W panelboard.

Figure 230.251

_____ 6. For the installation in Figure 230.251, the minimum NEMA size starter is ___.

a. 4 b. 1 c. 2 d. 5 e. 3

Objective 232.7 Worksheet

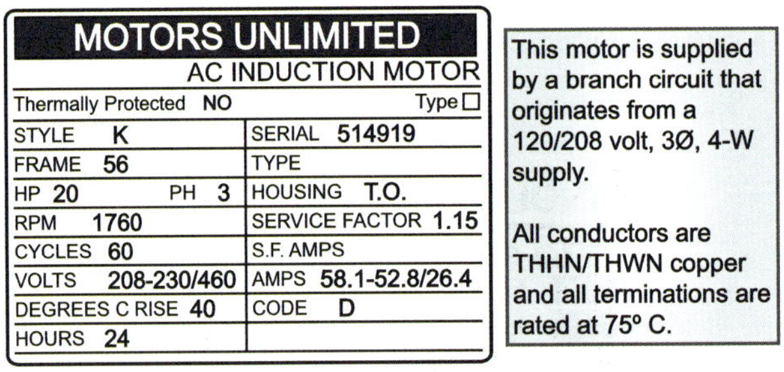

Figure 230.131

Motor nameplate: MOTORS UNLIMITED AC INDUCTION MOTOR; Thermally Protected NO; Type ☐; STYLE K; SERIAL 514919; FRAME 56; TYPE; HP 20; PH 3; HOUSING T.O.; RPM 1760; SERVICE FACTOR 1.15; CYCLES 60; S.F. AMPS; VOLTS 208-230/460; AMPS 58.1-52.8/26.4; DEGREES C RISE 40; CODE D; HOURS 24.

This motor is supplied by a branch circuit that originates from a 120/208 volt, 3Ø, 4-W supply. All conductors are THHN/THWN copper and all terminations are rated at 75° C.

_____ 1. Refer to the nameplate and information shown above in Figure 230.131. A Square D starter requires part number ___ heaters, with the appropriate prefix.

    a. 74.6   b. 94.0   c. 87.7   d. 81.5   e. none of these

_____ 2. Refer to the nameplate and information shown above in Figure 230.131. A Cutler-Hammer starter is used. The heater part number is FH___.

    a. 83   b. 82   c. 84   d. 85   e. none of these

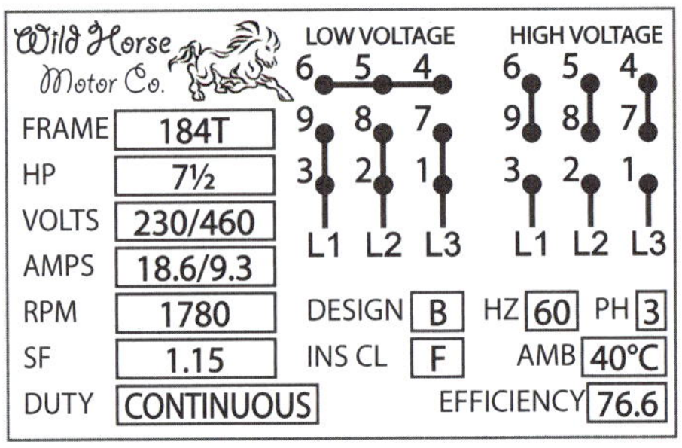

Figure 230.122

Wild Horse Motor Co. nameplate: FRAME 184T; HP 7½; VOLTS 230/460; AMPS 18.6/9.3; RPM 1780; SF 1.15; DUTY CONTINUOUS; DESIGN B; HZ 60; PH 3; INS CL F; AMB 40°C; EFFICIENCY 76.6. Low voltage and high voltage connection diagrams shown.

_____ 3. Refer to the nameplate shown above in Figure 230.122. For a correctly sized Square D starter, the heater catalog # for the heaters is ___. This motor is supplied power from a 480/277 volt, 3-phase 4-wire panelboard.

    a. B32   b. B17.5   c. B40   d. B14   e. none of these

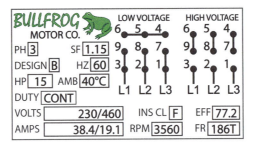

Figure 230.128

Motor: Branch Circuit Grounding Conductors, Disconnects, Starters, and Overload Protection

_____ 4. Refer to the nameplate shown on the previous page in Figure 230.128. The branch circuit for the motor is supplied from a 480/277 volt, 3Ø, 4-wire panelboard. An existing Cutler-Hammer NEMA size 3 starter will be used for this motor. You will select C-H # FH ___ heaters to install in this starter.

   a. 48   b. 47   c. 54   d. 79   e. none of these

_____ 5. Refer to the nameplate shown on the previous page in Figure 230.128. The branch circuit for the motor is supplied from a 480/277 volt, 3Ø, 4-wire panelboard. An existing Siemens NEMA size 3½ starter will be used for this motor. You will select Siemens # E___ heaters to install in this starter.

   a. 66   b. 62   c. 61   d. 65   e. none of these

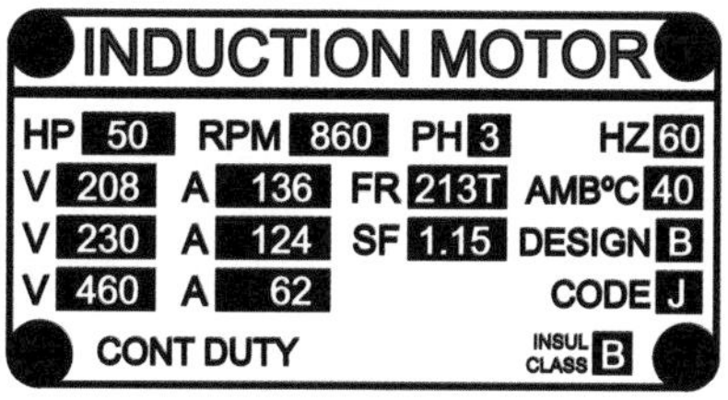

INDUCTION MOTOR
HP 50 RPM 860 PH 3 HZ 60
V 208 A 136 FR 213T AMB°C 40
V 230 A 124 SF 1.15 DESIGN B
V 460 A 62   CODE J
CONT DUTY   INSUL CLASS B

The branch circuit originates in a 120/240 volt, 3Ø, 4W panelboard.
All terminations are 75°C.
All conductors are THWN/THHN copper.
Use Exc 1 to 430.52(C)(1) to size OCPDs.

Figure 230.252

_____ 6. Refer to Figure 230.252, shown above. For a Cutler-Hammer starter, the part number for the heater(s) is FH___.

   a. 93   b. 92   c. 24   d. 25   e. none of these

_____ 7. Refer to Figure 230.252, shown above. When a Siemens starter is used, the part number for the heater(s) is E___.

   a. 99   b. 101   c. 34   d. 32   e. none of these

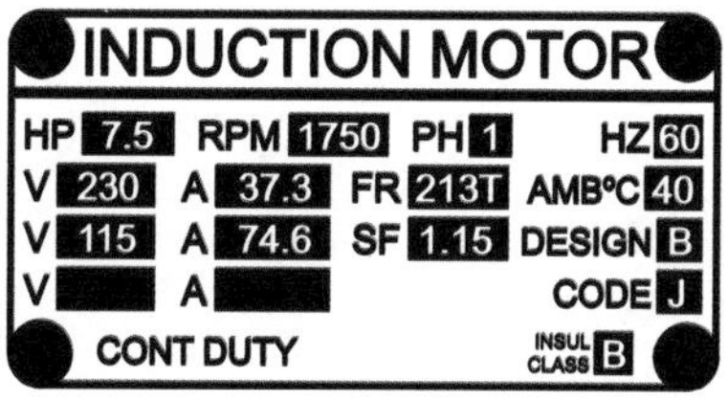

INDUCTION MOTOR
HP 7.5 RPM 1750 PH 1 HZ 60
V 230 A 37.3 FR 213T AMB°C 40
V 115 A 74.6 SF 1.15 DESIGN B
V   A   CODE J
CONT DUTY   INSUL CLASS B

Figure 230.108

_____ 8. Refer to the motor nameplate shown above in Figure 230.108. A NEMA starter will be used for this motor. The fewest number of heaters that can be installed in the starter per the NEC® is ___. The motor circuit is supplied from a 2-pole breaker in a 120/240 volt, 3Ø, 4-W panel.

   a. 1   b. 0   c. 3   d. 2

SER FAC	KVA CODE	HZ	INDUCTION MOTOR
1.15	H	60	
Horsepower	RPM	PH	MODEL
7½	3525	3	TEFC
TIME HRS	INS CL	FRAME	MAX AMB
24	B	256U	40° C
VOLTS		AMPS	
208-230/460		23.1-21/10.5	
Electric Motor Company			

Figure 230.127

_____ 9. Refer to the nameplate shown above in Figure 230.127. The branch circuit for the motor is supplied from a 277/480 volt, 3Ø, 4-wire panelboard. Although it may be too large, a Siemens Size 4 starter will be used for the motor. You will select E___ heaters to install in the starter.

   a. 56
   b. 57
   c. 60
   d. 55
   e. req'd heaters won't fit into this starter

_____ 10. Refer to the nameplate shown above in Figure 230.127. The branch circuit for the motor is supplied from a 120/208 volt, 3Ø, 4-wire panelboard. Although it may be larger than required, a Siemens Size 4 starter will be used for the motor. You will select E___ heaters to install in the starter.

   a. 67
   b. 65
   c. 70
   d. 71
   e. 69

Lesson 232 Safety Worksheet

Heavy Equipment

Many types and sizes of heavy equipment are commonly used on job sites of all varieties. This equipment comes in many types and sizes, from mini track-hoes to skid steer loaders to very large earth movers. All of this equipment creates hazards to which we need to be aware. The hazards can be struck-by type injuries or fatalities, or they may be caught-in or caught between hazards. Both struck-by and caught-in or between accidents continue to be two of OSHA's top four causes of death and serious injury. This makes them part of OSHA's "Focus Four Program".

The "Focus Four Program" places emphasis on falls, caught-in or between, struck-by and electrical hazards. These four types of accidents cause upwards of 90% of serious injuries and fatalities on construction sites.

The OSHA standard that applies to heavy equipment is subpart O of **CFR1926 Construction Industry Regulations**. The exact article starts at 1926.600 and run through 1926.606. (See **OSHA.gov** for this information.)

The general requirements of this regulation for equipment, as found in CFR1926.600, state that all equipment left unattended at night, adjacent to a highway in normal use, or adjacent to construction areas where work is in progress, shall have appropriate lights or reflectors, or barricades equipped with appropriate lights or reflectors, to identify the location of the equipment. Any heavy machinery, equipment, or parts of the equipment or machinery, which are suspended or raised by slings, hoists, or jacks shall be substantially blocked or cribbed to prevent these parts from falling or shifting before employees are permitted to work under or between these parts. Bulldozer and scraper blades, end-loader buckets, dump bodies, and similar equipment shall be lowered or blocked when being repaired or when not in use. Whenever equipment is parked, the parking brake shall be set. If the equipment is parked on a slope or incline, the parking brake shall be set and the wheels chocked. All cab glass shall be safety glass, or equivalent, that introduces no visible distortion affecting safe operation of any machine.

CFR1926.601 covers motor vehicles that operate within an off-highway jobsite, not open to public traffic have the following general requirements. All vehicles shall have a service brake system, an emergency brake system, and a parking brake system in operable condition. At least two headlights and two taillights are required if poor visibility is present. An adequate audible warning device (horn) shall be at the operator's station in working order. No employer shall use any motor vehicle equipment that has an obstructed view to the rear, unless the vehicle has a reverse signal alarm audible above the surrounding noise level, or the vehicle is backed up only when an observer signals that it is safe to do so. All vehicles with cabs shall be equipped with windshields and powered wipers. Cracked or broken glass shall be replaced. Defrosters are necessary in areas where conditions warrant.

All haulage vehicles loaded by cranes, loaders or similar equipment, shall have a cab shield or similar canopy adequate to protect the operator from shifting or falling materials. Tools and material shall be secured to prevent movement when transported in the same compartment with employees. Seat belts and anchorages shall be installed in all motor vehicles and comply with DOT (Department of Transportation) standards. Trucks with dump bodies shall be equipped with positive means of support, permanently attached, and capable of being locked in position to prevent accidental lowering of the body while maintenance or inspection is being done. **All vehicles in use** shall be checked at the beginning of each shift to assure that all of the following are in safe operating condition and free of damage that could cause failure in use: service brakes, including trailer connections, parking brake system, emergency stopping system, tires, horn, steering mechanism, seat belts, operating controls and all safety devices. All defects shall be corrected before the vehicle is placed in service.

Motor: Branch Circuit Grounding Conductors, Disconnects, Starters, and Overload Protection

The material handling equipment regulations are found in CFR1926.602. This section applies to earthmoving equipment, scrapers, loaders, crawler or wheel tractors, bulldozers, off-highway trucks, graders, agricultural and industrial tractors and similar equipment. Seat belts are required in this type of equipment, unless the equipment does not have a roll-over protection structure (ROPS) or is designed only for stand up operation. Roadways must be suitable for the equipment that will be operated on them. Brakes and fenders must be maintained in good operating order. Roll over protective structures (ROPS) must comply with OSHA subpart W. Material handling equipment are required to have audible alarms to alert others working in the area to any movement of these machines. An unobstructed view to the rear is required or an audible alarm or signal person to assure the backing move can be made safely. Scissor points on all front-end loaders, which constitute a hazard to the operator during normal operation, shall be guarded.

Lifting and hauling equipment have some additional requirements. No modifications of the equipment can be made which affect the capacity or safe operation of the equipment. Spinner knobs are prohibited on most of this equipment. Unauthorized personnel shall not be permitted to ride on powered industrial trucks. If workers are authorized to ride on equipment, a safe place to ride shall be provided. Use of any safety platform on lifting equipment is required to be securely fastened to the carriage or forks and a means provided to turn off the equipment from the platform.

Site clearing is covered under CFR1926.604. Employees engaged in site clearing shall be protected from hazards of irritant and toxic plants. All equipment used in site clearing operations shall be equipped with roll over guards that meet the specified requirements of the subpart. In addition, the overhead covering of the canopy shall be at least 1/8-inch steel plate or ¼-inch woven mesh with openings no larger than 1 inch. The ear opening of the canopy shall also have ¼-inch woven mesh with openings not exceeding 1 inch.

Pile driving equipment is covered in CFR1926.603 and marine operations are covered in CFR1926.605. If you are involved in this type of operations, please see the OSHA website for further information at **OSHA.gov**.

The required types of Personal Protective Equipment (PPE) when operating heavy equipment would be:

- Hardhats for any possible overhead impacts or electrical contacts
- Eye Protection (Safety Glasses) with side shields for debris and dust
- Gloves (Heavy Duty Leather) to protect our hands from cuts and abrasions
- Proper Footwear (Heavy Duty Boots) with Safety Toes if required
- Ear Protection to reduce exposure to high noise levels
- High Visibility Safety Apparel (vests or shirts and hard hats)

Traffic Control is required if work is on public streets or highways. This control must conform with OSHA's "MUTCD" Manual on Uniform Traffic Control Devices. This information outlines what is required in PPE, procedures for directing traffic, measures to protect the public and precautions to protect workers in the zones. See **OSHA.gov** for additional requirements or to download this information.

Year Two (Student Manual)

Questions

_____ 1. Heavy equipment is required to have ___ or similar audible device.

   a. a horn
   b. headlights
   c. a backup camera
   d. turn signals
   e. all of these

_____ 2. Backhoe blades or buckets should be lowered or blocked to prevent movement, when ___.

   a. performing maintenance or repairs
   b. left unattended for any period of time
   c. both a and b
   d. neither a nor b

_____ 3. When unattended heavy equipment is parked overnight along on the roadside, the equipment shall be made visible to traffic on the highway by ___.

   a. illuminating the equipment with lights
   b. having reflectors affixed to the equipment
   c. parking the equipment behind barricades that have reflectors
   d. any of these
   e. none of these

_____ 4. When backing up a piece of material-handling heavy equipment, which of the following is required?

   a. An audible, backup alarm must sound.
   b. The operator must have an unobstructed view to the rear.
   c. A signal person must direct the operation.
   d. Any one of these is acceptable.

_____ 5. When operating heavy equipment, ___ are required PPE.

   a. ear protection
   b. safety glasses with side shields
   c. both a and b
   d. neither a nor b

Motor: Branch Circuit Grounding Conductors, Disconnects, Starters, and Overload Protection

_____ 6. The wheels of heavy equipment should be chocked, when ___.

    a. performing maintenance or repairs
    b. left unattended for any period of time
    c. parked on an incline or sloping surface
    d. all of these
    e. none of these

E·T·N

EATON CORPORATION **CUSTOMER FOCUS**

An innovative electrical distribution solution for healthcare

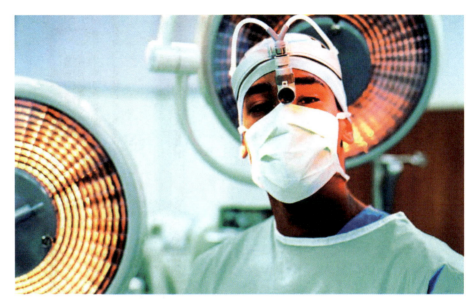

Customer
Healthcare Provider

Markets Served
Healthcare

The Value of PowerChain™ Management

PowerChain™ Management helps enterprises achieve a competitive advantage through proactive management of the power system as a strategic, integrated asset throughout its lifecycle.

 Enhanced Safety

 Greater Reliability

 Operating Cost Efficiencies

 Effective Use of Capital

 Risk Mitigation

To learn more, contact:

Celestine Huggins
Eaton Corporation
11 Corporate Circle
Sumter, SC 29151
803-481-6682
CelestineBHuggins@eaton.com

A reliable, high-quality power system is critical to every enterprise, but managing a power system effectively can be a challenge. Here is an example of how Eaton's PowerChain™ Management solutions helped a healthcare provider achieve:

Greater Reliability – Maintain vital operations with steady, high-quality power every minute of every day.

Operating cost efficiencies – Reduce operating costs with effective energy management and maintenance strategies.

Effective use of capital – Put your capital to work faster, and keep it working longer, with electrical designs that require less equipment and space, services that extend equipment life and preassembled equipment that can be installed faster.

Risk mitigation – Reduce the risk of construction delays and cost overruns with a coordinated approach to power system design, procurement, installation and maintenance.

Background

While the desire for a smaller footprint in electrical rooms has become virtually a universal demand across all market segments, it holds special significance within the healthcare industry.

Medical facilities commonly require four electrical systems, placing increased emphasis on space.

In northern California, two leading healthcare providers, undertaking separate projects at about the same time, explored solutions to make their electrical distribution equipment fit in the allotted space.

Challenges

Both hospital expansion projects — one a five-story women and children's center, the other a five-story bed tower — designated four electrical systems, all in parallel, to be located in the respective electrical rooms. The systems are: critical branch, equipment branch, life safety and normal.

Each system consists of:

- 480Y/277V supply panelboard
- 75kVA up to 150kVA dry type distribution transformer
- 42-circuit 208Y/120V panelboards (2-6 panels)
- transient voltage surge suppression (TVSS) required to protect sensitive electronic-based medical diagnostic equipment.

Eaton Corporation is a diversified industrial manufacturer with 2005 sales of $11.1 billion. Eaton is a global leader in electrical systems and components for power quality, distribution and control; fluid power systems and services for industrial, mobile and aircraft equipment; intelligent truck drivetrain systems for safety and fuel economy; and automotive engine air management systems, powertrain solutions and specialty controls for performance, fuel economy and safety. Eaton has 60,000 employees and sells products to customers in more than 125 countries. Learn more at www.eaton.com.

EATON CORPORATION CUSTOMER FOCUS

> "…in fact, saving more than 60% of floor space over a conventional layout."
>
> *Fred Paul*
> *Eaton Application Engineer, Sacramento, CA*

Installing conventional equipment would mean the delivery of countless components to designated floors of the structures, as well as painstaking installation of the panelboards, switchboards, transformers and TVSS.

The process would be time consuming and costly to contractors. It also would impact the construction schedule.

A specific challenge of providing TVSS as part of the distribution equipment forced one competitor to drop out of the running in the design stage.

Solution

The two healthcare providers happened to be working with the same local engineering consultant on their respective projects. The consultant, during a conversation with Eaton's Sacramento-based application engineer Fred Paul, immediately selected Cutler-Hammer® Integrated Facility Systems™ (IFS) as the way to proceed.

IFS was not entirely new to the consultant who had been introduced to it a few months earlier when Paul hosted a Lunch & Learn event for him and his staff. Impressed with the presentation, the consultant telephoned Paul and explained his dilemma. The consultant related that the architect already had informed him that expanding the severely restricted electrical room was out of the question on both projects.

With less than one week to work from a preliminary single line and develop an effective solution, Paul visited the consultant and demonstrated that IFS would fit comfortably in the room as designed, in fact saving more than 60% of floor space over a conventional layout.

While space was the major consideration for the consultant, he also discovered other favorable features. Among them:

- IFS door-to-door construction, allowing the panelboard to swing open rather than removing it for ease of maintenance
- door-mounted TVSS, which no competitor could match

For each project, Paul was able to send drawings generated from Bid Manager, an Eaton product configuration tool, directly to the IFS Solutions Center in Sumter, South Carolina. A quick three-day turnaround gave the consultant CAD-generated finish drawings with sufficient quality and detail to import to the electrical design drawings for submittal to OSHPOD (State of California Office of Statewide Health & Planning Development) for approval.

The bed tower project mirrors the other project in its IFS configuration. One exception is that still other components — Power Xpert™ meters and software — were added to the system as an internal energy and power quality auditing application.

Result

The contractor was able to install IFS immediately after the floor decks were poured and prior to the walls being built. In most cases, installation of the electrical distribution equipment could not have begun before the walls were constructed and sheet rocked.

Using IFS reduced the installation cycle time.

© 2007 Eaton Corporation
All Rights Reserved
Printed in USA
Publication No. IA01502002E
January 2007

Lesson 233
Locked Rotor Current and Phase Loss for Motors-A/C and Refrigeration Equipment- Generators and Fire Pumps

Electrical Curriculum

Year Two
Student Manual

Purpose

This lesson will discuss the currents required to start motors and the effects of single-phasing three-phase motors. This lesson also addresses NEC® requirements for air conditioning and refrigeration equipment, generators, and fire pumps.

Homework

(Due at the beginning of this class)

For this lesson, you should:
- Thoroughly read the material contained within Lesson 233.
- Read only the following in Electrical Systems
    A/C AND REFRIGERATION EQUIPMENT – ARTICLE 440:
        Introductory paragraphs
        Marking on Hermetic Refrigerant Motor – 440.4
        Markings on Controllers – 440.5
        Ampacity and Rating of Equipment and Compressor Motor – 440.6
        Highest Rated Motor – 440.7
        Single Machine – 440.8
        Disconnecting Means – 440.11
        Branch Circuit Fuses or Circuit Breakers
        Branch Circuit Conductors – Article 440, Part IV
        Branch Circuit Overload Protection– Article 440, Part VI
        Room A/Cs – Article 440, Part VII
        LCDI Protection for Room Air Conditioners – 440.65
    Generators – Article 445
    FIRE PUMPS – ARTICLE 695

- Refer to "Advisory Notes for IEC 2nd Year Apprenticeship Motor Studies" from Lessons 230 and 232.
- Complete Objective 233.1 Worksheet.
- Complete Objective 233.2 Worksheet.
- Complete Objective 233.3 Worksheet.
- Complete Objective 233.4 Worksheet.
- Read and complete Lesson 233 Safety Worksheet.

- Complete additional assignments or worksheets if available and directed by your instructor.

Objectives

By the end of this lesson, you should:

233.1 ....Understand motor locked rotor current and be able to perform calculations to determine approximate locked rotor current when a motor starts or stalls.

233.2 ....Understand the effects of phase loss on a motor and be able to perform calculations to determine approximate circuit current when a motor single phases.

233.3 ....Be able to determine the minimum sizes or ratings required for conductors, safety switches, and OCPDs used for air conditioning and refrigeration equipment branch circuits.

233.4 ....Understand the purpose of a fire suppression system and the NEC® requirements for fire pumps and generators.

233.1 Locked Rotor Current

Starting current and locked-rotor current (LRC or LRA) are equivalent terms. When a motor is started from a standstill, the motor will draw a greater amount of current, than indicated on the nameplate, to get it started and up to rated speed. This is *starting current* or *locked rotor current* and is the amount of current that flows in the circuit conductors, when a motor is first energized and the shaft starts to rotate. Similarly, if a motor stalls (is caused to stop without removing power), locked rotor current will flow in the circuit conductors.

Approximate LRC can be calculated. If the Code letter for a motor is known (from the motor's nameplate), NEC® Table 430.7(B) can be utilized. Figure 233.1 shows the nameplate for a typical 3Ø motor. This 20 hp motor has a Code letter G. NEC® Table 430.7(B) shows a range of 5.6 to 6.29 kVA per hp for Code Letter G. The worst case scenario is when the highest value (6.29) is used. Always use the highest value in your calculations. Use the following formula to calculate the approximate LRA for this motor. Use 230 volts as the line voltage for this example.

☞ $$LRA = \frac{(\text{Table kVA per hp} \times 1000 \times (\text{hp})}{E_{line} \times (1.732)}$$

$$LRA = \frac{(6.29) \times 1000 \times (20)}{230 \times (1.732)} = 315.8 \text{ amps}$$

Typical Three-Phase Motor Nameplate

MOTORS UNLIMITED
AC INDUCTION MOTOR

Thermally Protected NO			Type ☐
STYLE K		SERIAL 7142599	
FRAME 56		TYPE	
HP 20	PH 3	HOUSING T.O.	
RPM 1760		SERVICE FACTOR 1.15	
CYCLES 60		S.F. AMPS	
VOLTS 208-230/460		AMPS 57.1-51.6/25.8	
DEGREES C RISE 40		CODE G	
HOURS 24			

Figure 233.1

A "rule-of-thumb" method can be used for typical motors. To apply this method, simply multiply the nameplate FLA times 6. For the motor in Figure 233.1, this results in an approximate LRA of 309.6 amps (6 × 51.6 amps) at 230 volts. This method can be utilized for motors that have relatively low starting torques, such as the Code Letter G motor shown in Figure 233.1. For motors with Code letters higher or lower than G, the results of this method are less accurate.

For "Design Letter" or "NEMA Design" motors, NEC® Table 430.251(A) and (B) can be utilized to get *estimated* starting or locked-rotor currents. For example, the 10 horsepower motor shown at the right in Figure 233.2 is a NEMA Design B motor. NEC® Table 430.251(B) shows that this motor has an approximate LRA of 162 amps at 230 volts.

The great amount of starting current required for motors is why "lights dim" momentarily when a motor on the same circuit starts. This is also why large motors are not started "across-the-line". Starting methods that require less starting current must used for large motors. If alternative starting

Typical Three-Phase Motor Nameplate

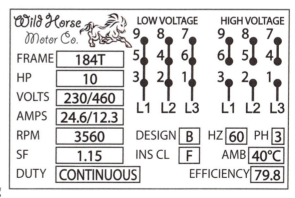

Figure 233.1

233.2 Single-Phasing of Three-Phase Motors

Single-Phasing When Running

Single-phasing is when one leg (phase) of a 3-phase system is lost. If this phase loss occurs while the motor is running, the motor slows down, causing a drop in torque. This reduction of torque causes the stator winding current to increase to the point of damaging the windings.

If a 3-phase motor is started with power available from only two of the three circuit conductors, very little torque is produced to turn the motor shaft. This causes excessive winding current and the windings will be damaged within a very short period of time.

Refer again to Figure 233.1 for the nameplate of a typical 3Ø motor. When operated at 230 volts, this motor should be drawing 51.6 amps of line current once it reaches its rated shaft speed (1760 rpm). If one of the circuit conductors loses power, comes apart in a junction box for example, the motor will single-phase. This 20 horsepower motor wants to deliver 20 horsepower, even though only two energized circuit conductors remain. Recall that, in their simplest form:

$$P_{1\emptyset} = I_{1\emptyset\text{-line}} \times E_{line} = HP \times 746 \quad \text{and}$$

$$P_{3\emptyset} = I_{3\emptyset\text{-line}} \times E_{line} \times 1.732 = HP \times 746$$

When a three-phase motor single phases, single-phase power equals three-phase power. That is, $P_{1\emptyset} = P_{3\emptyset}$.

These can be algebraically manipulated to become:

$$I_{1\emptyset\text{-line}} \times E_{line} = I_{3\emptyset\text{-line}} \times E_{line} \times 1.732$$

Since the line voltage hasn't changed, the line current when a motor single phases equals:

$$☞ I_{1\emptyset\text{-line}} = I_{3\emptyset\text{-line}} \times 1.732$$

For this 20 hp motor, after single-phasing at 230 volts, the line current equals:

$$I_{1\emptyset\text{-line}} = I_{3\emptyset\text{-line}} \times 1.732 = 51.6 \text{ amps} \times 1.732 = \mathbf{89.37} \text{ amps}$$

Since this motor is connected for low voltage, half of this line current (89.37 ÷ 2 = 44.69 amps) flows through each parallel-connected stator winding. This winding current is certainly in excess of the rated winding current and the windings will be damaged if the motor is allowed to operate at single-phase for very long.

Single-Phasing When Starting

For a three-phase motor that is started when only single-phase power (two "hots") is available, the LRA is multiplied by 1.732 to get the approximate current that would flow through the circuit conductors.

For the Code Letter G motor shown in Figure 233.1, the LRA was found to be about 315.8 amps when started at 230 volts. If this motor tried to start with single-phase power available, the starting current would be:

$$\text{single-phasing line current} = 315.8 \times 1.732 = \mathbf{547} \text{ amps}$$

Again, since this motor is connected for low voltage, half flows through each winding. If allowed to have this much current flowing through its windings for more than a few seconds, the motor would, most likely, have to be replaced.

233.3 Air Conditioning and Refrigeration

Most air conditioning and refrigeration equipment comes as a packaged unit with the MCA (minimum circuit ampacity) and MOCP (maximum overcurrent protection) on the nameplate. If the manufacturer's nameplate information is not used to size wire and OCPDs, the warranty may be voided. Remember that the NEC® at 110.3(B) requires that listed or labeled equipment be installed in accordance with any manufacturer instructions.

233.4 Generators and Fire Pumps

A fire pump is a stationary pump that is connected to a building's fire-protection, sprinkler piping system so that in the event of fire, water is dispersed through the sprinkler heads to extinguish the fire. A jockey pump is a small pump connected to this system and is intended to maintain pressure in the piping system to an artificially high level so that the operation of a single fire sprinkler will cause an appreciable pressure drop. This pressure drop is sensed by the fire pump automatic controller and causes the fire pump to start. Although not covered by the NEC®, a jockey (or pressure-maintenance) pump is essentially a portion of the fire pump's control system.

Most stationary fire pump assemblies are installed with devices that automatically sense a pressure loss in a fire protection system, and start running to supply water or boost pressure in that system. From time to time, though, small water leaks, unwanted pressure drops, or even temperature changes may "fool" the fire pump into starting when it isn't needed.

To prevent these false starts and maintain the fire pump's life expectancy, small pressure-maintenance or jockey pumps are installed to maintain a relatively constant pressure on the fire protection system.

Objective 233.1 Worksheet

```
┌─────────────────────────────────┐
│         MOTORS UNLIMITED         │
│          AC INDUCTION MOTOR      │
│ Thermally Protected  NO   Type □ │
│ STYLE   K        SERIAL  514919  │
│ FRAME   56       TYPE            │
│ HP  20      PH 3 HOUSING  T.O.   │
│ RPM   1760       SERVICE FACTOR 1.15 │
│ CYCLES  60       S.F. AMPS       │
│ VOLTS  208-230/460  AMPS 58.1-52.8/26.4 │
│ DEGREES C RISE 40   CODE  D      │
│ HOURS  24                        │
└─────────────────────────────────┘
```

This motor is supplied by a branch circuit that originates from a 120/208 volt, 3Ø, 4-W supply.

All conductors are THHN/THWN copper and all terminations are rated at 75° C.

Figure 230.131

_____ 1. Refer to the nameplate and information shown above in Figure 230.131. When this motor tries to start with 3Ø power available, it will pull ____ A. For this motor's Code Letter, use nameplate voltage and the range value that results in the highest current.

   a. 316.8   b. 356.4   c. 348.6   d. 249.3   e. 216.03

_____ 2. Refer to the nameplate and information shown above in Figure 230.131. When this motor is running normally, it is drawing ____ A.

   a. 100.6   b. 102.9   c. 91.45   d. 52.8   e. 58.1

```
┌─────────────────────────────────┐
│         MOTORS UNLIMITED         │
│          AC INDUCTION MOTOR      │
│ Thermally Protected  NO   Type □ │
│ STYLE   K        SERIAL  7142599 │
│ FRAME   56       TYPE            │
│ HP  20      PH 3 HOUSING  T.O.   │
│ RPM   1760       SERVICE FACTOR 1.15 │
│ CYCLES  60       S.F. AMPS       │
│ VOLTS  208-230/460  AMPS 58.1-52.8/26.4 │
│ DEGREES C RISE 40   CODE  J      │
│ HOURS  24                        │
└─────────────────────────────────┘
```

Figure 230.126

_____ 3. Refer to the nameplate shown above in Figure 230.126. When the circuit originates in a 120/240 volt, 3-phase, 4-wire panelboard, the motor should draw ____ A of current when it starts.

   a. 349   b. 439   c. 459   d. 317   e. 384

_____ 4. The locked rotor current of a 75 hp, 460-volt, three-phase, Design B motor is ____ amps.

   a. 166   b. 543   c. 450   d. 333   e. 340

Year Two (Student Manual)

_____ 5. Using the rule of thumb for a typical motor, about how much locked rotor current will a 75 hp, 460 volt, three-phase motor draw?

  a. 260   b. 333   c. 576   d. 435   e. 288

Objective 233.2 Worksheet

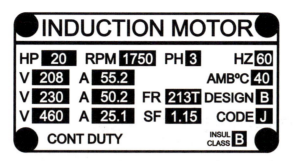

Figure 233.124

____ 1. Refer to the nameplate shown above in Figure 233.124. The motor is operating and supplied power from a 277/480 volt, 3-Ø, 4-wire panelboard. If a "hot" conductor breaks apart one of the several j-boxes in the branch circuit, the motor will single phase and ___ amps of current will flow through EACH of the motor windings.

   a. 47.8   b. 86.9   c. 33.8   d. 95.6   e. 43.5

____ 2. Refer to the nameplate shown above in Figure 233.124. The motor is operating and supplied power from a 120/240 volt, 3-Ø, 4-wire panelboard. If a "hot" conductor breaks apart in a j-box somewhere then the motor will single-phase and ___ amps of current will flow through the motor windings.

   a. 54   b. 43.5   c. 47.8   d. 86.9   e. 95.6

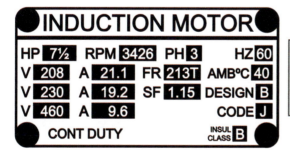

Figure 230.242

____ 3. Refer to the nameplate and information shown above in Figure 230.242. With this motor running normally, the service "loses a leg". The motor will continue to run for a while but will be damaged because it is drawing ___ A

   a. 15.6   b. 9.0   c. 16.6   d. 9.6   e. 11

4. Single-phasing is a term used to describe ___.

   a. the process of reversing direction of a 3Ø motor
   b. the calculation required to determine the equivalent size single-phase motor required to replace a three-phase motor
   c. the effect on equipment when one phase of a three-phase circuit is lost
   d. none of these

Figure 230.131

MOTORS UNLIMITED — AC INDUCTION MOTOR
Thermally Protected: NO   Type: ☐
STYLE K   SERIAL 514919
FRAME 56   TYPE
HP 20   PH 3   HOUSING T.O.
RPM 1760   SERVICE FACTOR 1.15
CYCLES 60   S.F. AMPS
VOLTS 208-230/460   AMPS 58.1-52.8/26.4
DEGREES C RISE 40   CODE D
HOURS 24

This motor is supplied by a branch circuit that originates from a 120/208 volt, 3Ø, 4-W supply.

All conductors are THHN/THWN copper and all terminations are rated at 75° C.

5. Refer to the nameplate and information shown above in Figure 230.131. This motor is not running. One of the 3Ø circuit conductors has come out of a wire nut in a junction box. It will draw ___ amps of starting current (locked rotor current), if it tries to start.

   a. 431.8   b. 548.7   c. 617.3   d. 374.2   e. 603.8

Locked Rotor Current and Phase Loss for Motors-A/C and Refrigeration Equipment- Generators and Fire Pumps   233-879

Objective 233.3 Worksheet

_____ 1. Refrigeration equipment requirements are provided in Article ___ in the NEC®.

   a. 460   b. 430   c. 445   d. 440   e. 470

> **ROOFTOP AIR CONDITIONING UNIT**
> Nameplate Information:
> MCA   68.8 amps (minimum ckt. amps)
> MOCP 90 amps (maximum OCP)
> Voltage 208   PH 3
> 3-ph motors: 1 @ 38.9 amps each
> 1-ph motors: 2 @ 9.8 amps each

Figure 233.135

_____ 2. Refer to the nameplate shown above in Figure 233.135. The largest fuse you could install in a fused safety switch is ___ A.

   a. 90   b. 60   c. 70   d. 80   e. fuses not permitted

_____ 3. Refer to the nameplate shown above in Figure 233.135. The largest circuit breaker you could install is ___ A.

   a. 90   b. 110   c. 100   d. circuit breaker not permitted   e. 80

_____ 4. Refer to the nameplate shown above in Figure 233.135. The smallest wire you could use for the circuit conductors is ___ AWG.

   a. 2   b. 8   c. 4   d. 6

_____ 5. Refer to the nameplate shown above in Figure 233.135. A nonfused safety switch with a voltage rating of 250 volts ___ to be used as the disconnecting means for this unit.

   a. is permitted   b. is not permitted

AC NAMEPLATE 3:
Lennox Industries   Model HS26-060-2P   Serial 5897F-30976
   Electrical Ratings: Compressor   Fan
                         VOLTS   240   VOLTS   240
                         PH   1   PH   1
                         RLA   23.9   FLA   1.5
                         LRA   32.9   HP   ¼

   MCA 32.9   Maximum fuse or circuit breaker-50

Year Two (Student Manual)

233-880   Locked Rotor Current and Phase Loss for Motors-A/C and Refrigeration Equipment- Generators and Fire Pumps

_____ 6. The nameplate information for a home air conditioning unit is shown above at AC NAMEPLATE 3. The minimum size Romex® that can be installed to this unit is ___ AWG.

    a. 8   b. 10   c. 4   d. 12   e. 6

_____ 7. HVAC is an abbreviation for ___.

    a. Heating-Ventilating-& Air Conditioning
    b. High Voltage Alternating Current
    c. High Volume Air Circulation
    d. none of these

ROOFTOP AIR CONDITIONING UNIT
Nameplate Information:
MCA   45.7 amps *(minimum ckt. amps)*
MOCP 50 amps *(maximum OCP)*
        HACR circuit breaker only
Voltage 480   PH 3
   3-ph motors: 2 @ 8.7 amps each
   1-ph motors: 3 @ 3.2 amps each

Figure 233.133

_____ 8. Refer to the nameplate shown above in Figure 233.133. The largest fuse you could install in a fused safety switch is ___ A.

    a. 40   b. 35   c. 50   d. 45   e. fuses not permitted

_____ 9. Refer to the nameplate shown above in Figure 233.133. The smallest wire you could use for the circuit conductors is ___ AWG.

    a. 4   b. 10   c. 8   d. 6

_____ 10. When a flexible cord is used to supply a 240-volt room air conditioner, the length of such cord shall not exceed ___.

    a. 4 feet   b. 6 feet   c. 10 feet   d. none of these

_____ 11. Refrigeration equipment requirements are provided in Article ___ in the NEC®.

    a. 460   b. 445   c. 430   d. 440   e. 470

Objective 233.4 Worksheet

_____ 1. The OCPDs protecting circuits for supervised fire pumps are selected to ___.

   a. carry LRC for an indefinite period
   b. permit the pump to burn up before the OCPDs open
   c. both a and b
   d. neither a nor b

_____ 2. A fire pump is permitted to be supplied by ___ the main distribution (service) panel.

   a. a fused safety switch connected ahead of
   b. a circuit breaker in
   c. either a or b
   d. neither a nor b

_____ 3. Article ___ in the NEC® provides requirements for the installation of generators.

   a. 445
   b. 440
   c. 470
   d. 460
   e. 430

4. Energized parts within control cabinets and other control enclosures used for fire pump systems shall be ___.

   a. located at least 12″ above the floor level
   b. separated from the fire pump by a wall with at least a 1-hour fire rating
   c. both a and b
   d. neither a nor b

5. When circuit conductors are installed in conduit between the disconnecting means and the fire pump that is located with the building, these conduits shall be ___.

   a. encased in at least 2″ of concrete
   b. painted red and properly labeled
   c. either a or b
   d. neither a nor b

6. The smallest THHN/THWN copper conductors permitted to be run, paralleled in four separate conduits, between a 3Ø, 350 KVA generator and a 208-volt distribution panelboard are ___ AWG or kcmil.

   a. 350   b. 500   c. 250   d. 300   e. 4/0

7. ___ kcmil are the minimum size copper THHN/THWN conductors permitted to be run between a generator and a distribution panelboard. The generator is 480-volt, 3Ø rated at 500 KVA. The conductors are to be run in **parallel** using **two** conduits.

   a. 350   b. 400   c. 300   d. 500   e. 600

8. Article 695 covers the installation of ___.

   a. fire pumps   b. jockey pumps   c. both a and b   d. neither a nor b

9. Article 695 includes voltage drop requirements that are ___.

   a. mandatory
   b. recommended

Lesson 233 Safety Worksheet

Weather Conditions

Weather conditions can affect your personal safety. It can be the extremes, such as the ice and snow, that are very common in our northern locations, to tornadoes in the heartland, to hurricanes in the southeast, to severe thunderstorms over most of our country or areas that are subjected to several of these extremes.

At the first warning signs of any of these conditions, you should immediately initiate a plan of action to protect yourself and those around you. You need to make plans to secure the jobsite to protect other personnel and materials from these conditions as well.

The most common occurrences are thunderstorms. Some of these may be severe thunderstorms with heavy rain and very dangerous lightning. Thunder is the sound of air being heated to extreme temperatures during thunderstorms which causes the air to expand and then contract as it cools. This creates sound waves: thunder. Although you may be able to hear the thunder, you may not see the lightning because it is hidden in the heavy clouds. Lighting may be within 10 miles, if you hear thunder. A good safety rule is: If you hear it, fear it. If you see it, flee it. What this means is that if you hear thunder, lightning is close enough to worry about.

The best shelter from lightning is in a permanent building. Trailers, sheds or small buildings do not provide as much safety. Vehicles with metal roofs offer some safety, but do not touch any metal surfaces of the vehicle. The best recommendation is to immediately head for safe shelter at the first sound of thunder.

The heavy rain that can accompany these storms can cause flash flooding and water damage to the jobsite. Obviously, water in energized electrical equipment can cause electrical arc conditions and equipment damage or injuries to anyone working on or near the equipment. Precautions must be taken to protect any electrical equipment from water. If this equipment is exposed to water, follow the manufacturer's recommendations for cleaning and restoration before attempting to energize or re-energize the equipment.

In many areas of the country, tornadoes are common. Tornado-producing storms are powerful and extremely dangerous. Many lives are lost each year due to these events. Tornadoes are spawned from thunderstorms and hurricanes. They appear as a rotating, funnel-shaped cloud that normally extends to the ground with whirling winds that can exceed 250 miles per hour. The destruction caused by these storms can be over a mile wide and fifty miles or more long. These extreme winds can destroy almost anything in their path. Most buildings will not survive a direct hit by these storms. Mobile homes, modular homes and wooden structures sustain the most damage. Concrete structures provide the best protection from the damaging winds and flying debris. Jobsite office or storage trailers **are not** a safe place to get away from the tornado.

To protect yourself from a tornado, take shelter in a safe place. A safe place is one that is on the lowest level of the structure, in the center of the building, in an area without any windows. Many times a hallway or corridor will be a more substantial part of the building. If you are at home and have a basement, go there. If you are in a mobile or modular home, go to a more sturdy structure nearby. Get under something sturdy, such as a heavy table, and stay there until the danger passes. If you are driving and a tornado is sighted or suspected, do not try to outrun or drive away from the tornado. This is because tornadoes do not always move in a straight line. You cannot always tell which direction it may be moving. The road you turn into may also have turns which could take you into the path of the storm. Again, the safest thing to do is find a sturdy building and go inside. If this is not possible, get out of your vehicle and lay flat in a low area that will not be subject to flooding. Protect your head and neck, and wait out the storm. <u>Do not seek shelter under an overpass</u>. These become areas of extremely high velocity wind and can be a place where large amounts of debris are blown, thus making them unsafe for you. People are often killed from doing this. Also, do not seek shelter in wooded areas.

Large hail often accompanies tornadoes. Wind-blown debris can become high-speed, flying missiles. Flying debris can cause more property damage and death, if you are struck by it. Severe thunderstorm watches and warnings and tornado watches and warnings can be similar. A **watch** means that in a defined area conditions are favorable for the formation of these storms. A **warning** means that a storm has been visually sighted or indicated on radar in the defined area. If you think a tornado is possible, act immediately. They are very dangerous.

Hurricanes pose severe threats primarily in the southeastern part of our country. These powerful storms can cause severe damage to very large areas. Unlike tornadoes or severe thunderstorms, they are forecast longer in advance. This provides us with longer time to prepare for them. One problem is that they do not always follow the exact path that forecasters predict. Pay particular attention to watches and warnings and know the differences between them. A hurricane watch means conditions are <u>possible</u> for the storms to form in the watch area usually within the next 36 hours. The areas are usually outlined on a weather map shown on television stations and internet based weather stations. A hurricane warning means that conditions are <u>expected</u> in the warning area usually within the next 24 hours. *These watches and warnings are different than those associated with severe thunderstorms and tornadoes. Learn the difference between them.*

Hurricanes may cause evacuation of large areas. Information is provided for evacuation routes and information regarding traffic restrictions and more. Have a plan of where you will go and a back-up plan. Make sure you have maps that may help you if your selected route of travel becomes too congested or closed. Take along phone numbers of family and friends as well as the places where you are traveling. Be sure to take any medications and supplies, some bedding (sleeping bags and pillows), bottled water, a flashlight with extra batteries, first aid kit, important papers and food.

Any of these weather conditions can create dangers and hardships and perhaps even death if not taken seriously. Pay attention to the watches and warnings that are given for your area. If instructions are given to take immediate shelter or to evacuate, do so without delay. Your life may be at risk. Don't become a victim or another statistic. Material things can be replaced a human life can not.

Much additional information is available at the following websites:

National Weather Service at www.weather.gov

National Hurricane Center at www.nhc.noaa.gov

American Red Cross at www.redcross.org

Federal Emergency Management Agency at www.fema.gov

Questions

_____ 1. If you hear thunder but don't see lightning, you are in no danger.

    a. False   b. True

_____ 2. A tornado warning is issued, in a defined area, when ___.

    a. conditions are favorable for the formation of tornadoes
    b. a tornado has been visually sighted or indicated on radar

_____ 3. The most common weather occurrences are ___.

    a. blizzards   b. thunderstorms   c. floods   d. hurricanes

_____ 4. A tornado watch is issued, in a defined area, when ___.

    a. a tornado has been visually sighted or indicated on radar
    b. conditions are favorable for the formation of tornadoes

_____ 5. The best shelter from lightning is in a ___.

    a. job trailer   b. storage shed   c. permanent building   d. portable building

_____ 6. The best shelter, from the flying debris and wind produced by a tornado, is ___.

      a. under an overpass
      b. laying flat in a low area
      c. in a concrete structure
      d. in a wood-frame building

_____ 7. The best shelter from lightning is in a ___.

      a. portable building
      b. storage shed
      c. job trailer
      d. permanent building

HALO

6" Remodel Housing   6" New Construction Housing   5" Remodel Housing   5" New Construction Housing

Halo Compact Fluorescent Downlights Provide Full Range Dimming!

Contractors have more choices with Halo's Energy Star/Title 24 products.

Meet the latest Energy Star and California Title 24 requirements with these compact fluorescent recessed downlights from Halo. They feature energy efficient electronic ballasts and the latest CFL lamp technology. These units combine high light output, pleasing color temperatures and full range dimming from 15% to 100%.

5" units utilize 26W lamps and 6" units utilize 26W and 32W lamps. All units are AIR-TITE™, IC rated and accept a wide range of trims guaranteeing a solution for any lighting need including wall wash and shower lights.

Installation features include Halo GOT-NAIL!™ bar hangers, Slide-N-Side™ junction boxes and quick connect wiring connectors. These features make Halo, from Cooper Lighting, the easiest and fastest downlight to install.

www.cooperlighting.com

Lesson 234
Motor Feeder Conductors, OCPDs, and Taps – Motor Branch Circuit Conductors and OCPDs

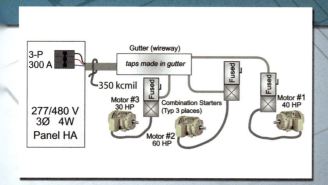

Electrical Curriculum

Year Two
Student Manual

IEC
PRIDE
NATIONAL

Purpose

This lesson will discuss the sizing of motor feeder conductors, overcurrent protective devices, and tap conductors. Additionally, this lesson will address the sizing of motor branch circuit conductors on the load side of overcurrent protective devices that are located at the termination of a motor feeder tap conductor.

Homework

(Due at the beginning of this class)

For this lesson, you should:

- Thoroughly read the material contained within Lesson 234.

- Read only the following in Electrical Systems
    MOTORS – ARTICLE 430:
    Sizing Feeder Conductors with More than One Motor – 430.24
    Sizing Feeder Fuses or Circuit Breakers – 430.62(A)

- Refer to "Advisory Notes for IEC 2nd Year Apprenticeship Motor Studies" from Lessons 230 and 232.

- Complete Objective 234.1 Worksheet.

- Complete Objective 234.2 Worksheet.

- Complete Objective 234.3 Worksheet.

- Complete Objective 234.4 Worksheet.

- Read and complete Lesson 234 Safety Worksheet.

- Complete additional assignments or worksheets if available and directed by your instructor.

Objectives

By the end of this lesson, you should:

234.1 ....Be able to answer general-type questions concerning motors.
234.2 ....Be able to perform the required calculations to determine the size of motor feeder conductors.
234.3 ....Be able to perform the required calculations to determine the size of motor feeder OCPDs.
234.4 ....Be able to determine the minimum permitted size for motor feeder tap conductors and to recognize that, when a tap conductor is terminated in an OCPD, conductors on the load size of tap conductor OCPDs are branch circuit conductors.

Motor Feeder Conductors, OCPDs, and Taps – Motor Branch Circuit Conductors and OCPDs

234.1 General Motor Knowledge

234.2 Motor Feeder Conductors

Determining the minimum size permitted for motor feeder conductors is a simple process, if basic steps are performed. These basic steps are as follows:

Step 1: Look up the NEC® Table FLA values, at the operating voltage of the feeder circuit, for all motors connected to the motor feeder.

Step 2: Select the motor with the largest Table FLA value and multiply this number by 1.25.

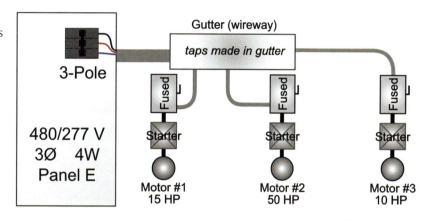

Figure 234.1

Step 3: Add the value from Step 2 to the FLAs of the remaining motors (other than the largest)

Step 4: Using NEC® Table 310.16, select a conductor (from the 75°C column) that has an ampacity that is greater than or equal to the value found in Step 3.

Example 1a – Sizing Feeder Conductors Using Figure 234.1

Step 1: Use NEC® Table 430.250, in the 460-volt column, to find the table FLAs for the three motors shown.
- T.FLA for Motor 1 = 21 amps
- T.FLA for Motor 2 = 65 amps *(largest motor in the group)*
- T.FLA for Motor 3 = 14 amps

Step 2: 65 amps × 1.25 = 81.25 amps

Step 3: 81.25 amps + 21 amps + 14 amps = 116.25 amps
 Since .25 is not a major fraction, 116.25 rounds down to 116 amps

Step 4: From NEC® Table 310.16, a 1 AWG conductor is the smallest that can carry 116 amps

☞ **1 AWG** is the smallest permitted size for the motor feeder.

234.3 Motor Feeder OCPDs

The key to determining the maximum permitted size or rating of a motor feeder OCPD is to first determine the maximum permitted size or rating of an OCPD, of the same type, that would be used for the largest motor connected to the feeder. Then the table FLAs of the remaining motors is added to this value. This is the calculation method set forth in NEC® 430.62(A).

Example 1b – Sizing Feeder Conductors Using Figure 234.1

Determine the maximum permitted rating for a circuit breaker protecting a feeder. Although the taps terminate in fused switches, the feeder OCPD in the panel is to be a circuit breaker. Therefore, the largest breaker size permitted for the largest motor (Motor 2) must be determined in order to determine the feeder breaker size. To find the maximum size OCPD, for single motor circuits, follow the procedure listed in Lesson 231.

> Motor 2 breaker size = 65 amps × 2.5 = 162.5 amps ⇒ 163 A *(major fraction)*
>> Since 163 A is not a standard size, go up to the next standard size. A 175 Amp breaker could be used for Motor 2.

Now the feeder breaker size can be determined. To do this, add together the size found for the largest motor (Motor 2 at 175 A) and the table FLAs from the remaining motors (Motors 1 and 3).

> Feeder breaker size = 175 + 21 + 14 = 210 Amps
>> Note that NEC® 430.62(A) is **not** followed by an exception that permits this breaker size to go up to the next standard rating. Therefore, the next standard size **below** 210 amps is a 200 A breaker. This is the largest permitted for the feeder breaker.

☞ **200 Amp** is the largest permitted for the motor feeder OCPD.

234.4 Motor Feeder Taps

The limitations shown in the in NEC® at 430.28 are specifically for *motor* feeder taps. These tap conductors sizes are limited so that, in the event of a fault (short-circuit or ground-fault) on the tap conductors, the OCPD protecting the feeder will open. The OCPDs at the termination of the tap conductors protect the tap conductors against overload.

Again, refer to Figure 234.1. Note that the tap conductors are connected (tapped) to the feeder inside the gutter. These tap conductors are terminated in the fused safety switches for each of the motors. Fuses are installed in the safety switches. These are the fuses that protect the tap conductors against overload. These fuses must also protect the motor against short circuits and ground faults, and are sized according to NEC® 430.52(C). These fuses are sized as was done previously for single motor circuits.

Motor Feeder Conductors, OCPDs, and Taps – Motor Branch Circuit Conductors and OCPDs   234-891

Example 1c – Sizing Feeder Tap Conductors Using Figure 234.1

Determine the minimum sizes permitted for the tap conductors to Motors 1, 2, and 3 given the following tap lengths.
Motor 1: The length from the tap point in the gutter to the line side of the safety switch is 8 feet.
Motor 2: The length from the tap point in the gutter to the line side of the safety switch is 18 feet.
Motor 3: The length from the tap point in the gutter to the line side of the safety switch is 28 feet.

Since the tap length for Motor 1 is 8 feet, the limitations of NEC® 430.28(1) are to be used.
Remember that, since all terminations have 75°C ratings, the 75°C column of Table 310.16 is to be used when determining conductor sizes.

☞ **10 foot rule: Tap Ampacity ≥ (Feeder OCP rating) ÷ 10**

In Example 1b it was found that a 200 Amp breaker is used for the feeder OCPD.

Motor 1-Tap Ampacity ≥ 200 A ÷ 10 = 20 amps ⇒ 14 AWG

Note that the minimum permitted size for the tap may not be large enough for the motor load. For Motor 1 compare the 14 AWG found for the tap to the size required for the motor. Then select the larger of the two.

Minimum conductor size required for Motor 1:
   Motor 1 Table FLA × 1.25 = 21 × 1.25 = 26.25 = 26 amps ⇒ 10 AWG

Since 10 AWG is larger than 14 AWG, 10 AWG is the smallest size permitted to serve the load of Motor 1.

☞ **10 AWG** conductors must be used for the tap conductors between the gutter and Motor 1 OCPDs.

The tap length for Motor 2 is 18 feet and the limitations of NEC® 430.28(2) are to be used.

☞ **25 foot rule: Tap Ampacity ≥ (Feeder Conductor Ampacity) ÷ 3**

In Example 1a it was found that 1 AWG, with an ampacity of 130 amps, is used for the feeder conductors.

Motor 2-Tap Ampacity ≥ (Feeder Conductor Ampacity) ÷ 3 = 130 A ÷ 3 = 43.33 amps
   Drop 0.33 (not a major fraction) = 43 amp wire ⇒ 8 AWG

Again, the minimum permitted size for the tap may not be large enough for the motor load. For Motor 2 compare the 8 AWG found for the tap to the size required for the motor. Then select the larger of the two.

Minimum conductor size required for Motor 2:
   Motor 2 Table FLA × 1.25 = 65 × 1.25 = 81.25 = 81 amps ⇒ 4 AWG

Since 4 AWG is larger than 8 AWG, 4 AWG is the smallest size permitted to serve the load of Motor 2.

☞ **4 AWG** conductors must be used for the tap conductors between the gutter and Motor 2 OCPDs.

Year Two (Student Manual)

The tap length for Motor 3 is 28 feet and the limitations of NEC® 430.28(3) are to be used.

☞ **greater than 25 foot rule: Tap Ampacity ≥ Feeder Conductor Ampacity**

In Example 1a it was found that 1 AWG, with an ampacity of 130 amps, is used for the feeder conductors. This is also the smallest permitted size that can be used for this tap that is longer than 25 feet.
Motor 3-Tap Ampacity ≥ Feeder Conductor Ampacity ⇒ 1 AWG

This, of course, is going to be large enough to serve Motor 3. But, just to be sure that the previous calculation of the feeder size is correct, compare this to the conductor size required for Motor 3.

Minimum conductor size required for Motor 3:
    Motor 3 Table FLA × 1.25 = 14 × 1.25 = 17.5 = 18 amps ⇒ 14 AWG

Since 1 AWG is larger than 14 AWG, 1 AWG is the smallest size permitted for the tap conductors between the gutter and the fuses protecting Motor 3.

☞ **1 AWG** conductors must be used for the tap conductors between the gutter and Motor 3 OCPDs.

Example 1d – *Sizing DETD fuses for Tapped Motor Circuits Using Figure 234.1*
    The OCPDs at the termination of the motor feeder tap conductors (the fuses in the fused safety switches) are required to be sized to protect the motors. This procedure is exactly the same as was used previously for single-motor branch circuits.

Motor 1 fuses: Table FLA × 1.75 = 21 × 1.75 = 36.75 ⇒ 37 amps ⇒ 40 A

☞ **40 Amp** DETD fuses must be used in the fused safety switch for Motor 1.

Motor 2 fuses: Table FLA × 1.75 = 65 × 1.75 = 113.75 ⇒ 114 amps ⇒ 125 A

☞ **125 Amp** DETD fuses must be used in the fused safety switch for Motor 2.

Motor 3 fuses: Table FLA × 1.75 = 14 × 1.75 = 24.5 ⇒ 25 amps ⇒ 25 A DETD Fuses

☞ **25 Amp** DETD fuses must be used in the fused safety switch for Motor 3.

Example 1e – *Sizing Motor Circuit Conductors Using Figure 234.1*
    The conductors on the **load** side of the fuses in the safety switch are, by definition, branch circuit conductors. These motor circuit conductors are protected against short circuits and ground faults by the fuses and do NOT have to comply with the tap rules of NEC® 430.28. These conductors only have to be sized to serve the motor load. These calculations were performed in Example 1c, when comparisons were made to the tap conductor sizes. These calculations are reprinted below.

Motor 1 Conductors: Table FLA × 1.25 = 21 × 1.25 = 26.25 ⇒ 26 amps ⇒ 10 AWG

☞ **10 AWG** conductors are the smallest permitted between Motor 1 and the load side of the fused safety switch for Motor 1.

Motor 2 Conductors: Table FLA × 1.25 = 65 × 1.25 = 81.25 ⇒ 81 amps ⇒ 4 AWG

☞ **4 AWG** conductors are the smallest permitted between Motor 2 and the load side of the fused safety switch for Motor 2.

Motor 3 Conductors: Table FLA × 1.25 = 14 × 1.25 = 17.5 ⇒ 18 amps ⇒ 14 AWG

☞ **14 AWG** conductors are the smallest permitted between Motor 3 and the load side of the fused safety switch for Motor 3.

Motor Feeder Conductors, OCPDs, and Taps – Motor Branch Circuit Conductors and OCPDs

Objective 234.1 Worksheet

_____ 1. The FLC from ___ is used to determine the branch circuit conductor size for a 3Ø motor.

   a. the nameplate   b. T.430.248   c. T.430.250   d. T.430.247

_____ 2. When time-delay fuses are used to protect motor branch circuit conductors, they would have a ___ ampere rating than would be required if nontime-delay (one-time) fuses were used.

   a. larger   b. smaller   c. the same size

_____ 3. Which formula below is used to size the conductors for a motor feeder?

   a. [(largest FLA) × (125%)] + [all other FLAs]
   b. [all FLAs] × 125%
   c. [(largest FLA) × (125%)] + [all FLAs]

_____ 4. For a 5 HP, single-phase motor operating at 208 volts, you would use a FLA of ___ amps to calculate wire size.

   a. 16.7   b. 38.5   c. 186   d. 30.8   e. 102

_____ 5. Generally, the current shown in Tables 430.247 to 430.250 will be ___ the actual current shown on the nameplate of a motor.

   a. more than   b. less than   c. the same as

_____ 6. The FLA shown on a motor nameplate is used to size ___.

   a. OL protection   b. conductors   c. SC&GF protection   d. all of these

Objective 234.2 Worksheet

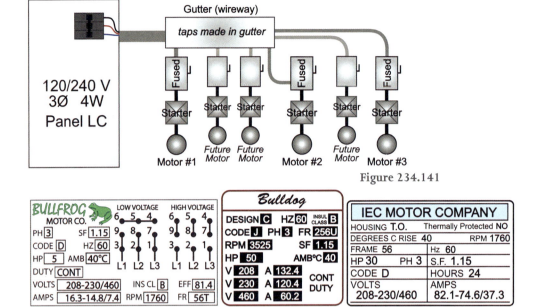

Figure 234.141

Feeder to serve these three motors. Use Exc. 1 to 430.52(C)(1) to determine individual motor OCPDs.

_____ 1. Refer to the information shown above in Figure 234.141. Without regard to the future motors, the smallest conductors permitted between Panel LC and the gutter are ___ AWG or kcmil.

    a. 350   b. 4/0   c. 250   d. 400   e. 300

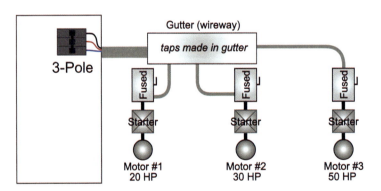

Figure 234.144

_____ 2. Refer to the information shown above in Figure 234.144. Where the breaker is in a 120/208 volt, 3Ø, 4-wire panel the smallest feeder conductors permitted between the 3-pole breaker and the wireway are ___ AWG or kcmil.

    a. 400   b. 4/0   c. 3/0   d. 350   e. 500

Motor Feeder Conductors, OCPDs, and Taps – Motor Branch Circuit Conductors and OCPDs   234-897

Scenario 234-401
A 3-pole breaker is installed in Panel K. A set of motor feeder conductors are then run from the breaker to a large junction box where taps are made to three 3Ø motors.

Motor #1 = 5 hp   Motor #2 = 25 hp   Motor #3 = 40 hp

Use 430.52(C)(1) Exc 1 to determine individual motor OCPDs.

_____ 3. Refer to the information shown above in Scenario 234-401. Panel K is a 120/208 volt, 3Ø, 4-wire panel. The minimum size conductors permitted to be installed between the panel and the large junction box are ___ AWG or kcmil.

   a. 250   b. 4/0   c. 350   d. 400   e. 300

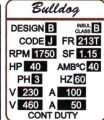

Feeder to serve these three motors. Use Exc. 1 to 430.52(C)(1) to determine individual motor OCPDs.   Figure 234.140

_____ 4. Refer to the information shown above in Figure 234.140. The smallest conductors permitted to be installed from a 120/240 volt, 3Ø, 4-wire panelboard for this motor feeder are ___ AWG or kcmil.

   a. 3/0   b. 2/0   c. 300   d. 4/0   e. 250

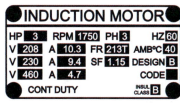 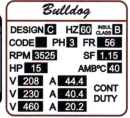

Figure 234.154

A motor feeder originates in a 120/208 volt, 3Ø, 4-W switchboard or panelboard. The motor circuits to continuously-operated motors 1, 2, and 3 (left-to-right) are tapped in the gutter and terminate in either a fused safety switch or a breaker enclosure (with an inverse time circuit breaker). The switch or breaker is used as the disconnecting means and is located near each motor.

_____ 5. Refer to the information shown above in Figure 234.154. The smallest feeder conductors permitted to be installed are ___ AWG or kcmil.

   a. 1/0   b. 3   c. 1   d. 3/0   e. none of these

Year Two (Student Manual)

Motor Feeder Conductors, OCPDs, and Taps – Motor Branch Circuit Conductors and OCPDs

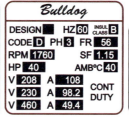
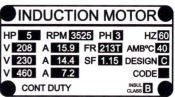

Figure 234.151

Left-to-right, motors 1, 2, and 3 are supplied by a feeder that enters a gutter. The feeder originates in a 277/480 volt, 3Ø, 4-W switchboard or panelboard. The motor circuits are tapped in the gutter and terminate in either a fused safety switch or a breaker enclosure (with an inverse time circuit breaker), located near each motor. The tap length from the gutter to the switch (or ITCB) for Motor #1 is 8 feet. The tap length from the gutter to the switch (or ITCB) for Motor #2 is 17 feet. The tap length from the gutter to the switch (or ITCB) for Motor #3 is 30 feet. All motors will be operated continuously.

Use Exc. 1 to 430.52(C)(1) to determine individual motor OCPDs.

_____ 6. Refer to the information shown above in Figure 234.151. The smallest feeder conductors permitted to be installed are ___ AWG or kcmil.

    a. 2   b. 1/0   c. 3   d. 2/0   e. none of these

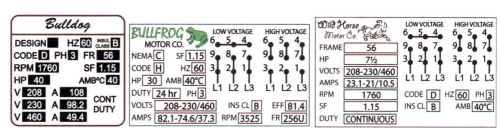

Figure 234.145

_____ 7. Refer to the nameplates & information shown above in Figure 234.145. The smallest feeder conductors permitted to be installed between the 120/240 volt, 3Ø, 4-wire panel and the gutter are ___ AWG or kcmil.

    a. 350   b. 4/0   c. 300   d. 250   e. 3/0

8. Size the minimum THWN/THHN copper 480-volt feeder conductors required to feed one each 50 hp, 40 hp, 30 hp, and 10 hp motors. Each motor is rated for 208-230/460 volts, three phase.

   a. 2/0 AWG   b. 250 kcmil   c. 500 kcmil   d. 1/0 AWG   e. 4/0 AWG   f. none of these

Objective 234.3 Worksheet

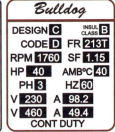

Figure 234.150

Left-to-right, motors 1, 2, and 3 are supplied by a feeder. The feeder originates in a 120/240 volt, 3Ø, 4-W switchboard or panelboard and feeds the gutter. The motor circuits are tapped in the gutter and terminate in either a fused safety switch or a breaker enclosure (with an inverse time circuit breaker), located near each motor. The tap length from the gutter to the switch (or ITCB) for Motor #1 is 8 feet. The tap length from the gutter to the switch (or ITCB) for Motor #2 is 17 feet. The tap length from the gutter to the switch (or ITCB) for Motor #3 is 30 feet.

All motors will be operated continuously.

_____ 1. Refer to the information shown above in Figure 234.150. The maximum permitted rating for an ITCB for the feeder would be ___ A.

   a. 250   b. 300   c. 225   d. 200   e. none of these

_____ 2. Refer to the information shown above in Figure 234.150. When the tap for Motor 2 terminates in a breaker enclosure, the maximum permitted size for an ITCB in this enclosure would be ___ A.

   a. 250   b. 200   c. 300   d. 150   e. none of these

_____ 3. Refer to the information shown above in Figure 234.150. The maximum permitted size for DETD fuses for the feeder would be ___ A.

   a. 200   b. 250   c. 300   d. 225   e. none of these

_____ 4. Refer to the information shown above in Figure 234.150. When the tap for Motor 1 terminates in a safety switch, the maximum permitted size for DETD fuses in this switch would be ___ A.

   a. 110
   b. 100
   c. 80
   d. 70
   e. none of these

Motor Feeder Conductors, OCPDs, and Taps – Motor Branch Circuit Conductors and OCPDs   234-901

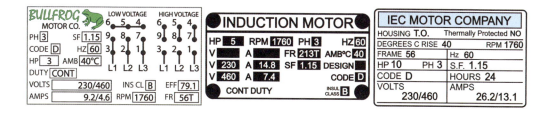

Figure 234.152

Left-to-right, motors 1, 2, and 3 are supplied by a feeder that enters a gutter and originates in a 120/240 volt, 3Ø, 4-W switchboard or panelboard. The motor circuits are tapped in the gutter and terminate in either a fused safety switch or a breaker enclosure (with an inverse time circuit breaker) that is located near each motor. The tap length from the gutter to the switch (or ITCB) for Motor #1 is 8 feet. The tap length from the gutter to the switch (or ITCB) for Motor #2 is 37 feet. The tap length from the gutter to the switch (or ITCB) for Motor #3 is 15 feet.

All motors will be operated continuously.   Use Exc. 1 to 430.52(C)(1) to determine individual motor OCPDs.

_____ 5. Refer to the information shown above in Figure 234.152. The maximum permitted size for DETD fuses for the feeder would be ___ A.

   a. 90   b. 70   c. 110   d. 60   e. none of these

_____ 6. Refer to the information shown above in Figure 234.152. The maximum permitted rating for an ITCB for the feeder would be ___ A.

   a. 100   b. 90   c. 125   d. 150   e. none of these

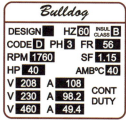 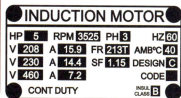

Figure 234.151

Left-to-right, motors 1, 2, and 3 are supplied by a feeder that enters a gutter. The feeder originates in a 277/480 volt, 3Ø, 4-W switchboard or panelboard. The motor circuits are tapped in the gutter and terminate in either a fused safety switch or a breaker enclosure (with an inverse time circuit breaker), located near each motor. The tap length from the gutter to the switch (or ITCB) for Motor #1 is 8 feet. The tap length from the gutter to the switch (or ITCB) for Motor #2 is 17 feet. The tap length from the gutter to the switch (or ITCB) for Motor #3 is 30 feet.

All motors will be operated continuously.
   Use Exc. 1 to 430.52(C)(1) to determine individual motor OCPDs.

_____ 7. Refer to the information shown on the previous page in Figure 234.151. The maximum permitted rating for an ITCB for the feeder would be ___ A.

    a. 250   b. 200   c. 225   d. 175   e. none of these

_____ 8. Refer to the information shown on the previous page in Figure 234.151. The maximum permitted size for DETD fuses for the feeder would be ___ A.

    a. 100   b. 125   c. 150   d. 175   e. none of these

_____ 9. Refer to the information shown on the previous page in Figure 234.151. When the tap for Motor 3 terminates in a safety switch, the maximum permitted size for DETD fuses in this switch would be ___ A.

    a. 15
    b. 35
    c. 25
    d. 45
    e. none of these

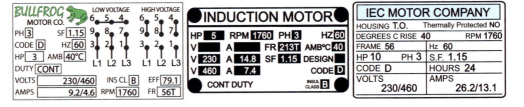

Figure 234.153

Left-to-right, continuously-operated motors 1, 2, and 3 are supplied by conductors that are tapped from a feeder. The feeder, inside the gutter, originates in a 277/480 volt, 3Ø, 4-W switchboard or panelboard. The tap conductors exit the gutter and terminate in either a fused safety switch or a breaker enclosure (with an inverse time circuit breaker), located near each motor. The tap lengths, between the gutter and the switch (or ITCB) for each motor, are as follows:

Motor #1 is 8 feet.   Motor #2 is 37 feet.   Motor #3 is 15 feet.

_____ 10. Refer to the information shown above in Figure 234.153. When the tap for Motor 2 terminates in a breaker enclosure, the maximum permitted size for an ITCB in this enclosure would be ___ A.

    a. 20
    b. 50
    c. 40
    d. 25
    e. 30
    f. none of these

Motor Feeder Conductors, OCPDs, and Taps – Motor Branch Circuit Conductors and OCPDs 234-903

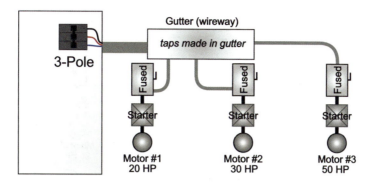

Figure 234.144

_____ 11. Refer to the information shown above in Figure 234.144. The breaker is in a 277/480 volt, 3Ø, 4-wire panel. The breaker is permitted to have a rating not greater than ___ amps where the feeder conductors are as small as permitted.

a. 300   b. 250   c. 175   d. 225   e. 200

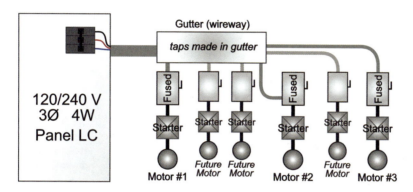

Figure 234.141

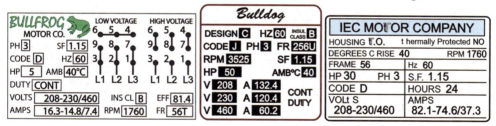

Feeder to serve these three motors. Use Exc. 1 to 430.52(C)(1) to determine individual motor OCPDs.

_____ 12. Refer to the information shown above in Figure 234.141. Without regard to the future motors, the maximum size circuit breaker permitted to be installed in Panel LC to protect the three motors is a ___ Amp.

a. 400   b. 300   c. 350   d. 450   e. 500

Year Two (Student Manual)

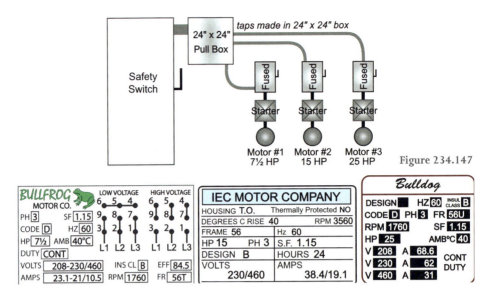

Figure 234.147

_____ 13. Refer to the nameplates & information shown above in Figure 234.147. The safety switch is supplied with 120/240 volt, 3Ø, 4-wire power. The fuses installed in the switch can have a rating not greater than ___ amps. The feeder conductors are as small as permitted for the motors shown.

a. 300   b. 225   c. 250   d. 200   e. 175

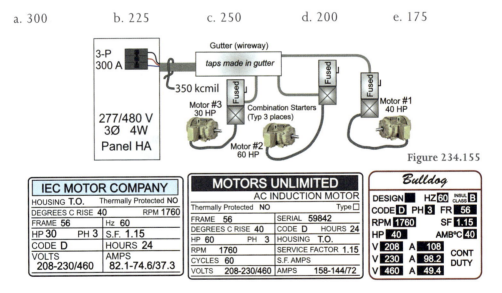

Figure 234.155

Although only 3 motors are shown, the feeder and the feeder OCPD were installed to supply power to (future) additional motors. The motor circuits are tapped in the gutter and the tap lengths are:
Motor #3 - 12 ft   Motor #2 - 23 ft   Motor #1 - 51 ft.

_____ 14. The maximum permitted size DETD fuses that can be installed in the combination starter for Motor 3 are ___ A, as determined from the information provided in Figure 234.155.

a. 100   b. 50   c. 45   d. 90   e. none of these

15. b. 350

Objective 234.4 Worksheet

Scenario 234-400

To allow for additional capacity, a set of 750 kcmil motor feeder conductors were installed between a 3-pole 450 A breaker in Panel LA and a gutter. Panel LA is 120/240 3Ø, 4-wire.

The tap conductor length from the gutter to the line side of the Motor #1 fused safety switch is 8 feet. Motor #1 is a 5 hp 230 volt, 3Ø motor.
The tap conductor length from the gutter to the line side of the Motor #2 fused safety switch is 16 feet. Motor #2 is a 30 hp 230 volt, 3Ø motor.
The tap conductor length from the gutter to the line side of the Motor #1 fused safety switch is 31 feet. Motor #3 is a 50 hp 230 volt, 3Ø motor. Don't forget to use Exc. 1 to 430.52(C)(1) to determine the rating of the largest motor's individual OCPD.

_____ 1. Refer to the information shown above in Scenario 234-400. For Motor #3 the minimum permitted size of the circuit conductors (from the fused switch to the motor) is ___ AWG or kcmil.

a. 3/0   b. 1   c. 2/0   d. 1/0   e. 2   f. 750

_____ 2. Refer to the information shown above in Scenario 234-400. For Motor #1 the minimum permitted size of the tap conductors (from the gutter to the switch) is ___ AWG.

a. 8   b. 4   c. 6   d. 3   e. 14   f. 10

_____ 3. Refer to the information shown above in Scenario 234-400. For Motor #3 the minimum permitted size of the tap conductors (from the gutter to the switch) is ___ AWG.

a. 1/0   b. 750   c. 2   d. 1   e. 2/0   f. 3/0

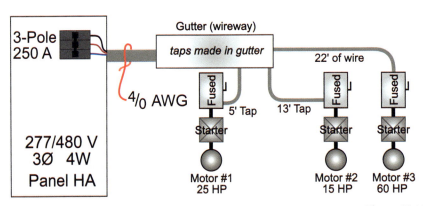

Figure 234.143

_____ 4. Refer to the information shown above in Figure 234.143. The smallest conductors permitted between the gutter and the switch for Motor 1 are ___ AWG.

a. 4   b. 14   c. 4/0   d. 10   e. 12   f. 8

_____ 5. Refer to the information shown on the previous page in Figure 234.143. The smallest conductors permitted between the gutter and the switch for Motor 3 are ___ AWG.

a. 4/0   b. 3   c. 4   d. 8   e. 10

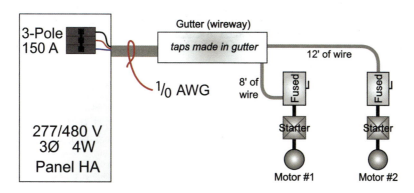

Figure 234.142

_____ 6. Refer to the information shown above in Figure 234.142. Motor 1 is rated at 30 hp and Motor 2 is rated at 7 ½ hp. The smallest conductors permitted between the fused switch and Motor 1 are ___ AWG. The fuses have been properly sized for the motor.

a. 1/0   b. 8   c. 14   d. 3   e. 6

_____ 7. Refer to the information shown above in Figure 234.142. Motor 1 is rated at 30 hp and Motor 2 is rated at 7½ hp. The smallest conductors permitted between the gutter and the switch for Motor 1 are ___ AWG.

a. 1/0   b. 14   c. 6   d. 8   e. 3

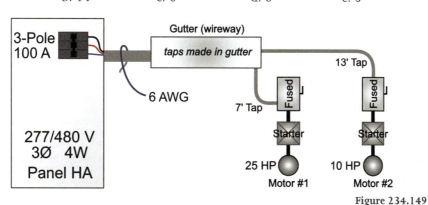

Figure 234.149

_____ 8. Refer to the information shown above in Figure 234.149. The minimum permitted size of the tap conductors (from the gutter to the switch) is ___ AWG for the 10 HP motor.

a. 6   b. 10   c. 8   d. 14   e. 12   f. none of these

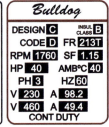

Figure 234.150

Left-to-right, motors 1, 2, and 3 are supplied by a feeder. The feeder originates in a 120/240 volt, 3Ø, 4-W switchboard or panelboard and feeds the gutter. The motor circuits are tapped in the gutter and terminate in either a fused safety switch or a breaker enclosure (with an inverse time circuit breaker), located near each motor. The tap length from the gutter to the switch (or ITCB) for Motor #1 is 8 feet. The tap length from the gutter to the switch (or ITCB) for Motor #2 is 17 feet. The tap length from the gutter to the switch (or ITCB) for Motor #3 is 30 feet.

All motors will be operated continuously.

_____ 9. Refer to the information shown above in Figure 234.150. The minimum tap conductor size for Motor 2 would be ___ AWG or kcmil. The feeder conductors are the minimum permitted size and are protected with an ITCB of the maximum permitted rating.

a. 4   b. 1   c. 2   d. 3   e. none of these

_____ 10. Refer to the information shown above in Figure 234.150. The minimum tap conductor size for Motor 1 would be ___ AWG or kcmil. The feeder conductors are the minimum permitted size and are protected with DETD fuses of the maximum permitted rating.

a. 10   b. 4   c. 6   d. 8   e. none of these

_____ 11. Refer to the information shown above in Figure 234.150. The minimum tap conductor size for Motor 1 would be ___ AWG or kcmil. The feeder conductors are the minimum permitted size and are protected with an ITCB of the maximum permitted rating.

a. 10   b. 6   c. 4   d. 8   e. none of these

_____ 12. A 300 amp, 240-volt, three-phase feeder consisting of 350 kcmil conductors is tapped to feed a 30 hp, three-phase motor. If the tap conductor length is 18 feet, what is the minimum size tap conductor permitted?

a. 3 AWG
b. 1 AWG
c. 8 AWG
d. 1/0 AWG
e. 2 AWG
f. none of these

13. A 300 amp, 480-volt, three-phase feeder consisting of 350 kcmil conductors is tapped to feed a 30 hp, three-phase motor. If the tap conductor length is 18 feet, what is the minimum size tap conductor permitted?

    a. 6 AWG
    b. 3 AWG
    c. 8 AWG
    d. 1 AWG
    e. 1/0 AWG
    f. none of these

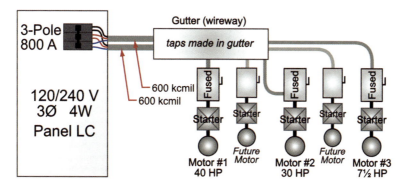

Figure 234.146

14. Refer to the information shown above in Figure 234.146. When the conductor length from the gutter to the line side of the motor #3 fused safety switch is 48 feet, the minimum permitted size of the tap conductors (from the gutter to the switch) is ___ AWG or kcmil.

    a. 1/0   b. 10   c. 2/0   d. 2 sets of 600se. 3

15. Refer to the information shown above in Figure 234.146. When the conductor length from the gutter to the line side of the motor #1 fused safety switch is 9 feet, the minimum permitted size of the tap conductors (from the gutter to the switch) is ___ AWG.

    a. 3   b. 1   c. 1/0   d. 2   e. 4

Objective 234 Safety Worksheet

New Employees

New employees present several challenges and opportunities for us. The opportunities are being able to teach them how to be safe and what to be aware of on our jobsites. Our obligations are to make them understand what can injure or kill them and how they can protect themselves from these dangers. Another opportunity is that we have a chance to teach them how to do things safely before they learn the short-cuts that can get them in trouble.

We can never assume that a new employee understands or has ever been exposed to what we see everyday on our jobsites. Things that we now take for granted because we have been working for a while, they may have never seen or experienced for themselves. These facts contribute to a large number of fatalities and severe injuries to employees who have had little or no real safety training and little time on the job. We can reduce this possibility by providing safety training to these individuals. OSHA states that "each employer shall provide to each of his employees, employment and a place of employment which are free from recognized hazards that are causing or are likely to cause death or serious physical harm to his employees". From this we understand the employer has the obligation to provide a safe workplace.

We must help these new employees be safe until they fully understand what can hurt them and what they must do to protect themselves.

The supervisor has a responsibility to explain the safety requirements for the jobsite and the work to be performed. Some of the topics that need to be discussed are:

- Where are 1st aid supplies and who is certified to administer the aid?

- What PPE is required and who will train the new employee in the proper use and care?

- Hand and power tools require training in safety use and care.

- Electrical requirements, grounded cords, GFCI protection for employees

- Ladders are common, is the employee trained in the proper use and care?

- Where are the containers for trash and debris?

- Where are the required fire extinguishers?

- HazCom – do we have any hazardous materials onsite we need to be aware of?

- Some other topics could include crane and overhead hazards, heavy equipment, excavations and trenching, motor vehicles, scaffolds, confined spaces and fall protection as they apply.

Making sure safety items are discussed and training is completed will help any new employees get a good safe start in their new job or assignment.

Questions

_____ 1. New employees must be advised as to what PPE requirements apply to the job to which they are assigned.

   a. False
   b. True

_____ 2. Safety topics that need to be discussed with new employees include____.

   a. hand and power tools
   b. 1st aid
   c. HazCom
   d. GFCI protection
   e. ladders
   f. all of these

_____ 3. Making sure that safety issues are discussed and training is completed will help any new employees get a good safe start in their new job or assignment.

   a. True
   b. False

Lesson 235
Final Exam Review

Electrical Curriculum

**Year Two
Student Manual**

IEC
NATIONAL

Purpose

The second year final exam review is scheduled for this lesson.

Homework

(Due at the beginning of this class)

For this lesson, you should:

- Review all your worksheets, homework, and quizzes.
- Complete your "Practice Exam."

- Be prepared to clarify any questions or problem areas you encountered during your review.

Objectives

This lesson will determine your proficiency in the subject matter from the previous lessons. The instructor should review next week's material beforehand to determine if the students will require additional instruction before assigning the worksheets for that lesson.

Year Two (Student Manual)

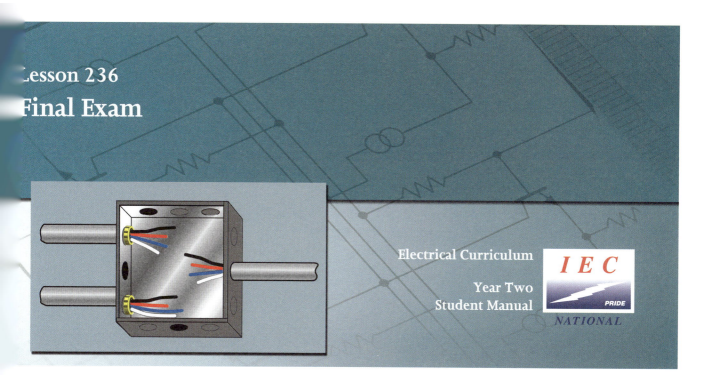

Lesson 236
Final Exam

Electrical Curriculum

Year Two
Student Manual

Purpose
The second year final exam is scheduled for this lesson.

Homework
(Due at the beginning of this class)
For this lesson, you should:

- Review all your worksheets, homework, and quizzes.
- Complete your "Practice Exam."

- Be prepared to clarify any questions or problem areas you encountered during your review.

Objectives
This lesson will determine your proficiency in the subject matter from the previous lessons. The instructor should review next week's material beforehand to determine if the students will require additional instruction before assigning the worksheets for that lesson.

236-920

Final Exam

Year Two (Student Manual)

Yr-2 Annex A – Glossary of Transformer Terms

Accurate Ratio Transformer	A small transformer which transforms at an accurate ratio to allow an attached instrument to gauge the current or voltage without actually running full power through the instrument.
Ammeter	An instrument used to measure current.
Autotransformer	A special type of power transformer, consisting of a single, continuous winding that is tapped on one side to provide either a step-up or step-down function.
Basic Impulse Level	A measure of the ability of a transformer's insulation system to withstand very high-voltage, short-time surges Burden Term applied to the tiny load that an instrument's delicate moving elements places on an accurate-ratio transformer.
Coils	See "Winding."
Copper Loss	The energy wasted as heat in the copper (or aluminum) windings, as copper/aluminum is not a perfect conductor of electricity.
Core	A component of a transformer. The iron or steel core provides a controlled path for the magnetic flux generated in the transformer by the current flowing through the windings.
Core and Coils	A non-enclosed, dry type transformer, mounted on a base as one unit.
Core Type	A type of core where the windings surround the laminated iron core.
Current Transformer	A type of instrument transformer used to measure current.
Deflection	Literally, the amount of movement the indicator of an instrument makes when sensing. Also called the reading of the instrument.
Distribution Transformer	A transformer used to supply relatively small amounts of power to residences. It is used at the end of the electrical utility's delivery system. Often mounted on a pole.
Delta	A three-phase transformer connection where the phases are connected in a manner which resembles the Greek letter Delta.
Dry Type Transformer	A transformer designed to operate in air. The design does not require the assistance of a liquid to dissipate excess heat. Natural or fan-assisted circulation through ventilation openings is all that is required to meet temperature classification requirements.
Eddy Current	Induced voltage in the core as a result of transformer operation. The currents move through the core in circular paths.
Eddy-Current Loss	The energy wasted by eddy currents creating heat in the core, as this does not aid in the induction process.
Efficiency	A rating of the percentage of input power transmitted through the transformer. This number will never be 100% in the real world due to copper losses, eddy-current losses and other inefficiencies.

Encapsulated Transformer	A specialty dry type transformer, sealed in an enclosure. It is capable of moving excessive heat away from the core and coils without ventilation openings.
Full Capacity Above Normal Tap	A special voltage tap used to account for voltage fluctuations on the input side. Allows for fine-tuning of the output voltage when the input voltage is higher than expected.
Full Capacity Below Normal Tap	A special voltage tap used to account for voltage fluctuations on the input side. Allows for fine-tuning of the output voltage when the input voltage is lower than expected.
Input	The voltage source coming into the transformer.
Instrument Transformer	A small transformer which transforms at an accurate ratio to allow an attached instrument to gauge the current or voltage without actually running full power through the instrument.
Insulation System Temperature Classification	A statement of the maximum temperature permitted in the hottest spot in the winding, at a specified ambient temperature, usually 40°C. Exceeding this figure will likely result in an insulation failure.
Liquid-Filled Transformer	A type of transformer cooled by mounting in a sealed tank filled with liquid. The liquid is normally oil, but silicone and other liquids may also be used.
Magnetic Coupling	The method by which one circuit is linked to another circuit by a common magnetic field.
Magnetic Flux	Lines of magnetic force surrounding a magnet or electromagnet.
NEMA	Abbreviation for National Electrical Manufacturers Association. An organization of manufacturers of electrical products.
Output	The transformed voltage exiting the transformer and going out to the load.
Potential Transformer	A type of instrument transformer used to measure voltage.
Power Transformer	A transformer used primarily to couple electrical energy from a power supply line to a circuit system.
Primary	The winding(s) connected to the leads or lugs marked with "H".
Secondary	The winding(s) connected to the leads or lugs marked with "X".
Shell Type	A type of core where the core surrounds the windings.
Step-Up Transformer	A transformer in which the output voltage is higher than the input voltage. The secondary winding has more turns of wire than the primary winding.
Step-Down Transformer	A transformer in which the output voltage is lower than the input voltage. The secondary winding has fewer turns of wire than the primary winding.
Three-Phase Transformer	A transformer used to transform voltage provided by a three-phase power system.
Transformer	A device that transfers electrical energy from one electric circuit to another, without changing the frequency, by the principles of electromagnetic induction. The energy transfer usually takes place with a change of voltage.

Transformer Bank	An arrangement of three individual single-phase transformers, configured to transform three-phase voltage.
Turns	The number of times the wire of a winding actually goes around the core.
Turns Ratio	A comparison of the number of turns in the primary versus the number of turns in the secondary. Directly related to the voltage ratio.
Voltage Ratio	A comparison of the voltage entering the primary versus the voltage exiting the secondary. Directly related to the turns ratio.
Voltage Tap	An additional connection to a winding, which permits use of only a specific part of the winding. This allows the same winding to handle multiple voltage levels.
Voltmeter	An instrument used to measure voltage.
Wattmeter	An instrument used to measure wattage.
Winding	Turns of wire around the core of the transformer. Connects the core to either the input, in the case of the primary winding, or the output, in the case of the secondary winding.
Wye	A three-phase transformer connection where the phases are connected in a manner which resembles the letter "Y." Often called a "star connection."

Yr-2 Annex B – Single-Phase Transformer Data

Group I – Single Phase

240 x 480 PRIMARY VOLTS — 120/240 SECONDARY VOLTS — FOUR WINDINGS — 1Ø, 60 Hz

KVA Rating	Catalog # Terminations: Lugs or Wire Leads (WL)		Approximate Dimensions (inches)			Approximate Shipping Weight (lb)	Type Mtg. W – Wall F - Floor	Knockouts (inches)	Weather Shield Req'd	Wiring Diagram – Enclosure Type
			Height	Width	Depth					
.05	1101	WL	6.41	3.14	3.05	4	W	0.875	NO	1–A
.10	1102	WL	7.16	3.89	3.67	5	W	0.875	NO	1–A
.15	1103	WL	7.16	3.89	3.67	7	W	0.875	NO	1–A
.25	1104	WL	8.68	4.08	3.88	10	W	0.50-0.75	NO	2–B
.50	1105	WL	9.06	4.37	4.20	15	W	0.50-0.75	NO	2–B
.75	1106	WL	9.68	4.75	4.50	19	W	0.50-0.75	NO	2–B
1.00	1107	WL	10.50	5.50	5.13	24	W	0.50-0.75	NO	2–B
1.50	1108	WL	11.62	5.50	5.13	30	W	0.50-0.75	NO	2–B
2.00	1109	WL	13.00	5.50	5.13	38	W	0.50-0.75	NO	2–B
3.00	1110	WL	11.50	10.31	7.13	55	W	0.75-1.25	NO	2–C
3.00	1111	WL	11.50	10.31	7.13	55	W	0.75-1.25	NO	3–C
5.00	1112	WL	14.38	10.31	7.13	75	W	0.75-1.25	NO	2–C
5.00	1113	WL	14.38	10.31	7.13	75	W	0.75-1.25	NO	3–C
7.50	1114	WL	15.19	13.50	10.84	115	W	0.75-1.25	NO	4–D
10.00	1115	WL	15.19	13.50	10.84	125	W	0.75-1.25	NO	4–D
15.00	1116	Lugs	16.94	14.12	11.59	170	F or W①	1.00-1.50	YES	4–E
25.00	1117	Lugs	18.44	16.13	13.34	250	F or W①	1.00-1.50	YES	4–E
37.50	1118	Lugs	25.50	24.39	19.37	280	F or W①	NA	YES	5–E
50.00	1119	Lugs	25.50	24.39	19.37	350	F or W①	NA	YES	5–E
75.00	1120	Lugs	35.47	31.90	26.88	430	F	NA	YES	5–E
100.00	1121	Lugs	41.52	32.90	29.87	525	F	NA	YES	5–E
167.00	1122	Lugs	45.60	39.50	35.50	1050	F	NA	YES	5–E
250.00	1123	Lugs	45.60	39.50	35.50	1440	F	NA	YES	5–E

① Wall mounting brackets must be ordered separately for these sizes.

Wall mounting brackets are used to wall mount most 15 through 50 kVA and some 75 kVA transformers. This bracket allows for 6-inch clearance from the wall as recommended by Eaton and most other manufacturers.

WALL MOUNTING BRACKETS

Year Two (Student Manual)

WEATHERSHIELDS

A weathershield kit consisting of a front and rear cover shield must be installed on all ventilated dry-type distribution transformers when the unit is located outdoors. The shields protect the transformer top ventilation openings against rain but allow for proper ventilation. Field installation hardware is not required. Proper installation provides a NEMA 3R rating.

Group III – Single Phase

120 x 240 PRIMARY VOLTS — 120/240 SECONDARY VOLTS — FOUR WINDINGS — 1Ø, 60 Hz

KVA Rating	Catalog # Terminations: Lugs or Wire Leads (WL)		Approximate Dimensions (inches)			Approximate Shipping Weight (lb)	Type Mtg. W – Wall F - Floor	Knockouts (inches)	Weather Shield Req'd	Wiring Diagram – Enclosure Type
			Height	Width	Depth					
1.0	1301	WL	10.50	5.50	5.13	24	W	0.50-0.75	NO	13–B
1.5	1302	WL	11.62	5.50	5.13	30	W	0.50-0.75	NO	13–B
2.0	1303	WL	13.00	5.50	5.13	38	W	0.50-0.75	NO	13–B
3.0	1304	WL	11.50	10.31	7.13	55	W	0.75-1.25	NO	13–C
5.0	1305	WL	14.38	10.31	7.13	75	W	0.75-1.25	NO	13–C
7.5	1306	WL	15.19	13.50	10.84	115	W	0.75-1.25	NO	13–D
10.0	1307	WL	15.19	13.50	10.84	125	W	0.75-1.25	NO	13–D
15.0	1308	Lugs	16.94	14.12	11.59	170	W	1.00-1.50	YES	13–E
25.0	1309	Lugs	18.44	16.13	13.34	250	W	1.00-1.50	YES	13–E

Group V – Single Phase

208 PRIMARY VOLTS — 120/240 SECONDARY VOLTS — THREE WINDINGS — 1Ø, 60 Hz

KVA Rating	Catalog # Terminations: Lugs or Wire Leads (WL)		Approximate Dimensions (inches)			Approximate Shipping Weight (lb)	Type Mtg. W – Wall F - Floor	Knockouts (inches)	Weather Shield Req'd	Wiring Diagram – Enclosure Type
			Height	Width	Depth					
1.0	1501	WL	10.50	5.50	5.13	24	W	0.50-0.75	NO	6–B
1.5	1502	WL	11.62	5.50	5.13	30	W	0.50-0.75	NO	6–B
2.0	1503	WL	13.00	5.50	5.13	38	W	0.50-0.75	NO	6–B
3.0	1504	WL	11.50	10.31	7.13	55	W	0.75-1.25	NO	6–C
5.0	1505	WL	14.38	10.31	7.13	75	W	0.75-1.25	NO	6–C
7.5	1506	WL	15.19	13.50	10.84	115	W	0.75-1.25	NO	6–D
10.0	1507	WL	15.19	13.50	10.84	125	W	0.75-1.25	NO	6–D
15.0	1508	Lugs	16.94	14.12	11.59	170	W	1.00-1.50	YES	6–E
25.0	1509	Lugs	18.44	16.13	13.34	250	W	1.00-1.50	YES	6–E
37.5	1510	Lugs	25.48	24.39	19.37	257	F or W①	N/A	YES	58–E
50.0	1511	Lugs	25.48	24.39	19.37	340	F or W①	N/A	YES	17–E
75.0	1512	Lugs	35.40	31.90	26.88	420	F or W①	N/A	YES	17–E

① Wall mounting brackets must be ordered separately for these sizes.

Single-Phase Transformer Data

Group VI – Single Phase

277 PRIMARY VOLTS — 120/240 SECONDARY VOLTS — THREE WINDINGS — 1Ø, 60 Hz

KVA Rating	Catalog #	Terminations: Lugs or Wire Leads (WL)	Approximate Dimensions (inches)			Approximate Shipping Weight (lb)	Type Mtg. W – Wall F - Floor	Knockouts (inches)	Weather Shield Req'd	Wiring Diagram – Enclosure Type
			Height	Width	Depth					
1.0	1601	WL	10.50	5.50	5.13	24	W	0.50-0.75	NO	7–B
1.5	1602	WL	11.62	5.50	5.13	30	W	0.50-0.75	NO	7–B
2.0	1603	WL	13.00	5.50	5.13	38	W	0.50-0.75	NO	7–B
3.0	1604	WL	11.50	10.31	7.13	55	W	0.75-1.25	NO	7–C
5.0	1605	WL	14.38	10.31	7.13	75	W	0.75-1.25	NO	7–C
7.5	1606	WL	15.19	13.50	10.84	115	W	0.75-1.25	NO	7–D
10.0	1607	WL	15.19	13.50	10.84	125	W	0.75-1.25	NO	7–D
15.0	1608	Lugs	16.94	14.12	11.59	170	W	1.00-1.50	YES	7–E
25.0	1609	Lugs	18.44	16.13	13.34	250	W	1.00-1.50	YES	7–E

Group VII – Single Phase

120/208/240/277 PRIMARY VOLTS — 120/240 SECONDARY VOLTS — 1Ø, 60 Hz

KVA Rating	Catalog #	Terminations: Lugs or Wire Leads (WL)	Approximate Dimensions (inches)			Approximate Shipping Weight (lb)	Type Mtg. W – Wall F - Floor	Knockouts (inches)	Weather Shield Req'd	Wiring Diagram – Enclosure Type
			Height	Width	Depth					
1.0	1701	WL	10.50	5.50	5.13	23	W	0.50-0.75	NO	23–B
1.5	1702	WL	11.62	5.50	5.13	30	W	0.50-0.75	NO	23–B
2.0	1703	WL	13.00	5.50	5.13	37	W	0.50-0.75	NO	23–B
3.0	1704	WL	11.50	10.31	7.13	55	W	0.75-1.25	NO	23–C
5.0	1705	WL	14.38	10.31	7.13	75	W	0.75-1.25	NO	23–C
7.5	1706	WL	15.19	13.50	10.84	105	W	0.75-1.25	NO	63–D
10.0	1707	WL	15.19	13.50	10.84	124	W	0.75-1.25	NO	63–D
15.0	1708	Lugs	16.94	14.12	11.59	171	W	1.00-1.50	YES	63–E
25.0	1709	Lugs	18.44	16.13	13.34	261	W	1.00-1.50	YES	63–E

Many other transformers are available. These include those that have two sets of windings and are rated to change 277/480 volts to 208/277 1Ø. For example, if you have available only a 2-wire 277 volt circuit, but a piece of equipment requires 208 volts, you could use this transformer. You could also use this transformer if your system (or circuit) is 480 volts (no neutral) but your lighting needs to operate at 277 volts.

SINGLE-PHASE

CONNECTION DIAGRAMS

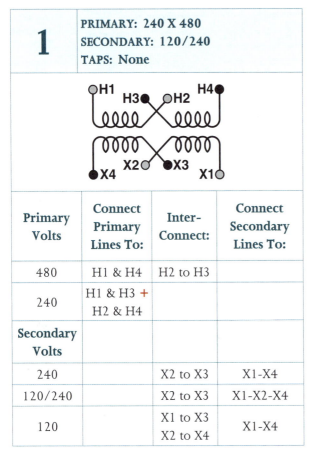

1   PRIMARY: 240 X 480
SECONDARY: 120/240
TAPS: None

Primary Volts	Connect Primary Lines To:	Inter-Connect:	Connect Secondary Lines To:
480	H1 & H4	H2 to H3	
240	H1 & H3 + H2 & H4		
Secondary Volts			
240		X2 to X3	X1-X4
120/240		X2 to X3	X1-X2-X4
120		X1 to X3 X2 to X4	X1-X4

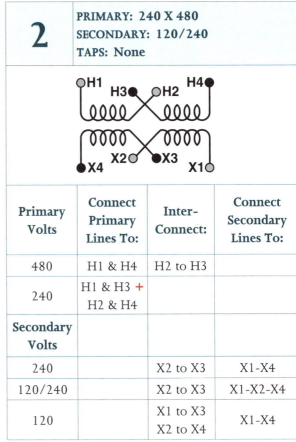

2   PRIMARY: 240 X 480
SECONDARY: 120/240
TAPS: None

Primary Volts	Connect Primary Lines To:	Inter-Connect:	Connect Secondary Lines To:
480	H1 & H4	H2 to H3	
240	H1 & H3 + H2 & H4		
Secondary Volts			
240		X2 to X3	X1-X4
120/240		X2 to X3	X1-X2-X4
120		X1 to X3 X2 to X4	X1-X4

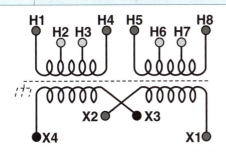

3
PRIMARY: 240 X 480
SECONDARY: 120/240
TAPS: 2-2½% FCAN & 2-2½% FCBN

Primary Volts	Connect Primary Lines To:	Inter-Connect:	Connect Secondary Lines To:
252	H1 & H8	H1 to H5 H4 to H8	
240	H1 & H7	H1 to H5 H3 to H7	
228	H1 & H6	H1 to H5 H2 to H6	
504	H1 & H8	H4 to H5	
492	H1 & H8	H3 to H5	
480	H1 & H7	H3 to H5	
468	H1 & H7	H2 to H5	
456	H1 & H6		
Secondary Volts			
240		X2 to X3	X1-X4
120/240		X2 to X3	X1-X2-X4
120		X1 to X3 X2 to X4	X1-X4

4
PRIMARY: 240 X 480
SECONDARY: 120/240
TAPS: 2-2½% FCAN & 2-2½% FCBN

Primary Volts	Connect Primary Lines To:	Inter-Connect:	Connect Secondary Lines To:
216	H1 & H10	H1 to H9 H10 to H2	
228	H1 & H10	H1 to H8 H10 to H3	
240	H1 & H10	H1 to H7 H10 to H4	
252	H1 & H10	H1 to H6 H10 to H5	
432	H1 & H10	H2 to H9	
444	H1 & H10	H3 to H9	
456	H1 & H10	H3 to H8	
468	H1 & H10	H4 to H8	
480	H1 & H10	H4 to H7	
492	H1 & H10	H5 to H7	
504	H1 & H10	H5 to H5	
Secondary Volts			
240		X2 to X3	X1-X4
120/240		X2 to X3	X1-X2-X4
120		X1 to X3 X2 to X4	X1-X4

5

PRIMARY: 240 X 480
SECONDARY: 120/240
TAPS: 2-2½% FCAN & 2-2½% FCBN

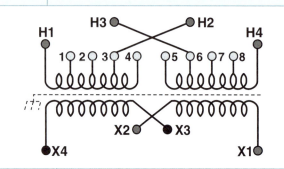

Primary Volts	Connect Primary Lines To:	Inter-Connect:	Connect Secondary Lines To:
216	H1 & H4	H1 to H3 to 8 H2 to H4 to 1	
228	H1 & H4	H1 to H3 to 7 H2 to H4 to 2	
240	H1 & H4	H1 to H3 to 6 H2 to H4 to 3	
252	H1 & H4	H1 to H3 to 5 H2 to H4 to 4	
432	H1 & H4	H2 to 1 & H3 to 8	
444	H1 & H4	H2 to 2 & H3 to 8	
456	H1 & H4	H2 to 2 & H3 to 7	
468	H1 & H4	H2 to 3 & H3 to 7	
480	H1 & H4	H2 to 3 & H3 to 6	
492	H1 & H4	H2 to 4 & H3 to 6	
504	H1 & H4	H2 to 4 & H3 to 5	
Secondary Volts			
240		X2 to X3	X1-X4
120/240		X2 to X3	X1-X2-X4
120		X1 to X3 X2 to X4	X1-X4

6

PRIMARY: 208
SECONDARY: 120/240
TAPS: 2-5% FCBM

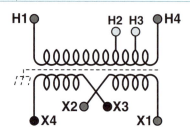

Primary Volts	Connect Primary Lines To:	Inter-Connect:	Connect Secondary Lines To:
208	H1 & H4		
198	H1 & H3		
187	H1 & H2		
Secondary Volts			
240		X2 to X3	X1-X4
120/240		X2 to X3	X1-X2-X4
120		X1 to X3	X1-X4

Single-Phase Transformer Data   Yr-2 Annex B-931

7
PRIMARY: 277
SECONDARY: 120/240
TAPS: 2-5% FCBN

Primary Volts	Connect Primary Lines To:	Inter-Connect:	Connect Secondary Lines To:
277	H1 & H4		
263	H1 & H3		
250	H1 & H2		
Secondary Volts			
240		X2 to X3	X1-X4
120/240		X2 to X3	X1-X2-X4
120		X1 to X3 / X2 to X4	X1-X4

13
PRIMARY: 120 X 240
SECONDARY: 120/240
TAPS: None

Primary Volts	Connect Primary Lines To:	Inter-Connect:	Connect Secondary Lines To:
240	H1 & H4	H2 to H3	
120	H1 & H3	H2 & H4	
Secondary Volts			
240		X2 to X3	X1-X4
120/240		X2 to X3	X1-X2-X4
120		X1 to X3 / X2 to X4	X1-X4

Year Two (Student Manual)

23

PRIMARY: 120/280/240/277
SECONDARY: 120/240

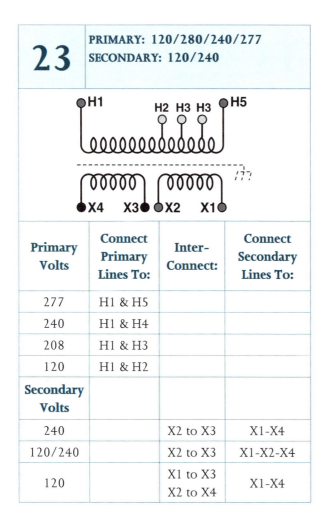

Primary Volts	Connect Primary Lines To:	Inter-Connect:	Connect Secondary Lines To:
277	H1 & H5		
240	H1 & H4		
208	H1 & H3		
120	H1 & H2		
Secondary Volts			
240		X2 to X3	X1-X4
120/240		X2 to X3	X1-X2-X4
120		X1 to X3 X2 to X4	X1-X4

58

PRIMARY: 208
SECONDARY: 120/240
TAPS: 2-5% FCBN

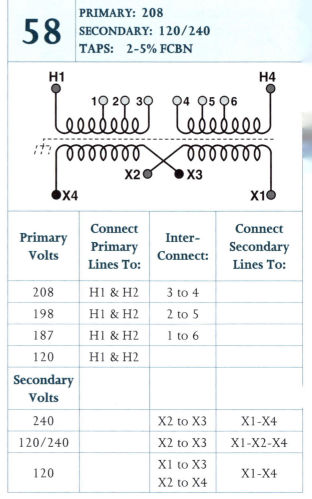

Primary Volts	Connect Primary Lines To:	Inter-Connect:	Connect Secondary Lines To:
208	H1 & H2	3 to 4	
198	H1 & H2	2 to 5	
187	H1 & H2	1 to 6	
120	H1 & H2		
Secondary Volts			
240		X2 to X3	X1-X4
120/240		X2 to X3	X1-X2-X4
120		X1 to X3 X2 to X4	X1-X4

63 — PRIMARY: 240 X 480 / SECONDARY: 120/240

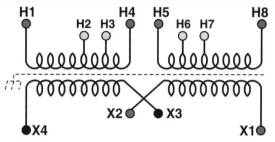

Primary Volts	Connect Primary Lines To:	Inter-Connect:	Connect Secondary Lines To:
120	H1 & H8	H1 to H6 H3 to H8	
208	H1 & H8	H2 to H7	
240	H1 & H8	H3 to H6	
277	H1 & H8	H4 to H5	
Secondary Volts			
240		X2 to X3	X1-X4
120/240		X2 to X3	X1-X2-X4
120		X1 to X3 X2 to X4	X1-X4

Transformer Enclosure Types

Enclosure types shown are for reference only.
Types A, B, C, D, F and I are totally enclosed and encapsulated.

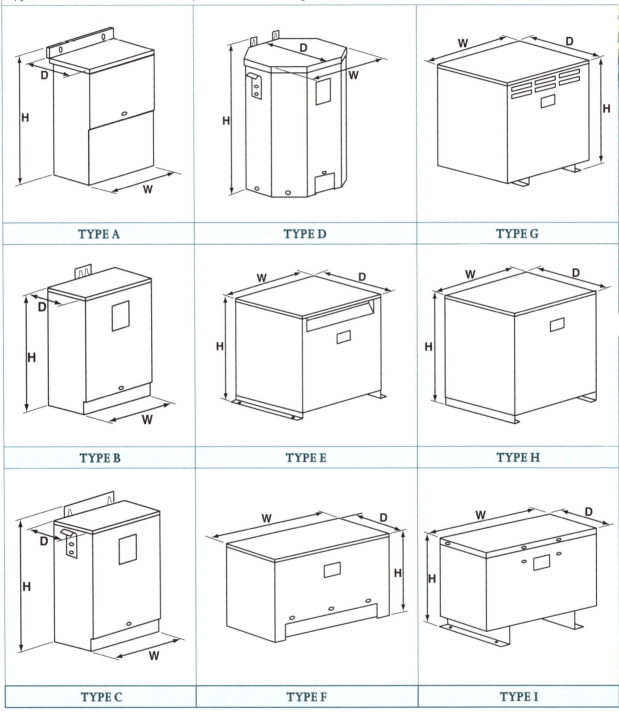

Three-Phase Transformer Data

Group D – Three Phase

480 Delta PRIMARY VOLTS — 208Y/120 SECONDARY VOLTS 3Ø - 60 Hz
may be used on 480Y/277V supply

KVA Rating	Catalog # Terminations: Lugs or Wire Leads (WL)	Approximate Dimensions (inches)			Approximate Shipping Weight (lb)	Type Mtg. W – Wall F - Floor	Knockouts (inches)	Weather Shield Req'd	Wiring Diagram – Enclosure Type
		Height	Width	Depth					
3	3101	WL	10.38	12.37	7.47	75	W	0.75-1.25	NO	21–F
6	3102	WL	11.83	14.17	8.82	140	W	0.75-1.25	NO	21–F
9	3103	WL	14.03	17.77	11.52	180	W	0.75-1.25	NO	21–F
15	3104	Lugs	18.86	20.30	9.03	250	F ①	NA	NO	21–I
25	3105	Lugs	25.48	24.39	19.37	290	F ①	NA	YES	22–E
30	3106	Lugs	25.48	24.39	19.37	290	F ①	NA	YES	22–E
37.5	3107	Lugs	25.48	24.39	19.37	400	F ①	NA	YES	22–E
45	3108	Lugs	25.48	24.39	19.37	400	F ①	NA	YES	22–E
50	3109	Lugs	29.41	28.15	22.37	475	F ①	NA	YES	22–E
75	3110	Lugs	29.41	28.15	22.37	500	F ①	NA	YES	22–E
112.5	3111	Lugs	35.47	31.90	26.88	750	F	NA	YES	22–E
150	3112	Lugs	41.52	32.90	29.87	970	F	NA	YES	22–E
225	3113	Lugs	41.52	32.90	29.87	1200	F	NA	YES	22–E
300	3114	Lugs	45.60	39.50	35.50	1550	F	NA	YES	22–E
500	3115	Lugs	57.80	45.60	41.50	2480	F	NA	YES	22–G
750	3116	Lugs	62.80	54.00	41.50	3600	F	NA	YES	22–G
1000	3117	Lugs	62.80	54.00	41.50	4300	F	NA	YES	80–G

① Wall mounting brackets are available for these sizes. Refer to the brackets shown in Yr-2 Annex B.

Group H – Three Phase

480 Delta PRIMARY VOLTS — 240Δ/120 TAP SECONDARY VOLTS 3Ø - 60 Hz
NON-VENTILATED TRANSFORMER *may be used on 480Y/277V supply*

KVA Rating	Catalog # Terminations: Lugs or Wire Leads (WL)	Approximate Dimensions (inches)			Approximate Shipping Weight (lb)	Type Mtg. W – Wall F - Floor	Knockouts (inches)	Weather Shield Req'd	Wiring Diagram – Enclosure Type
		Height	Width	Depth					
30	3301	Lugs	29.41	28.15	22.37	600	F ①	NA	NO	26–H
45	3302	Lugs	35.47	31.90	26.88	750	F	NA	NO	26–H
75	3303	Lugs	41.52	32.90	29.87	1125	F	NA	NO	26–H
112.5	3304	Lugs	45.59	39.50	35.50	1150	F	NA	NO	26–H

① Wall mounting brackets are available for these sizes. Refer to the brackets shown in Yr-2 Annex B.

Weather Shields and Enclosure Types are shown in Yr-2 Annex B.

Group G – Three Phase

480 Delta PRIMARY VOLTS — 240Δ/120 TAP SECONDARY VOLTS 3Ø - 60 Hz

may be used on 480Y/277V supply

KVA Rating	Catalog #	Terminations: Lugs or Wire Leads (WL)	Approximate Dimensions (inches)			Approximate Shipping Weight (lb)	Type Mtg. W – Wall F - Floor	Knockouts (inches)	Weather Shield Req'd	Wiring Diagram – Enclosure Type
			Height	Width	Depth					
3	3201	WL	10.38	12.37	7.47	75	W	0.75-1.25	NO	25–F
6	3202	WL	11.83	14.17	8.82	140	W	0.75-1.25	NO	25–F
9	3203	WL	14.03	17.77	11.52	180	W	0.75-1.25	NO	25–F
15	3204	Lugs	18.86	20.30	9.03	250	F ①	NA	NO	25–I
30	3205	Lugs	25.50	24.39	19.37	325	F ①	NA	YES	26–E
45	3206	Lugs	25.50	24.39	19.37	400	F ①	NA	YES	26–E
75	3207	Lugs	29.41	28.15	22.37	500	F ①	NA	YES	26–E
112.5	3208	Lugs	35.47	31.90	26.88	750	F	NA	YES	26–E
150	3209	Lugs	41.52	32.90	29.87	1125	F	NA	YES	26–E
225	3210	Lugs	41.52	32.90	29.87	1200	F	NA	YES	26–E
300	3211	Lugs	45.60	39.50	35.50	1550	F	NA	YES	26–G
500	3212	Lugs	62.00	54.00	42.00	2675	F	NA	YES	27–G
750	3213	Lugs	62.80	54.00	41.50	3406	F	NA	YES	26–G

① Wall mounting brackets are available for these sizes. Refer to the brackets shown in Yr-2 Annex B.

Group H1 – Three Phase

480 Delta PRIMARY VOLTS — 480Y/277 TAP SECONDARY VOLTS 3Ø - 60 Hz

may be used on 480Y/277V supply

KVA Rating	Catalog #	Terminations: Lugs or Wire Leads (WL)	Approximate Dimensions (inches)			Approximate Shipping Weight (lb)	Type Mtg. W – Wall F - Floor	Knockouts (inches)	Weather Shield Req'd	Wiring Diagram – Enclosure Type
			Height	Width	Depth					
15	3401	Lugs	18.86	20.30	9.03	250	F ①	NA	NO	31–I
30	3402	Lugs	25.50	24.39	19.37	325	F ①	NA	YES	31–E
45	3403	Lugs	25.50	24.39	19.37	400	F ①	NA	YES	31–E
75	3404	Lugs	29.41	28.15	22.37	600	F ①	NA	YES	31–E
112.5	3405	Lugs	35.47	31.90	26.88	710	F	NA	YES	31–E
150	3406	Lugs	41.52	32.90	29.87	1155	F	NA	YES	31–E
225	3407	Lugs	41.52	32.90	29.87	1210	F	NA	YES	31–E
300	3408	Lugs	45.60	39.50	35.50	1600	F	NA	YES	31–E
500	3409	Lugs	62.00	54.00	42.00	2620	F	NA	YES	32–G

① Wall mounting brackets are available for these sizes. Refer to the brackets shown in Yr-2 Annex B.

Weather Shields and Enclosure Types are shown in Yr-2 Annex B.

Three-Phase Connection Diagrams

21 PRIMARY: 480 Volts Delta SECONDARY: 208Y/120 Volts TAPS: 2-5% FCBN

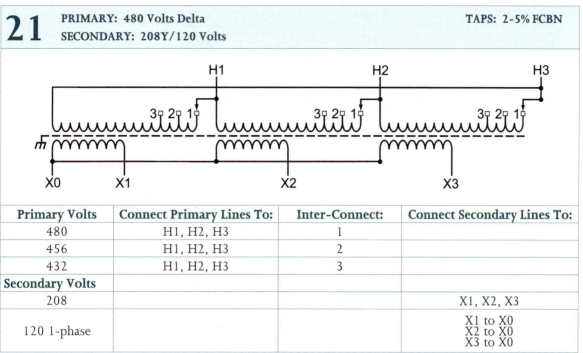

Primary Volts	Connect Primary Lines To:	Inter-Connect:	Connect Secondary Lines To:
480	H1, H2, H3	1	
456	H1, H2, H3	2	
432	H1, H2, H3	3	
Secondary Volts			
208			X1, X2, X3
120 1-phase			X1 to X0 X2 to X0 X3 to X0

22 PRIMARY: 480 Volts Delta SECONDARY: 208Y/120 Volts TAPS: 2-2½% FCAN & 4-2½% FCBN

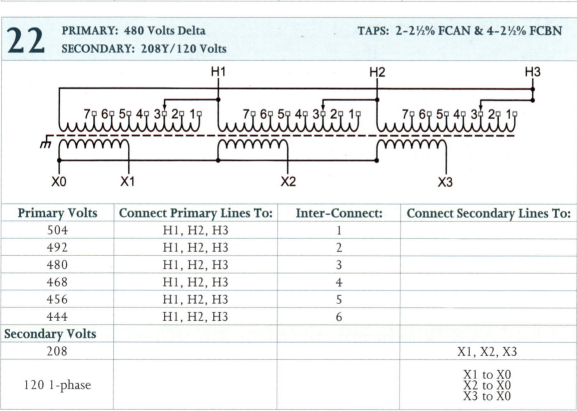

Primary Volts	Connect Primary Lines To:	Inter-Connect:	Connect Secondary Lines To:
504	H1, H2, H3	1	
492	H1, H2, H3	2	
480	H1, H2, H3	3	
468	H1, H2, H3	4	
456	H1, H2, H3	5	
444	H1, H2, H3	6	
Secondary Volts			
208			X1, X2, X3
120 1-phase			X1 to X0 X2 to X0 X3 to X0

25

PRIMARY: 480 Volts Delta
SECONDARY: 208Y/120 Volts
TAPS: 2-5% FCBN

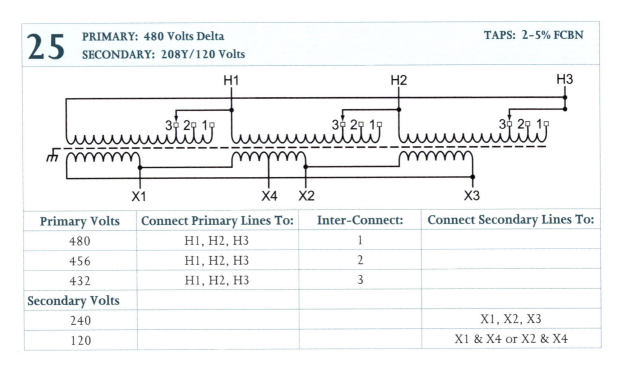

Primary Volts	Connect Primary Lines To:	Inter-Connect:	Connect Secondary Lines To:
480	H1, H2, H3	1	
456	H1, H2, H3	2	
432	H1, H2, H3	3	
Secondary Volts			
240			X1, X2, X3
120			X1 & X4 or X2 & X4

26

PRIMARY: 480 Volts Delta
SECONDARY: 240 Volts Delta
TAPS: 2-2½% FCAN & 4-2½% FCBN

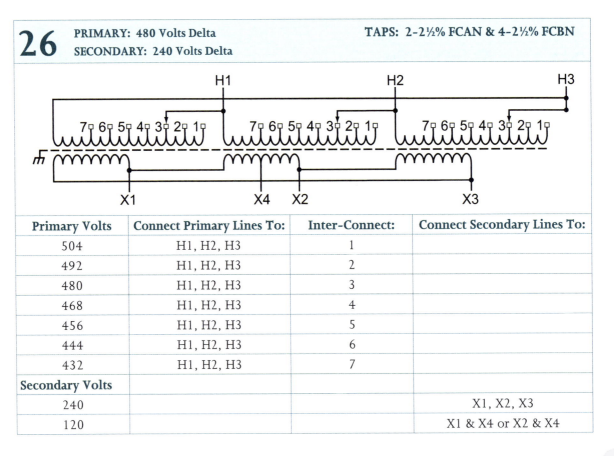

Primary Volts	Connect Primary Lines To:	Inter-Connect:	Connect Secondary Lines To:
504	H1, H2, H3	1	
492	H1, H2, H3	2	
480	H1, H2, H3	3	
468	H1, H2, H3	4	
456	H1, H2, H3	5	
444	H1, H2, H3	6	
432	H1, H2, H3	7	
Secondary Volts			
240			X1, X2, X3
120			X1 & X4 or X2 & X4

27

PRIMARY: 480 Volts Delta
SECONDARY: 240 Volts Delta
TAPS: 2-2½% FCAN & 4-2½% FCBN

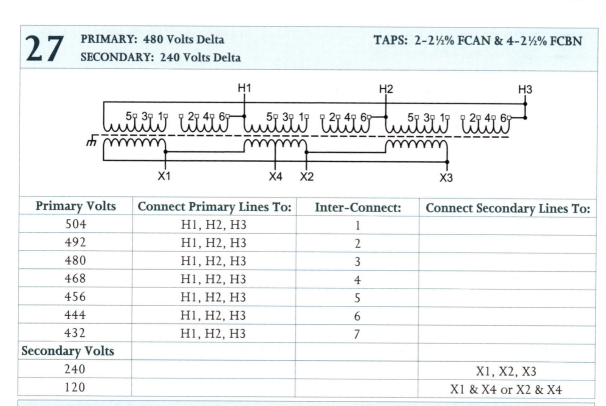

Primary Volts	Connect Primary Lines To:	Inter-Connect:	Connect Secondary Lines To:
504	H1, H2, H3	1	
492	H1, H2, H3	2	
480	H1, H2, H3	3	
468	H1, H2, H3	4	
456	H1, H2, H3	5	
444	H1, H2, H3	6	
432	H1, H2, H3	7	
Secondary Volts			
240			X1, X2, X3
120			X1 & X4 or X2 & X4

31

PRIMARY: 480 Volts Delta
SECONDARY: 480Y/277 Volts
TAPS: 2-2½% FCAN & 4-2½% FCBN

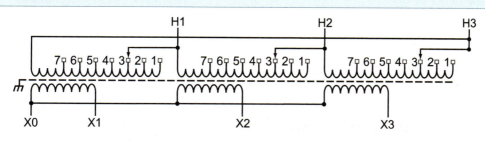

Primary Volts	Connect Primary Lines To:	Inter-Connect:	Connect Secondary Lines To:
504	H1, H2, H3	1	
492	H1, H2, H3	2	
480	H1, H2, H3	3	
468	H1, H2, H3	4	
456	H1, H2, H3	5	
444	H1, H2, H3	6	
432	H1, H2, H3	7	
Secondary Volts			
480			X1, X2, X3
277 1-phase			X1 to X0 X2 to X0 X3 to X0

32   PRIMARY: 480 Volts Delta   TAPS: 2-2½% FCAN & 4-2½% FCBN
SECONDARY: 480Y/277 Volts

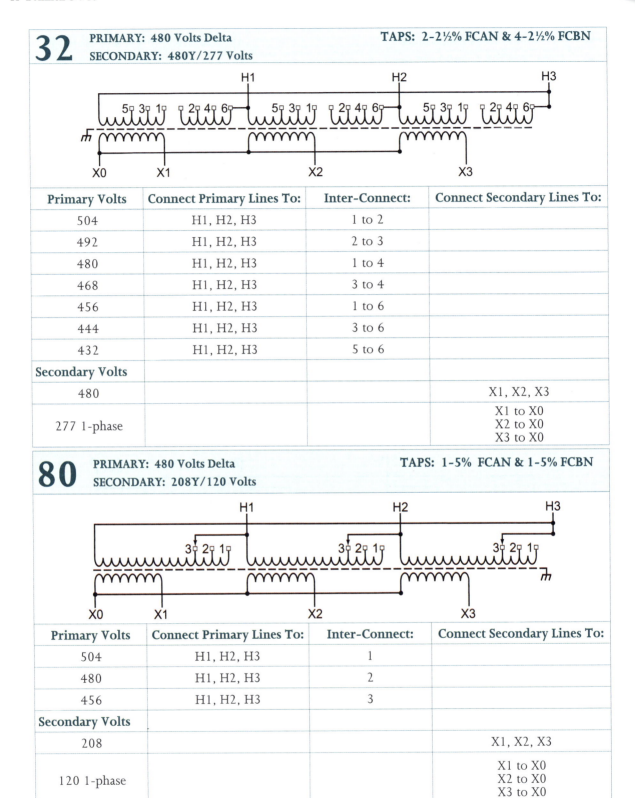

Primary Volts	Connect Primary Lines To:	Inter-Connect:	Connect Secondary Lines To:
504	H1, H2, H3	1 to 2	
492	H1, H2, H3	2 to 3	
480	H1, H2, H3	1 to 4	
468	H1, H2, H3	3 to 4	
456	H1, H2, H3	1 to 6	
444	H1, H2, H3	3 to 6	
432	H1, H2, H3	5 to 6	
Secondary Volts			
480			X1, X2, X3
277 1-phase			X1 to X0 X2 to X0 X3 to X0

80   PRIMARY: 480 Volts Delta   TAPS: 1-5% FCAN & 1-5% FCBN
SECONDARY: 208Y/120 Volts

Primary Volts	Connect Primary Lines To:	Inter-Connect:	Connect Secondary Lines To:
504	H1, H2, H3	1	
480	H1, H2, H3	2	
456	H1, H2, H3	3	
Secondary Volts			
208			X1, X2, X3
120 1-phase			X1 to X0 X2 to X0 X3 to X0

Yr-2 – Annex D – Buck-Boost Transformers

Included in this annex are the following:

- Enclosure Type Illustrations
- Specification and Dimension Tables
- Selection Tables for Single-Phase Installations
- Insulating Xfmr Connection Diagrams
- Single-Phase Autotransformer Connection Diagrams
- Selection Tables for Three-Phase Installations
- Three-Phase Autotransformer Connection Diagrams
- 3Ø, 4-Wire WYE-WYE Selection Table
- 3Ø, 4-Wire WYE-WYE Autotransformer Connection Diagrams

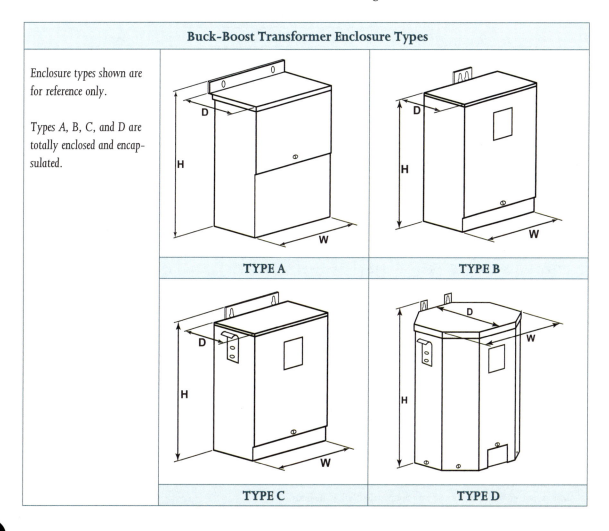

Buck-Boost Transformer Enclosure Types

Enclosure types shown are for reference only.

Types A, B, C, and D are totally enclosed and encapsulated.

TYPE A   TYPE B   TYPE C   TYPE D

SPECIFICATIONS & DIMENSIONS
GROUP I — 120 x 240 PRIMARY VOLTS – 12/24 SECONDARY VOLTS – 60 Hz

Catalog Number	Insulating XFMR KVA Rating	Secondary Maximum Current Output		Approximate Dimensions Inches (cm)			Approx. Net Wt. Lbs (kg)	Enclosure Type
		at 12 V	at 24 V	Height	Width	Depth		
11101	0.050	4.16	2.08	6.41 (16.3)	3.14 (8.0)	3.05 (7.7)	4 (1.8)	A
11102	0.100	8.32	4.16	7.16 (18.2)	3.89 (9.9)	3.67 (9.3)	5 (2.3)	A
11103	0.150	12.52	6.25	7.16 (18.2)	3.89 (9.9)	3.67 (9.3)	7 (3.2)	A
11104	0.250	20.80	10.40	8.68 (22.0)	4.08 (10.4)	3.88 (9.9)	10 (4.5)	B
11105	0.500	41.60	20.80	9.06 (23.0)	4.37 (11.1)	4.20 (10.7)	15 (6.8)	B
11106	0.750	62.50	31.25	9.68 (24.6)	4.75 (12.1)	4.51 (11.5)	19 (8.6)	B
11107	1.0	83.20	41.60	10.50 (26.7)	5.50 (14.0)	5.13 (13.0)	24 (10.9)	B
11108	1.5	125.00	62.50	11.62 (29.5)	5.50 (14.0)	5.13 (13.0)	30 (13.6)	B
11109	2.0	166.00	83.20	13.00 (33.0)	5.50 (14.0)	5.13 (13.0)	38 (17.2)	B
11110	3.0	250.00	125.00	11.50 (29.2)	10.31 (26.2)	7.13 (18.1)	55 (24.9)	C
11111	5.0	416.60	208.00	14.38 (36.5)	10.31 (26.2)	7.13 (18.1)	75 (34.0)	C
11112*	7.5	625.00	312.50	20.81 (52.9)	11.12 (28.2)	10.84 (27.5)	125 (56.7)	D
11113*	10.0	833.00	416.60	20.81 (52.9)	11.75 (29.8)	11.59 (29.4)	160 (72.6)	D

All units have ground studs for use with nonmetallic wiring methods.

SPECIFICATIONS & DIMENSIONS
GROUP II — 120 x 240 PRIMARY VOLTS – 16/32 SECONDARY VOLTS – 60 Hz

Catalog Number	Insulating XFMR KVA Rating	Secondary Maximum Current Output		Approximate Dimensions Inches (cm)			Approx. Net Wt. Lbs (kg)	Enclosure Type
		at 16 V	at 32 V	Height	Width	Depth		
21101	0.050	3.12	1.56	6.41 (16.3)	3.14 (8.0)	3.05 (7.7)	4 (1.8)	A
21102	0.100	6.25	3.12	7.16 (18.2)	3.89 (9.9)	3.67 (9.3)	5 (2.3)	A
21103	0.150	9.38	4.69	7.16 (18.2)	3.89 (9.9)	3.67 (9.3)	7 (3.2)	A
21104	0.250	15.60	7.80	8.68 (22.0)	4.08 (10.4)	3.88 (9.9)	10 (4.5)	B
21105	0.500	31.20	15.60	9.06 (23.0)	4.37 (11.1)	4.20 (10.7)	15 (6.8)	B
21106	0.750	46.90	23.40	9.68 (24.6)	4.75 (12.1)	4.51 (11.5)	19 (8.6)	B
21107	1.0	62.50	31.20	10.50 (26.7)	5.50 (14.0)	5.13 (13.0)	24 (10.9)	B
21108	1.5	93.70	46.90	11.62 (29.5)	5.50 (14.0)	5.13 (13.0)	30 (13.6)	B
21109	2.0	125.00	62.50	13.00 (33.0)	5.50 (14.0)	5.13 (13.0)	38 (17.2)	B
21110	3.0	187.50	93.80	11.50 (29.2)	10.31 (26.2)	7.13 (18.1)	55 (24.9)	C
21111	5.0	312.00	156.00	14.38 (36.5)	10.31 (26.2)	7.13 (18.1)	75 (34.0)	C
21112*	7.5	468.00	234.00	20.81 (52.9)	11.12 (28.2)	10.84 (27.5)	125 (56.7)	D
21113*	10.0	625.00	312.00	20.81 (52.9)	11.75 (29.8)	10.84 (27.5)	160 (72.6)	D

All units have ground studs for use with nonmetallic wiring methods.

SPECIFICATIONS & DIMENSIONS

GROUP III — 240 x 480 PRIMARY VOLTS — 24/48 SECONDARY VOLTS — 60 Hz

Catalog Number	Insulating XFMR KVA Rating	Secondary Maximum Current Output		Approximate Dimensions Inches (cm)			Approx. Net Wt. Lbs (kg)	Enclosure Type
		at 16 V	at 32 V	Height	Width	Depth		
31101	0.050	2.08	1.04	6.41 (16.3)	3.14 (8.0)	3.05 (7.7)	4 (1.8)	A
31102	0.100	4.16	2.08	7.16 (18.2)	3.89 (9.9)	3.67 (9.3)	5 (2.3)	A
31103	0.150	6.24	3.12	7.16 (18.2)	3.89 (9.9)	3.67 (9.3)	7 (3.2)	A
31104	0.250	10.40	5.20	8.68 (22.0)	4.08 (10.4)	3.88 (9.9)	10 (4.5)	B
31105	0.500	20.80	10.40	9.06 (23.0)	4.37 (11.1)	4.20 (10.7)	15 (6.8)	B
31106	0.750	31.20	15.60	9.68 (24.6)	4.75 (12.1)	4.51 (11.5)	19 (8.6)	B
31107	1.0	41.60	20.80	10.50 (26.7)	5.50 (14.0)	5.13 (13.0)	24 (10.9)	B
31108	1.5	62.40	31.20	11.62 (29.5)	5.50 (14.0)	5.13 (13.0)	30 (13.6)	B
31109	2.0	83.20	41.60	13.00 (33.0)	5.50 (14.0)	5.13 (13.0)	38 (17.2)	B
31110	3.0	125.00	62.50	11.50 (29.2)	10.31 (26.2)	7.13 (18.1)	55 (24.9)	C
31111	5.0	208.00	104.00	14.38 (36.5)	10.31 (26.2)	7.13 (18.1)	75 (34.0)	C
31112	7.5	312.00	156.00	20.81 (52.9)	11.12 (28.2)	10.84 (27.5)	135 (61.2)	D
31113*	10.0	416.00	208.00	20.81 (52.9)	11.75 (29.4)	11.59 (29.4)	160 (72.6)	D

All units have ground studs for use with nonmetallic wiring methods.

GROUP I		
120 × 240 V Input — 12/24 V Output		
INPUT	OUTPUT	Connection Diagram
120	12	K
120	24	L
240	12	M
240	24	N

GROUP II		
120 × 240 V Input — 16/32 V Output		
INPUT	OUTPUT	Connection Diagram
120	16	K
120	32	L
240	16	M
240	32	N

GROUP III		
120 × 240 V Input — 24/48 V Output		
INPUT	OUTPUT	Connection Diagram
240	24	K
240	48	L
480	24	M
480	48	N

BUCK-BOOST Insulating Transformer Selection Tables

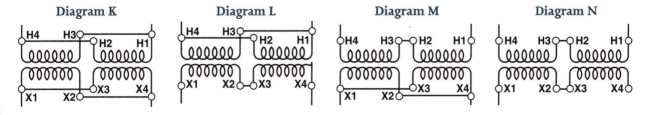

Diagram K   Diagram L   Diagram M   Diagram N

Yr-2 Annex D-944

INSULATING TRANSFORMER – CONNECTION DIAGRAMS

Selection Table BB1-1 — SINGLE-PHASE
GROUP I — 120 x 240 PRIMARY VOLTS – 12/24 SECONDARY VOLTS

SINGLE PHASE			BOOSTING								BUCKING					
Available Line Voltage (input)			95	100	105	110	189	208	215	220	125	132	230	245	250	252
Load Voltage (output)			114	120	115	120	208	230	237	242	113	120	208	222	227	240
Cat #																
11101	LOAD	KVA	0.24	0.25	0.48	0.50	0.43	0.48	0.49	0.50	0.52	0.54	0.47	0.50	0.52	1.02
		Amps	2.08	2.08	4.17	4.17	2.08	2.08	2.08	2.08	4.60	4.60	2.28	2.28	2.28	4.37
.050 KVA	Max OCPD		6	6	10	10	6	6	6	6	10	10	6	6	6	10
11102	LOAD	KVA	0.47	0.50	0.96	1.01	0.87	0.96	0.99	1.01	1.04	1.08	0.95	1.00	1.04	2.04
		Amps	4.17	4.17	8.33	8.33	4.17	4.17	4.17	4.17	9.20	9.20	4.56	4.56	4.58	8.75
.100 KVA	Max OCPD		10	10	15	15	10	10	10	10	15	15	10	10	10	15
11103	LOAD	KVA	0.71	0.75	1.43	1.51	1.30	1.43	1.48	1.51	1.56	1.62	1.42	1.50	1.56	3.00
		Amps	6.25	6.25	12.5	12.5	6.25	6.25	6.25	6.25	13.8	13.8	6.86	6.86	6.86	13.1
.150 KVA	Max OCPD		15	15	20	20	15	15	15	15	20	20	15	15	15	15
11104	LOAD	KVA	1.19	1.25	2.40	2.50	2.16	2.39	2.46	2.52	2.60	2.75	2.37	2.50	2.60	5.10
		Amps	10.4	10.4	20.8	20.8	10.4	10.4	10.4	10.4	22.8	22.8	11.4	11.4	11.4	21.8
.250 KVA	Max OCPD		25	25	40	30	15	15	15	15	30	30	15	15	15	30
11105	LOAD	KVA	2.37	2.50	4.80	5.00	4.33	4.79	4.93	5.04	5.20	5.40	4.47	5.00	5.20	10.2
		Amps	20.8	20.8	41.7	41.7	20.8	20.8	20.8	20.8	46.8	46.8	22.8	22.8	22.8	43.7
.500 KVA	Max OCPD		35	35	60	60	30	30	30	30	60	60	30	30	30	60
11106	LOAD	KVA	3.56	3.75	7.17	7.56	6.50	7.19	7.41	7.56	7.80	8.15	7.10	7.50	7.80	15.3
		Amps	31.3	31.3	83.3	83.3	31.3	31.3	31.3	31.3	68.5	69.5	34.4	34.4	34.4	65.5
.750 KVA	Max OCPD		50	50	90	90	45	45	45	45	80	80	40	40	40	80
11107	LOAD	KVA	4.75	5.00	9.58	10.0	8.66	9.58	9.87	10.0	10.4	10.8	9.5	10.0	10.0	20.4
		Amps	41.7	41.7	83.3	83.3	41.7	41.7	41.7	41.7	91.5	91.5	45.8	45.8	45.8	87.5
1.0 KVA	Max OCPD		70	70	125	125	60	60	60	60	110	110	60	60	50	110
11108	LOAD	KVA	7.12	7.5	14.4	15.1	13.0	14.3	14.8	15.1	15.0	16.2	14.2	15.0	15.6	30.6
		Amps	62.5	62.5	125	125	62.5	62.5	62.5	62.5	138	138	68.6	68.6	68.6	132
1.5 KVA	Max OCPD		100	100	175	175	90	90	90	90	150	175	80	80	80	175
11109	LOAD	KVA	9.5	10.0	19.2	20.2	17.3	19.2	19.7	20.1	20.8	21.6	19.0	20.0	20.3	40.8
		Amps	83.3	83.3	167	167	83.3	83.3	83.3	83.3	183	183	91.6	91.6	91.2	175
2.0 KVA	Max OCPD		125	125	250	250	125	125	125	125	225	225	110	110	110	225
11110	LOAD	KVA	14.2	15.0	28.8	30.0	26.0	28.7	29.6	30.3	31.2	32.5	28.5	30.0	31.2	61.0
		Amps	125	125	250	250	125	125	125	125	275	275	137	137	137	263
3.0 KVA	Max OCPD		200	200	350	350	175	175	175	175	350	350	175	175	175	350
11111	LOAD	KVA	23.7	25.0	47.9	50.0	43.3	47.8	49.3	50.3	52.0	54.0	47.4	50.0	52.0	102
		Amps	208	208	417	417	208	208	208	208	457	457	228	228	228	437
5.0 KVA	Max OCPD		350	350	600	600	300	300	300	300	600	600	300	300	300	600
11112*	LOAD	KVA	35.6	37.5	71.9	75.6	65.0	71.8	74.0	75.6	78.0	81.0	71.0	76.0	78.0	153
		Amps	313	313	625	625	313	313	313	313	688	688	344	344	344	655
7.5 KVA	Max OCPD		500	500	1000	1000	450	450	450	450	800	800	400	400	400	800
11113*	LOAD	KVA	47.5	50.0	95.8	100	86.6	95.8	98.7	101	104	108	95.0	100	104	204
		Amps	417	417	833	833	417	417	417	417	915	915	458	458	458	875
10 KVA	Max OCPD		700	700	1200	1200	600	600	600	600	1200	1200	600	600	600	1200
CONNECTION DIAGRAM			D	D	C	C	H	H	H	H	F	F	I	I	I	E

* These larger KVA transformers have two X1, two X2, two X3, and two X4 leads.

Year Two (Student Manual)

Selection Table BB1-2 — SINGLE-PHASE
GROUP II — 120 x 240 PRIMARY VOLTS – 16/32 SECONDARY VOLTS

SINGLE PHASE			BOOSTING							BUCKING					
Available Line Voltage (input)			95	100	105	208	215	215	220	225	135	240	240	245	250	255
Load Voltage (output)			120	114	119	240	244	230	235	240	119	208	225	230	234	239
Cat #																
21101	LOAD	KVA	0.19	0.36	0.37	0.38	0.38	0.72	0.73	0.75	0.52	0.54	0.47	0.50	0.52	1.02
		Amps	1.56	3.13	3.13	1.56	1.56	3.13	3.13	3.13	3.54	1.77	3.33	3.33	3.33	3.33
.050 KVA		Max OCPD	6	6	6	6	6	6	6	6	6	6	6	6	6	6
21102	LOAD	KVA	0.38	0.71	0.74	0.75	0.76	1.44	1.47	1.50	0.84	0.74	1.50	1.53	1.56	1.59
		Amps	3.13	6.25	6.25	3.13	3.13	6.25	6.25	6.25	7.08	3.54	6.67	6.67	6.67	6.67
.100 KVA		Max OCPD	10	10	15	15	10	10	10	10	15	15	10	10	10	15
21103	LOAD	KVA	0.56	1.07	1.12	1.13	1.14	2.16	2.20	2.25	1.26	1.11	2.25	2.30	2.34	2.39
		Amps	4.69	9.38	9.38	4.69	4.69	9.38	9.38	9.38	10.6	5.31	10.0	10.0	10.0	10.0
.150 KVA		Max OCPD	10	15	15	10	10	15	15	15	15	6	15	15	15	15
21104	LOAD	KVA	0.94	1.78	1.86	1.88	1.91	3.59	3.67	3.75	2.11	1.84	3.75	3.83	3.90	3.98
		Amps	7.81	15.6	15.6	7.81	15.6	15.6	15.6	15.6	17.7	8.85	16.7	16.7	16.7	16.7
.250 KVA		Max OCPD	15	25	25	15	15	25	25	25	20	15	20	20	20	20
21105	LOAD	KVA	1.88	3.56	3.72	3.75	3.81	7.19	7.34	7.50	4.21	3.68	7.50	7.67	7.80	7.97
		Amps	15.6	31.3	31.3	15.6	15.6	31.3	31.3	31.3	35.2	17.7	33.3	33.3	33.3	33.3
.500 KVA		Max OCPD	25	45	45	25	25	45	45	45	40	20	40	40	40	40
21106	LOAD	KVA	2.81	5.34	5.58	5.63	5.72	10.8	11.0	11.3	6.32	5.53	11.3	11.5	11.7	11.9
		Amps	23.4	46.9	46.9	23.4	23.4	46.9	46.9	46.9	53.2	26.6	50.0	50.0	50.0	50.0
.750 KVA		Max OCPD	40	70	70	40	40	70	70	70	60	30	60	60	60	60
21107	LOAD	KVA	3.75	7.13	7.44	7.50	7.63	14.4	14.7	15.0	8.43	7.37	15.0	15.3	15.6	15.9
		Amps	31.3	62.5	62.5	31.3	31.3	62.5	62.5	62.5	70.8	35.4	66.7	66.7	66.7	66.7
1.0 KVA		Max OCPD	50	90	90	50	50	90	90	90	80	40	80	80	80	80
21108	LOAD	KVA	5.63	10.7	11.2	11.3	11.4	21.6	22.0	22.5	12.6	11.1	22.5	23.0	23.4	23.9
		Amps	46.9	93.8	93.8	46.9	46.9	93.8	93.8	93.8	106	53.1	100	100	100	100
1.5 KVA		Max OCPD	80	150	150	70	70	125	125	125	125	60	125	125	125	125
21109	LOAD	KVA	7.5	14.3	14.9	15.0	15.3	28.8	29.4	30	16.9	14.7	30	30.7	31.2	31.7
		Amps	62.5	125	125	62.5	62.5	125	125	125	142	70.8	133	133	133	133
2.0 KVA		Max OCPD	100	200	200	90	90	175	175	175	175	80	175	175	175	175
21110	LOAD	KVA	11.3	21.9	22.3	22.5	22.9	43.1	44.1	45	25.3	22.1	45	46	46.8	47.8
		Amps	93.8	188	188	93.8	93.8	188	188	188	213	106	200	200	200	200
3.0 KVA		Max OCPD	150	300	300	150	150	250	250	250	250	125	250	250	250	250
21111	LOAD	KVA	18.8	35.6	37.2	37.5	38.1	71.9	73.4	75	42.2	36.8	75	76.7	78	79.7
		Amps	156	313	313	156	156	313	313	313	354	177	333	333	333	333
5.0 KVA		Max OCPD	250	450	450	225	225	450	450	450	400	200	400	400	400	400
21112*	LOAD	KVA	28.1	53.4	55.8	56.3	57.2	108	110	113	63.2	55.3	113	115	117	120
		Amps	234	469	469	234	234	469	469	469	513	257	500	500	500	500
7.5 KVA		Max OCPD	400	700	700	350	350	700	700	700	600	300	600	600	600	600
21113*	LOAD	KVA	37.5	71.3	74.4	75	76.3	144	147	150	84.3	73.7	150	153	156	159
		Amps	313	625	625	313	313	625	625	625	708	354	667	667	667	667
10 KVA		Max OCPD	500	1000	1000	450	450	1000	1000	1000	800	400	800	800	800	800
CONNECTION DIAGRAM			D	D	C	C	H	H	H	H	F	F	I	I	I	E

* These larger KVA transformers have two X1, two X2, two X3, and two X4 leads.

Selection Table BB1-3 — SINGLE-PHASE
GROUP III — 240 x 480 PRIMARY VOLTS – 24/48 SECONDARY VOLTS

SINGLE PHASE			BOOSTING										BUCKING			
Available Line Voltage (input)			230	380	416	425	430	435	440	440	450	460	277	480	480	504
Load Voltage (output)			277	420	457	467	473	457	462	484	472	483	230	436	456	480
Cat #																
31101	LOAD	KVA	0.29	0.44	0.48	0.49	0.49	0.95	0.96	0.50	0.98	1.01	0.29	0.50	1.05	1.10
		Amps	1.04	1.04	1.04	1.04	1.04	2.08	2.08	1.04	2.08	2.08	1.25	1.15	2.29	2.29
.050 KVA		Max OCPD	3	3	3	3	3	6	6	3	6	6	3	3	6	6
31102	LOAD	KVA	0.58	0.87	0.95	0.97	0.99	1.90	1.93	1.01	1.97	2.01	0.58	1.00	2.09	2.20
		Amps	2.08	2.08	2.08	2.08	2.08	4.17	4.17	2.08	4.17	4.17	2.50	2.29	4.58	4.58
.100 KVA		Max OCPD	6	6	6	6	6	10	10	6	10	10	6	6	10	10
31103	LOAD	KVA	0.87	1.31	1.43	1.46	1.48	2.86	2.89	1.51	2.95	3.02	0.86	1.50	3.14	3.30
		Amps	3.13	3.13	3.13	3.13	3.13	6.25	6.25	3.13	6.25	6.25	3.75	3.44	6.88	6.88
.150 KVA		Max OCPD	10	6	6	6	6	15	15	6	15	15	6	6	15	15
31104	LOAD	KVA	1.44	2.19	2.38	2.43	2.46	4.76	4.81	2.52	4.92	5.03	1.44	2.50	5.23	5.50
		Amps	5.21	5.21	5.21	5.21	5.21	10.4	10.4	5.21	10.4	10.4	6.25	5.73	11.5	11.5
.250 KVA		Max OCPD	15	10	10	10	10	15	15	10	15	15	10	10	15	15
31105	LOAD	KVA	2.89	4.38	4.76	4.86	4.93	9.52	9.62	5.04	9.83	10.1	2.88	5.0	10.5	11.0
		Amps	10.4	10.4	10.4	10.4	10.4	20.8	20.8	10.4	20.8	20.8	12.5	11.5	22.9	22.9
.500 KVA		Max OCPD	20	15	15	15	15	30	30	15	30	30	15	15	30	30
31106	LOAD	KVA	4.33	6.56	7.14	7.3	7.39	14.3	14.4	7.56	14.8	15.1	4.31	7.49	15.7	16.5
		Amps	15.6	15.6	15.6	15.6	15.6	31.3	31.3	15.6	31.3	31.3	18.8	17.2	34.4	34.4
.750 KVA		Max OCPD	25	25	25	25	25	45	45	25	45	45	20	20	45	45
31107	LOAD	KVA	5.77	8.57	9.52	9.73	9.85	19.0	19.3	10.1	19.7	20.1	5.75	9.99	20.9	22
		Amps	20.8	20.8	20.8	20.8	20.8	41.7	41.7	20.8	41.7	41.7	25	22.9	45.8	45.8
1.0 KVA		Max OCPD	35	30	30	30	30	60	60	30	60	60	30	30	60	60
31108	LOAD	KVA	8.66	13.1	14.3	14.6	14.8	28.6	28.9	15.1	29.5	30.2	8.63	15.0	31.4	33
		Amps	31.3	31.3	31.3	31.3	31.3	62.5	62.5	31.3	62.5	62.5	37.5	34.4	68.8	68.8
1.5 KVA		Max OCPD	50	50	45	45	45	90	90	45	90	90	40	40	90	90
31109	LOAD	KVA	11.5	17.5	19.0	19.5	19.7	38.1	38.5	20.2	39.3	40.3	11.5	20.0	41.8	44
		Amps	41.7	41.7	41.7	41.7	41.7	83.3	83.3	41.7	83.3	83.3	50	45.8	91.7	91.7
2.0 KVA		Max OCPD	70	60	60	60	60	110	110	60	110	110	60	60	110	110
31110	LOAD	KVA	17.3	26.3	28.6	29.2	29.6	57.1	57.8	30.3	59	60.4	17.3	30.0	62.7	66
		Amps	62.5	62.5	62.5	62.5	62.5	125	125	62.5	125	125	75	68.8	138	138
3.0 KVA		Max OCPD	100	90	90	90	90	175	175	90	175	175	80	80	175	175
31111	LOAD	KVA	28.9	43.8	47.6	48.6	49.3	95.2	96.2	50.4	98.3	101	28.8	50	105	110
		Amps	104	104	104	104	104	208	208	104	208	208	125	115	229	229
5.0 KVA		Max OCPD	175	150	150	150	150	300	300	150	300	300	150	150	300	300
31112	LOAD	KVA	43.3	65.6	71.4	73.0	73.9	143	144	75.6	148	151	43.1	74.9	157	165
		Amps	156	156	156	156	156	313	313	156	313	313	250	229	458	458
7.5 KVA		Max OCPD	250	225	225	225	225	450	450	225	450	450	200	200	450	450
31113*	LOAD	KVA	57.7	87.5	95.2	97.3	98.5	190	193	101	197	201	57.5	99.9	209	220
		Amps	208	208	208	208	208	417	417	208	417	417	250	229	458	458
10 KVA		Max OCPD	350	300	300	300	300	600	600	300	600	600	300	300	600	600
	CONNECTION DIAGRAM		D	H	H	H	H	G	G	H	G	G	J	I	E	E

* This larger KVA transformer has two X1, two X2, two X3, and two X4 leads.

Single-Phase Autotransformer CONNECTION DIAGRAMS

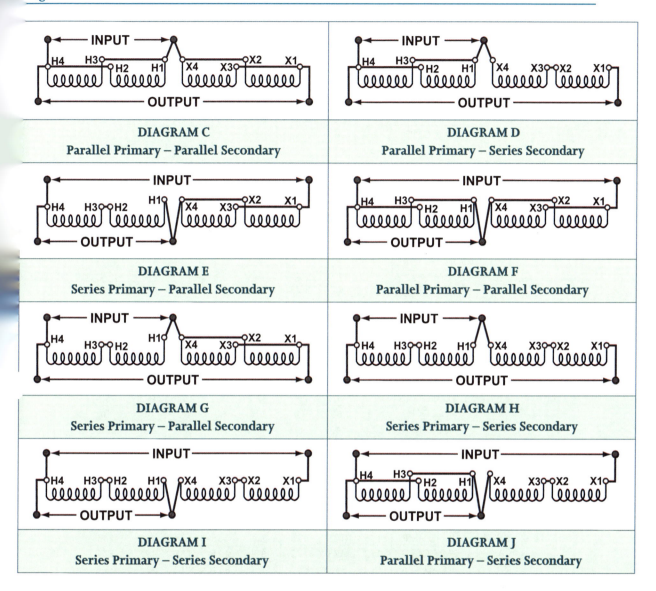

Selection Table BB3-1 — THREE PHASE

GROUP I 3Ø Transformer Arrangements — 120 x 240 PRIMARY VOLTS – 12/24 SECONDARY

THREE PHASE			BOOSTING			BUCKING				
Available Line Voltage (input)			189	208	220	219	230	250	255	264
Load Voltage (output)			208	230	242	208	208	227	232	240
Cat #	RATING									
11101	.050 KVA	LOAD KVA	0.75	0.83	0.87	1.58	0.83	0.90	0.92	0.95
		LOAD AMPS	2.08	2.08	2.08	4.39	2.30	2.29	2.29	2.29
		Max OCPD	6	6	6	10	6	6	6	6
11102	.100 KVA	LOAD KVA	1.50	1.66	1.75	3.16	1.66	1.80	1.84	1.91
		LOAD AMPS	4.17	4.17	4.17	8.77	4.61	4.59	4.58	4.58
		Max OCPD	10	10	10	15	10	10	10	10
11103	.150 KVA	LOAD KVA	2.25	2.49	2.62	4.74	2.49	2.71	2.76	2.86
		LOAD AMPS	6.25	6.25	6.25	13.16	6.91	6.88	6.87	6.88
		Max OCPD	15	15	15	20	15	15	15	15
11104	.250 KVA	LOAD KVA	3.75	4.15	4.37	7.9	4.15	4.51	4.6	4.76
		LOAD AMPS	10.4	10.4	10.4	21.9	11.5	11.5	11.5	11.5
		Max OCPD	15	15	15	30	15	15	15	15
11105	.500 KVA	LOAD KVA	7.51	8.3	8.73	15.8	8.3	9.02	9.2	9.53
		LOAD AMPS	20.8	20.8	20.8	43.9	23.0	22.9	22.9	22.9
		Max OCPD	30	30	30	60	30	30	30	30
11106	.750 KVA	LOAD KVA	11.3	12.5	13.1	23.7	12.5	13.5	13.8	14.3
		LOAD AMPS	31.3	31.3	31.3	65.8	34.6	34.4	34.4	34.4
		Max OCPD	45	45	45	80	40	40	40	40
11107	1.0 KVA	LOAD KVA	15.0	16.6	17.5	31.6	16.6	18.0	18.4	19.1
		LOAD AMPS	41.7	41.7	41.7	87.7	46.1	45.9	45.8	45.8
		Max OCPD	60	60	60	110	60	60	60	60
11108	1.5 KVA	LOAD KVA	22.5	24.9	26.2	47.4	24.9	27.1	27.6	28.6
		LOAD AMPS	62.5	62.5	62.5	132	69.1	68.8	68.7	68.8
		Max OCPD	90	90	90	175	80	80	80	80
11109	2.0 KVA	LOAD KVA	30.0	33.2	34.9	63.2	33.2	36.1	36.8	38.1
		LOAD AMPS	83.3	83.3	83.3	175	92.2	91.8	91.6	91.7
		Max OCPD	125	125	125	225	110	110	110	110
11110	3.0 KVA	LOAD KVA	45.0	49.8	52.4	94.8	49.8	54.1	55.2	57.2
		LOAD AMPS	125	125	125	263	138	138	137	138
		Max OCPD	175	175	175	350	175	175	175	175
11111	5.0 KVA	LOAD KVA	75.1	83.0	87.3	158	83.0	90.2	92.0	95.3
		LOAD AMPS	208	208	208	439	230	229	229	229
		Max OCPD	300	300	300	600	300	300	300	300
11112*	7.5 KVA	LOAD KVA	113	125	131	237	124	135	138	143
		LOAD AMPS	313	313	313	658	346	344	343	344
		Max OCPD	450	450	450	800	400	400	400	400
11113*	10 KVA	LOAD KVA	150	166	175	316	166	180	184	191
		LOAD AMPS	417	417	417	877	461	459	458	458
		Max OCPD	600	600	600	1200	600	600	600	600
Number of Transformers Required			2	2	2	2	2	2	2	2
CONNECTION DIAGRAM			B-B	B-B	B-B	C-C	E-E	E-E	E-E	E-E

* These larger KVA transformers have two X1, two X2, two X3, and two X4 leads.

Selection Table BB3-2 — THREE PHASE
GROUP II 3Ø Transformer Arrangements 120 x 240 PRIMARY VOLTS – 16/32 SECONDARY VOLTS

THREE PHASE			BOOSTING			BUCKING					
Available Line Voltage (input)			195	208	225	240	245	250	256	265	272
Load Voltage (output)			208	240	240	208	230	234	240	234	240
Cat #	RATING										
21101	.050 KVA	LOAD KVA	1.13	0.63	1.30	0.56	1.33	1.35	1.39	0.72	0.74
		LOAD AMPS	3.13	1.56	3.13	1.56	3.33	3.34	3.33	1.77	1.77
		Max OCPD	6	3	6	3	6	6	6	3	3
21102	.100 KVA	LOAD KVA	2.25	1.27	2.60	1.13	2.65	2.71	2.77	1.43	1.47
		LOAD AMPS	6.25	3.13	6.25	3.13	6.66	6.68	6.67	3.54	3.54
		Max OCPD	15	6	15	10	15	15	15	6	6
21103	.150 KVA	LOAD KVA	3.38	1.90	3.90	1.69	3.98	4.06	4.16	2.15	2.21
		LOAD AMPS	9.38	4.69	9.38	4.69	9.99	10.0	10.0	5.31	5.31
		Max OCPD	15	10	15	10	15	15	15	10	10
21104	.250 KVA	LOAD KVA	5.63	3.17	6.50	2.81	6.63	6.77	6.93	3.59	3.68
		LOAD AMPS	15.6	7.81	15.6	7.81	16.6	16.7	16.7	8.85	8.85
		Max OCPD	25	15	25	15	20	20	20	15	15
21105	.500 KVA	LOAD KVA	11.3	6.33	13.0	5.63	13.3	13.5	13.9	7.17	7.36
		LOAD AMPS	31.3	15.6	31.3	15.6	33.3	33.4	33.3	17.7	17.7
		Max OCPD	45	25	45	20	40	40	40	20	20
21106	.750 KVA	LOAD KVA	16.9	9.50	19.5	8.44	19.9	20.3	20.8	10.7	11.0
		LOAD AMPS	46.9	23.4	46.9	23.4	49.9	50.1	50.0	26.5	26.5
		Max OCPD	70	35	70	30	60	60	60	30	30
21107	1.0 KVA	LOAD KVA	22.5	12.7	26.0	11.3	26.5	27.1	27.7	14.3	14.7
		LOAD AMPS	62.5	31.3	62.5	31.3	66.6	66.7	66.7	35.4	35.4
		Max OCPD	90	45	90	35	80	80	80	40	40
21108	1.5 KVA	LOAD KVA	33.8	19.0	39.0	16.9	39.9	40.6	41.6	21.5	22.1
		LOAD AMPS	93.8	46.9	93.8	46.9	99.9	100	100	53.1	53.1
		Max OCPD	125	70	125	60	125	125	125	60	60
21109	2.0 KVA	LOAD KVA	45.0	25.3	52.0	22.5	53.0	54.1	55.4	28.7	29.4
		LOAD AMPS	125	62.5	125	62.5	133	134	133	70.8	70.8
		Max OCPD	175	90	175	70	175	175	175	80	80
21110	3.0 KVA	LOAD KVA	67.6	38.0	77.9	33.8	79.6	81.2	83.1	43.0	44.2
		LOAD AMPS	187	93.8	188	93.8	200	200	200	106	106
		Max OCPD	250	150	250	110	250	250	250	125	125
21111	5.0 KVA	LOAD KVA	113	63.3	130	56.3	133	135	139	71.7	73.5
		LOAD AMPS	313	156	313	156	333	334	333	177	177
		Max OCPD	450	225	450	175	400	400	400	200	200
21112*	7.5 KVA	LOAD KVA	169	95.0	195	84.4	199	203	208	108	110
		LOAD AMPS	469	234	469	234	499	501	500	265	265
		Max OCPD	700	350	700	300	600	600	600	300	300
21113*	10 KVA	LOAD KVA	225	127	260	113	265	271	277	143	147
		LOAD AMPS	625	313	625	313	666	668	667	354	354
		Max OCPD	1000	450	1000	350	800	800	800	400	400
Number of Transformers Required			2	2	2	2	2	2	2	2	2
CONNECTION DIAGRAM			G-G	B-B	G-G	D-D	C-C	C-C	C-C	E-E	E-E

* These larger KVA transformers have two X1, two X2, two X3, and two X4 leads.

Selection Table BB3-3 — THREE PHASE

GROUP III — 3Ø Transformer Arrangements — 240 x 480 PRIMARY VOLTS – 24/48 SECONDARY VOLTS

THREE PHASE			BOOSTING							BUCKING						
Available Line Voltage (input)			380	430	440	460	460	480	480	440	440	460	460	480	480	500	500
Load Voltage (output)			420	473	462	506	483	528	504	400	419	438	418	457	436	455	477
Cat # & Rating																	
31101	.050 KVA	LOAD KVA	0.86	0.76	0.85	1.66	0.91	1.74	0.95	0.79	1.58	1.66	0.83	1.73	0.86	0.9	1.8
		LOAD AMPS	1.04	1.04	1.04	2.08	1.04	2.08	1.04	1.14	2.18	2.18	1.14	2.18	1.14	1.14	2.18
		Max OCPD	3	3	3	6	3	6	3	3	6	6	3	6	3	3	6
31102	.100 KVA	LOAD KVA	1.73	1.51	1.70	3.33	1.82	3.48	1.90	1.59	3.17	3.31	1.66	3.46	1.73	1.80	3.61
		LOAD AMPS	2.08	2.08	2.08	4.16	2.08	4.16	2.08	2.29	4.37	4.37	2.29	4.37	2.29	2.29	4.37
		Max OCPD	6	6	6	10	6	10	6	6	10	10	6	10	6	6	10
31103	.150 KVA	LOAD KVA	2.60	2.27	2.56	4.99	2.73	5.22	2.85	2.38	4.75	4.97	2.48	5.19	2.59	2.7	5.41
		LOAD AMPS	3.12	3.12	3.12	6.24	3.12	6.25	3.12	3.42	6.55	6.55	3.43	6.55	3.43	3.43	6.55
		Max OCPD	10	6	6	15	6	15	6	6	15	15	6	15	6	6	15
31104	.250 KVA	LOAD KVA	4.33	3.78	4.26	8.32	4.56	8.7	4.76	3.96	7.92	8.28	4.14	8.64	4.32	4.51	9.02
		LOAD AMPS	5.2	5.2	5.2	10.4	5.2	10.4	5.2	5.72	10.9	10.9	5.72	10.9	5.72	5.72	10.9
		Max OCPD	15	10	10	15	10	15	10	10	15	15	10	15	10	10	15
31105	.500 KVA	LOAD KVA	8.6	7.56	8.52	16.6	9.11	17.4	9.51	7.93	15.9	16.6	8.28	17.3	8.64	9.02	18.1
		LOAD AMPS	10.4	10.4	10.4	20.8	10.4	20.8	10.4	11.4	21.8	21.8	11.4	21.8	11.4	11.4	21.8
		Max OCPD	20	15	15	30	15	30	15	15	30	30	15	30	15	15	30
31106	.750 KVA	LOAD KVA	12.9	11.3	12.8	25.0	13.7	26.1	14.3	11.9	23.8	24.9	12.4	25.9	13.0	13.5	27.1
		LOAD AMPS	15.6	15.6	15.6	31.2	15.6	31.2	15.6	17.2	32.8	32.8	17.2	32.8	17.2	17.2	32.8
		Max OCPD	25	25	25	45	25	45	25	20	40	40	20	40	20	20	40
31107	1.0 KVA	LOAD KVA	17.3	15.1	17.0	33.3	18.2	34.8	19.0	15.9	31.7	33.1	16.6	34.6	17.3	18.0	36.1
		LOAD AMPS	20.8	20.8	20.8	41.6	20.8	41.6	20.8	22.9	43.7	43.7	22.9	43.7	22.9	22.9	43.7
		Max OCPD	35	30	30	60	30	60	30	30	60	60	30	60	30	30	60
31108	1.5 KVA	LOAD KVA	25.9	22.7	25.6	49.9	27.3	52.2	28.5	23.8	47.6	49.7	24.9	51.8	25.9	27.1	54.1
		LOAD AMPS	31.2	31.2	31.2	62.4	31.2	62.4	31.2	34.3	65.5	65.5	34.3	65.5	34.3	34.3	65.5
		Max OCPD	50	45	45	90	45	90	45	40	80	80	40	80	40	40	40
31109	2.0 KVA	LOAD KVA	34.6	30.3	34.0	66.6	36.5	69.6	38.0	31.7	63.4	66.3	33.1	69.2	34.6	36.1	72.2
		LOAD AMPS	41.6	41.6	41.6	83.2	41.6	83.2	41.6	45.8	87.4	87.4	45.8	87.4	45.8	45.8	87.4
		Max OCPD	70	60	60	110	60	110	60	60	110	110	60	110	60	60	110
31110	3.0 KVA	LOAD KVA	52	45.5	51.2	100	54.7	105	57.1	47.6	95.3	99.6	49.8	104	51.9	54.2	108
		LOAD AMPS	62.5	62.5	62.5	125	62.5	125	62.5	68.8	131	131	68.8	131	68.8	68.8	131
		Max OCPD	100	90	90	175	90	175	90	80	175	175	80	175	80	80	175
31111	5.0 KVA	LOAD KVA	86.1	75.6	85.2	166	91.2	174	95.1	79.3	159	166	82.8	173	86.4	90.1	180
		LOAD AMPS	104	104	104	208	104	208	104	114	218	218	114	218	114	114	218
		Max OCPD	175	150	150	300	150	300	150	150	300	300	150	300	150	150	300
31112	7.5 KVA	LOAD KVA	129	113	128	250	137	261	143	119	238	249	124	259	130	135	271
		LOAD AMPS	156	156	156	312	156	312	156	172	328	328	172	328	172	172	328
		Max OCPD	250	225	225	450	225	450	225	200	400	400	200	400	200	200	400
31113*	10 KVA	LOAD KVA	173	151	170	333	182	348	190	159	317	331	166	346	173	180	361
		LOAD AMPS	208	208	208	416	208	416	208	229	437	437	229	437	229	229	437
		Max OCPD	350	300	300	600	300	600	300	300	600	600	300	600	300	300	600
# of Transformers Required			2	2	2	2	2	2	2	2	2	2	2	2	2	2	2
CONNECTION DIAGRAM			B-B	B-B	G-G	B-B	G-G	B-B	G-G	E-E	C-C	C-C	E-E	C-C	E-E	E-E	C-C

* This larger KVA transformer has two X1, two X2, two X3, and two X4 leads.

Three-Phase Autotransformer CONNECTION DIAGRAMS

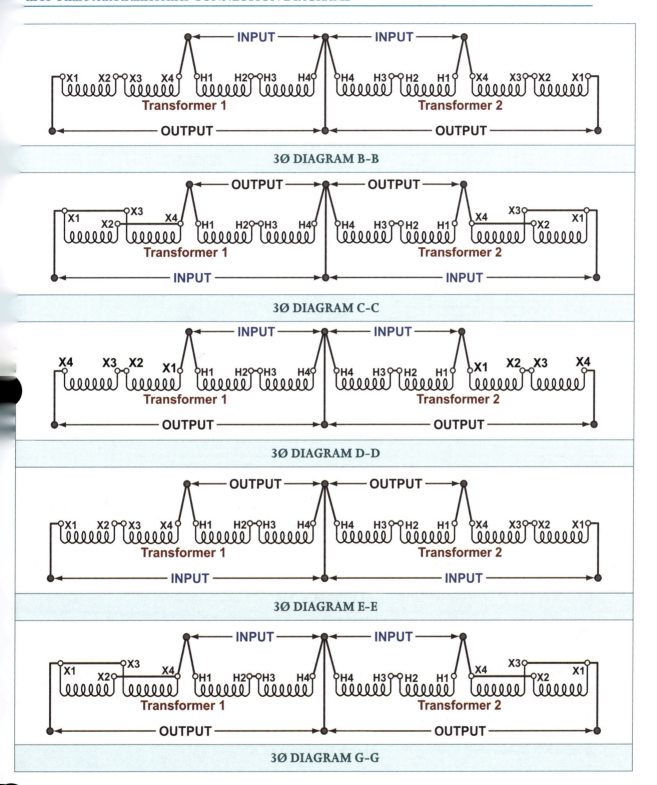

3Ø, 4-Wire WYE-WYE Selection Table

Cat #	KVA RATING	Available Line Voltage (input)	BOOSTED to 480Y/277 Volt Output to Load			
			424Y/245	436Y/252	450Y/260	457Y/264
11101	.050	Load kVA		1.5		3
21101	.050	Load kVA	1.12		2.25	
11102	.100	Load kVA		3		6
21102	.100	Load kVA	2.25		4.5	
11103	.150	Load kVA		4.5		9
21103	.150	Load kVA	3.38		6.76	
11104	.250	Load kVA		7.5		15
21104	.250	Load kVA	5.62		11.25	
11105	.500	Load kVA		15		30
21105	.500	Load kVA	11.25		22.5	
11106	.750	Load kVA		22.5		45
21106	.750	Load kVA	16.9		33.8	
11107	1.0	Load kVA		30		60
21107	1.0	Load kVA	22.5		45	
11108	1.5	Load kVA		45		90
21108	1.5	Load kVA	33.8		67.6	
11109	2.0	Load kVA		60		120
21109	2.0	Load kVA	45		90	
11110	3.0	Load kVA		90		180
21110	3.0	Load kVA	67.6		135	
		CONNECTION DIAGRAM	Y-Y-10	Y-Y-10	Y-Y-9	Y-Y-9

3Ø, 4-Wire WYE-WYE Autotransformer CONNECTION DIAGRAMS

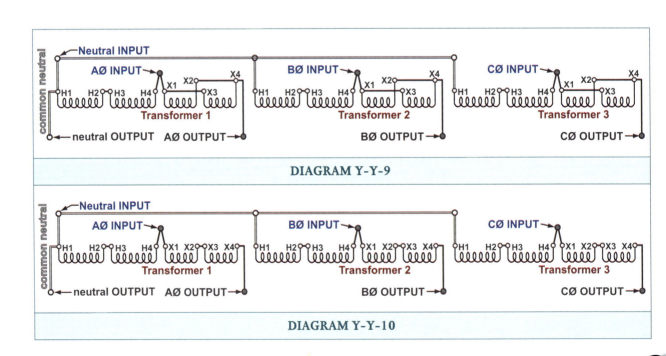

DIAGRAM Y-Y-9

DIAGRAM Y-Y-10

Yr-2 – Annex L

Lesson 216   Single-Phase Transformer Lab

Objectives
There are several objectives for this lab – Each student should be able to:

General
- explain the purpose of the main "kill" switch and the purpose of the lockout rings with the instructor's padlock and the use of the additional padlocks
- correctly replace the lockout ring and padlock each time the trainer is de-energized
- twist stranded wire and get all the strands under the terminal screws. You may have to show the students how to twist the wire, stick it under the screw, then use their screwdrivers to curve it before tightening the terminal screw. Don't be surprised if they don't know to curve the wire in the direction of tightening!
- determine nameplate rated voltage of a transformer or load (lamps here) and do not apply a voltage that exceeds this rating
- properly ground the secondary of a transformer as per the NEC. There are several spools of green wire around the lab. This wire should be used for the system bonding jumper between the secondaries and the "grounding electrode terminal point" that is located at each station. These ground terminals are also bonded to each transformer enclosure and to the safety switches.

Part A
- locate primary H terminals and secondary X terminals
- use an ohmmeter
- see that each of the two primary windings is separate from the other
  [ie ∞ Ω resistance or perfect insulation]
- see that each of the two secondary windings is separate from the other
  [ie ∞ Ω resistance or perfect insulation]
- see that the windings are not connected to the enclosure or the core around which the windings are wound.

Part B1
- locate the proper power source, use a voltmeter, and determine actual line-to-line voltage of the power source
- know when to connect the windings in series or in parallel so that applied voltage doesn't exceed the rated voltage of each primary winding

Part B2
- recognize that the output voltage of each winding SET (X1-X2 and X3-X4) has a relationship [follows the nameplate* voltage ratio] with the actual primary voltage applied to the primary winding

- know when to connect the secondary windings in series or in parallel to match the required output to the rated voltage of a load
- recognize when transformer windings and the elements of a load are connected in series or in parallel or in series/parallel

NOTE: A sufficient supply of extra lamps should be available so if you misconnect the loads your instructor will let you blow them out. It's better to do it here with a 50 cent lamp than with a load that would cost the contractor much more to replace!

* NO LOAD output voltages are slightly higher than expected because the secondary windings of small transformers are "compensated". Once a significant load is connected (as would be in the field) the voltage ratio will more closely match the nameplate voltage ratio. If these small transformers did not have compensated windings the VR would be less than the nameplate rating when loaded.

Part B3
- know when to connect the secondary windings in series or in parallel to match the required output to the rated voltage of a load – this seems like repetition however students don't always grasp that although the input voltage has not changed the secondary can be changed from 240 to get 120 volts out.

Part C1
- this is more repetition to reinforce that the voltage ratios remain unchanged for each individual set of windings within a dual voltage transformer so that the primary supply voltage and primary connections can be changed without affecting the secondary.

Part C2
- recognize the results of an open neutral in a multiwire branch circuit with balanced loads. Here the midpoint of the secondary is "grounded". Rarely are we lucky enough to have balanced loads on a multiwire branch circuit so that if the neutral opens there would be no significant damage
- recognize the results of an open neutral in a multiwire branch circuit with **un**balanced loads. This should drive home that generally we are NOT lucky and don't have balanced loads on multiwire branch circuits.

Part D
- use Ohm's Law to explain open neutral projects done in the lab. These are to be done in class after they complete the lab projects.

Lesson 16 Single-Phase Transformer Lab

> This lab can be a demonstration by the instructor or can be completed by the students if equipment is available. For each lab station the following should be available:
>
> 1............480 volt 1Ø power supply with lockout capability
> 1............240 volt 1Ø power supply with lockout capability
> 1............208 volt 1Ø power supply with lockout capability
> 1............volt-ohmmeter
> 4............120-Volt lampholders with low-wattage lamps (4 to 15 W)
> Qty........extra lamps to allow for misconnection
> 1............100 VA 240/480 to 120/240 Volt transformer (rating dependent on lamp load)
> ....(fused protection is recommended for each winding)
> Qty........18 AWG wire
> Qty........wire nuts

Single-Phase Transformer Lab #1 - Activities

All secondaries are to be **grounded** (as required by the NEC) to the available grounding electrode terminal point.

A. Before connecting your transformer, and using your ohmmeter to make measurements:

1. The actual resistance (not impedance) between X1 and X2 is _____ Ω.

2. The actual resistance (not impedance) between X3 and X4 is _____ Ω.

3. The actual resistance (not impedance) between X1 and X4 is _____ Ω.

4. The actual resistance (not impedance) between X2 and X3 is _____ Ω.

5. The actual resistance (not impedance) between H1 and H2 is _____ Ω.

6. The actual resistance (not impedance) between H3 and H4 is _____ Ω.

7. The actual resistance (not impedance) between H1 and H4 is _____ Ω.

8. The actual resistance (not impedance) between H2 and H3 is _____ Ω.

9. The actual resistance (not impedance) between X1 and the metal enclosure, which is connected to the metal core, is _____ Ω.

10. The actual resistance (not impedance) between H3 and the metal enclosure, which is connected to the metal core, is _____ Ω.

11. The actual resistance (not impedance) between X1 and H3 is _____ Ω.

B1. Connect 480 volts as your source voltage to your transformer **primary**.
Draw your connections on the illustration shown below.

12. Again, using your **ohmmeter**, the resistance between H1 & H4 is now _____ Ω.

13. You connected your primary windings in ___.
    a. series   b. parallel

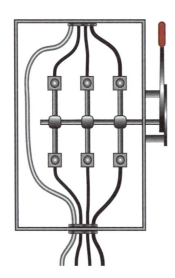
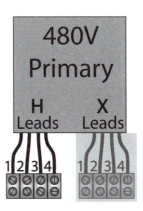

Now apply voltage and using your *voltmeter*:

14. The actual voltage between H1 and H2 is _____ volts.

15. The actual voltage between H3 and H4 is _____ volts.

16. The actual voltage between H1 and H4 is _____ volts.

17. The actual voltage between X1 and X2 is _____ volts.

18. The actual voltage between X3 and X4 is _____ volts.

Leave your **primary** connected to the 480 volt supply!
Disconnect power and continue your connections.

B2. Connect your transformer **secondary** to produce a 240-volt output. Draw your secondary connections on the illustration shown at the right and connect to all four lamps so that **rated** voltage is applied to the lamps.

19. You connected your secondary windings in ___.
    a. series   b. parallel

20. Your lamp load should be connected as in drawing ___ below.

Figure A

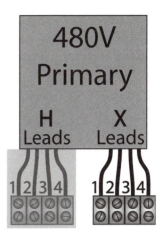

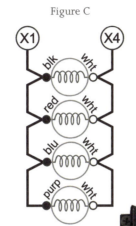

21. Your lamp bank is connected in ___.
    a. series   b. parallel   c. series/parallel

Now apply voltage and using your **voltmeter:**

22. The actual voltage between X1 and X2 is _____ volts.

23. The actual voltage between X3 and X4 is _____ volts.

24. The actual voltage between X1 and X4 is _____ volts.

25. The actual voltage measured at the Lamp Bank T-Strip terminals for each lamp is:
_____ volts/_____ volts/_____ volts/ & _____ volts.

Leave your primary connected to the 480 volt supply!
Disconnect power and continue your connections

B3. Connect your transformer **secondary** to produce a 120 volt output. Draw your secondary connections on the illustration shown at the right and connect to all four lamps so that **rated** voltage is applied to the lamps.

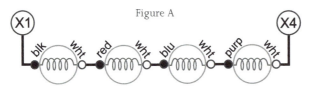

26. You connected your secondary windings in ___.
    a. series   b. parallel

27. Your lamp load should be connected as in drawing ___ below.

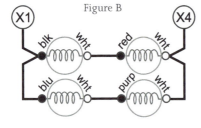

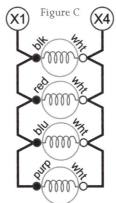

28. Your lamp bank is connected in ___.
    a. series   b. parallel   c. series/parallel

29. Does the wire connected to X1 or X4 meet the NEC definition of a grounded conductor?
    a. yes   b. no

30. If you answered "no", why is it not?

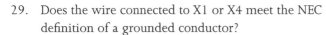

Now apply voltage and using your **voltmeter:**

31. The actual voltage between X1 and X2 is _____ volts.

32. The actual voltage between X3 and X4 is _____ volts.

Lesson 16 Single-Phase Transformer Lab   Yr-2 Annex L-959

33. The actual voltage between X1 and X4 is _____ volts.
34. The actual voltage dropped across each lamp at the Lamp Bank T-Strip terminals is:
 _____ volts/ _____ volts/ _____ volts/ & _____ volts.

35. The actual voltage between your "neutral" and the grounded safety switch is _____ volts.

36. Disconnect and cap off the connection to the grounding electrode point. Now the voltage between the "neutral" on the transformer and the grounded safety switch is _____ volts.

37. Could this transformer's secondary be connected to produce a 120/240 volt 1Ø, 3-wire output with a 480 volt input on the primary?
    a. yes   b. no

Leave your secondary connected for a 120 volt OUTPUT then Disconnect power and continue your connections.

C1. **Connect** 240 volts as your source voltage to your transformer **primary**. Your transformer **secondary** should still be connected to produce a 120 volt output. Draw both secondary and primary connections on the illustrations shown at the right.

38. Using your **ohmmeter** measure the resistance from H1 to H4. It is _____ Ω.

39. You connected your primary windings in ___.
    a. series   b. parallel

40. Your secondary windings are still connected in ___.
    a. series   b. parallel

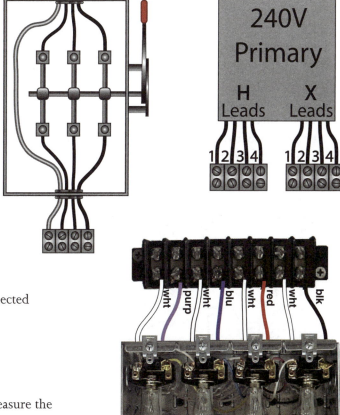

Now apply voltage and using your **voltmeter** measure the actual voltages:

41. between H1 & H2 is _____ volts.

Year Two (Student Manual)

42. between H3 & H4 is _____ volts.

43. between H1 & H4 is _____ volts.

44. between X1 & X2 is _____ volts.

45. between X3 & X4 is _____ volts.

46. between X1 & X4 is _____ volts.

47. across each lamp at the Lamp Bank T-Strip terminals is:

    _____ volts/ _____ volts/ _____ volts/ & _____ volts.

48. Are these secondary and load measurements very close to the same voltage measurements you got at Questions 31 through 34?
    a. yes   b. no

Leave your primary connected to the 240 volt supply!
Disconnect power and continue your connections.

C2. **Re-Connect** your transformer **secondary** to produce a 120/240 volt output. Draw your secondary connections on the illustration shown at the right. Connect your lights to the secondary terminals as per the schematic shown at the right.

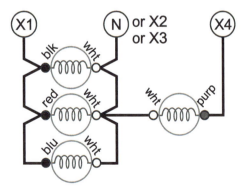
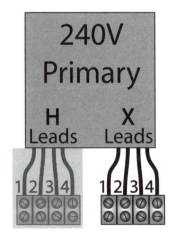

Run one "hot" from X1 to three lights and the other "hot" from X4 to one light. **Remember** to run only one extra longneutral (leave about **18" of slack**) from X2/X3 to your first lamp on the T-strip and "jumper the neutrals" together on the T-strip!

Lesson 16 Single-Phase Transformer Lab   Yr-2 Annex L-961

Now apply voltage and using your voltmeter make the following actual voltage measurements:

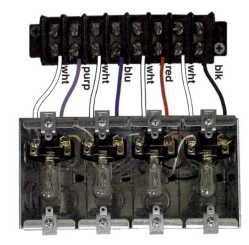

49. between X1 & X2 is _____ volts.

50. between X3 & X4 is _____ volts.

51. between X1 & X4 is _____ volts.

52. across the T-Strip terminals of the three lamps that are connected to X1 & the neutral are:
_____ volts/ _____ volts/and _____ volts.

53. The actual voltage across the T-Strip terminals of the single lamp and its neutral is _____ volts.

Your **instructor** will now *carefully* cut your neutral:

54. The "3-lamp set" is ___ than the "single lamp".
    a. dimmer   b. brighter

55. The "single lamp" is ___ than the "3-lamp set".
    a. dimmer   b. brighter

56. Using your *voltmeter* the actual voltage dropped across the T-Strip terminals of the lamps in the "3-lamp set" are: _____ volts/ _____ volts/ & _____ volts.

57. Using your *voltmeter* the actual voltage measured across the T-Strip terminals of the single lamp is _____ volts.

Leave your secondary connected for a 120/240 volt OUTPUT then Disconnect power and continue your connections.

C3. **Reconnect** your neutral that was cut in **Part C2**. Now move your supply voltage connections from the 240-volt supply to the 208-volt supply.

Now apply voltage and using your voltmeter:

58. The actual voltage between H1 and H2 is _____ volts.

59. The actual voltage between H3 and H4 is _____ volts.

60. The actual voltage between H1 and H4 is _____ volts.

Year Two (Student Manual)

Yr-2 Annex L-962

Lesson 16 Single-Phase Transformer Lab

61. The actual voltage between X1 and X2 is _____ volts.

62. The actual voltage between X3 and X4 is _____ volts.

63. The actual voltage between X1 and X4 is _____ volts.

64. The actual voltage across the T-Strip terminals of any of the lamps in the "3-lamp set" is _____ volts.

65. The actual voltage across the T-Strip terminals of the single lamp is _____ volts.

66. Energize your 277/480 volt safety switch and measure the voltages between its T-strip and the 120/208 volt T-strip.
    480 AØ to 208 AØ is _____ volts
    480 AØ to 208 BØ is _____ volts
    480 AØ to 208 CØ is _____ volts
    480 BØ to 208 AØ is _____ volts
    480 BØ to 208 BØ is _____ volts
    480 BØ to 208 CØ is _____ volts
    480 BØ to 208 AØ is _____ volts
    480 BØ to 208 BØ is _____ volts
    480 BØ to 208 CØ is _____ volts

67. A 120 volt switch is supplied from Circuit 4 in a 120/208 volt 3Ø, 4-wire panel. A 277 volt switch is supplied from Circuit 5 in a 277/480 volt 3Ø, 4-wire panel. Both of these switches are installed in a 2-gang 1900 box with a plaster ring. Does the NEC require you to install a partition between the two switches?
    a. yes   b. no

You should now **Disconnect power**,
    rig down,
        clean your area,
            return your tool bin as you received it (no more – no less),
                return to class,
                    and complete **Part D**.

When you return to class **Answer** the following questions.

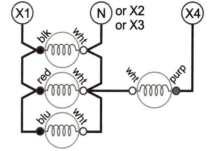

68. If you are using 6 watt lamps and had measured the current in the wire between X1 and the first lamp of the "3-lamp set" at 150 mA and the current in the wire between X4 and the lamp at 50 mA there should be _____ mA of current in the neutral. A drawing of this circuit is shown at the right.

69. The fixed ("hot") resistance of a 6 watt 120-volt lamp is _____ Ω.

70. The rated current of a 6 watt 120-volt lamp is _____ mA.

Refer to the drawing of the "open neutral" circuit shown at the right to answer Questions 71 through 75.

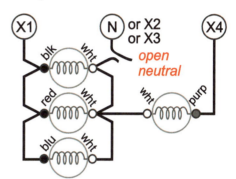

71. With the neutral cut you should now get about _____ mA of current between X1 and the first lamp of the "3-lamp set".

72. With the neutral cut you should now get about _____ mA of current between X4 and the single lamp.

73. The answer you gave to Question #71 can be explained because there should be _____ mA of current flowing through each lamp in the "3-lamp set" instead of the rated current you calculated in Question #70.

74. This is ___ than the rated current.
    a. higher   b. lower

The answer you gave to Question #72 can be explained because there should be _____ mA of current flowing through the lamp in the "1-lamp set" instead of the rated current you calculated in Question #70.

75. This is ___ than the rated current.
    a. higher   b. lower

Yr-2 – Annex M

Lesson 229 Lab   Identifying the Leads of an Unmarked 3-Phase Motor

Objectives

- Be able to determine and mark the leads of an unmarked 9-lead, wye-wound, 3-phase motor.
- Understand how to determine the leads of an unmarked 9-lead, delta-wound, 3-phase motor.

Materials and Equipment Needed for the Lab

- 9-lead, single-speed, wye-wound, 3-phase motor with the lead markings masked (covered)
- Two Analog DC voltmeters
- Analog or digital ohmmeter
- Four alligator clips
- Small 12-volt battery – If a 6-volt battery is used, the voltage test results will be half that found with a 12-volt battery.
- Masking tape
- Fine point permanent marking pen, such as a Sharpie®
- Numeric wire markers (1 – 9)

9-Lead, Delta-Wound Motor

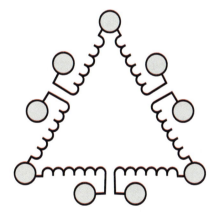

9-Lead, Wye-Wound Motor

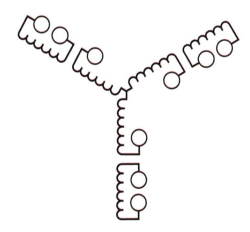

Yr-2 Annex M-966

Procedures

1. First, make certain that all 9 motor leads are separated (insulated) from each other.

2. Determine if the motor is wye or delta wound. Resistance measurements will be made with an ohmmeter to determine winding sets. Refer to the drawings of the delta and wye-wound motor windings shown on the previous page. A resistance measurement of somewhere between 0.2 and 20 ohms will indicate that a winding has continuity and has been identified.

    a. Select one of the nine leads and measure the resistance between it and the other leads. If continuity is indicated between the selected lead and **two** other leads, a **3**-lead winding set has been identified.

    b. If the lead selected in Step 2a has continuity to only one of the other leads, tape these two leads together and select another lead. Make continuity measurements between the newly selected lead and the remaining leads. As before, if this lead has continuity to **two** other leads, a **3**-lead set has been identified.

    c. If the lead selected in Step 2b has continuity to only one of the other leads, tape these two leads together and select another lead. Make continuity measurements between the newly selected lead and the remaining leads. As before, if this lead has continuity to **two** other leads, a **3**-lead set has been identified.

    d. By this point, the following winding sets should have been identified:

    - Three separate **3**-lead winding sets have been identified

        ☞ This is a delta motor – Skip to Procedure Step 6

    OR

    - Three separate **2**-lead winding sets have been identified

        ☞ This is a wye motor – Proceed to Procedure Step 3

    OR

    - One **3**-lead winding set and two separate **2**-lead winding sets have been identified

        ☞ This is a wye motor – Proceed to Procedure Step 4

3. This is a Y-wound motor and three separate 2-lead winding sets have been identified. The remaining three leads must be 7-8-9. Verify that continuity exists among these three leads and permanently mark them as 7, 8, and 9. Skip to Procedure Step 5.

4. This is a Y-wound motor. One 3-lead winding set and two separate 2-lead winding sets have been identified. Permanently mark the three-lead set as 7, 8, and 9. Verify that continuity exists between the remaining two leads and tape these together. Continue to Procedure Step 5.

5. For a wye-wound motor:

    a. Select any one of the 2-lead winding set leads. Select a DC volt scale of about 10 volts and clip the analog voltmeter leads to these two motor leads. Make certain that the other 7 motor leads are insulated from each other. Connect the negative battery terminal to the motor lead marked "8". Watch for deflection (movement) of the needle on the voltmeter, and momentarily connect the positive battery terminal to the motor lead marked "9". If no deflection occurred, motor leads "1" and "4" have been located. Verify this by moving the voltmeter leads to another 2-lead winding set and repeat the procedure. If motor leads 1 and 4 were first selected, meter deflection should be observed for this set. Repeat, and look for deflection, on the remaining 2-lead winding set.

b. It must now be determined which motor lead is "1" and which is "4". Connect the voltmeter to the motor leads marked 1 and 4. Make certain that the other seven motor leads are insulated from each other. Connect the *positive* battery terminal to the motor lead marked "7". Watch for positive voltmeter deflection when the negative battery terminal is momentarily connected to motor lead 8. If the meter deflected in the positive direction when the connection to lead 8 was "made", the motor lead connected to the "positive" voltmeter lead should be permanently marked as "1" and the motor lead connected to the "common" voltmeter lead should be permanently marked as "4". If a negative deflection was seen on the voltmeter, reverse the meter lead connections at 1 and 4. Repeat the momentary battery connection to lead 8.

c. It can now be determined which motor lead is "3" and which is "6". Connect the voltmeter to the motor leads marked 3 and 6. Make certain that the other seven motor leads are insulated from each other. Connect the *positive* battery terminal to the motor lead marked "9". Watch for positive voltmeter deflection when the negative battery terminal is momentarily connected to motor lead 7. If the meter deflected in the positive direction when the connection to lead 7 was "made", the motor lead connected to the "positive" voltmeter lead should be permanently marked as "3" and the motor lead connected to the "common" voltmeter lead should be permanently marked as "6". If a negative deflection was seen on the voltmeter, reverse the meter lead connections at 3 and 6. Repeat the momentary battery connection to lead 7.

d. Finally, motor leads 2 and 5 can now be determined. Connect the voltmeter to the motor leads marked 2 and 5. Make certain that the other seven motor leads are insulated from each other. Connect the *positive* battery terminal to the motor lead marked "8". Watch for positive voltmeter deflection when the negative battery terminal is momentarily connected to motor lead 9. If the meter deflected in the positive direction when the connection to lead 9 was "made", the motor lead connected to the "positive" voltmeter lead should be permanently marked as "2" and the motor lead connected to the "common" voltmeter lead should be permanently marked as "5". If a negative deflection was seen on the voltmeter, reverse the meter lead connections at 2 and 5. Repeat the momentary battery connection to lead 9.

e. All nine of the wye-wound motor leads should now be permanently marked.

6. For a delta-wound motor:

   a. Make certain that all nine motor leads are insulated from each other.

   b. Select any one of the 3-lead winding sets and mark the leads as X, Y, and Z.

   c. Connect the 12-volt battery terminals to leads X and Z and measure the voltage between leads X and Y. If the voltage is 6 VDC, the Y lead is the corner lead of this 3-lead winding set. The Y lead can be permanently marked as "1". Skip to Procedure Step 6e.

d. If zero volts was measured between X and Y, move the battery connections to Y and Z. Now measure the voltage between leads X and Y. If the voltage between X and Y is 6 volts, permanently mark lead X as "1".

e. The corner of the next 3-lead winding set must be identified and marked. Mark this set as U-V-W.

f. Connect the 12-volt battery terminals to leads U and W and measure the voltage between leads U and V. If the voltage is 6 VDC, the V lead is the corner lead of this 3-lead winding set. The V lead can be permanently marked as "2". Skip to Procedure Step 6h.

g. If zero volts was measured between U and V, move the battery connections to V and W. Now measure the voltage between leads U and V. If the voltage between U and V is 6 volts, permanently mark this lead U as "2".

h. The corner of the remaining 3-lead winding set must be identified and marked. Mark this set as R-S-T.

i. Connect the 12-volt battery terminals to leads R and T and measure the voltage between leads R and S. If the voltage is 6 VDC, the S lead is the corner lead of this 3-lead winding set. The S lead can be permanently marked as "3". Skip to Procedure Step 6k.

j. If zero volts was measured between R and S, move the battery connections to S and T. Now measure the voltage between leads R and S. If the voltage between R and S is 6 volts, permanently mark this lead R as "3".

k. Leads 5 and 7 can now be identified. Using a DC voltmeter, connect one of its leads to Lead 2 and the other meter lead to one of the letter-marked (U or V or W) leads. Connect the leads from a second DC voltmeter to Lead 2 and the remaining letter-marked (U or V or W) lead. Twist together the two letter-marked leads (X or Y or Z) associated with Lead 1. Connect one 12-volt battery terminal to the twisted leads (X or Y or Z). Now, momentarily connect the other battery terminal to Lead 1 and note which voltmeter yields the greater deflection. The letter-marked lead (U or V or W) connected to the meter with the greater deflection should be marked as 7. The remaining letter-marked lead should be marked as 5.

l. Follow the same procedure used in Step 6k to identify Leads 6 and 8. The two voltmeters will be moved to the Lead 3 set. This set consists of Lead 3 plus the two remaining letter-marked leads (R or S or T). The battery will be kept with the Lead 1 set. Again, determine which voltmeter has the greater deflection when the battery is connected to Lead 1. The letter-marked lead connected to the meter with the greater deflection can be marked as 6, and the one with the lesser deflection can be marked as 8.

m. All that remains is to identify Leads 4 and 9. Connect one of voltmeter leads to Lead 1 and the other voltmeter lead to one of the letter-marked (X or Y or Z) leads. Connect the leads from the second voltmeter to Lead 1 and the remaining letter-marked (X or Y or Z) lead. Twist together the two letter-marked leads (U or V or W) associated with Lead 1. Connect one 12-volt battery terminal to the twisted leads (U or V or W). Now, momentarily connect the other battery terminal to Lead 2 and note which voltmeter yields the greater deflection. The letter-marked lead (X or Y or Z) connected to the meter with the greater deflection should be marked as 4. The remaining letter-marked lead, the one with the lesser deflection, should be marked as 9.

n. All nine of the delta-wound motor leads should now be permanently marked.

Yr-2 Annex M-969

Student Notes

Year Two (Student Manual)

Yr-2 Annex M-970

Year Two (Student Manual)

Annex N - Tables

Complete this table for **1Ø** Motor Branch Circuit Calculations

Are the BC conductors to be installed in parallel conduits? If **YES** use the table on the **back** of this page. Continue here **ONLY** if you are using only one set of conductors in one conduit.									
Again, make certain that the motor is **single**-phase!!!!					Using Table 430.52				
Go to Table 430.248				Table 310.16	DETD	ITCB		OCPD TO BE USED	
E_{line}	HP	T. FLA	T. FLA × 1.25	75° wire size	T.FLA × **1.75**	T.FLA × **2.5**		FUSE	CKT BKR
				→					
				Select next largest wire if calc not std size				See Article 240 & go up to next std size if calc not std size	

Complete this table for **3Ø** Motor Branch Circuit Calculations

Are the BC conductors to be installed in parallel conduits? If **YES** use the table on the **back** of this page. Continue here **ONLY** if you are using only one set of conductors in one conduit.									
Again, make certain that the motor is **three**-phase!!!!					Using Table 430.52				
Go to Table 430.250				Table 310.16	DETD	ITCB		OCPD TO BE USED	
E_{line}	HP	T. FLA	T. FLA × 1.25	75° wire size	T.FLA × **1.75**	T.FLA × **2.5**		FUSE	CKT BKR
				→					
				Select next largest wire if calc not std size				See Article 240 & go up to next std size if calc not std size	

Yr-2 Annex N-**972**

Complete this table for **1Ø** Motor Branch Circuit Calculations for **PARALLEL** conductors

Are the BC conductors to be installed in parallel conduits?									
If **NO** use the table on the **back** of this page. Continue here **ONLY** if you are using PARALLEL sets of conductors in two or more conduits.									
Again, make certain that the motor is **single**-phase!!!!						Using Table 430.52				
Go to Table 430.248					Table 310.16	DETD	ITCB	OCPD TO BE USED	
E_{line}	HP	T. FLA	T. FLA × 1.25	(T.FLA × 1.25) / # of conduits	75° wire size	T.FLA × 1.75	T.FLA × 2.5	FUSE	CKT BKR
				→	→				
					Select next largest wire if calc not std size			See Article 240 & go up to next std size if calc not std size	

Complete this table for **3Ø** Motor Branch Circuit Calculations in **PARALLEL** conductors

Are the BC conductors to be installed in parallel conduits?									
If **NO** use the table on the **back** of this page. Continue here **ONLY** if you are using PARALLEL sets of conductors in two or more conduits.									
Again, make certain that the motor is **three**-phase!!!!						Using Table 430.52				
Go to Table 430.250					Table 310.16	DETD	ITCB	OCPD TO BE USED	
E_{line}	HP	T. FLA	T. FLA × 1.25	(T.FLA × 1.25) / # of conduits	75° wire size	T.FLA × 1.75	T.FLA × 2.5	FUSE	CKT BKR
				→	→				
					Select next largest wire if calc not std size			See Article 240 & go up to next std size if calc not std size	

Year Two (Student Manual)

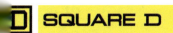

Thermal Unit Selection

For 1.15 to 1.25 service factor motors, use 100% of motor FLC

For 1.0 to 1.10 service factor motors, use 90% of motor FLC

Table 14 — Size 00

Motor FLC (Amps)			Thermal Unit Number
1 T.U.	2 T.U.	3 T.U.	
0.43-0.47	0.41-0.45	0.40-0.41	A .49
0.48-0.51	0.46-0.50	0.42-0.46	A .54
0.52-0.56	0.51-0.55	0.47-0.51	A .59
0.57-0.64	0.56-0.62	0.52-0.57	A .65
0.65-0.69	0.63-0.67	0.58-0.62	A .71
0.70-0.76	0.68-0.72	0.63-0.67	A .78
0.77-0.84	0.73-0.81	0.68-0.75	A .86
0.85-0.91	0.82-0.88	0.76-0.80	A .95
0.92-1.01	0.89-0.97	0.81-0.89	A 1.02
1.02-1.15	0.98-1.08	0.90-1.02	A 1.16
1.16-1.23	1.09-1.18	1.03-1.09	A 1.25
1.24-1.37	1.19-1.32	1.10-1.21	A 1.39
1.38-1.45	1.33-1.40	1.22-1.29	A 1.54
1.46-1.56	1.41-1.48	1.30-1.37	A 1.63
1.57-1.67	1.49-1.60	1.38-1.48	A 1.75
1.68-1.77	1.61-1.72	1.49-1.58	A 1.86
1.78-1.92	1.73-1.84	1.59-1.72	A 1.99
1.93-2.09	1.85-2.00	1.73-1.85	A 2.15
2.10-2.31	2.01-2.22	1.86-2.05	A 2.31
2.32-2.56	2.23-2.45	2.06-2.29	A 2.57
2.57-2.92	2.46-2.82	2.30-2.62	A 2.81
2.93-3.16	2.83-3.08	2.63-2.84	A 3.61
3.17-3.48	3.09-3.39	2.85-3.10	A 3.95
3.49-3.83	3.40-3.75	3.11-3.46	A 4.32
3.84-4.24	3.76-4.16	3.47-3.85	A 4.79
4.25-4.62	4.17-4.51	3.86-4.16	A 5.30
4.63-4.92	4.52-4.83	4.17-4.46	A 5.78
4.93-5.61	4.84-5.49	4.47-5.08	A 6.20
5.62-5.85	5.50-5.67	5.09-5.35	A 6.99
5.86-6.36	5.68-6.16	5.36-5.82	A 7.65
6.37-6.99	6.17-6.75	5.83-6.34	A 8.38
7.00-7.67	6.76-7.00	6.35-6.95	A 9.25
7.68-8.15	...	6.96-7.00	A 9.85
8.16-9.00	...	...	A 11.0

Table 15 — Size 0

Motor FLC (Amps)			Thermal Unit Number
1 T.U.	2 T.U.	3 T.U.	
0.31-0.33	0.31-0.33	0.29-0.31	B 0.44
0.34-0.36	0.34-0.36	0.32-0.36	B 0.51
0.37-0.40	0.37-0.40	0.37-0.38	B 0.57
0.41-0.48	0.41-0.48	0.39-0.46	B 0.63
0.49-0.57	0.49-0.57	0.47-0.55	B 0.71
0.58-0.64	0.58-0.64	0.56-0.61	B 0.81
0.65-0.70	0.65-0.70	0.62-0.66	B 0.92
0.71-0.77	0.71-0.77	0.67-0.75	B 1.03
0.78-0.85	0.78-0.85	0.76-0.83	B 1.16
0.86-0.99	0.86-0.99	0.84-0.93	B 1.30
1.00-1.10	1.00-1.10	0.9-1.06	B 1.45
1.11-1.28	1.11-1.28	1.07-1.18	B 1.67
1.29-1.41	1.29-1.41	1.19-1.31	B 1.88
1.42-1.58	1.42-1.58	1.32-1.47	B 2.10
1.59-1.80	1.59-1.80	1.48-1.67	B 2.40
1.81-2.03	1.81-2.03	1.68-1.83	B 2.65
2.04-2.25	2.04-2.25	1.84-2.04	B 3.00
2.26-2.51	2.26-2.51	2.05-2.38	B 3.30
2.52-2.83	2.52-2.83	2.39-2.60	B 3.70
2.84-3.29	2.84-3.29	2.61-3.13	B 4.15
3.30-3.75	3.30-3.75	3.14-3.59	B 4.85
3.76-4.22	3.76-4.22	3.60-3.94	B 5.50
4.23-4.65	4.23-4.65	3.95-4.19	B 6.25
4.66-5.16	4.66-5.16	4.20-4.72	B 6.90
5.17-5.53	5.17-5.53	4.73-5.21	B 7.70
5.54-6.09	5.54-6.09	5.22-5.51	B 8.20
6.10-6.80	6.10-6.80	5.52-6.17	B 9.10
6.81-7.60	6.81-7.60	6.18-7.07	B 10.2
7.61-8.35	7.61-8.35	7.08-8.05	B 11.5
8.36-9.04	8.36-9.04	8.06-8.69	B 12.8
9.05-9.99	9.05-9.99	8.70-9.32	B 14
10.0-11.1	10.0-11.1	9.33-10.2	B 15.5
11.2-12.3	11.2-12.3	10.3-11.3	B 17.5
12.4-13.7	...	11.4-12.0	B 19.5
13.8-15.4	...	...	B 22
15.5-18.0	...	...	B 25

Following Sections for Size 1 Only

1 T.U.	2 T.U.	3 T.U.	Thermal Unit Number
...	11.2-12.3	...	B 17.5
...	12.4-13.7	11.4-12.1	B 19.5
...	13.8-15.4	12.2-13.7	B 22
15.5-17.2	15.5-17.2	13.8-15.2	B 25
17.3-19.4	17.3-19.4	15.3-17.2	B 28.0
19.5-21.7	19.5-21.7	17.3-18.9	B 32
21.8-23.9	21.8-23.9	19.0-21.4	B 36
24.0-26.0	24.0-26.0	21.5-23.7	B 40
...	...	23.8-26.0	B 45

Table 58 — Size 2

Motor FLC (Amps)		Thermal Unit Number
1 or 2 T.U.	3 T.U.	
3.37-3.82	3.28-3.51	B 4.85
3.83-4.33	3.52-3.89	B 5.50
4.34-4.79	3.90-4.14	B 6.25
4.80-5.33	4.15-4.73	B 6.90
5.34-5.79	4.74-5.22	B 7.70
5.80-6.27	5.23-5.53	B 8.20
6.28-7.03	5.54-6.21	B 9.10
7.04-7.88	6.22-7.17	B 10.2
7.89-8.73	7.18-8.19	B 11.5
8.74-9.55	8.20-8.90	B 12.8
9.56-10.6	8.91-9.57	B 14
10.7-11.8	9.58-10.6	B 15.5
11.9-13.1	10.7-11.8	B 17.5
13.2-14.9	11.9-12.7	B 19.5
15.0-16.9	12.8-14.4	B 22
17.0-18.8	14.5-16.1	B 25
18.9-21.5	16.2-18.2	B 28.0
21.6-24.1	18.3-20.2	B 32
24.2-26.8	20.3-22.8	B 36
26.9-29.9	22.9-25.6	B 40
30.0-35.5	25.7-28.8	B 45
35.6-36.5	28.9-30.6	B 50
36.6-39.6	30.7-32.4	B 56
39.7-41.5	32.5-34.6	B 62
41.6-45.0	34.7-38.6	B 70
...	38.7-45.0	B 79

Table 16 — Size 3

Motor FLC (Amps)			Thermal Unit Number
1 T.U.	2 T.U.	3 T.U.	
16.2-17.5	15.1-16.2	14.3-15.4	CC 20.9
17.6-18.8	16.3-17.3	15.5-16.4	CC 22.8
18.9-20.5	17.4-19.5	16.5-18.5	CC 24.6
20.6-22.2	19.6-20.7	18.6-19.6	CC 26.3
22.3-23.7	20.8-22.3	19.7-21.1	CC 28.8
23.8-25.4	22.4-24.0	21.2-22.7	CC 31.0
25.5-27.3	24.1-25.7	22.8-24.4	CC 33.3
27.4-29.3	25.8-27.5	24.5-26.1	CC 36.4
29.4-31.5	27.6-29.6	26.2-28.1	CC 39.6
31.6-33.9	29.7-31.7	28.2-30.0	CC 42.7
34.0-36.2	31.8-33.9	30.1-32.1	CC 46.6
36.3-39.3	34.0-36.6	32.2-34.7	CC 50.1
39.4-42.3	36.7-39.3	34.8-37.3	CC 54.5
42.4-45.3	39.4-42.3	37.4-40.1	CC 59.4
45.4-48.3	42.4-44.9	40.2-42.6	CC 64.3
48.4-52.0	45.0-48.3	42.7-45.8	CC 68.5
52.1-54.9	48.4-50.9	45.9-48.3	CC 74.6
55.0-59.7	51.0-55.5	48.4-52.6	CC 81.5
59.8-65.4	55.6-59.9	52.7-56.8	CC 87.7
65.5-69.6	60.0-64.2	56.9-60.9	CC 94.0
69.7-74.8	64.3-68.7	61.0-65.1	CC 103
74.9-79.7	68.8-71.4	65.2-67.7	CC 112
79.8-83.1	71.5-74.8	67.8-70.9	CC 121
83.2-86.0	74.9-78.0	71.0-73.9	CC 132
...	78.1-80.7	74.0-76.5	CC 143
...	80.8-86.0	76.6-80.2	CC 156
...	...	80.3-83.1	CC 167
...	...	83.2-86.0	CC 180

Table 61 — Size 4

Motor FLC (Amps)		Thermal Unit Number
2 T.U.	3 T.U.	
46.8-50.0	45.3-48.2	CC 64.3
50.1-54.2	48.3-52.4	CC 68.5
54.3-58.3	52.5-56.4	CC 74.6
58.4-63.6	56.5-61.2	CC 81.5
63.7-68.5	61.3-66.1	CC 87.7
68.6-74.0	66.2-71.4	CC 94.0
74.1-79.8	71.5-77.0	CC 103
79.9-83.0	77.1-79.0	CC 112
83.1-88.9	79.1-84.7	CC 121
89.0-95.6	84.8-91.1	CC 132
95.7-102	91.2-98.1	CC 143
103-109	98.2-104	CC 156
110-119	105-113	CC 167
120-133	114-123	CC 180
...	124-133	CC 196

SQUARE D *by* **Schneider Electric**

Thermal Unit Selection

For 1.15 to 1.25 service factor motors, use 100% of motor FLC

For 1.0 to 1.10 service factor motors, use 90% of motor FLC

Motor and controller in *same ambient temperature*

Thermal Unit Selection

For 1.15 to 1.25 service factor motors, use 100% of motor FLC

For 1.0 to 1.10 service factor motors, use 90% of motor FLC

Size 5

Table 24

Motor FLC (Amps)	Thermal Unit Number
88.2-95.1	DD 112
95.2-101	DD 121
102-111	DD 128
112-119	DD 140
120-131	DD 150
132-149	DD 160
150-170	DD 185
171-180	DD 220
181-187	DD 240
198-204	DD 250
205-213	DD 265
214-237	DD 280
238-243	DD 300
244-266	DD 320

Size 6

Table 20

Motor FLC (Amps)	Thermal Unit Number
133-148	B 1.30
149-174	B 1.45
175-195	B 1.67
196-219	B 1.88
220-239	B 2.10
240-271	B 2.40
272-308	B 2.65
309-348	B 3.00
349-397	B 3.30
398-429	B 3.70
430-495	B 4.15
496-520	B 4.85

NEMA Size Motor Starters					
Maximum Horsepower Ratings @ Nominal Voltage					
NEMA Size	Single Phase		Three Phase		
	120 V	240 V	208 V	240 V	480 V
00	1/3	1	1 1/2	1 1/2	2
0	1	2	3	3	5
1	2	3	7 1/2	7 1/2	10
1P	3	5	—	—	—
2	3	7 1/2	10	15	25
3	7 1/2	15	25	30	50
4	—	—	40	50	100
5	—	—	75	100	200
6	—	—	150	200	400
7	—	—	—	300	600
8	—	—	—	450	900
9	—	—	—	800	1600

The heaters shown below provide a maximum trip rating of 125% of minimum motor amperes for 40°C motors (service factor 1.15). **For other motors** (service factor 1.0), **select the next lower** listed heater code number within the designated table which provides a maximum trip rating of approximately 115%.

SIEMENS Table 233

NEMA Size Starter							Heater Code
00,0,1	1 3/4	2, 2 1/2	3, 3 1/2	4(JB)	4 1/2, 5	6	
.38-.40							E6
.41-.43							E7
.44-.48							E8
.49-.53							E9
.54-.57							E11
.58-.62							E12
.63-.66							E13
.67-.72							E14
.73-.80							E16
.81-.85							E17
.86-.92							E18
.93-.99							E19
1.00-1.08							E23
1.09-1.23							E24
1.24-1.37							E26
1.38-1.54					88.0-98.0	166-195	E27
1.55-1.69					98.1-108	196-217	E28
1.70-1.80					109-114	218-229	E29
1.81-1.94					115-122	230-245	E31
1.95-2.07					123-130	246-261	E32
2.08-2.26					131-140	262-281	E33
2.27-2.54	2.27-2.54				141-155	282-311	E34
2.55-2.69	2.55-2.69				156-166	312-331	E36
2.70-2.88	2.70-2.88				167-177	332-355	E37
2.89-3.14	2.89-3.14				178-193	356-387	E38
3.15-3.40	3.15-3.40				194-209	388-419	E39
3.41-3.81	3.41-3.81				210-233	420-467	E41
3.82-4.25	3.82-4.25				234-248	468-515	E42
4.26-4.62	4.26-4.62					516-563	E44
4.63-5.09	4.63-5.09						E46
5.10-5.61	5.10-5.61						E47
5.62-5.91	5.62-5.91						E48
5.92-6.15	5.92-6.15						E49
6.16-6.70	6.16-6.70						E50
6.71-7.54	6.71-7.54						E51
7.55-8.29	7.55-8.29						E52
8.30-8.99	8.30-8.99						E53

Continued on the next page.

The heaters shown below provide a maximum trip rating of 125% of minimum motor amperes for 40°C motors (service factor 1.15). **For other motors** (service factor 1.0), **select the next lower** listed heater code number within the designated table which provides a maximum trip rating of approximately 115%.

continued, SIEMENS   Table 233

00,0,1	1 3/4	2, 2 1/2	3, 3 1/2	4(JB)	4 1/2, 5	6	Heater Code
9.00-9.85	9.00-9.85						E54
9.86-10.4	9.86-10.4	9.1-10.4					E55
10.5-12.0	10.5-12.0	10.5-12.0	14.8-15.8				E56
12.1-13.6	12.1-13.6	12.1-13.6	15.9-17.2				E57
13.7-15.6	13.7-15.6	13.7-15.6	17.3-18.6				E60
15.7-17.0	15.7-17.0	15.7-17.0	18.7-20.4				E61
17.1-18.4	17.1-19.4	17.1-19.4	20.5-22.5				E62
18.5-19.4	19.5-20.9	19.5-20.9	22.6-24.7				E65
19.5-21.3	21.0-22.2	21.0-22.2	24.8-27.2				E66
21.4-24.4	22.3-25.3	22.3-25.3	27.3-29.9				E67
24.5-25.9	25.4-26.9	25.4-26.9	30.0-33.5	19.4-21.2			E69
26.0-27.0	27.0-30.2	27.0-30.2	33.6-36.4	21.3-23.3			E70
			36.5-39.6	23.4-25.7			E71
	30.3-33.3	30.3-33.3		25.8-28.2			E72
	33.4-36.0	33.4-35.3	39.7-43.6	28.3-31.0			E73
			43.7-46.5				E73A
		35.4-41.5	46.6-51.6	31.1-34.1			E74
		41.6-45.0	51.7-54.4	34.2-37.5			E76
		45.1-52.3	54.5-58.0				E77
		52.4-55.7	58.1-63.0	37.6-37.5			E78
		55.8-60.0	63.1-67.7	41.4-45.3			E79
			67.8-72.4	45.4-49.9			E80
				45.0-55.9			E88
				56.0-60.9			E89
				61.0-65.9			E91
				66.0-69.9			E92
				70.0-75.9			E93
			72.5-80.0	76-81.9			E94
			80.1-88.1	82-86.9			E96
			88.2-91.5	87-92.9			E97
			91.6-96.8	93-97.9			E98
			96.9-99.0	98.0-107.9			E99
			99.1-108	108.0-113.9			E101
				114.0-132.5			E102
							E103
							E104

Cutler-Hammer (Westinghouse) NEMA Size Starter			Non Compensated Open Starters		Non-Compensating ENCLOSED Starters		Heater Catalog Number
			Block Type Overload Using 3 Heaters	Single Pole Type Overload	Block Type Overload Using 3 Heaters	Single Pole Type Overload	
			Full Load Current of Motor (Amps.)				
FOR SIZE 2 STARTERS	FOR SIZE 1 STARTERS	FOR SIZE 0 STARTERS	.25 - .27	.29 - .31	.24 - .25	.28 - .30	FH03
			.28 - .31	.32 - .35	.26 - .28	.31 - .34	FH04
			.32 - .34	.36 - .39	.29 - .31	.35 - .37	FH05
			.35 - .38	.40 - .43	.32 - .35	.38 - .42	FH06
			.39 - .42	.44 - .48	.38 - .39	.43 - .47	FH07
			.43 - .46	.49 - .53	.40 - .43	.48 - .52	FH08
			.47 - .50	.54 - .58	.44 - .47	.53 - .56	FH09
			.51 - .55	.59 - .64	.48 - .51	.57 - .63	FH10
			.56 - .62	.65 - .71	.52 - .57	.64 - .70	FH11
			.63 - .68	.72 - .79	.58 - .63	.71 - .77	FH12
			.69 - .75	.80 - .87	.64 - .70	.78 - .85	FH13
			.76 - .83	.88 - .96	.71 - .77	.86 - .94	FH14
			.84 - .91	.97 - 1.06	.78 - .85	.95 - 1.03	FH15
			.92 - 1.00	1.07 - 1.16	.86 - .93	1.04 - 1.13	FH16
			1.01 - 1.11	1.17 - 1.28	.94 - 1.03	1.14 - 1.25	FH17
			1.12 - 1.22	1.29 - 1.41	1.04 - 1.16	1.26 - 1.38	FH18
			1.23 - 1.34	1.42 - 1.55	1.14 - 1.25	1.39 - 1.52	FH19
			1.35 - 1.47	1.56 - 1.71	1.26 - 1.37	1.53 - 1.67	FH20
			1.48 - 1.62	1.72 - 1.87	1.38 - 1.51	1.68 - 1.83	FH21
			1.63 - 1.78	1.88 - 2.06	1.52 - 1.65	1.84 - 2.01	FH22
			1.79 - 1.95	2.07 - 2.26	1.66 - 1.81	2.02 - 2.21	FH23
			1.96 - 2.15	2.27 - 2.48	1.82 - 1.99	222 - 2.43	FH24
			2.16 - 2.35	2.49 - 2.72	2.00 - 2.19	2.44 - 2.66	FH25
			2.36 - 2.58	2.73 - 2.99	2.20 - 2.39	2.67 - 2.92	FH26
			2.59 - 2.83	3.00 - 3.28	2.40 - 2.63	2.93 - 3.21	FH27
			2.84 - 3.11	3.29 - 3.60	2.64 - 2.89	3.22 - 3.53	FH28
			3.12 - 3.42	3.61 - 3.95	2.90 - 3.17	3.54 - 3.87	FH29
			3.43 - 3.73	3.96 - 4.31	3.18 - 3.47	3.88 - 4.22	FH30
			3.74 - 4.07	4.32 - 4.71	3.48 - 3.79	4.23 - 4.61	FH31
			4.08 - 4.39	4.72 - 5.14	3.80 - 4.11	4.62 - 4.9	FH32
			4.40 - 4.87	5.15 - 5.6	4.12 - 4.55	5.0 - 5.5	FH33
			4.88 - 5.3	5.7 - 6.2	4.56 - 5.0	5.6 - 6.0	FH34
			5.4 - 5.9	6.3 - 6.8	5.1 - 5.5	6.1 - 6.6	FH35
			6.0 - 6.4	6.9 - 7.5	5.6 - 5.9	6.7 - 7.3	FH36
			6.5 - 7.1	7.6 - 8.2	6.0 - 6.6	7.4 - 8.0	FH37
			7.2 - 7.8	8.3 - 9.0	6.7 - 7.2	8.1 - 8.7	FH38
			7.9 - 8.5	9.1 - 9.9	7.3 - 7.9	8.8 - 9.7	FH39
			8.6 - 9.4	10.0 - 10.8	8.0 - 8.7	9.8 - 10.5	FH40
			9.5 - 10.3	10.9 - 11.9	8.8 - 9.5	10.6 - 11.7	FH41
			10.4 - 11.3	12.0 - 13.1	9.6 - 10.5	11.8 - 12.7	FH42
			11.4 - 12.4	13.2 - 14.3	10.6 - 11.5	12.8 - 14.0	FH43
			12.5 - 13.5	14.4 - 15.7	11.6 - 12.6	14.1 - 15.3	FH44
			13.6 - 14.9	15.8 - 17.2	12.7 - 13.8	15.4 - 16.6	FH45
			15.0 - 16.3	17.3 - 18.9	13.9 - 15.1	16.7 - 18.3	FH46
			16.4 - 18.0	19.0 - 20.8	15.2 - 16.7	18.4 - 20.0	FH47
			18.1 - 19.8	20.9 - 22.9	16.8 - 18.3	20.1 - 21.9	FH48
			19.9 - 21.7	23.0 - 25.2	18.4 - 20.2	22.0 - 23.9	FH49
			21.8 - 23.9	25.3 - 27.6	20.3 - 22.2	24.0 - 26.2	FH50
			24.0 - 26.2		22.3 - 24.3	26.3 - 28.6	FH51
			26.3 - 28.7	30.4 - 33.3	24.4 - 26.6	28.7 - 31.4	FH52
			28.8 - 31.4	33.4 - 36.4	26.7 - 29.1	31.5 - 34.5	FH53
			31.5 - 34.4	36.5 - 39.9	29.2 - 32.0	34.6 - 37.9	FH54
			34.5 - 37.9	40.0 - 43.9	32.1 - 35.2	38.0 - 41.7	FH55
			38.0 - 41.5		35.3 - 38.5	42.0 - 45.0	FH56
			41.6 - 45.0		38.6 - 42.2		FH57

Cutler-Hammer (Westinghouse) NEMA Size Starter	Ambient Compensated Encl. Starters	Non-Compensating Enclosed Starters	Cat. No.
		All Applications	
	Full Load Current of Motor (Amps.)		
FOR SIZE 3 STARTERS	12.8 - 14.1	11.9 - 13.0	FH68
	14.2 - 15.5	13.1 - 14.3	FH69
	15.6 - 17.1	14.4 - 15.9	FH70
	17.2 - 18.9	16.0 - 17.4	FH71
	19.0 - 20.8	17.5 - 19.1	FH72
	20.9 - 22.9	19.2 - 21.1	FH73
	23.0 - 25.2	21.2 - 23.2	FH74
	25.3 - 27.8	23.3 - 25.6	FH75
	27.9 - 30.6	25.7 - 28.1	FH76
	30.7 - 33.5	28.2 - 30.8	FH77
	33.6 - 37.5	30.9 - 34.5	FH78
	37.6 - 41.5	34.6 - 38.2	FH79
	41.6 - 46.3	38.3 - 42.6	FH80
	46.4 - 50	42.7 - 46	FH81
	51 - 55	47 - 51	FH82
	56 - 61	52 - 56	FH83
	62 - 66	57 - 61	FH84
	67 - 73	62 - 67	FH85
	74 - 78	68 - 72	FH86
	79 - 84	73 - 77	FH87
	85 - 92	78 - 84	FH88
FOR SIZE 4 STARTERS	93 - 101	85-91	FH89
	102 - 110	92-99	FH90
	111 - 122	100-110	FH91
	123 - 129	111-122	FH92
	130 - 133	123-128	FH93
		129-133	FH94
FOR SIZE 5 STARTERS	118 - 129	118 - 129	FH24
	130 - 141	130 - 141	FH25
	142 - 155	142 - 155	FH26
	156 - 170	156 - 170	FH27
	171 - 187	171 - 187	FH28
	188 - 205	188 - 205	FH29
	206 - 224	206 - 224	FH30
	225 - 244	225 - 244	FH31
	245 - 263	245 - 263	FH32
	264 - 292	264 - 270	FH33
	293 - 300	-	FH34

240 Volt — Light Duty — Class 3130

www.us.schneider-electric.com
FOR CURRENT INFORMATION

General Duty Safety Switches

General Duty—Up To 100 kA Short Circuit Current Rating With Proper Current Limiting Fusing

General duty safety switches are designed for residential and commercial applications where economy is a prime consideration. Typical loads are lighting, air conditioning, and appliances. They are suitable for use as service equipment when equipped with a factory-installed neutral assembly or a field-installed service grounding kit.

General duty safety switches are UL Listed, File E2875, and meet or exceed the NEMA Standard KS1.

400 and 600 A general duty switches (NEMA 1 only) will accept Class J fuses and are UL Listed for use on systems with up to 100 kA available fault current. 600 A requires Class J fuse kit—GDJK600 (page 3-3). 400 A requires moving load base.

Class T 400–800 A general duty safety switches use 300 Vac Class T fuses and are UL Listed for use on systems with up to 100 kA available fault current.

Table 3.1:   Fusible

UL Listed Short Circuit Withstand Rating		
Switch Type	Fuse Class	Short Circuit Rating
Fusible	Plug	10 kA
	H	10 kA
	K	10 kA
	J	100 kA
	R	100 kA
	T	100 kA
Non-Fusible ▲	H	10 kA
	K	10 kA
	J	100 kA
	R	100 kA ■
	T	100 kA

▲ The UL Listed short-circuit current rating for Square D general duty, not fusible switches is based on the switch being used in conjunction with fuses. Evaluation of non-fusible switches in conjunction with molded case circuit breakers has not been performed. If a UL Listed short-circuit current rating is required, this non-fusible switch must be replaced with a Square D general duty fusible safety switch equipped with the appropriate class and size fusing. The UL Listed short-circuit current rating of the fusible switch is typically as follows: when used with Class H & K fuses—10,000 A, Class R and J fuses—100,000 A. Consult the wiring diagram of the switch to verify the UL Listed short-circuit current rating.

■ 50 kA for 60 A non-fusible switch.

System	A	Fuse	NEMA 1 Indoor Cat. No.	$ Price	NEMA 3R△ Rainproof Cat. No.	$ Price	Class R Fuse Kits Field-Installable ■ Cat. No.	$ Price	Std. (Fast Acting One-Time Fuses) 1Ø	3Ø	Max. (Dual Element Time-Delay Fuses) 1Ø	3Ø
2 Wire (1 Blade and Fuseholder, 1 Neutral)—120 Vac												
	30	Plug	Use Light Duty Device for This Application (see below)						—	—	—	—
	30	Cart.	Use 3 Wire Devices for this application.						—	—	—	—
3 Wire (2 Blades and Fuseholders, 1 Neutral)—120/240 Vac (Plug), 240 Vac (Cart.) Maximum												
	30	Plug	D211N	60.00	D211NRB	118.00			1-1/2	—	3	—
	30	Cart.	D221N	81.00	D221NRB	125.00	DRK30	17.10	1-1/2	3♦	3	7-1/2♦
	60	Cart.	D222N	137.00	D222NRB	217.00	RFK03H	17.00	3	7-1/2♦	10	15♦
	100	Cart.	D223N	284.00	D223NRB	320.00	RFK10	31.80	7-1/2	15♦	15	30♦
	200	Cart.	D224N▼	589.00	D224NRB▼	800.00	HRK1020	31.80	15	25♦	—	60♦
	400	Cart.	D225N	1703.00	D225NR	2306.00	DRK40	74.00	—	—	—	—
	600	Cart.	D226N	3406.00	D226NR	4379.00	DRK600	74.00	—	—	—	—
4 Wire (3 Blades and Fuseholders, 1 Neutral)—240 Vac Maximum												
	30	Cart.	D321N	125.00	D321NRB	195.00	DRK30	17.10	1-1/2	3	3	7-1/2
	60	Cart.	D322N	217.00	D322NRB	294.00	RFK03H	17.00	3	7-1/2★	10	15★
	100	Cart.	D323N	376.00	D323NRB	544.00	RFK10	31.80	7-1/2	15★	15	30★
	200	Cart.	D324N▼	801.00	D324NRB▼	974.00	HRK1020	31.80	15	25★	—	60★
	400	Cart.	D325N	2075.00	D325NR	2595.00	DRK40	74.00	—	50	—	125
	400	Class T	D325NT	1996.00	D325NTR	2494.00			—	50	—	—
	600	Cart.	D326N	3882.00	D326NR	5251.00	DRK600	74.00	—	75	—	150
	600	Class T	D326NT	3732.00	D326NTR	5046.00			—	75	—	—
	800	Class T	T327N	6481.00	T327NR	8292.00			—	100	—	—

▲ Bolt-on hubs—Refer to page 3-9.
■ When installed, this kit rejects all but Class R fuses. When installed with this kit and Class R fuses, the switch is UL Listed for use on systems with up to 100 kA available fault current.
♦ For corner grounded delta systems only. Use switching poles for ungrounded conductors.
★ If corner grounded delta, use outer switching poles for ungrounded conductors.
▼ For 200% neutral, order (1) additional neutral kit SN20A at $133. and (1) neutral jumper kit SN20NI at $18.40.

Table 3.2:   Non-Fusible

System	A	NEMA 1 Indoor Cat. No.	$ Price	NEMA 3R Rainproof △ Cat. No.	$ Price	Horsepower Ratings (Max.) 1Ø	3Ø
2 Wire (2 Blades)—240 Vac Maximum							
	30	—	—	DU221RB●	118.00	3	—
	60	—	—	DU222RB●	235.00	10	—
	60	QO260NATS□◊	107.00	QO200TR□◊★	107.00	10	—
	100	QO2000NS□◊	184.00	QO2000NRB□★	225.00	20	—
	200	Use 3P Switch	—	Use 3P Switch	—	—	—
	400	Use 3P Switch	—	Use 3P Switch	—	—	—
	600	Use 3P Switch	—	Use 3P Switch	—	—	—
3 Wire (3 Blades)—240 Vac Maximum							
	30	DU321●	103.00	DU321RB●	195.00	3	7-1/2
	60	DU322●	137.00	DU322RB●	295.00	10	15
	100	DU323▽	318.00	DU323RB▽	544.00	15	30
	200	DU324●	589.00	DU324RB●	974.00	15	60✳
	400	DU325●	1465.00	—	—	—	125
	600	DU326✳	2794.00	—	—	—	150

△ Bolt-on hubs—Refer to page 3-9.
□ Enclosed molded case switch—Refer to 1-22.
◊ Includes factory-installed grounding kit.
★ Not service entrance rated—Refer to page 1-19 for more information.
▽ Includes factory-installed neutral. For service equipment use or when neutral is required, order part number SN0610 at $71.00.
● Suitable for use as service equipment, requires field installation of neutral assembly SN20A at $133.00.
✳ If corner grounded delta, install neutral and use outer switching poles for ungrounded conductors.

Light Duty—Visible Blades 10 kA Short Circuit Current Rating

The Square D light duty enclosed switch is ideal for home applications in disconnecting power to workshops, hobby rooms, furnaces and garages.

Table 3.3:   Fusible

Heavy Duty Safety Switches

240 Volt

Class **3110**

◻ SQUARE D
www.us.schneider-electric.com
FOR CURRENT INFORMATION

Visible blade heavy duty safety switches are designed for application where maximum performance and continuity of service are required. All heavy duty safety switches feature quick-make, quick-break operating mechanism, a dual cover interlock and a color coded indicator handle. They are suitable for use as service equipment when equipped with a field- or factory-installed neutral assembly or equipment grounding kit, unless a 600Y/347 V or 480 Y/277 V, 1000 A or greater, solidly grounded WYE system is used, per NEC 215-10. Heavy duty safety switches are UL Listed (except as noted), File E2875 & 154828 and meet or exceed the NEMA Standard KS1. For UL Listed short circuit current ratings, see page 3-6.

FUSED

Table 3.10: 240 Volt—Single Throw Fusible

System	Amperes	NEMA 1 Indoor Cat. No.	$ Price	NEMA 3R Rainproof (Bolt-on Hubs, page 3-9) Cat. No.	$ Price	NEMA 4, 4X, 5, ▲ 304 Stainless Steel (for 316 stainless, see page 3-7) Dust tight, Watertight, Corrosion Resistant (Watertight Hubs, page 3-9) Cat. No.	$ Price	NEMA 12K With Knockouts (Watertight Hubs, page 3-9) Cat. No.	$ Price	NEMA 12, 3R ■ Without Knockouts (Watertight Hubs, page 3-9) Cat. No.	$ Price	Horsepower Ratings ♦ 240 Vac Std. (Using Fast Acting, One Time Fuses) 1Ø	3Ø	Max. (Using Dual Element, Time Delay Fuses) 1Ø	3Ø	250 Vdc ★
2 Wire (2 Blades and Fuseholders)—240 Vac, 250 Vdc																	
	30	Use 3 Wire Devices For 2 Wire Applications					H221DS	1298.00	H221A	336.00	H221AWK	315.00	1-1/2	3 ▼	3	7-1/2 ▼	5
	30					—	—	—	—	H2212AWK△	392.00	1-1/2	—	3	—	5
	60					H222DS	1558.00	H222A	431.00	H222AWK	431.00	3	7-1/2 ▼	10	15 ▼	10
	100					H223DS	3396.00	H223A	672.00	H223AWK	632.00	7-1/2	15 ▼	15	30 ▼	20
	200					H224DS	4640.00	H224A	1158.00	H224AWK	1095.00	15	25 ▼	—	60 ▼	40
	400	H225	1819.00	H225R	2589.00	H225DS	9654.00	—	—	H225AWK	2775.00	—	—	—	—	50
	600	H226	3616.00	H226R	4854.00	H226DS	13848.00	—	—	H226AWK	4362.00	—	75 ▼	—	200 ▼	50
	800	H227	5639.00	H227R◻	7655.00	—	—	—	—	H227AWK	6883.00	50	—	50	—	50
	1200	H228	7788.00	H228R◻	10324.00	—	—	—	—	H228AWK	10543.00	50	—	50	—	50
3 Wire (2 Blades and Fuseholders, 1 Neutral)—240 Vac, 250 Vdc																	
	30	H221N	157.00	H221NRB	298.00	Use 2 Wire Devices, Field-Installable Solid Neutral Assemblies Order Separately See page 3-10.					1-1/2	3 ▼	3	7-1/2 ▼	5
	60	H222N	314.00	H222NRB	561.00							3	7-1/2 ▼	10	15 ▼	10
	100	H223N	477.00	H223NRB	724.00							7-1/2	15 ▼	15	30 ▼	20
	200	H224N	859.00	H224NRB	1041.00							15	25 ▼	—	60 ▼	40
	400	H225N	2061.00	H225NR	2830.00	H225NDS	9858.00	—	—	H225NAWK	2869.00	—	50 ▼	—	125 ▼	50
	600	H226N	3879.00	H226NR	5118.00	H226NDS	14054.00	—	—	H226NAWK	4624.00	—	75 ▼	—	200 ▼	50
	800	H227N	6711.00	H227NR◻	8144.00	—	—	—	—	H227NAWK	8225.00	50	—	50	—	50
	1200	H228N	8281.00	H228NR◻	11110.00	—	—	—	—	H228NAWK	11456.00	50	—	50	—	50
3 Wire (3 Blades and Fuseholders)—240 Vac, 250 Vdc																	
	30	Use 4 Wire Devices For 3 Wire Applications					H321DS	1366.00	H321A	426.00	H321AWK	403.00	1-1/2	3	3	5	
	60					H322DS	1688.00	H322A	609.00	H322AWK	576.00	3	7-1/2	10	15	10
	100					H323DS	3564.00	H323A	941.00	H323AWK	887.00	7-1/2	15	15	30	20
	200					H324DS	4997.00	H324A	1360.00	H324AWK	1284.00	15	25	—	60	40
	400	H325	2283.00	H325R	2650.00	H325DS	9974.00	—	—	H325AWK	2835.00	—	50	—	125	50
	600	H326	4113.00	H326R	5524.00	H326DS	14266.00	—	—	H326AWK	4910.00	—	75	—	200	50
	800	H327	7637.00	H327R◻	9899.00	—	—	—	—	H327AWK	9685.00	50	100	50	250	50
	1200	H328	9678.00	H328R◻	12485.00	—	—	—	—	H328AWK	11633.00	50	100	50	250	50
4 Wire (3 Blades and Fuseholders, 1 Neutral)—240 Vac, 250 Vdc																	
	30	H321N	209.00	H321NRB	370.00	Use 3 Wire Devices, Field-Installable Solid Neutral Assemblies Order Separately. See page 3-10					1-1/2	3	3	7-1/2	5
	60	H322N	352.00	H322NR	594.00							3	7-1/2	10	15	10
	100	H323N	561.00	H323NRB	852.00							7-1/2	15	15	30	20
	200	H324N	967.00	H324NRB	1165.00							15	25	—	60	40
	400	H325N	2525.00	H325NR	2881.00	H325NDS	10214.00	—	—	H325NAWK	3090.00	—	50	—	125	50
	600	H326N	4346.00	H326NR	5748.00	H326NDS	14506.00	—	—	H326NAWK	5171.00	—	75	—	200	50
	800	H327N	8126.00	H327NR◻	10375.00	—	—	—	—	H327NAWK	10586.00	50	100	50	250	50
	1200	H328N	10209.00	H328NR◻	13139.00	—	—	—	—	H328NAWK	13343.00	50	100	50	250	50
4 Wire (4 Blades and Fuseholders)																	
	30	Use 600 Vac Devices. See page 3-5.														
	60															
	100															
	200															
	400															
	600															

▲ Complete rating is NEMA 3, 3R, 4, 4X, 5 and 12. For NEMA 3R applications, remove drain screw from bottom endwall.
■ Also suitable for NEMA 3R application by removing drain screw from bottom endwall.
♦ Refer to page 7-31 for additional motor application data. The starting current of motors of more than standard horsepower may require the use of fuses with appropriate time delay characteristics.
★ For switching dc, use two switching poles.
▼ For corner grounded delta systems only and with neutral assembly installed. Use switching poles for ungrounded conductors.
△ 60 ampere switch with 30 ampere fuse spacing and clips. Must use 60 A enclosure accessories including electrical interlocks.
◻ Suitable for NEMA 5 applications with drain screw installed.

Heavy Duty Safety Switches — FUSED — 600 Volt

SQUARE D — schneider-electric.com

Class 3110

Table 3.11: 600 Volts—Single Throw Fusible

System	Amperes	NEMA 1 Indoor Cat. No.	$ Price	NEMA 3R Rainproof (Bolt-on Hubs, page 3-9) Cat. No.	$ Price	NEMA 4, 4X, 5▲ 304 Stainless Steel (for 316 stainless, see page 3-7) Dust tight, Watertight, Corrosion Resistant (Watertight Hubs, page 3-9) Cat. No.	$ Price	NEMA 12K With Knockouts (Watertight Hubs, page 3-9) Cat. No.	$ Price	NEMA 12, 3R■ Without Knockouts (Watertight Hubs, page 3-9) Cat. No.	$ Price	480 Vac Std. (Using Fast Acting, One Time Fuses) 3Ø	480 Vac Max. (Using Dual Element, Time Delay Fuses) 3Ø	600 Vac Std. (Using Fast Acting, One Time Fuses) 3Ø	600 Vac Max. (Using Dual Element, Time Delay Fuses) 3Ø	dc▼ 250	dc▼ 600

2 Wire (2 Blades and Fuseholders)—600 Vac, 600 Vdc

	30					Use 3 Wire Devices For 2 Wire Applications						—	—	—	—	—	—
	60											—	—	—	—	—	—
	100											—	—	—	—	—	—
	200											—	—	—	—	—	—
	400	H265	2804.00	H265R	3616.00	H265DS	9974.00	—	—	H265AWK	3350.00	100★	250★	—	—	50	50
	600	H266	4435.00	H266R	7124.00	H266DS	14266.00	—	—	H266AWK	4894.00	150★	400★	—	—	50	50
	800	H267	6910.00	H267R□	10923.00	—	—	—	—	H267AWK	10184.00	—	—	—	—	50	50
	1200	H268	9713.00	H268R□	11994.00	—	—	—	—	H268AWK	12029.00	—	—	—	—	50	50

3 Wire (3 Blades and Fuseholders)—600 Vac, 600 Vdc▼

	30	H361	352.00	H361RB	599.00	H361DS	1680.00	H361A	676.00	H361AWK	637.00	5	15	7-1/2	20	5	15
	30	H361-2△	411.00	H3612RB△	699.00	—	—	H361-2A△	690.00	H3612AWK△	651.00	5	15	7-1/2	20	—	15
	60	H362	425.00	H362RB	703.00	H362DS	1847.00	H362A	698.00	H362AWK	656.00	15	30	15	50	—	30
	100	H363	792.00	H363RB	1096.00	H363DS	3662.00	H363A	1084.00	H363AWK	1026.00	25	60	30	75	50	50
	200	H364	1138.00	H364RB	1506.00	H364DS	5123.00	H364A	1696.00	H364AWK	1600.00	50	125	60	150	40	50
	400	H365	3034.00	H365R	3688.00	H365DS	10214.00	—	—	H365AWK	3641.00	100	250	125	350	50	50
	600	H366	5099.00	H366R	7266.00	H366DS	14056.00	—	—	H366AWK	6135.00	150	400	200	500	50	50
	800	H367	8879.00	H367R□	11000.00	—	—	—	—	H367AWK	10901.00	200	500	250	500	50	50
	1200	H368	11671.00	H368R□	13339.00	—	—	—	—	H368AWK	13137.00	200	500	250	500	50	50

3 Wire (3 Blades and Fuseholders, 1 Neutral)—600 Vac, 600 Vdc▼

	30	H361N	411.00	H361NRB	657.00	Use 3 Wire Devices Field-Installable Solid Neutral Assemblies. Order Separately. See page 3-10						5	15	7-1/2	20	—	15
	60	H362N	473.00	H362NRB	756.00							15	30	15	50	—	30
	100	H363N	852.00	H363NRB	1158.00							25	60	30	75	—	50
	200	H364N	1246.00	H364NRB	1605.00	H364NDS	5247.00	H364NA	1810.00	H364NAWK	1705.00	50	125	60	150	40	50
	400	H365N	3265.00	H365NR	3843.00	H365NDS	10445.00	—	—	H365NAWK	3882.00	100	250	125	350	50	50
	600	H366N	5346.00	H366NR	7369.00	H366NDS	14748.00	—	—	H366NAWK	6400.00	150	400	200	500	50	50
	800	H367N	9362.00	H367NR□	11470.00	—	—	—	—	H367NAWK	11502.00	200	500	250	500	50	50
	1200	H368N	12076.00	H368NR□	13995.00	—	—	—	—	H368NAWK	13880.00	200	500	250	500	50	50

4 Wire (4 Blades and Fuseholders)—600 Vac, 600 Vdc◊   2Ø / 2Ø / 2Ø / 2Ø

	30	H461	609.00	—	—	H461DS	1958.00	—	—	H461AWK	743.00	7-1/2	20	10	25	5	15
	60	H462	710.00	—	—	H462DS	2046.00	—	—	H462AWK	838.00	15	40	20	50	10	30
	100	H463	1185.00	—	—	H463DS	5563.00	—	—	H463AWK	1288.00	25	50	30	75	20	30
	200	H464	1971.00	—	—	H464DS	8397.00	—	—	H464AWK	2148.00	50	—	50	—	40	50
	400	H465	4140.00	—	—	—	—	—	—	H465AWK	4538.00	100	250	125	350	50	50
	600	H466	6736.00	—	—	—	—	—	—	—	—	150	400	200	500	50	50

6 Wire (6 Blades and Fuseholders)—600 Vac ◊   3Ø / 3Ø / 3Ø / 3Ø

| | 100 | — | — | — | — | H663DS | 17309.00 | — | — | H663AWK | 3408.00 | 25 | 60 | 30 | 75 | — | — |
| | 200 | — | — | — | — | H664DS | 23595.00 | — | — | H664AWK | 8148.00 | For applications requiring motor disconnect capability, use electrical interlock. Refer to page 3-9. | | | | | |

♦ Complete rating is NEMA 3, 3R, 4, 4X, 5 and 12.
■ Also suitable for NEMA 3R application by removing drain screw from bottom endwall.
♦ Refer to page 7-35 for additional motor application data. The starting current of motors of more than standard horsepower may require the use of fuses with appropriate time delay characteristics.
▼ For corner grounded delta systems only and with neutral assembly installed. Use switching poles for ungrounded conductors.
◊ On 3P devices, use two outside poles for switching dc.
△ 60 A switch with 30 A fuse spacing and clips. Must use 60 A enclosure accessories including electrical interlocks.
▲ Suitable for NEMA 5 applications with drain screw installed.
□ Not suitable for use as service equipment.

Class H Fuse Provisions:

Fusible Square D 30 through 600 A heavy duty safety switches accept Class H fuses as standard. With Class H fuses installed, the switch is UL Listed for use on systems with up to 10 kA available fault current.

Class R Fuse Provisions:

Fusible Square D 30–600 A heavy duty safety switches will accept Class R fuses as standard. A field-installable rejection kit is available which, when installed, rejects all but Class R fuses. With the installation of the rejection kit and Class R fuses, the switch is UL Listed for use on systems with up to 200 kA available fault current. See Class R fuse kits on page 3-9.

Class J Fuse Provisions:

Provisions for installing Class J fuses are included in 30 through 400 A 600 Volt, and 100 through 400 A 240 Volt, fusible heavy duty safety switches. Conversion to Class J fuse spacing requires relocating the load side fuse base assembly from the standard Class H fuse location to an alternate position as marked in the enclosure. With Class J fuses installed, the switch is UL Listed for use on systems with up to 200 kA available fault current. Switches rated 600 A, 240 or 600 Volt, require the addition of an adapter kit, H600J at **$304**. One kit per 3P switch.

Class L Fuse Provisions:

Heavy Duty Safety Switches — No-Fuse — 600 Volt

Table 3.12: 600 Volt—Single Throw Non-Fusible

System	Rating (A)	NEMA 1 Indoor		NEMA 3R Rainproof (Bolt-on Hubs, page 3-9)		NEMA 4, 4X, 5 ▲ 304 Stainless Steel (for 316 stainless, see page 3-7) Dust tight, Watertight Corrosion Resistant (Watertight Hubs, page 3-9)		NEMA 12K With Knockouts (Watertight Hubs, page 3-9)		NEMA 12, 3R ■ Without Knockouts (Watertight Hubs, page 3-9)		Horsepower Ratings (Max.) ♦							
												Volts ac						dc★
												240		480		600		
		Cat. No.	$ Price	Cat. No.	$ Price	Cat. No.	$ Price	Cat. No.	$ Price	Cat. No.	$ Price	1Ø	3Ø	1Ø	3Ø	1Ø	3Ø	250	600
2 Wire (2 Blades)—600 Vac, 600 Vdc																			
	30					Use 3 Wire Devices For 2 Wire Applications						—	—	—	—	—	—	—	—
	60											—	—	—	—	—	—	—	—
	100											—	—	—	—	—	—	—	—
	200											—	—	—	—	—	—	—	—
	400	HU265	1833.00	HU265R	2509.00	HU265DS	8541.00	—	—	HU265AWK	2141.00	—	—	—	—	—	—	50	50
	600	HU266	3264.00	HU266R	5022.00	HU266DS	12303.00	—	—	HU266AWK	3605.00	—	—	—	—	—	—	50	50
	800	HU267	4978.00	HU267R▼	8589.00	—	—	—	—	HU267AWK	8638.00	50	—	50	—	50	—	—	50
	1200	HU268	6817.00	HU268R▼	11595.00	—	—	—	—	HU268AWK	11681.00	50	—	50	—	—	—	50	50
3 Wire (3 Blades)—600 Vac, 600 Vdc																			
	30	HU361	186.00	HU361RB	325.00	HU361DS	1413.00	HU361A	459.00	HU361AWK	431.00	5	10	7-1/2	20	10	30	5	15
	30	HU361EI△	425.00	HU361RBEI△	564.00	HU361DSEI△	1653.00	HU361AEI△	698.00	HU361AWKEI△	671.00	5	10	7-1/2	20	10	30	5	15
	30	HU3612□	246.00	HU3612RB□	425.00	—	—	—	—	HU3612A□	473.00	HU3612AWK□	444.00	5	10	7-1/2	20	10	30	5	15
	60	HU362	325.00	HU362RB	584.00	HU362DS	1680.00	HU362A	583.00	HU362AWK	555.00	10	20	25	50	30	60	10	30
	60	—	—	—	—	HU362DSEI△	1981.00	—	—	—	—	10	20	25	50	30	60	10	30
	100	HU363	522.00	HU363RB	817.00	HU363DS	3401.00	HU363A	843.00	HU363AWK	796.00	20	40	40	75	40	100	20	50
	200	HU364	806.00	HU364RB	990.00	HU364DS	4640.00	HU364A	1131.00	HU364AWK	1069.00	15	60	50	125	50	150	40	50
	400	HU365	1869.00	HU365R	2560.00	HU365DS	9529.00	—	—	HU365AWK	2682.00	—	125	—	250	—	350	50	50
	600	HU366	3328.00	HU366R	5122.00	HU366DS	12708.00	—	—	HU366AWK	4474.00	—	200	—	400	—	500	50	50
	800	HU367	6652.00	HU367R▼	8700.00	—	—	—	—	HU367AWK	8731.00	50	250	50	500	50	500	50	50
	1200	HU368	8947.00	HU368R▼	11911.00	—	—	—	—	HU368AWK	11960.00	50	250	50	500	50	500	50	50
4 Wire (4 Blades)—600 Vac, 600 Vdc ⊖												2Ø	3Ø	2Ø	3Ø	2Ø	3Ø		
	30	HU461◊	551.00	—	—	HU461DS	1724.00	—	—	HU461AWK★	610.00	10	10	20	25	30	10▽	15▽
	60	HU462◊	609.00	—	—	HU462DS	2018.00	—	—	HU462AWK	672.00	20	20	40	50	50	60	10	30
	100	HU463◊	1098.00	—	—	HU463DS	4934.00	—	—	HU463AWK	1194.00	30	40	50	75	50	75	20	30
	200	HU464◊	1599.00	—	—	HU464DS	7496.00	—	—	HU464AWK	1888.00	50	60	50	125	50	150	40	50
	400	HU465	3467.00	—	—	—	—	—	—	HU465AWK	3781.00	—	125	—	250	—	350	50	50
	600	HU466	6048.00	—	—	—	—	—	—	—	—	—	200	—	400	—	500	50	50
6 Wire (6 Blades)—600 Vac ⊖												3Ø		3Ø		3Ø			
	30	—	—	—	—	HU661DS	7935.00	—	—	HU661AWK✱	2238.00	—	10	—	20	—	30	—	—
	60	—	—	—	—	HU662DS	8836.00	—	—	HU662AWK✱	2589.00	—	20	—	50	—	60	—	—
	100	—	—	—	—	HU663DS	13762.00	—	—	HU663AWK✱	3195.00	—	40	—	75	—	75	—	—
	200	—	—	—	—	HU664DS	18877.00	—	—	HU664AWK✱	7025.00	—	60	—	125	—	150	—	—

▲ Complete rating is NEMA 3, 3R, 4, 4X, 5 and 12.
■ Also suitable for NEMA 3R application by removing drain screw from bottom endwall.
♦ Refer to page 7-32 for additional motor application data.
★ For switching dc, use two switching poles.
▼ Suitable for NEMA 5 applications with drain screw installed.
△ Switches with EI suffix are stocked with factory-installed electrical interlocks with one normally open and one normally closed contact.
□ Use 60 A enclosure accessories, including electrical interlocks.
◊ No knockouts are provided.
✱ Check series number on switch for correct accessory. See page 3-13.
▽ HU461AWK (Series E1) is rated 5 hp@250 Vdc, 10 hp@600 Vdc.
⊖ Not suitable for use as service equipment.
✱ One enclosure for NEMA 1, 3, 3R or 12 applications. UL Listed.

UL Listed Maximum Short Circuit Current Ratings—AC only

NOTE: Consult the wiring diagram of the switch to verify the UL Listed short circuit current rating.

Table 3.13: Fusible Safety Switches
For the short circuit current rating, refer to the table below.

Heavy Duty Safety Switch Type	UL Listed Fuse Class	UL Listed Short Circuit Current Ratings
Fusible	H, K	10 kA
	R, J, L	200 kA ♦

♦ On 600 V, 200 A switches, 100,000 A max. on corner grounded delta when protected by Class J or R fuses.

Table 3.14: Maximum I²t and Ip Ratings of Heavy Duty Switches ⊖

Switch Rating (A)	Max. I²t Rating (Amp² sec.)	Max Ip Rating (A)
30	50,000	14,000
60	200,000	26,000
100	500,000	32,000
200	2,000,000	50,000

Any brand of circuit breaker or fuse not exceeding the ampere rating of the switch may be used in conjunction with an non-fused safety switch when there is up to 10 kA short circuit current available. (See table below.)

Above 10, kA—When applied on systems with greater than 10 kA short circuit current available, the UL Listed short circuit current rating for Square D non-fused switches is based upon the switch being used in conjunction with fuses or Square D circuit breakers or Mag-Gard motor circuit protectors.

Table 3.15: Non-Fusible Safety Switches

Heavy Duty Safety Switch Type	Switch Rating (A) ◊	Fuse or Circuit Breaker Type	UL Listed Short Circuit Current Rating
Non-Fused Switches	All	Any Brand Circuit Breaker	Up to 10 kA
		H, K	
		R,T,J,L	200 kA
	30–100	FA	14 kA
	30–100	FH	18 kA
	200	KA	
	400	LA	22 kA

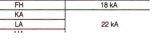

Yr-2 Annex N-**984**

Year Two (Student Manual)